Y0-BXJ-158

Modern Machine Shop's

Guide to Engineering Materials

By
Woodrow Chapman

Hanser Gardner Publications
Cincinnati, Ohio

Library of Congress Cataloging-in-Publication Data

Chapman, Woodrow W. (Woodrow Wilson), 1944-
Modern machine shop guide to engineering materials / Woodrow Chapman.
p. cm.
Includes index.
ISBN 1-56990-358-1
1. Materials—Handbooks, manuals, etc. I. Title.
TA403.4.C463514 2004
620.1'1—dc22

2004000741

A **MODERN MACHINE SHOP** book

Published by

Gardner Publications, Metalworking's Premier Publisher

www.mmsonline.com

Hanser Gardner Publications

6915 Valley Avenue

Cincinnati, OH 45244-3029

Preface

Since the publication of *Modern Machine Shop's Handbook for the Metalworking Industries* two years ago, we have had several inquiries and suggestions about offering a number of selections from the handbook in shorter, focused volumes. This book, which covers materials used in engineering and manufacturing, is the first of these more concise editions, printed in an enlarged format in order to enhance readability.

The text, tables, and diagrams contained within reveal the properties and characteristics of a wide range of both metallic and nonmetallic materials, plus detailed information about their heat treatment, chemical composition, and testing methods. In addition to common metals such as alloy steel, aluminum, and copper, extensive data on plastics, fiber reinforced composites, and heat resistant "superalloys" will be found, alongside those for specialized magnesium and titanium alloys. Keeping up with advances in material science can be challenging, due to developments in metallurgy, polymers, and fibers, but every effort has been made to provide current, useful, and practical knowledge that an engineer, designer, or machinist normally consults in order to predict a material's suitability for a particular application.

Finally, a word of thanks to the companies and associations listed on the following page whose cooperation and contributions have made this book possible. Every effort has been made to acknowledge the copyrighted material supplied by each contributor within the text, and any lapses of identification are regretted.

AISI
American Iron and Steel Institute
1000 16th Street NW
Washington, DC 20006
www.steel.org

ANSI
American National Standards Institute
11 West 42nd Street
New York, NY 10036
www.ansi.org

ASME
American Society of Mechanical Engineers
345 East 47th Street
New York, NY 10017
www.asme.org

ASTM International
100 Bar Harbor Drive
West Conshohocken, PA 19428-2959
www.astm.org

AWS
American Welding Society
550 NW Lejeune Road
Miami, FL 33135
www.aws.org

The Aluminum Association
818 Connecticut Avenue NW
Washington, DC 20006
www.aluminum.org

Bethlehem Steel Corporation
1170 Eighth Avenue
Bethlehem, PA 18016
www.bethsteel.com

Copper Development Association
260 Madison Avenue
New York, NY 10016
www.copper.org

DoAll Co.
254 N. Laurel Avenue
Des Plaines, IL 60016-4398
www.doall.com

Kennametal Inc.
1600 Technology Way
Latrobe, PA 15650
www.kennametal.com

MODERN MACHINE SHOP Magazine
6915 Valley Avenue
Cincinnati, OH 45244
www.mmsonline.com

Norton Company
1 New Bond Street
Worcester, MA 01615-0008
www.nortonabrasives.com

PLASTICS TECHNOLOGY Magazine
29 West 34th Street
New York, NY 10001
www.plasticstechnology.com

Quadrant Engineering Plastic Products
2120 Fairmont Avenue
Reading, PA 19612-4235
www.quadrantepp.com

SAE
Society of Automotive Engineers
400 Commonwealth Drive
Warrendale, PA 15096
www.sae.org

SSINA
Specialty Steel Industry of North America
3050 K St. NW
Washington, DC 20007
www.ssina.org

Timken Latrobe Steel
2626 Ligonier Street
Latrobe, PA 15650
www.timken.com/latrobe

Properties of Materials

Selection of a material for a specific application requires a full understanding of its mechanical properties including tensile properties, elastic and plastic behavior, brittle fracture strength, and hardness. In addition, physical properties such as corrosion resistance, thermal characteristics, and density must be considered. Choosing the correct material for use in a specific application may be a simple task or a complex one, depending on the requirements of the application. In most cases, a material is selected for use in a particular application for either or both of the following reasons. 1) The part or structure made from the material will satisfy, as completely as possible, all of the essential requirements of the application. 2) The part or structure made from the material can be produced and maintained for a lower total cost than is possible with any other material.

The raw material is an important economic factor in any selection criteria, as initial purchase costs, manufacturing costs, inspection costs, maintenance costs, etc., will in some ways be a direct result of the chosen material. The effect of these costs should be analyzed before final selection, as raw material costs alone very often do reflect the total life cycle expenses of any given material. However, in some instances, requirements are so stringent or unique that it is necessary to conduct a test program to evaluate various materials that may satisfy the design profile.

As with any engineering problem, a logical, organized approach is beneficial in solving selection problems. This section examines many of the characteristics of materials and explains the standard tests that are used to evaluate materials. By testing and using the comparative data, it is possible to assure that the end product will meet the essential functional requirements of the application.

Mechanical Properties

Elasticity

Tensile Stress. The rod illustrated in *Figure 1* is subjected to tensile loading, due to applied loads in opposite directions on each end. The intensity of this load creates stress in the rod, and the magnitude of the stress can be calculated with the following equation.

$$f = P \div A_0$$

where f = tensile stress
P = applied load (force)
A_0 = original cross-sectional area of the rod.

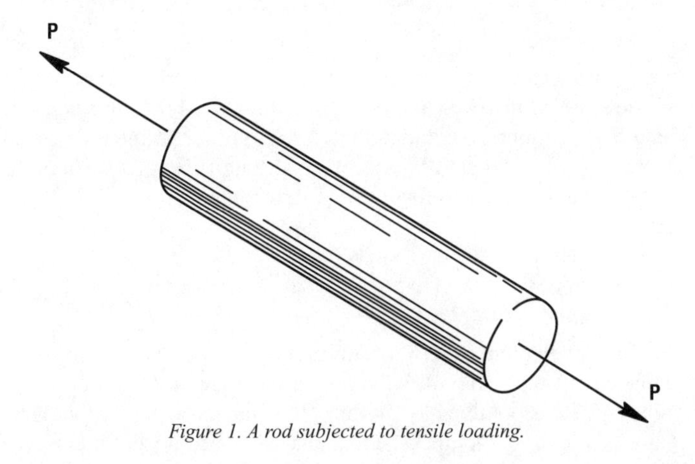

Figure 1. A rod subjected to tensile loading.

Shear Stress. Tensile stress, as shown on the rod in the figure, would make it longer. Compressive stress would be directed in the opposite directions, and would make it shorter. If a force acts normal to a given cross section, the resulting stress is normal stress. If, however, the force acts parallel to a cross section, as shown in *Figure 2*, the resultant stress is shear stress, and its magnitude can be calculated with the following equation.

$$f_S = P \div A$$

where f_S = shear stress
P = applied force acting parallel to the cross section
A = cross-sectional area.

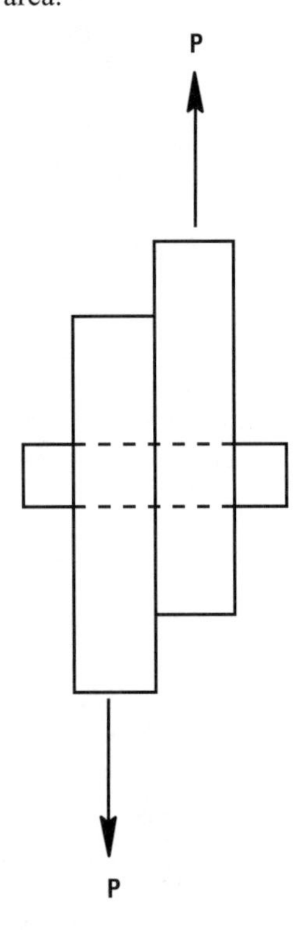

Figure 2. Shear loading on a pin.

Axial Strain. Every stress is accompanied by a corresponding deflection, or change in dimension, in the stressed member. For the tensile loading shown in *Figure 3*, the rod will increase in length an amount δ, provided that the magnitude of the load remains constant. When the deflection (δ) is divided by the original length, a value defined as the normal conventional strain can be calculated with the following equation.

$$e = \delta \div L_0$$

where e = axial strain
δ = length of deflection
L_0 = original length.

Shear Strain. If an element of uniform thickness is subjected to pure shear stress, there will be a displacement on each side of the element relative to the opposite side (shear strain). The shear strain is obtained by dividing this displacement by the distance between the sides of the element. It should be noted that shear strain is obtained by dividing a

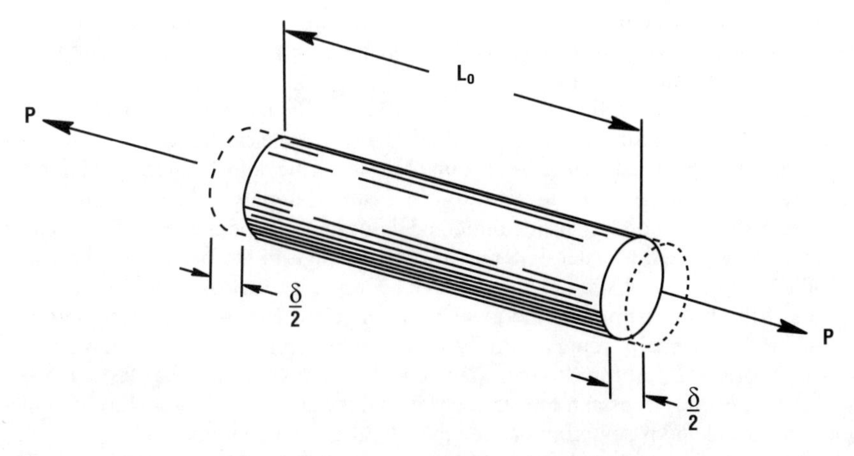

Figure 3. Deflection in a rod subjected to tensile load.

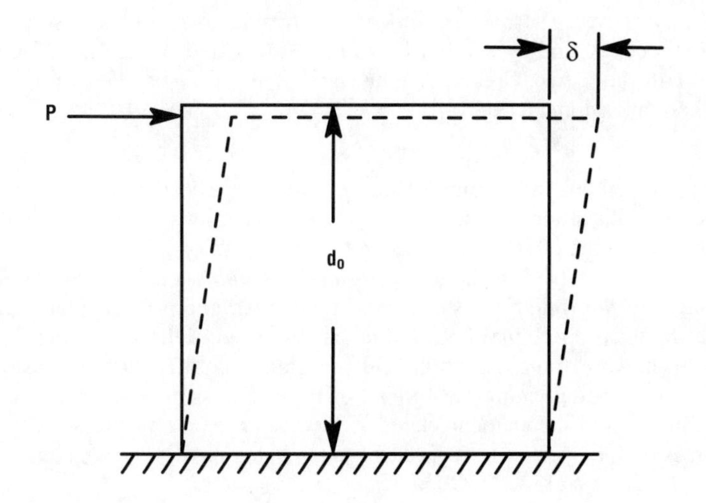

Figure 4. Shear deflection.

displacement by a distance at right angles to the displacement, whereas axial strain is obtained by dividing the deformation by a length measured in the same direction as the deformation. *Figure 4* illustrates shear strain (deflection), which can be calculated by the following equation.

$$e_S = \delta \div d_0$$

where e_S = shear strain
δ = length of deflection
d_0 = cross-sectional length.

Normal and Principal Strains; Poisson's Ratio. Normal strain is the strain associated with a normal stress, and it results in a change in length. Increases in length are denoted as positive strains and decreases in length are called negative strains. Positive strains are identified with the "+" sign and negative with the "–" sign. If one projects three mutually perpendicular planes through a point, there is always some orientation of this system such that only normal strains exist, and all shear strains are zero. The normal strains are called principal strains. Their direction coincides with the direction of the principal *stresses* only for materials that have uniform properties in all directions, which are known as isotropic materials.

In uniaxial loading, strain in the direction of the applied stress varies with that stress. The ratio of stress to strain has a constant value (E) within the elastic range of the material, but it decreases when plastic strain is encountered (in elastic deformation, any deformation caused by stress disappears upon removal of that stress; in plastic deformation, the stress strains the material beyond its elastic limit, and deformation is permanent). The axial strain is always accompanied by lateral strains of opposite sign in the two directions mutually perpendicular to the axial strain. Under uniaxial conditions, the absolute value of the ratio of either of the lateral strains to the axial strain is called *Poisson's ratio*. Simply stated, Poisson's ratio (normally represented by μ) describes unit lateral deflection and is the relationship of the lateral force to the axial force. For stresses within the elastic range, this ratio is approximately constant. For stresses beyond the elastic limit, this ratio is a function of the axial strain and is sometimes called the lateral contraction ratio.

Poisson's ratio can be viewed as a tensile stress acting along the x-axis of the beam, as shown in *Figure 5*. This force will produce a strain, e_X. The force will also produce transverse strains e_y and e_Z. According to Hooke's Law (see below), $e_X = f \div E$. Forces e_y and e_z will not be as large as e_X and will be negative as the bar will contract in these two directions. The ratio of e_y and e_z to e_X is Poisson's ratio, which is a nondimensional term.

$$\mu = e_y \div e_X = e_Z \div e_X \quad \text{or} \quad e_y = e_Z = \mu\, e_X.$$

In multiaxial loading, the strains resulting from the application of each of the stresses are additive: thus, the strains in each of the principal directions must be calculated, taking into account each of the principal stresses and Poisson's ratio.

Modulus of Elasticity. Also known as *Young's Modulus*, the modulus of elasticity is a measurement of the rigidity or stiffness of a material, and it is calculated as the ratio of stress below the proportional limit (which is discussed in the section on Stress-Strain Diagrams) to the corresponding strain. This constant value (E) of the ratio of stress to strain is determined by dividing the amount of tensile stress that an element is subjected to by the elongation of the element, stated as a percentage, at that stress. If, for example, a tensile stress of 4 ksi (27.6 MPa) produces an elongation of 2% in a material, E for that material will be 200 ksi (1380 MPa).

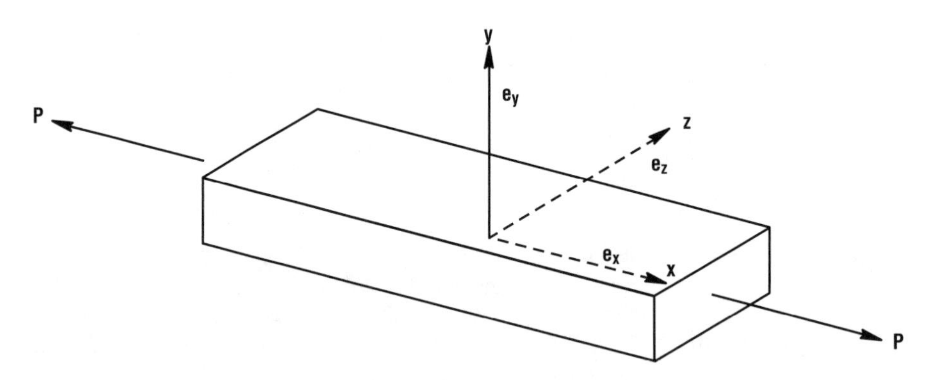

Figure 5. Three-dimensional strain (Poisson's ratio).

Modulus of Rigidity. This is sometimes referred to as the modulus of elasticity in shear, or simply the shear modulus, and it is the ratio of the shear stress to the shear strain at shear stresses below the proportional limit of the material. It can be calculated with the following equation.

$$G = E \div (2\,[1 + \mu])$$

where G = modulus of rigidity
E = modulus of elasticity
μ = Poisson's ratio.

Hooke's Law and Stress-Strain Diagrams. Hooke's Law states that stress is directly proportional to strain, and it is expressed as:

$$f \div e = E.$$

This law is applicable to all solid materials below the proportional limit where materials lose their elasticity and become plastic. This can be demonstrated by subjecting a specimen to an increasing tensile load until failure of that specimen occurs. Depending on the instrumentation of the test, a load deflection or load strain curve is usually obtained. The curves shown in *Figure 6* are typical, and it will be noted that the first part of the diagram is typically a straight line ascending vertically. This indicates a constant ratio between stress and strain over that range. The numerical value or the ratio is the modulus of elasticity for the material, and, up to the point where the line begins to curve, the material is in its elastic region. In the elastic region, stress and strain are related by $f = Ee$, or stress is equal to the elasticity modulus times the strain.

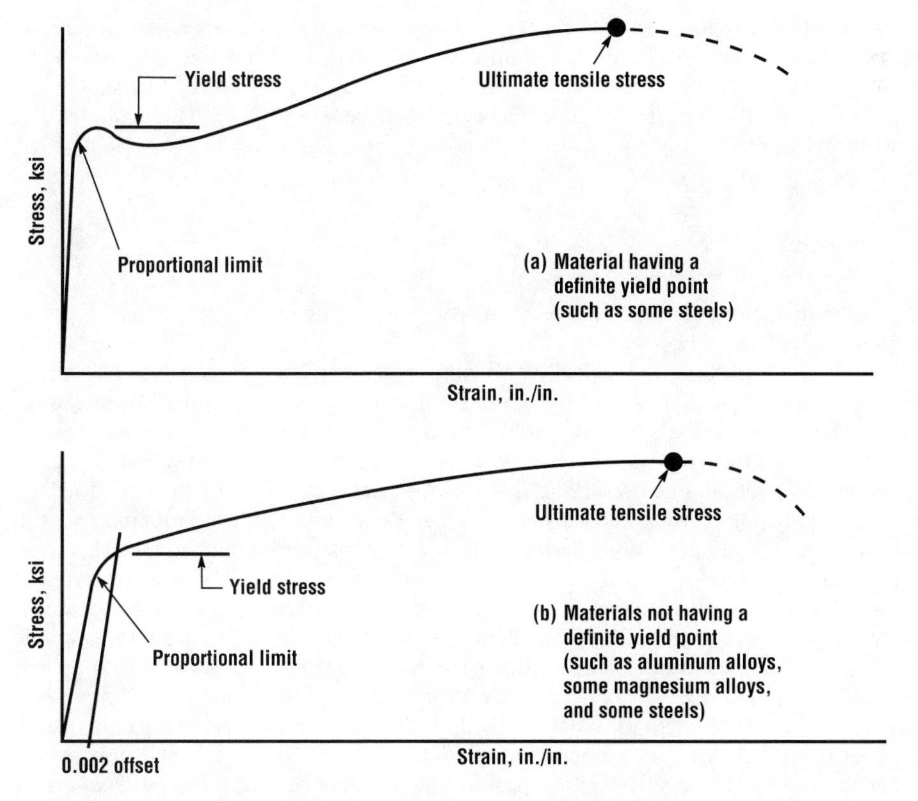

Figure 6. Typical stress-strain diagrams.

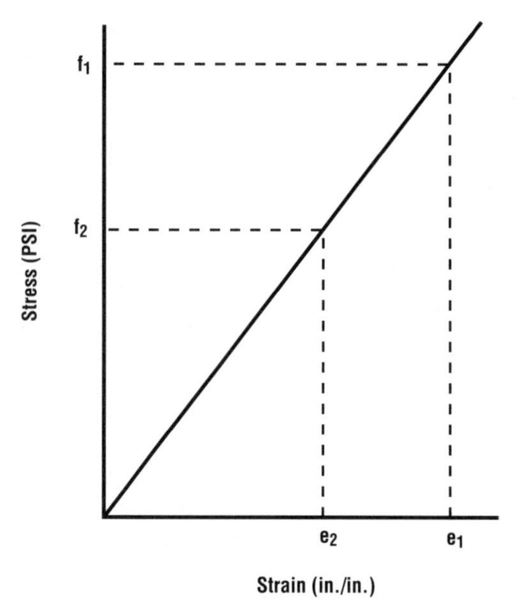

Figure 7. The elastic segment of a stress-strain diagram.

If a material is loaded to a stress f_1 in the elastic region, a corresponding strain e_1 will occur, as shown in *Figure 7*. If the load is removed, the total elastic strain will be recovered. If, rather than removed, the load is reduced to f_2, then the total existing strain will be e_2, and the amount of strain recovered will be $e = e_1 - e_2$. This strain can be calculated by the equation $f = Ee$.

$$f_1 = e_1E \quad \text{or} \quad e_1 = f_1 \div E$$
$$f_2 = e_2E \quad \text{or} \quad e_2 = f_2 \div E$$
therefore, $$e_1 - e_2 = (f_1 \div E) - (f_2 \div E)$$
$$\Delta e = (f_1 - f_2) \div E.$$

No permanent deflection or strain will occur if a material is stressed in the elastic region and the load is later released.

A material will behave elastically only until a characteristic stress is reached. At this stress, the straight line portion of the stress-strain curve ends. This point is known as the proportional limit, or the tensile proportional limit, and it is the maximum stress in which strain remains directly proportional to stress. The value that shows how much stress a material can withstand before plastic deformation occurs is its yield strength. However, except in the softer ferrous metals this point can be very difficult to pinpoint, and it is almost impossible to precisely determine this point on a stress-strain diagram. Therefore, it is customary to assign a small value of permanent strain and identify the corresponding stress value (at the intersection of the stress-strain curve) as the proportional limit. The selected permanent strain offset value should be stated when using the proportional limit. Ultimate tensile stress ratings, on the other hand, recognize the stress at the maximum load reached in a test. They are not necessarily an indication of the proportional limit of the material. Ultimate strength of a material is the maximum stress a material can withstand without fracturing. After the proportional limit has been exceeded, stress and strain are no longer related by Hooke's Law. In this plastic region, the total strain is composed of elastic strain plus plastic strain. Plastic strain is permanent strain that cannot be recovered after the load has been removed. Even though plastic strain is occurring, the amount of elastic strain can still be measured by Hooke's Law. As shown in *Figure 8*, a material loaded to point f_1 will follow the stress-strain curve (O-A-B in the figure) and the total strain will be e_2. When unloaded, it will unload along the line B-C, which is parallel to the original elastic line O-A. The remaining plastic strain at $f = 0$ will be e_1. The recovered elastic strain will be $e_2 - e_1$.

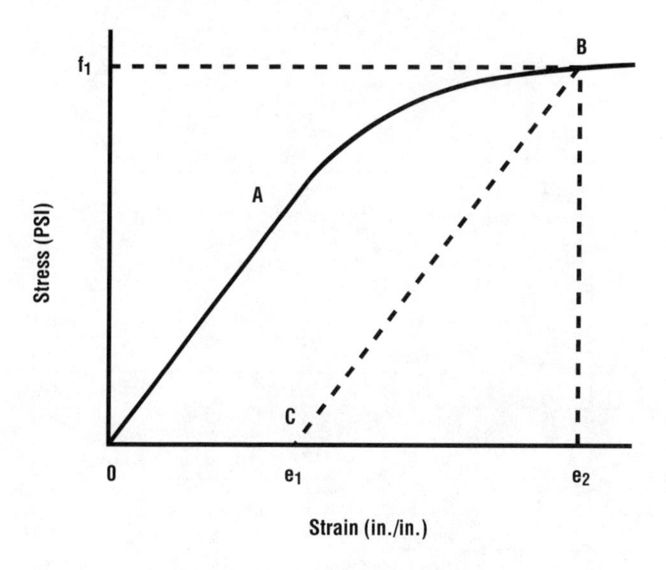

Figure 8. Material behavior in the plastic zone.

Mechanical Properties–Plastic Strain Characteristics

When a metal is deformed beyond its proportional limit, the magnitude of the stress required for further deformation increases. For example, during a tensile test, as a tensile specimen elongates, the load will increase due to strain hardening. Strain hardening (or work hardening) is the point at which a metal, which has become harder and harder under progressive deformation, cannot deform further without fracturing. Tensile strength and hardness are at their maximum and ductility is at its minimum. During strain hardening, the cross-sectional area of the specimen will begin to decrease. During the first part of the test, strain hardening predominates and the load increases. During the later stages of the test, strain hardening becomes less pronounced and the load reaches a maximum. Then it progressively decreases until fracture occurs. Strain will be uniform along the entire test section until the maximum is reached. At this load, some cross section of the bar that is infinitesimally weaker than the rest of the specimen will stretch, as plastic flow takes over, under constant load, while the other portions of the specimen will not. Because of this, localized strain will occur and the specimen will get thinner or "neck" at this particular location. The load will then begin to decrease, the neck will continue to grow thinner, and fracture will eventually occur at the neck.

Testing Mechanical Properties

As shown in **Table 1**, several tests are used to determine the mechanical properties of materials. Most of the properties discussed up to this point are measured with one or more of several common tension tests.

Table 1. Standard Testing Procedures to Determine Mechanical Properties of Materials.
(Source, Bethlehem Steel.)

Mechanical Property	Test Name	Units: English (Metric)
Cold Bending	Cold-bend	Angular Degrees (radians)
Compressive Strength	Compression	psi or ksi (MPa)
Corrosion-fatigue Limit	Corrosion-fatigue	psi or ksi (MPa)
Creep Strength	Creep	psi or ksi (MPa) per time and temperature
Elastic Limit	Tension; Compression	psi or ksi (MPa)
Elongation	Tension	Percent of a specific specimen gage length
Endurance Limit	Fatigue	psi or ksi (MPa)
Hardness	Static: Brinell; Rockwell; Vickers Dynamic: Shore (Scleroscope)	Empirical Numbers Empirical Numbers
Impact, Bending	Bend	ft-lb (Joule)
Impact, Torsional	Torsion-impact	ft-lb (Joule)
Modulus of Rupture	Bend	psi or ksi (MPa)
Proof Stress	Tension; Compression	psi or ksi (MPa)
Proportional Limit	Tension; Compression	psi or ksi (MPa)
Reduction of Area	Tension	percent
Shear Strength	Shear	psi or ksi (MPa)
Tensile Strength	Tension	psi or ksi (MPa)
Torsional Strength	Torsion	psi or ksi (MPa)
Yield Point	Tension	psi or ksi (MPa)
Yield Strength	Tension	psi or ksi (MPa)

Tension Tests

The requirements of this test are set down in ASTM specification E8. The basic methodology of the test is as follows.

There are two standard tensile test specimens: one with a rectangular cross section, and the other round. The grips can be of any configuration that will allow tensile loads to be applied axially through the specimen. Typical tensile test specimens are shown in *Figure 9*. General requirements for a specimen are as follows.

The gage section should be of uniform cross section along its length. It may taper slightly so that at the center of the gage length the width of the flat specimen may be reduced to a width that is 0.010 inch (0.254 mm) less than the width of the specimen at the ends of the gage length. For round specimens, the permissible reduction is limited to 1% of the diameter of the specimen at the center of the gage length.

The gage section should be free from burrs, scratches, pits, or other surface defects.

The ratio of the gage section length to specimen width or diameter should be at least 4:1.

The grip length should be long enough to prevent slippage or fracture in the grips.

The properties most often obtained from a tension test are yield strength, ultimate strength, percent elongation, and reduction of area. Other properties, less frequently measured, are the modulus of elasticity, Poisson's ratio, and the strain hardening exponent. The typical sequence of events in a tension test is as follows.

The cross-sectional area of the gage section is measured.

The specimen is marked with standard gage lengths for subsequent elongation

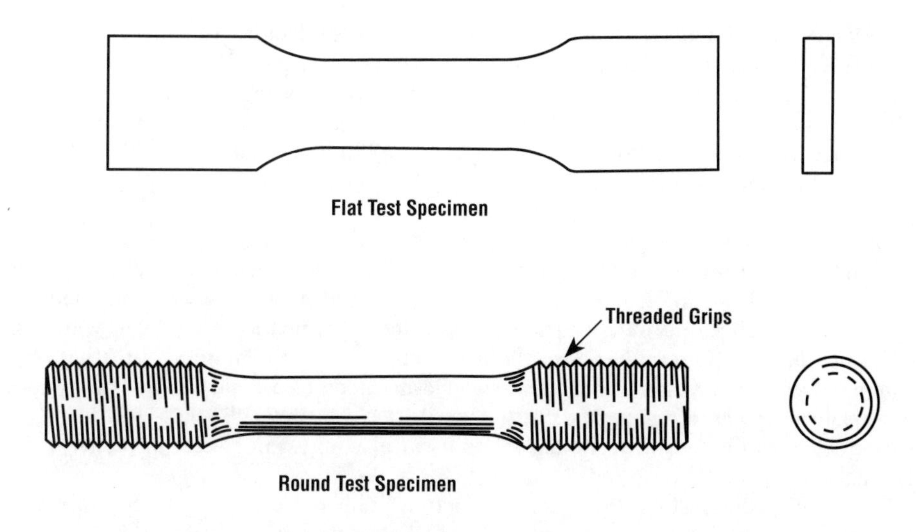

Figure 9. Flat and round tensile test specimens.

measurements, or fitted with an extensometer that will precisely measure changes in length.

The specimen is placed into a tensile machine and a suitable strain measuring device is attached.

The specimen is loaded to fracture.

For those alloys having a distinct yield point, the stress at which the yield point occurs is the yield strength. For those alloys without a distinct yield point, the yield strength is commonly designated as the stress at which 0.2% plastic strain has occurred. The ultimate strength is obtained by dividing the maximum load that the specimen carried by the original cross-sectional area. This load can normally be read from a dial on the tensile machine.

Percent elongation is the ratio of the increase in length of the gage section of the specimen to its original length expressed as a percent. It can be measured by indexing a gage length—usually 1 or 2 inches long—on the specimen gage section and then measuring this length after fracture. The percent elongation can be determined with the following equation.

$$\% \text{ elongation} = (\text{change in length} \div \text{original length}) \times 100.$$

Elongation measurements made in this manner are only an approximation of the true uniform strain that an alloy can undergo before necking. The measurement is made across the fracture, introducing an error, and localized strain occurring in the necked portion introduces additional error.

Reduction in area is the ratio of the decrease in area at the fracture cross section (the neck) to the original area. It can be measured with the following equation.

$$\% \text{ reduction in area} = (\text{original cross-sectional area} - \text{final cross-sectional area}) \div \text{original cross-sectional area} \times 100.$$

The application of tensile properties to design is straightforward. The yield and ultimate strengths designate how much load a structure can carry under ideal conditions. However, in many applications another criteria, such as fatigue characteristics or sheer stress, may determine the actual working stress of a structure.

Elongation and reduction in area are measurements of ductility (ductility, commonly expressed as percentage elongation in gage length, is the degree of extension that takes place in a material in tension before fracture) and have primary applications in fabrications and forming work. Elongation values are the main consideration in stretching or drawing operations, while the reduction rate is important in roll forming. Elongation values are also a qualitative indication of brittleness of an alloy or temper.

Compression Tests

Compression properties are determined by subjecting a specimen to an increasing compressive load until general yielding has occurred. Compression tests measure malleability (malleability is a measure of the extent to which a material can withstand deformation in compression before failure occurs), and in compressive testing only compressive yield strength and compressive elastic modulus are measured. This is done in a manner similar to the methods of the tensile test. Theoretically, these values should be the same as the tensile yield and modulus of elasticity values but, in reality, there is usually a small difference.

In compression testing, the specimen "barrels" rather than necks. Because different factors are at work in barreling, malleable specimens flow in response to the load and actually compress rather than fracture. As the material yields, it swells out (barrels), so that its increasing area continues to support the increasing load. Therefore, there is no measurement point for pinpointing the ultimate tensile strength of the material. Brittle specimens, on the other hand, will fracture into two or more pieces, making it possible to establish an ultimate tensile strength for these materials.

Compression tests are perhaps most useful for predicting how materials will flow in forging, extruding, rolling, and other operations. Generally, no significant data are obtained from a compression test that cannot be obtained from a tension test. Since a material fails by crack formation and propagation, a crack cannot form in malleable materials under compressive loading conditions. Failures under compressive loading can normally be attributed to buckling instability and in some cases shear or bending stresses.

Torsion Tests

Shear properties are usually determined through the torsion test. In this test, shear modulus, shear yield strength, shear ultimate strength (modulus of rupture), and the shear modulus are measured. Although the testing criteria for torsion testing have not been standardized, and it is almost never specified, the data is normally presented as a torque-twist curve on which the degree of twist in a specimen is plotted against the applied torque.

Torsion tests are made by restraining a cylindrical specimen at one end and subjecting it to a twisting movement at the other. The following equation can be used to predict the maximum shear stress of a thin walled tube. When using this and the following equation, a torque angle twist curve is plotted and the shear yield strength is determined at the proportional limit or at a specified permanent twist such as 0.001 radians per inch. It should be noted that this equation is slightly in error, but provides a useful approximation.

$$f_S = T \div (2r^2 \times t)$$

where T = torque
r = outer radius of tube
t = tube wall thickness.

For a solid cylinder, the equation is:

$F_S = 16T \div (\pi \times d^3)$, where d is the diameter of the cylinder. Like the above equation, this can only be applied when strain is proportional to stress. The modulus of rigidity (or modulus of elasticity in shear) can also be determined once the maximum shear stress is known.

$$G = (f_S \times l) \div (r \times \theta)$$

where l = length of test specimen
r = radius of the cylinder
θ = angle of twist (in radians) in length (l).

Additional information on torsion and other methods of materials testing appears in the Statics and Stress Analysis technical units that preceded this section.

Brittle Fracture Tests

The catastrophic failures of welded ships and tankers during World War II brought to the attention of engineers that structural steels can fail at very low stress levels in some environments. These failures occurred at very low ambient temperatures, and there was generally a notch, crack, or other defect present at the failure origin. This type of failure has appropriately been named brittle fracture, because the failure is preceded by little or no plastic strain. However, it is not the brittleness of the fracture that is important. The important consideration is that the failure stress may be considerably less than the yield strength of the material. The engineering methods used to prevent such low strength failures have taken both quantitative and qualitative approaches.

Quantitative Methods

Quantitative methods allow for establishing limiting parameters, such as minimum or maximum operating temperatures; maximum permissible defect size, or maximum safe operating stress for a particular alloy in a given application. These data initiating these parameters were all easily converted to design and operational reference values. In contrast, when using qualitative methods, data are not readily converted to design parameters.

Transition temperature method. The drop weight test described in ASTM E208-63T is a test method that was developed to determine the nil-ductility transition temperature of ferritic steels. The test relies on the concept that ferritic steels in the notched condition are markedly affected by temperature so that there is a characteristic temperature below which a given steel will fail in a brittle manner, but above which brittle fracture will not occur. The nil-ductility transition temperature as determined by the drop weight test is defined as the temperature at which, in a series of tests conducted under controlled conditions, specimens break, while at a temperature 10° F (5.5° C) higher, under duplicate test conditions, no break performance is obtained from near identical specimens.

The drop weight test is simple and inexpensive, and its significant feature of all sizes of specimens is the weld bead deposited on the tension side of the specimen along its longitudinal centerline, as shown in *Figure 10*. The test is conducted by positioning the specimen in a fixture, where it is struck by 60 or 100 pound weights dropped from a predetermined height. The weight is sufficient to deflect the specimen until it impinges on the anvil (about 5°) as shown in *Figure 11*. A cleavage crack forms in the weld bead as soon as incipient yield occurs, at about 3° deflection. A series of specimens are tested over a range of temperatures, and from these tests the nil-ductility transition temperature (which is the temperature at which the steel, in the presence of a cleavage crack, will not deform plastically before fracturing, but will fracture at the moment of yielding) is determined. A specimen is considered broken if it fractures to one or both edges of the tensile surfaces. Complete separation at the compression side is not required If a specimen develops a crack that does not extend to either edge of the tensile surface, it is regarded as a no-break performance.

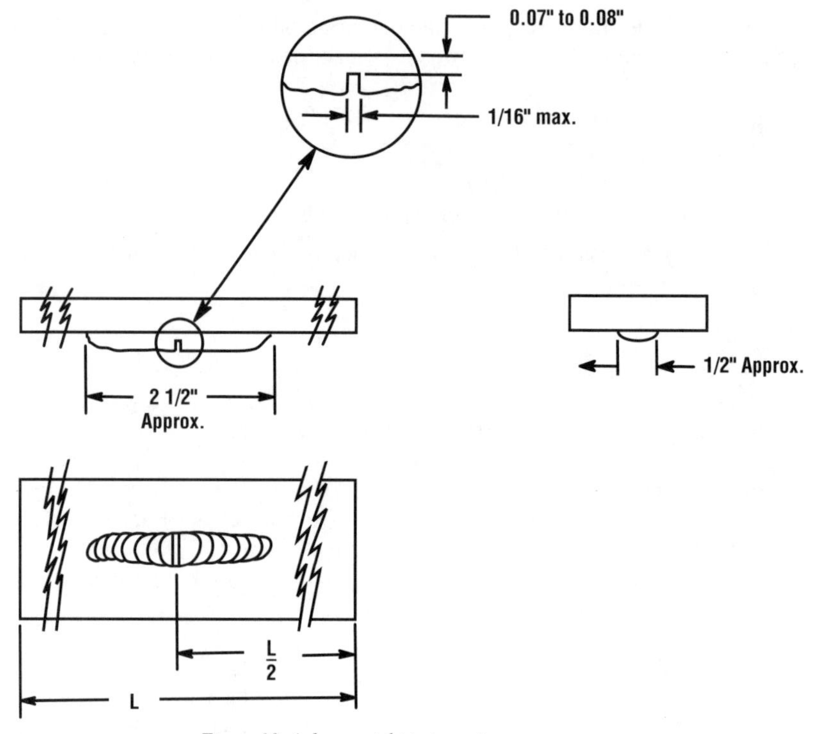

Figure 10. A drop weight test specimen.

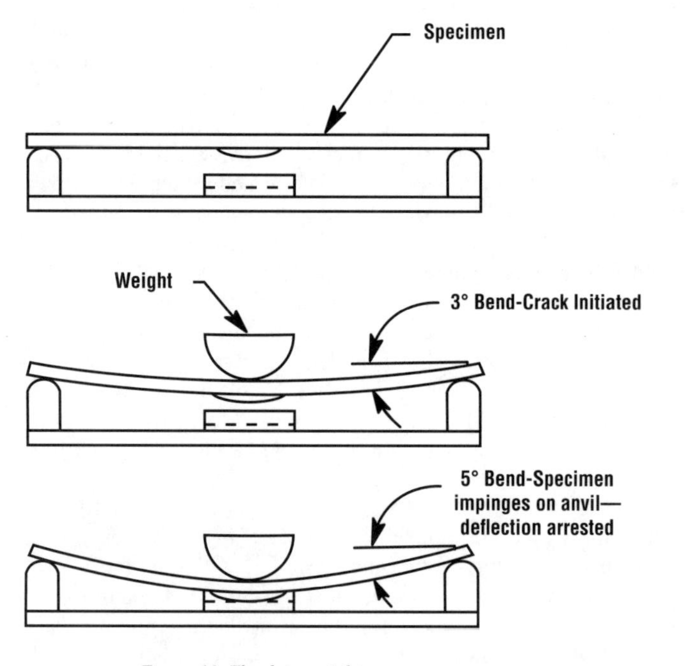

Figure 11. The drop weight test.

After the nil-ductility transition temperature of an alloy has been established, brittle fracture can be avoided by specifying the alloy for use only where it will be in service above the transition temperature. Or, if the operating temperature is fixed, an alloy can be selected that has a transition temperature below the lowest operating temperature. There are, however, two major disadvantages of the transition temperature approach to avoiding brittle fracture. First, it is sometimes necessary to use an alloy at a temperature below its transition temperature, and in these instances it would be desirable to know what level of stress can be considered safe. Secondly, the higher strength steel alloys do not have definite transition temperatures. To satisfy these two problems, the fracture mechanics method can be used.

Fracture mechanics method. This approach to brittle fracture analysis determines the fracture strength of an alloy in the presence of a defect of known geometry. It was developed from Griffith's elastic analysis of the fracture of brittle materials. A. A. Griffith was an English physicist who, in 1924, developed a method of putting crack prediction on an analytical basis. It is based on the concept that small elliptical cracks in a material act as local stress risers that cause stresses to exceed the strength of the material, even though the nominal stress across the section may be quite low. Furthermore, a crack will begin to propagate when the elastic energy released by propagation is equal to, or greater than, the energy of formation of the two new surfaces. Since the elastic energy increases with increasing stress, it is apparent that, at some value of stress, the strain energy released by the crack will exceed the energy of formation of the surfaces, and the crack will become self-propagating under its own stress concentration. Griffith's theory was developed for completely brittle materials that do not deform plastically, and it does not strictly apply to metals, which do undergo plastic deformation.

To account for plastic deformation, a modification of Griffith's theory allows the fracture stress of an alloy containing a crack of known size to be accurately predicted. The fracture stress is characterized by the strain energy release rate (G), an experimentally determined parameter. G increases with crack length, and the stress at which a crack of known size becomes self-propagating is called the critical strain energy rate (denoted by the subscript $_C$, as in G_C or K_C). G_C denotes the critical G determined under plane stress conditions, and G_{IC} is the critical G determined under plane strain conditions. G can also be expressed in terms of the stress intensity factor, K.

$$K_C = \sqrt{G_C \times E}$$

and, $K_{IC} = \sqrt{(G_{IC} \times E) \div (1 - \mu^2)}$

where E = modulus of elasticity

μ = Poisson's ratio.

G and K are both referred to as fracture toughness values. Since these values vary considerably from each other, it is important to specify which is being discussed.

By definition, a plane stress condition is one in which the stress in at least one direction is zero, as can be illustrated by a thin-walled pressure vessel or a thin sheet loaded in tension. In each instance, the stress through the thickness is zero. As applied to fracture mechanics, plane stress actually describes the stress state or restraint at the leading crack edge. Therefore, a through crack in thin material is in plane stress conditions because the stress in the thickness direction at the crack tip is zero or very small. For thicker material, the restraint at the crack front increases until full restraint exists and the stress in the thickness direction is quite high. The fully restrained condition is the plane strain condition. The thickness at which full restraint is reached differs for different materials, but is in the neighborhood of $^1/_4$ to $^1/_2$ inch.

The significance of the plane stress to plane strain transition is that a much lower fracture stress is required for a given defect size when plane strain conditions exist. This is apparent from the difference in magnitude between K_C and K_{IC}. As the thickness increases, the fracture toughness in the plane stress region (K_C) decreases to a constant value of K_{IC}, corresponding to plane strain or full restraint.

There are three basic types of cracks. 1) Through cracks which extend completely through the thickness of the material. 2) Surface cracks, which can be viewed on the outside of the material but do not pass through. 3) Embedded cracks which cannot be seen from the outside but are in the interior of the material. A fracture initiated by a through crack may occur under either plane stress or plane strain conditions, depending on the material thickness. The initial propagation of both the surface and embedded cracks is always under plane strain conditions. However, once a crack travels through the thickness it will be identical to a through crack and may be in either the plane stress or plane strain stress state depending, again, on the thickness.

The stress at which a tension member fails in the presence of a crack is dependent on the fracture toughness of the material, and the defect size. This relationship, sometimes called the inverse square root law, is the basis of fracture mechanics and is expressed by the following equation.

$$S_f = A \div \sqrt{c_l}$$

where S_f = fracture stress
A = constant expressing the fracture toughness
c_l = crack length.

For example, using this expression, if a stress of 100,000 PSI will cause failure in a material in which a 1/4 inch long crack is present, then it can be determined that a stress of 50,000 PSI would cause failure if the crack were 1 inch long.

$$S_{f1} = A \div \sqrt{c_{lA}} = 100{,}000 = A \div \sqrt{0.25}$$

therefore, $S_{f2} = (A \div \sqrt{0.25}) \div \sqrt{1} = 100{,}000 \times 0.5$,
$S_{f2} = 50{,}000$ PSI.

For the case of a crack in an infinitely wide solid, such as a small crack in a pressure vessel, this equation takes the forms:

$$S_f = K_C \div \sqrt{\pi \times 0.5\ c}$$

where K_C = fracture toughness in the plane stress region,

or $S_f = K_{IC} \div \sqrt{(\pi \times 0.5\ c)} \times (1 - \mu^2)$

where K_{IC} = fracture toughness in the plane strain region.

The above equations are standard ways of presenting fracture strength as a function of defect size. As discussed previously, the K_C values for thin material are considerably higher than K_{IC} values. Plotting the lower K_{IC} values in place of the K_C values would show that it takes a considerably longer crack to cause failure under plane stress conditions than it does under plane strain conditions at a given stress level.

Fracture toughness testing. The fracture toughness of a material is determined by loading a fatigue cracked specimen in tension and recording the load at which the crack begins to propagate, and the load at which it fails. Fracture toughness testing methods are sometimes very complex, but typical specimens used in the tests are shown in *Figure 12*. Specimen widths usually range from 1 to 4 inches, with the thickness normally being equal to that of the part that is to be made from the material. The major restriction on testing is that fracture toughness data may only be obtainable in the elastic region of the tensile curve. If yielding of the specimen occurs, then the equations that are used to determine K are no longer valid, since they are based on the elastic theory of stress analysis.

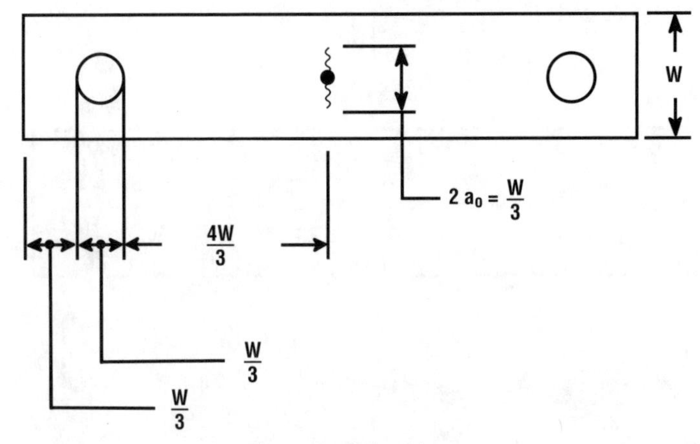

a. Center-cracked specimen.

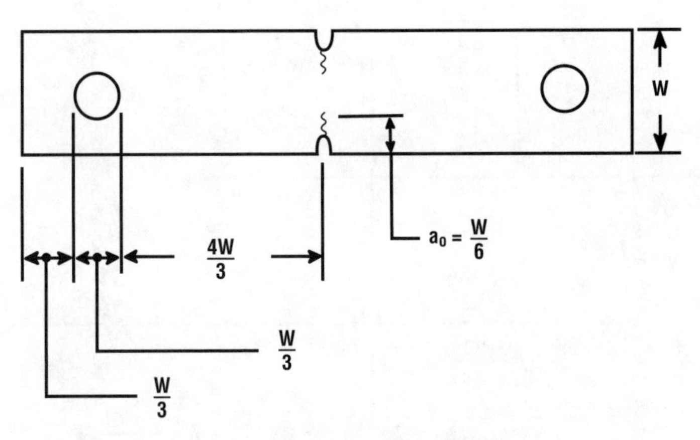

b. Double-edge-notched specimen.

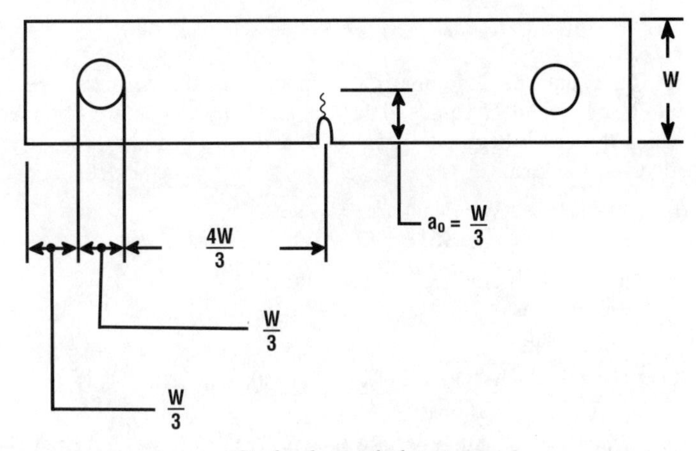

c. Single-edge-notched specimen.

Figure 12a-c. Common fracture toughness test specimens.

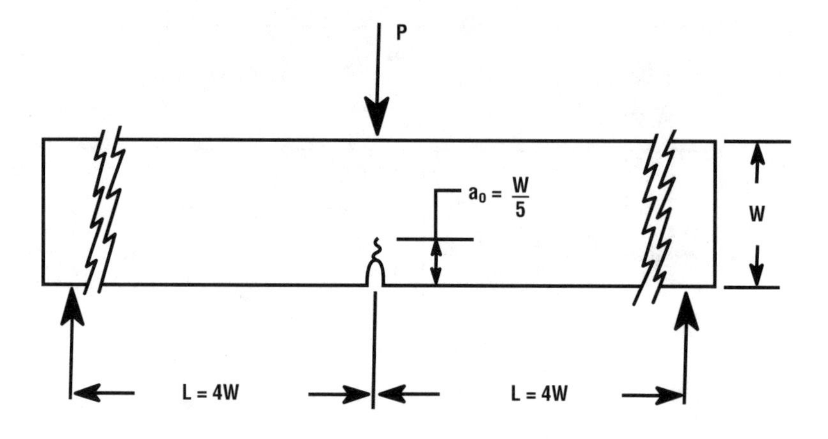

d. Three point bend specimen.

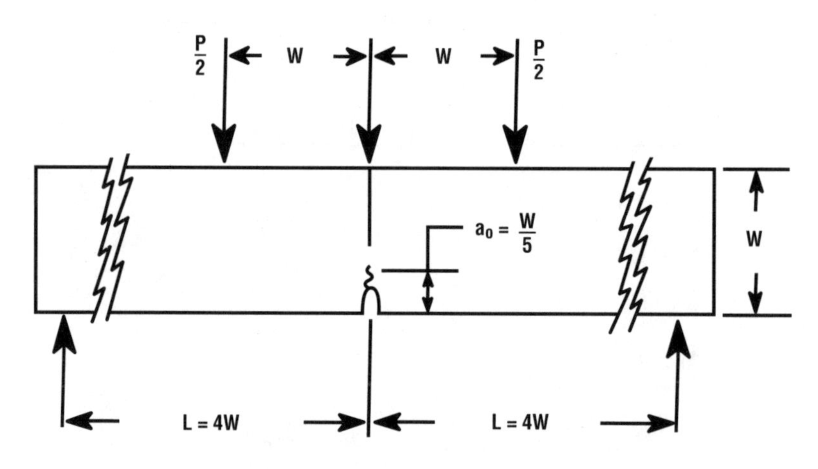

e. Four point bend specimen.

Figure 12d-e. Common fracture toughness test specimens.

The following equations accompany the specimens shown in *Figure 12*, and are used to calculate the K and G values. The K_{IC} and G_{IC} values are obtained by using the load at which the crack begins to propagate, and the K_C and G_C are determined by using the load at fast fracture.

Figure 12a: $G = (P^2 \div EWt^2) \times \tan(\pi a \div W)$
$EG_C = (1 - \mu^2) \times K_{IC}^2 = EG_{IC}$

where a = one-half crack length at fracture
t = thickness
W = width

Figure 12b: $G = (P^2 \div EWt^2) \times (\tan[\pi a \div W] + 0.1 \sin[2\pi a \div W])$
$EG_C = (1 - \mu^2) \times K_{IC}^2 = EG_{IC}$.

Figure 12c: $G_{IC} = (P^2 \div EWt^2) \times (7.59[a \div W] - 32[a \div W]^2 + 117[a \div W]^3)$
$EG_C = (1 - \mu^2) \times K_{IC}^2 = EG_{IC}$.

Figure 12d: $G_{IC} = (P^2 L^2 \div Et^2W^3) \times (31.7 [a \div W] - 64.8 [a \div W]^2 + 211 [a \div W]^3)$
$EG_C = (1 - \mu^2) \times K_{IC}^2 = EG_{IC}$.

Figure 12e: $G_{IC} = (P^2 L^2 \div Et^2W^3) \times (34.7 [a \div W] - 55.2 [a \div W]^2 + 196 [a \div W]^3)$
$EG_C = (1 - \mu^2) \times K_{IC}^2 = EG_{IC}$.

The load at which the crack first begins to propagate (the "pop-in" point) is measured by recording the strain occurring across the crack, in much the same way as in the tensile stress. A load-strain curve (as shown in *Figure 13*) can be obtained from the data collected, and the deviation from linearity is the pop-in point. Two types of behavior are demonstrated. In the first, pop-in occurs and then the load increases until failure under plane stress condition. In the second case, failure occurs immediately as pop-in initiates, indicating that at one-half inch thickness full restraint exists and the plane strain conditions prevail.

Designing against low strength failures. There are three basic fracture mechanics philosophies to designing against low strength failures: proof testing, leak before failure, and stress analysis method.

Proof testing postulates that if a structure contains a crack and it is loaded to a particular stress and the crack does not propagate, then it may be safely used at a slightly lower static operating stress. This is the basis of implementing fracture toughness through proof testing. For example, if a material has a characteristic fracture stress flaw curve such as the one shown in *Figure 14*, and is proof tested at point Sp, then the largest flaw that can possibly be present is slightly smaller than c_1. The static operating stress in this case is So, and the curve shows that for a failure to occur at So, a crack of size c_2 must be present. However, it has already been shown that the largest possible flaw that can exist is c_1, and since c_2 is larger than c_1, the structure can safely operate at So.

Leak before failure is primarily used in pressure vessel design. The success of this method is dependent on the strain-plane stress transition. A surface or embedded flaw in a pressure vessel may propagate at the proof, or operating, pressure of the vessel. The material to be used is selected on the basis that it will be able to tolerate a crack having a length of at least twice the wall thickness of the vessel at the required stress level, under plane strain conditions. Then, if a surface or embedded crack exists, it may propagate at the operating stress. In this case, it will pop through the thickness when its length is approximately twice the wall thickness. Provided the wall is thin enough, the crack

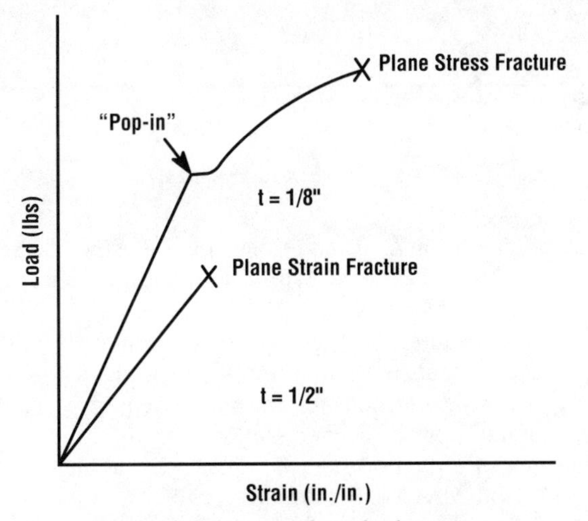

Figure 13. Fracture toughness load curves.

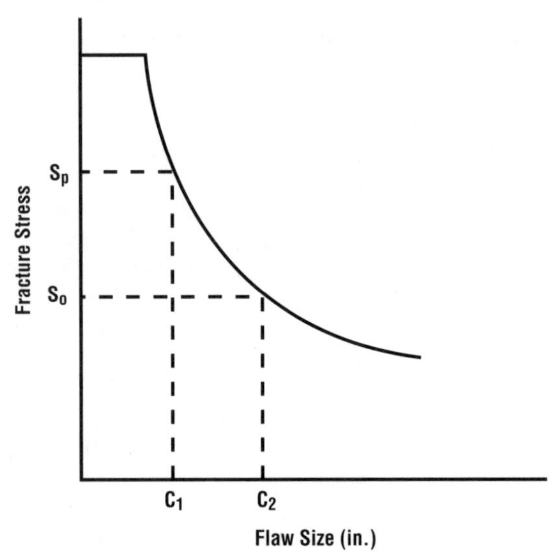

Figure 14. Fracture stress flaw size curve.

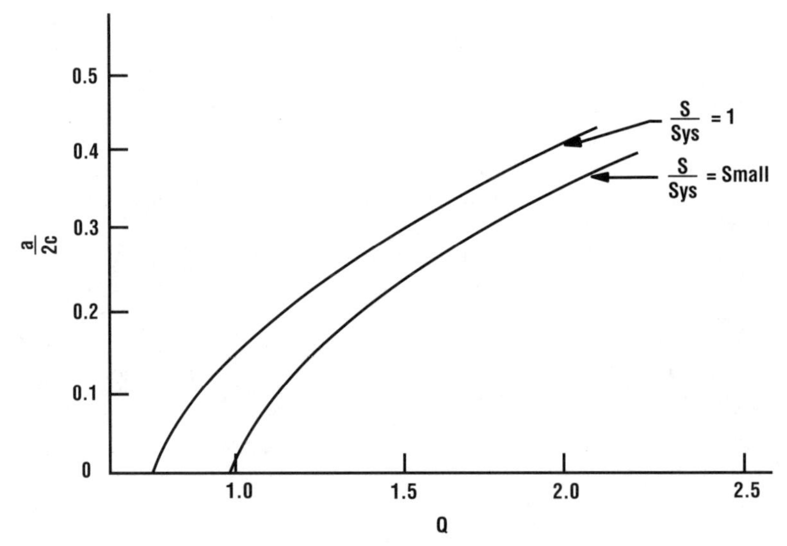

Figure 15. Flaw shape parameter curve.

is under plane stress conditions and additional load or stress is required for further crack propagation. Therefore, the vessel will leak and the defect can be easily detected before brittle failure can occur.

Stress analysis uses fracture toughness of the material and the defect size that can be detected by nondestructive inspection to limit the working stress of a part. The defect shape and size are both considered, and are expressed in terms of the defect size parameter, *a/Q*. The parameter is the ratio of *a*, the crack depth, to *Q*, the flaw shape parameter. *Q*, in turn, is a function of the ratio of crack depth to crack length: *Q* is equal to tensile stress times the crack depth, divided by the crack length. This function is plotted in *Figure 15*. To determine the fracture strength of a part having a known defect size, the value of Q is determined from the curve on the figure, and fracture stress is then determined with the following equation.

$$S = K_{IC} \div \sqrt{1.21} \times (a \div Q)$$

where S = applied stress

K_{IC} = plain strain fracture toughness.

Qualitative Methods

Charpy impact tests, the Izod test, and notched tensile tests are classified as qualitative brittle fracture tests. Essentially, the Charpy is a simple beam test, and the Izod is a cantilever beam test. They measure fracture toughness, and are useful for evaluating the transition from brittle to ductile structure in steels. Both are easy to conduct. The Charpy has gained wide recognition and acceptance because it can be done quickly, easily, and with reliable results.

In the Charpy test, a swinging pendulum is released from a known height to strike a specially prepared notched specimen (as shown in *Figure 16*) that is positioned in the anvil of an impact machine. The pendulum's knife-edged striker contacts the specimen at the bottom of the swing, at which point the kinetic energy of the pendulum reaches a maximum value of 25 to 240 ft lbs (35 to 325 J) on impact. After breaking the specimen, the pendulum continues in its swing to a height that is measured. Since the pendulum is released from a known height, its kinetic energy at impact is a known quantity. The energy expended in breaking the specimen can be expressed as a function of the difference in the release height of the pendulum and the height it attains after the specimen is broken. Therefore,

$$E_A = W (h_o - h_f)$$

where E_A = energy absorbed by the specimen in foot pounds or Joules

W = weight of the pendulum

h_o = release height

h_f = final height.

Normally, swinging pendulum impact machines are equipped with scales from which the energy absorbed by the specimen can be read directly in foot pounds or Joules.

The Charpy test is a convenient, though very definitely destructive, test method for determining the change in the fracture mode of a steel as a function of temperature. A series of specimens can be tested over a range of temperatures and the data thus obtained can be plotted to develop a brittle to ductile transition curve. The transition from a brittle to a ductile material usually occurs over a range of temperatures. Likewise, the fracture of a specimen changes from 100% cleavage (a bright, faceted appearance), to 100% shear (a silky, fibrous appearance). The temperature at which specimens show a fracture of 50% cleavage and 50% shear is frequently defined as the transition temperature.

The Charpy test is also a convenient method for comparing the notch toughness characteristics of various materials.

Methods and procedures for conducting Charpy tests are specified in ASTM E23. It should also be noted that the NATO nations use the Charpy test as the standard impact test for steel for guns.

The Izod test supports the specimen on one end only, leaving usually up to 75% of its area projecting from a fixture. The notched side of the specimen faces the anvil of the pendulum, and the force required to fracture the specimen is recorded, much as in a Charpy test. The singular drawback of the Izod test is that setup can be lengthy, making it impractical for testing materials at different temperatures.

Notched tensile tests are performed on cylindrical specimens having an annular notch,

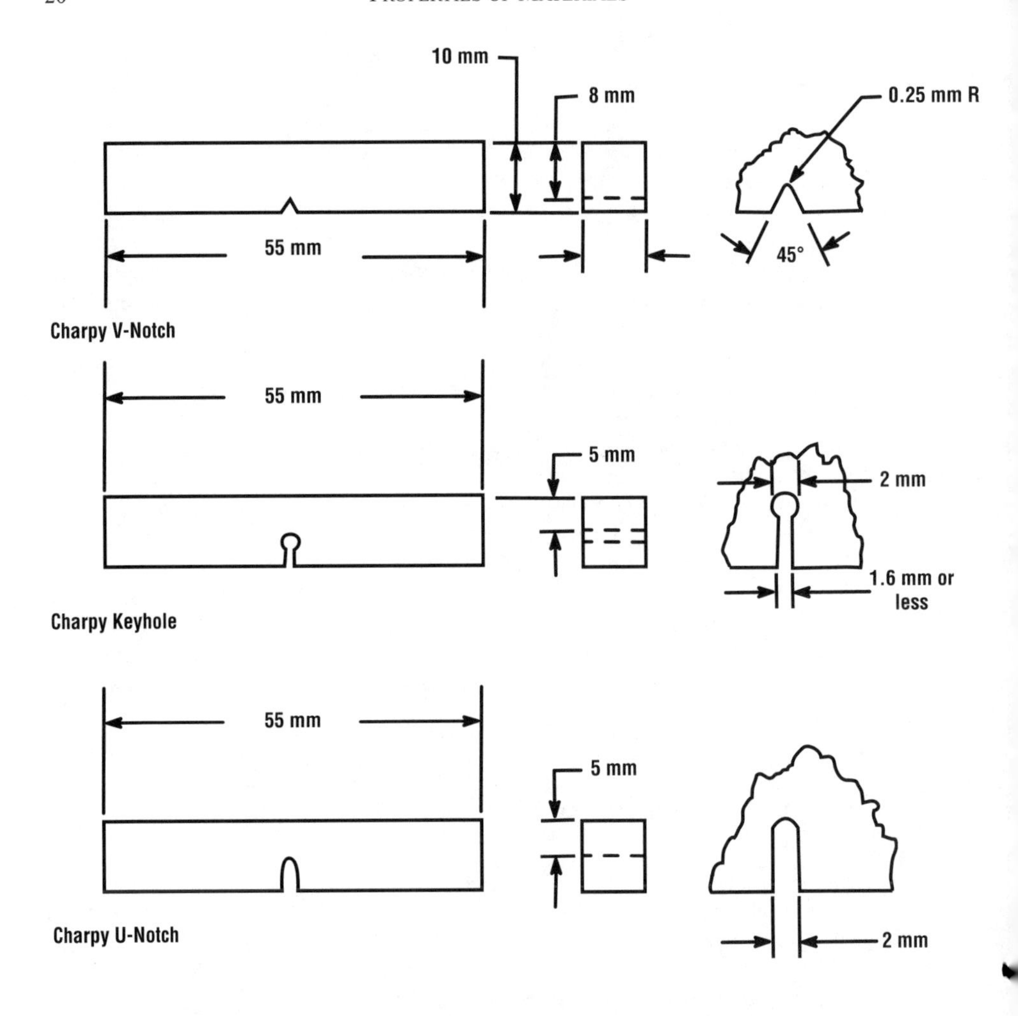

Figure 16. Charpy impact test specimens.

and on flat specimens having double edge notches. The notches act as stress concentrators and affect the load carrying capabilities of the specimens. The stress concentration, due to a notch or an abrupt change of section, is described in terms of K_t, the theoretical stress concentration factor. K_t is the ratio of the maximum stress, due to the notch, to the nominal stress across the area beneath the notch. It should not be confused with K_C or K_{IC}, the fracture toughness values. Values for K_t have been established for various geometries, the most common of which are shown in *Figure 17*. K_t is a function of both the specimen size and the notch radius, and is often approximated with the equation $K_t = \sqrt{a/r}$.

In the most commonly performed notch tests, a round specimen with a 50% notch depth is used. The restraint caused by the notch sets up a state of triaxial tension near the notch root. This stress state increases the flow resistance of the metal and thus decreases the ductility at failure, but it may also increase the fracture strength. The notch strength ratio (NSR—the ratio of notched strength to unnotched strength) has been shown to increase with increasing notch sharpness, at a constant notch depth, to a maximum value at which

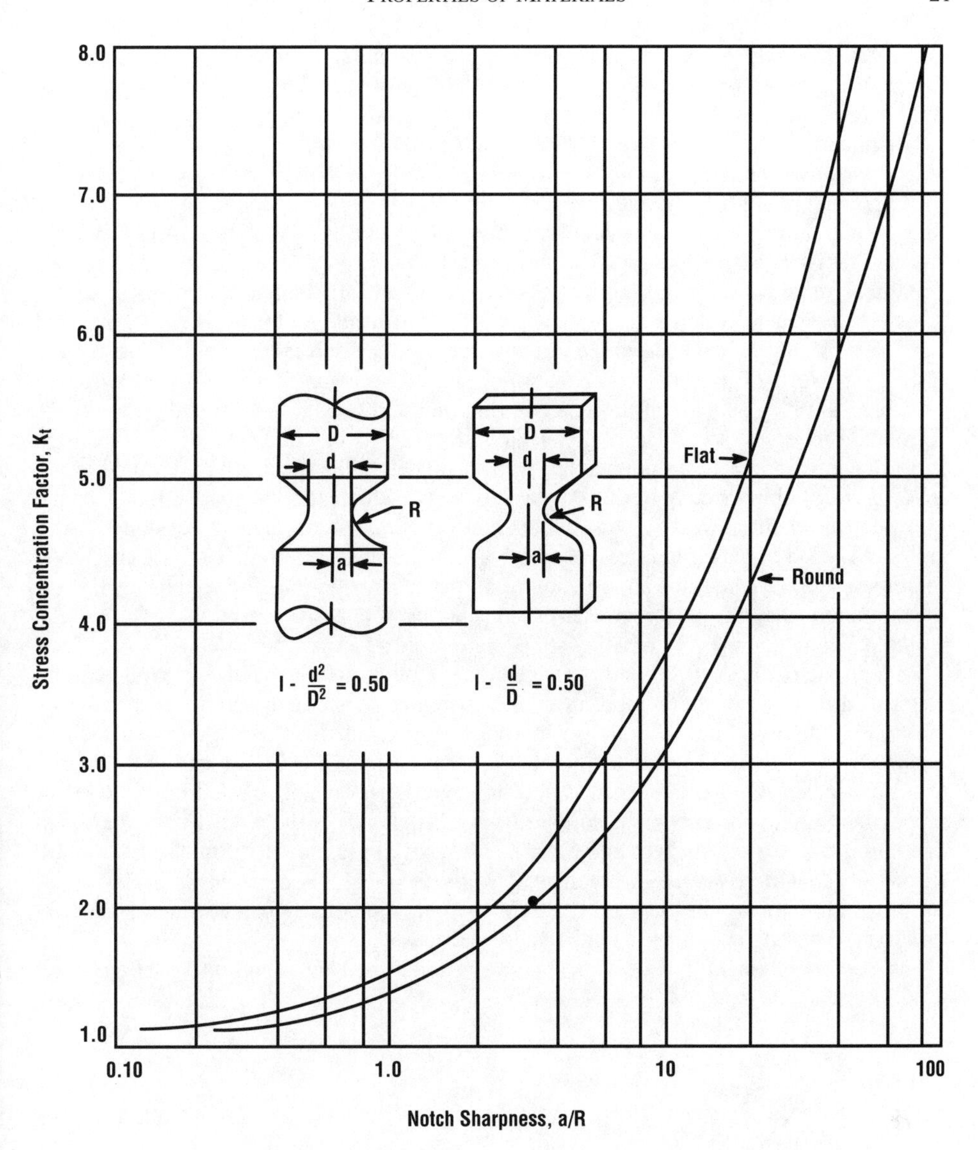

Figure 17. Effect of notch sharpness on the stress concentration factor.

the triaxiality becomes constant. Increasing the sharpness beyond this point causes a reduction in the NSR for notch-sensitive materials. Increasing the notch depth will increase the triaxiality, and will increase the notch strength of the specimen. The primary use of notched tensile tests is to screen materials.

Fatigue Tests

Fatigue has been defined as the "process of progressive localized permanent structural change occurring in material subjected to conditions that produce fluctuating stresses and strains at some point or points, and which may culminate in cracks or complete fracture after a sufficient number of fluctuations. Traditionally, tests have been performed on specimens having simple geometries in attempts to characterize the failure properties of particular materials. Fatigue tests have also been conducted for many reasons, including fatigue life information for design purposes, to evaluate the differences between materials, or to investigate the influences of heat treatment or mechanical working.

A material that is repeatedly subjected to tensile loading may fail even though the stresses imposed on the specimen are below yield strength. Failure may occur after a few load applications or after years of repeated applications. This phenomenon is known as fatigue, and fatigue life is dependent on many variables such as maximum stress, average stress, alternating stress, the ratio of alternating to average stress, yield strength, surface condition, and environment. Since the first indication of fatigue in a material is cracking, fatigue tests generally fall into two categories: those designed to initiate cracking, and those that propagate cracks to determine their growth rate and failure point. Most fatigue tests are performed by repeatedly rotating a simple beam until cracking is determined or it breaks, or by subjecting the specimen to alternating axial tension and strain loads. In rotating tests, the specimen is subjected to 1800 RPM or more. The pressures on the specimen provide axial-loads and axial-strains, subjecting the specimen to repeated cyclic loading until failure occurs, or until the testers and machine admit defeat and discontinue the test. The loading cycles are classed as tension-compression, tension-tension, or zero stress-tension. The maximum stress, *S max*, is the highest tension stress in the cycle, and the minimum stress, *S min*, is the largest compressive stress for compression-tension cycling, or the minimum tensile stress for tension-tension cycling. The mean (or average) of *S max* and *S min* is *Sm*, and *Sa* is the alternating stress, calculated as *S max* minus *Sm*.

The basic method of presenting fatigue data is on an S-N curve that plots maximum stress as a function of the number of cycles to failure. When reading an S/N curve, it is necessary to identify the "A" ratio (alternating stress, *Sa*, divided by the mean stress, *Sm*) or the stress ratio, "R" (*S min* divided by *S max*), because the fatigue life of a material is dependent on the maximum stress and the mean stress of the applied cyclic load. At a specified maximum stress, the fatigue life of a material is the lowest under reversed bending, when *Sm* equals zero and *Sa* = T. For a given maximum stress, the fatigue life increases as *Sm* increases and *Sa* decreases. On the example S-A chart shown in *Figure 18*, the curved horizontal lines labeled "endurance limit" are stresses below which a material will not fail by fatigue, under the indicated conditions.

A Soderberg diagram is another, and more useful, method of presenting fatigue data.

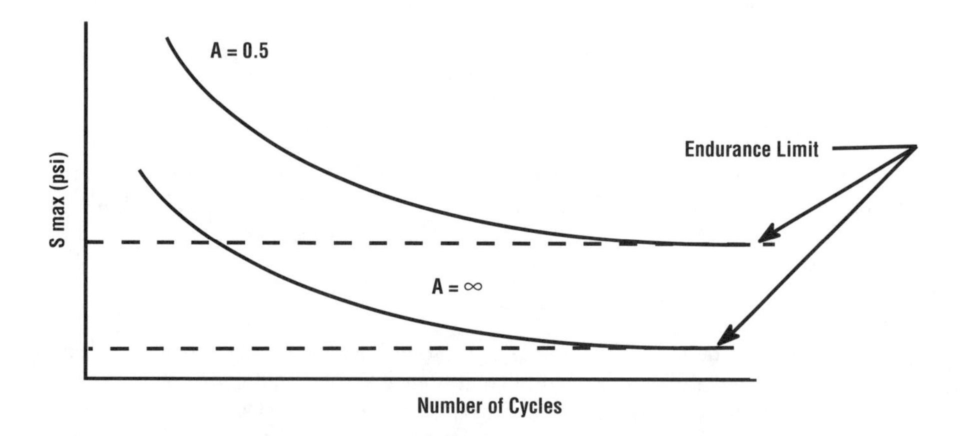

Figure 18. An S-N curve for predicting fatigue life.

Also called a "modified Goodman diagram," it plots the alternating stress (*Sa*) against the mean stress (*Sm*). *Figure 19* illustrates a sample Soderberg diagram, which shows the dependence of fatigue life on the alternating and mean stress, as well as on the A ratio. If the fatigue life is to be determined under the conditions of $Sa = Sm$ (A = 1) at a given maximum stress, and if $Sa = S\,max - Sm = Sa_1$, then the stress Sa_1 is read off the vertical axis and the fatigue life at the proper A ratio is read from the diagram—in this case 10^4 cycles. It can also be noted that at an A ratio of 5, the alternating stress that could be endured would be Sm_2 for 10^4 cycles. Ideally, the Soderberg diagram should be constructed by determining the S-N curve for several A ratios and then plotting these data. The diagram can, however, be determined by testing with only one A ratio. To do this, the S-N curve is established and the points are plotted on the appropriate A ratio line. As an example, data might be obtained from an A ratio of 1. These data are plotted on the 45° (A = 1) line as shown in *Figure 20*, and then a straight line can be drawn connecting these points and the material ultimate strength value plotted on the *Sm* axis. This will provide a Soderberg diagram that is conservative, but accurate enough for many engineering applications. This method offers considerable utility because most of the fatigue data found in the literature is in the form of S-N curves for reversed bending (A ratio = T). Using this method, it is possible to use such data for applications in which the A ratio is other than T. The dashed line in *Figures 19* and *20* that connects the yield strengths plotted on each axis limits the range over which the Soderberg diagram may be used. If values beyond this line are selected, the yield strength of the material is exceeded. Since yielding is usually a criteria for failure, fatigue tests in which the maximum stress is greater than yield are seldom needed.

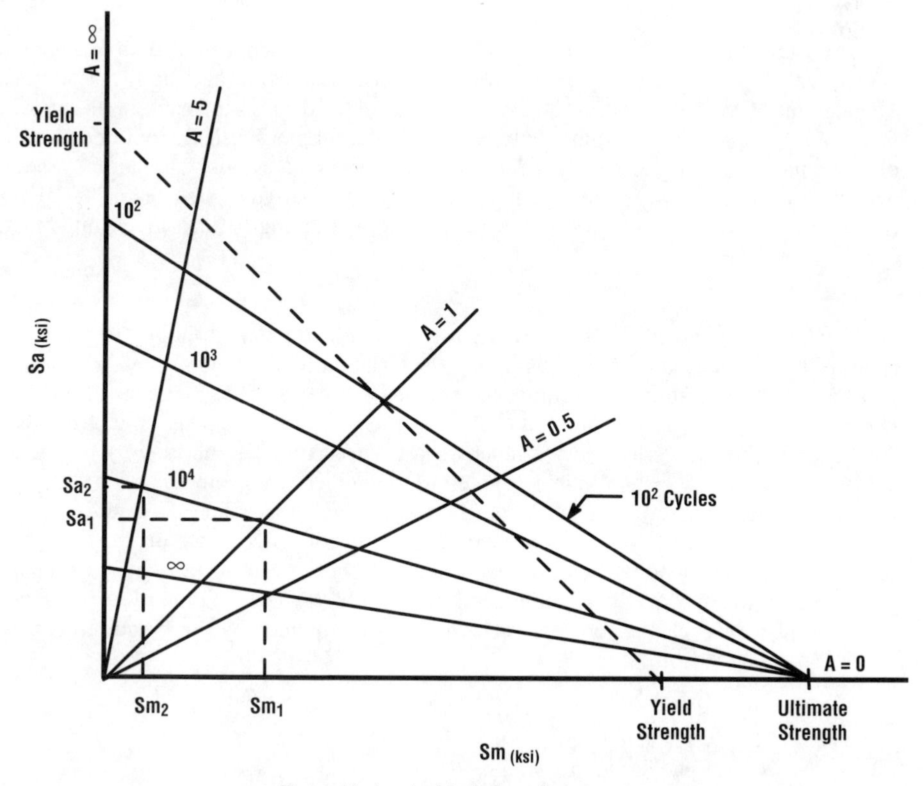

Figure 19. A typical Soderberg diagram.

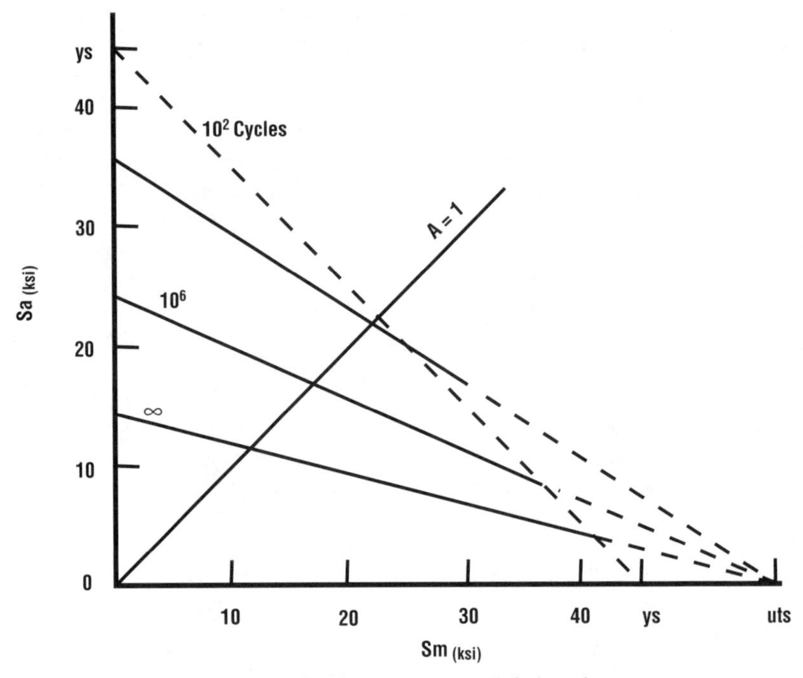

Figure 20. A method for constructing a Soderberg diagram.

There are many factors that affect fatigue life, and the presence of a notch or any other defect that causes a stress concentration can seriously reduce the fatigue life of a metal. The amount of the reduction in fatigue strength is dependent on the stress concentration factor, K_t. For a notched specimen with a K_t of 2, the fatigue strength would be one-half of the unnotched fatigue strength for a given number of cycles. The actual amount that the fatigue strength is reduced is expressed by the fatigue notch factor, K_f, which is the ratio of the unnotched strength to the notched strength for a given number of cycles. Therefore,

$$K_f = Su \div Sn,$$

where *Su* is the unnotched fatigue strength and *Sn* is the notched fatigue strength. For small values, K_f is equal to K_t, but as K_t increases, the ratio of the two values increases. K_f is determined by obtaining S-N curves for notched and unnotched specimens of the same material. Then, for a given number of cycles, the ratio of $Su \div Sn$ is determined from these curves. K_f is also dependent on the notch radius and sharpness, and may have different values for the same K_t, when K_t is developed from nonidentical geometries.

For loadings where the cyclic loading does not vary (i.e., where *Sa* and *Sm* remain constant), the fatigue life can be read directly from the S-N curve or the Soderberg diagram. When the loading cycle varies, a method such as the linear damage principle is used. The linear damage principle can be stated as "the total fatigue life is the sum of the number of cycles experienced at a given stress, divided by the fatigue life at that stress." In equation form, it is:

$$(M_1 \div N_1) + (M_2 \div N_2) + (M_3 \div N_3) + (Mi \div Ni) = 1,$$

where M = cycles at a given stress
N = fatigue life at that stress.

For example, if a material having the fatigue characteristics shown below had been subjected to 20,000 cycles at 20,000 PSI, how many additional cycles could it withstand at 50,000 PSI before it would fail?

Sample characteristics

S max (PSI)	Cycles to Failure
10,000	∞
20,000	70,000,000
30,000	20,000,000
50,000	6,000,000

In this example, M_1 = 20,000,000 for S = 20,000 PSI and N_1 = 70,000,000. M_2 is unknown for S = 50,000 PSI. N_2 = 6,000,000.

$$M_1 \div N_1 + M_2 \div N_2 = 1,$$
$$20.000{,}000 \div 70{,}000{,}000 + M_2 \div 6{,}000{,}000 = 1, \text{ and}$$
$$M_2 = 6{,}000{,}000 \times 1 - (20{,}000{,}000 \div 70{,}000{,}000),$$
$$1 = 7/7, \text{ and } 20{,}000{,}000 \div 70{,}000{,}000 = 2/7$$

therefore, $M_2 = 6{,}000{,}000 \times 5/7 = 4{,}280{,}000$.

As can be determined by the calculation, the part could be expected to withstand an additional 4,280,000 cycles at 50,000 PSI.

It is good design practice to eliminate stress concentrations. This applies not only to stress concentrations designed into a part, but also to those resulting from surface roughness and defects in the material. The effect of surface finish on fatigue life, as reported by G. Dieter Jr. in his book *Mechanical Metallurgy*, can be seen from the following readings obtained on 3130 steel specimens tested under completely reversed stress at 95,000 PSI.

Condition	Surface Roughness	Average Fatigue Life (Cycles)
As machined by lathe	105	24,000
Light hand polishing	6	91,000
Hand-polished	5	137,000
Superfinished	7	212,000
Ground	7	217,000
Ground and polished	2	234,000

Other surface properties can be adjusted to improve fatigue life. Carburizing, nitriding, and cold working will also increase fatigue life. Conversely, decarburization on the surface is particularly damaging to fatigue properties. Shot-peening introduces compressive residual stresses into the surface, as well as increasing strength by cold work. Both of these factors increase fatigue life, but shot-peening also increases surface roughness which of course decreases fatigue life.

Corrosion in metal

Most briefly stated, corrosion is the deterioration of a metal due to chemical or electrochemical action. Because it is so prevalent, corrosion losses are one of the most expensive problems faced in the application of metals.

The galvanic cell, in its simplest form, consists of two electrodes immersed in a conducting solution (the electrolyte). The electrodes may be two dissimilar metals, a metal and a conducting nonmetal (e.g., carbon), or a metal and an oxide. In each case, an electrical potential, which may produce current when the electrodes are joined by a suitable conductor, is induced between the electrodes. When current is flowing, reactions take place at each electrode as seen in *Figure 21*.

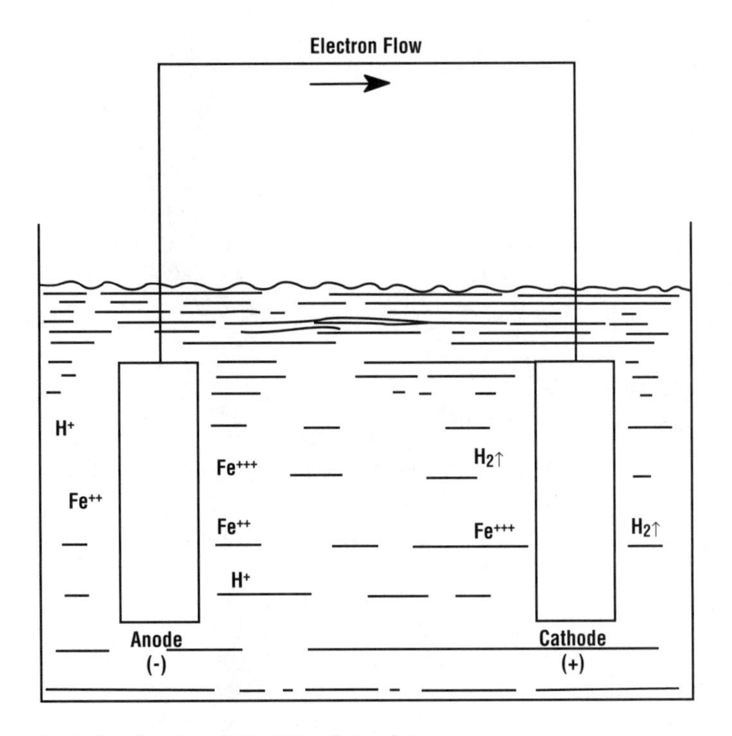

Anode Reaction: $Fe \rightarrow Fe^{++} + 2e^{-} \rightarrow Fe^{+++} + 3e^{-}$
Cathode Reaction: $2H^{+} + 2e^{-} \rightarrow H_2$ (gas)

Figure 21. A galvanic cell.

In a cell, dissolution of the metal occurs at the negative electrode (the anode), while hydrogen is evolved at the positive electrode (the cathode). The anodic metal is corroded and the cathode is protected. The rate of corrosion of the anodic metal depends on the degree of separation of the two metals concerned in the practical galvanic series of metals and alloys (see **Table 2**). The corrosion rate is also dependent on the conductivity and composition of the electrolyte and the relative area of the two metals. Galvanic corrosion in service applications occurs through the formation of composition cells or concentration cells.

Composition cells may occur between dissimilar metals or in a single metal that has areas of differing electrode potential. An example of the dissimilar metal cell is galvanized (Zn coated) steel. The zinc coating is anodic to the steel base metal. Therefore, when an electrolyte, such as water, is present between the Fe and Zn, a galvanic cell is produced. Since Zn is anodic to steel, it is preferentially attacked, thereby protecting the steel. A composition cell may be set up in a single metal because of the differences in electrode potential between different phases, or between the grain boundaries and the matrix. An example of this is intergranular corrosion. This type of attack takes place when the grain boundaries are anodic to the matrix. The occurrence of intergranular corrosion is strongly dependent on the thermal and mechanical treatment given the metal. Generally, the stronger the alloy is made through heat treatment or cold work, the less corrosion resistant it becomes.

Table 2. Galvanic Series of Metals and Alloys.

Anodic End	Magnesium
	Magnesium Alloys
	Zinc
	Alclad
	Aluminum 6053
	Cadmium
	Aluminum 2024
	Cast Iron
	Wrought Iron
	Mild Steel
	13% Chromium Steel Type 410 (Active)
	18-8-3 Chromium Nickel Stainless Steel Type 316
	18-8 Chromium Nickel Stainless Steel Type 304 (Active)
	Tin
	Lead
	Lead Tin Solders
	Naval Brass
	Manganese Bronze
	Muntz Metal
	76 Ni-16 Cr-7 Fe Alloy (Active)
	Nickel (Active)
	Silicon Bronze
	Copper
	Red Brass
	Aluminum Brass
	Admiralty Brass
	Yellow Brass
	76 Ni-16 Cr-7 Fe Alloy (Passive)
	Nickel (Passive)
	Silver Solder
	70-30 Cupro-Nickel
	Monel
	Titanium
	13% Chromium Steel Type 410 (Passive)
	18-8-3 Chromium Nickel Stainless Steel Type 316 (Passive)
	18-8 Chromium Nickel Stainless Steel Type 304 (Passive)
	Silver
	Graphite
	Gold
Cathodic End	Platinum

Concentration cells are formed when there is a difference in the concentration of the electrolyte between the areas in contact. The area with the weakest concentration of electrolyte is attacked. The same type of cell, but of considerably more importance, is the oxygen concentration cell, which is characterized by areas that have different levels of oxygen concentration. The area having the lowest oxygen concentration is anodic and is attacked. Corrosion caused by an oxygen concentration cell can usually be found under surface dirt, mill scale, or other areas that may be oxygen deficient.

Stress corrosion may occur in a susceptible material when it is subjected to residual or applied surface tensile stresses while exposed to a corrosive environment. Stress corrosion cracks initiate and propagate transverse to the loading direction, and a low strength failure results. The time required for stress corrosion to occur is dependent on several variables.

Strength level. As with general corrosion, the higher the strength level obtained through heat treatment or cold working of a given alloy, the lower its resistance to stress corrosion.

Grain orientation. The direction of the applied stress with reference to the grain orientation is of prime importance. The stress corrosion resistance is lowest in the short transverse grain direction and highest in the longitudinal direction, with the long transverse being intermediate. The stress corrosion resistance of a material in the short transverse may be only 10% as good as that in the longitudinal. The resistance in the long transverse direction is dependent on the grain geometry. A material with narrow elongated grains would have a long transverse susceptibility to stress corrosion nearly as poor as that in the short transverse direction. For equiaxed grains (a grain structure with approximately the same dimensions in all directions), such as found in cross rolled material, there is no difference between the longitudinal and the long transverse direction.

Stress level. At a high stress level, a susceptible alloy, when exposed to a corrosive medium, may fail in a matter of minutes. As the stress is decreased, the time to failure increases. This is the basis of the threshold stress concept, which postulates that for stress corrosion susceptible materials, there is a stress value (called threshold stress) below which stress corrosion cracking will not occur, while at higher levels stress corrosion cracking should be expected to occur.

Corrosion protection is afforded by two basic methods: surface protection and galvanic protection. Surface protection involves keeping corrosive environments away from the metal and is commonly achieved by plating, cladding, or painting. Galvanic protection is achieved by electrically connecting the metal with a sacrificial anode, such as zinc coating steels or burying magnesium bars in contact with underground pipelines. Stress corrosion is not usually a problem with carbon steel and wrought iron, but the higher strength alloy steels may be susceptible. It is prevented primarily with surface protection, proper grain direction exposure, and by control of the magnitude of applied stresses.

Physical properties

Thermal conductivity, designated by the letter K, is the measure of the ability of a material to carry or conduct heat. It is analogous to electrical conductivity, and is used in the analysis of heat transfer calculations.

Thermal expansion measures the change in dimension in a solid when it is subjected to a change in temperature. *(Also see p. 366.)* It can be expressed as:

$$\alpha = (\Delta L \div L_0) \times (1 \div T)$$

where α = thermal expansion, expressed as in./in./°F *or* mm/mm/°C
ΔL = change in dimension with temperature
L_0 = original dimension
T = temperature, in °F or °C.

As can be seen from the equation, thermal expansion is actually the strain in a solid due to a temperature change of 1° F or 1° C. The coefficient of thermal expansion depends on the temperature at which the measurement is taken, so the range of temperatures over which α is valid is always specified. The coefficient of thermal expansion is used to calculate volume changes and to determine stresses in restrained members that are subjected to temperature changes. As an example, if a restrained rod is heated from 68 to 268° F (20 to 131° C), which is a temperature change of 200° F, or 111° C, the amount it would increase in length is:

$$\Delta L = L_0 \times \alpha \times \Delta T,$$

or, $\Delta L \div L_0 = \alpha \times \Delta T = e$

where e = strain

ΔT = temperature change.

Since the rod is restrained, it cannot increase in length and a compressive stress (S) is induced into the rod. By definition,

$$S = E \times e.$$

The two equations above, when combined, show compressive stress as:

$$S = E \times \alpha \times \Delta T.$$

Using the above example and equations, determine the compressive strain on a restrained rod made of steel with a coefficient of thermal expansion of 9.9 in./in./°F × 10^{-6} (17.82 m/m/°K × 10^{-6}) and a modulus of elasticity of 28,000 ksi (190 GPa). The change in temperature is 200° F (111° C).

$$S = 28{,}000 \times (9.9 \times 10^{-6}) \times 200 = 55.4 \text{ ksi},$$

or, $S = 190 \times (17.82 \times 10^{-6}) \times 111 = 0.3578$ GPa (375,823,800 Pascals).

Therefore, a compressive force of 55.4 ksi (0.3578 GPa) would be induced in the rod by the indicated temperature change.

Density is the mass per unit volume of a solid, but it is commonly expressed as weight per unit volume, as pounds per cubic foot, pounds per cubic inch, kilograms per cubic meter, and grams per cubic centimeter. **Table 3** provides weights for selected materials. (It should be mentioned that mass density is measured in slugs/ft^3 in U.S. Customary Units and in kg/m^3 in the metric system.) Density becomes a design consideration when the weight of a structure must be controlled. The selection of an alloy is then made on the basis of a high yield strength to density ratio (also called the strength to weight ratio).

Hardness tests

In simple terms, hardness is a material's ability to resist surface abrasion. The relative hardness of materials can be seen on Mohs' hardness scale, which is based on fifteen materials, each of which is capable of scratching the one below it. Any material can be positioned on Mohs' scale according to which of the other materials it is able to scratch.

Mohs' Hardness Scale

Mineral	*Hardness index*
Diamond	15
Boron Carbide	14
Silicon Carbide	13
Fused Alumina	12
Fused Zirconia	11
Garnet	10
Topaz	9
Quartz or Stellite	8
Vitreous Silica	7

(Continued)

Mohs' Hardness Scale *(Continued)*

Mineral	*Hardness index*
Orthoclase (feldspar)	6
Apatite	5
Fluorite	4
Calcite	3
Gypsum	2
Talc	1

While this scale can be useful in determining the approximate relative hardness of a material, it is obviously not capable of recognizing the comparative hardness of steels and other metals. In material selection, hardness plays a large part in determining the machinability and strength, so several tests have been devised for accurately cataloging hardness values.

Table 3. Weights of Materials.

Material	Weight, Expressed in			
	Pounds per Cubic Foot	Pounds per Cubic Inch	Kilograms per Cubic Meter	Grams per Cubic Centimeter
Aluminum	168.5	.0975	2699.11	2.6988
Brass, 80% C, 20% Z	536.6	.3105	8595.51	8.5946
Brass, 70% C, 30% Z	526.7	.3048	8436.92	8.4368
Brass, 60% C, 40% Z	521.7	.3019	8356.83	8.3538
Brass, 50% C, 50% Z	511.7	.2961	8196.65	8.1960
Brick, common	112	.0648	1794.07	1.7937
Brick, fire	143	.0827	2290.64	2.2891
Brick, pressed	137	.0793	2194.53	2.1950
Brick, hard	125	.0723	2002.31	2.0013
Bronze, 90% C, 10% T	547.9	.3171	8776.51	8.7773
Cement, portland, loose	90	.0521	1441.66	1.4421
Cement, portland, set	183	.1059	2931.38	2.9313
Chromium	432.4	.2502	6926.38	6.9255
Clay, loose	63	.0365	1009.16	1.0103
Coal broken, loose, anthracite	54	.0313	864.99	0.8664
Coal broken, loose, bituminous	49	.0294	784.90	0.7861
Concrete	137	.0793	2194.53	2.1950
Copper	554.7	.3210	8885.44	8.8852
Earth, common, loam	75	.0434	1201.38	1.2013
Earth, packed	100	.0579	1601.85	1.6027
Glass	162	.0938	2594.99	2.5964
Gravel, dry, loose	90 to 106	-	1441.66 to 1697.96	-
Gravel, well shaken	99 to 117	-	1585.83 to 1874.16	-
Gold	1204.3	.6969	19291.03	19.2901
Ice	56	.0324	897.03	0.8968
Iron, cast	450	.2604	7208.31	7.2078

(Continued)

Table 3. Weights of Materials. *(Continued)*

Material	Weight, Expressed in			
	Pounds per Cubic Foot	Pounds per Cubic Inch	Kilograms per Cubic Meter	Grams per Cubic Centimeter
Iron, wrought	486.7	.2817	7796.18	7.7974
Lead	707.7	.4095	11336.26	11.3349
Lime	53	.0307	848.98	0.8498
Magnesium	108.6	.0628	1739.60	1.7383
Masonry	150	.0868	2402.77	2.4026
Masonry, dry rubble	138	.0799	2210.55	2.2116
Molybdenum	636.5	.3683	10195.75	10.1945
Mortar, set	103	.0590	1649.90	1.6497
Nickel	549	.3177	8794.13	8.7939
Petroleum, benzene	46	.0266	736.85	0.7363
Petroleum, gasoline	42	.0243	672.78	0.6726
Plaster of Paris	112	.0648	1794.07	1.7937
Platinum	1333.5	.7717	21360.62	21.3606
Quartz	162	.0938	2594.99	2.5964
Salt, common	48	.0278	768.88	0.7695
Sand, dry, loose	90 to 106	-	1441.66 to 1697.96	-
Sand, well shaken	99 to 106	-	1585.83 to 1874.16	-
Silver	657	.3802	10524.13	10.5239
Snow, freshly fallen	5 to 12	-	80.09 to 192.22	-
Snow, wet and compacted	15 to 50	-	240.28 to 800.92	-
Steel	490	.2836	7849.05	7.8500
Stone, gneiss	168	.0972	2691.10	2.6905
Stone, granite	168	.0972	2691.10	2.6905
Stone, limestone	162	.0938	2594.99	2.5964
Stone, marble	168	.0972	2691.10	2.6905
Stone, sandstone	143	.0828	2290.64	2.2919
Stone, shale	162	.0938	2594.99	2.5964
Stone, slate	175	.1013	21803.23	2.8040
Tar	75	.0434	1201.38	1.2013
Tin	455	.2632	7288.40	7.2881
Titanium	280.1	.1621	4486.77	4.4869
Tungsten	1192	.6898	19094.00	19.0936
Water, fresh	62.5	.0362	1001.15	1.0020
Water, sea water	64	.0370	1025.18	1.0242
Zinc	439.3	.2542	7036.91	7.0362
Wood, dry				
Ash, black	28	.0162	448.52	0.4484
Ash, white	41	.0237	656.76	0.6560

(Continued)

Table 3. Weights of Materials. *(Continued)*

Material	Weight, Expressed in			
	Pounds per Cubic Foot	Pounds per Cubic Inch	Kilograms per Cubic Meter	Grams per Cubic Centimeter
Wood, dry				
Beech	45	.0260	720.83	0.7197
Birch	44	.0255	704.81	0.7058
Birch, paper	38	.0220	608.70	0.6090
Cedar, Alaska	31	.0179	496.57	0.4955
Cedar, eastern red	33	.0191	528.61	0.5287
Cedar, southern white	23	.0133	368.42	0.3681
Cedar, western red	23	.0133	368.42	0.3681
Cherry	42	.0203	672.78	0.6726
Cherry, black	35	.0203	560.65	0.5619
Chestnut	41	.0237	656.76	0.6560
Cypress	30	.0174	480.55	0.4816
Elm	45	.0260	720.83	0.7197
Hemlock	29	.0168	464.54	0.4650
Hickory	49	.0284	784.90	0.7861
Locust	46	.0266	736.85	0.7363
Mahogany	53	.0307	848.98	0.8498
Maple, hard	43	.0249	688.79	0.6892
Maple, white	33	.0191	528.61	0.5287
Oak, chestnut	54	.0313	864.99	0.8664
Oak, live	59	.0341	954.09	0.9439
Oak, red, black	41	.0237	656.76	0.6560
Oak, white	46	.0266	738.85	0.7363
Pine, white	26	.0150	416.48	0.4152
Pine, yellow, longleaf	44	.0255	656.76	0.7058
Pine, yellow, shortleaf	36	.0208	576.66	0.5757
Poplar	28	.0162	448.52	0.4484
Redwood, California	26	.0150	416.48	0.4152
Spruce, white, black	27	.0156	432.50	0.4318
Sycamore	37	.0214	592.68	0.5923
Walnut, black	38	.0220	608.70	0.6089
Walnut, white	26	.0150	416.48	0.4152

There are three basic types of hardness tests: indention hardness, scratch hardness, and rebound hardness. As the name implies, indention hardness is measured by indenting the metal with a suitable load and indenting mechanism. The hardness value obtained is actually a measure of the material's resistance to plastic deformation, but through empirical calculations, hardness gives the designer an indication of the strength of the alloy or metal being tested. The advantage of indention hardness testing, which is the hardness test commonly applied to metals, is that it is fast, inexpensive, and virtually nondestructive. In the majority of cases, a small indentation less than $^1/_{16}$ inch in diameter is produced. The three common indention tests are Brinell, Rockwell, and Vickers. Conversion charts for relative hardnesses on these three scales, plus Scleroscope hardness, are given in **Table 4**, which provides Rockwell C conversions, and **Table 5**, which gives Rockwell B scale conversions.

Table 4. Hardness Conversion Scale for Brinell, Rockwell C, Vickers, Scleroscope, and Knoop Hardness Scales.

Brinell Hardness	Rockwell C Hardness	Vickers Hardness	Scleroscope Hardness	Knoop Hardness	Tensile Strength ksi	Tensile Strength MPa
3000 kg	150 kg					
767	66.5	882	93	870	-	-
745	65.3	842	91	584	-	-
722	64	800	88	822	-	-
705	63	772	85	799	-	-
688	62	746	84	776	-	-
670	61	720	83	754	-	-
654	60	697	81	732	-	-
634	59	674	80	710	351	2420
615	58	653	79	690	338	2330
595	57	633	77	670	325	2241
577	56	613	75	650	313	2158
560	55	595	74	630	301	2075
543	54	577	72	612	292	2013
525	53	560	71	594	283	1951
512	52	544	70	576	273	1882
496	51	528	68	558	264	1820
481	50	513	67	542	255	1758
469	49	498	66	526	246	1696
455	48	484	65	510	237	1634
443	47	471	63	495	229	1579
432	46	458	62	480	221	1524
421	45	446	60	466	215	1482
409	44	434	59	452	208	1434
400	43	423	58	438	201	1386
390	42	412	56	426	195	1344
381	41	402	55	414	188	1296
371	40	392	54	402	182	1255
362	39	382	52	391	177	1220
353	38	372	51	380	171	1179
344	37	363	50	370	166	1145
336	36	354	49	360	161	1110

(Continued)

Table 4. *(Continued)* **Hardness Conversion Scale for Brinell, Rockwell C, Vickers, Scleroscope, and Knoop Hardness Scales.**

Brinell Hardness	Rockwell C Hardness	Vickers Hardness	Scleroscope Hardness	Knoop Hardness	Tensile Strength ksi	Tensile Strength MPa
3000 kg	150 kg					
327	35	345	48	351	155	1069
319	34	336	47	342	152	1048
311	33	327	46	334	149	1027
301	32	318	45	326	146	1007
294	31	310	44	318	141	972
286	30	302	42	311	138	951
279	29	294	41	304	135	931
271	28	286	40	297	131	903
264	27	279	39	290	128	883
258	26	272	38	284	126	869
253	25	266	-	278	123	848
247	24	260	37	272	119	820
243	23	254	36	265	117	807
237	22	248	35	261	115	793
231	21	243	34	256	112	772
226	20	238	34	251	110	758

Brinell ratings 500 and above obtained with tungsten carbide ball. All Brinell with 3,000 kg load.
Rockwell C ratings obtained with diamond indenter and 150 kg load.

Table 5. Hardness Conversion Scale for Brinell, Rockwell B, Vickers, Scleroscope, and Knoop Hardness Scales.

Brinell Hardness	Brinell Hardness	Rockwell B Hardness	Vickers Hardness	Scleroscope Hardness	Knoop Hardness	Tensile Strength ksi	Tensile Strength MPa
3000 kg	500 kg	100 kg					
240	201	100	240	36	261	116	800
234	195	99	234	35	258	114	786
228	189	98	228	34	253	109	752
222	184	97	222	-	247	105	724
216	179	96	216	33	240	102	703
210	175	95	210	32	233	100	689
206	171	94	206	31	225	98	676
200	167	93	200	30	218	94	648
195	163	92	195	29	211	92	634
190	160	91	190	-	206	90	621
185	157	90	185	28	201	89	614
180	154	89	180	27	196	88	607
176	151	88	176	-	192	86	593

(Continued)

Table 5. *(Continued)* **Hardness Conversion Scale for Brinell, Rockwell B, Vickers, Scleroscope, and Knoop Hardness Scales.**

Brinell Hardness	Brinell Hardness	Rockwell B Hardness	Vickers Hardness	Scleroscope Hardness	Knoop Hardness	Tensile Strength ksi	Tensile Strength MPa
3000 kg	500 kg	100 kg					
172	148	87	172	26	188	84	579
169	145	86	169	-	184	83	572
165	142	85	165	25	180	82	565
162	140	84	162	-	176	81	558
159	137	83	159	24	173	80	552
156	135	82	156	-	170	76	524
153	133	81	153	23	167	73	503
150	130	80	150	-	164	72	496
147	128	79	147	22	161	70	483
144	126	78	144	-	158	69	476
141	124	77	141	21	155	68	469
139	122	76	139	-	152	67	462
137	120	75	137	-	150	66	455
135	118	74	135	-	147	65	448
132	116	73	132	-	145	64	441
130	114	72	130	20	143	63	434
127	112	71	127	-	141	62	427
125	110	70	125	19	139	61	421
123	109	69	123	-	137	60	414
121	108	68	121	18	135	59	407
119	106	67	119	-	133	58	400
117	104	66	117	17	131	57	393
116	102	65	116	16	129	56	386
114	100	64	114	15	127	-	-
112	99	63	112	-	125	-	-
110	98	62	110	-	124	-	-
108	96	61	108	-	122	-	-
107	95	60	107	-	120	-	-
106	94	59	106	-	118	-	-
104	92	58	104	-	117	-	-
103	91	57	103	-	115	-	-
101	90	56	101	-	114	-	-
100	89	55	100	-	112	-	-
-	87	54	-	-	111	-	-
-	86	53	-	-	110	-	-
-	85	52	-	-	109	-	-
-	84	51	-	-	108	-	-

(Continued)

Table 5. *(Continued)* **Hardness Conversion Scale for Brinell, Rockwell B, Vickers, Scleroscope, and Knoop Hardness Scales.**

Brinell Hardness	Brinell Hardness	Rockwell B Hardness	Vickers Hardness	Scleroscope Hardness	Knoop Hardness	Tensile Strength ksi	Tensile Strength MPa
3000 kg	500 kg	100 kg					
-	83	50	-	-	107	-	-
-	82	49	-	-	106	-	-
-	81	48	-	-	105	-	-
-	80	47	-	-	104	-	-
-	80	46	-	-	103	-	-
-	79	45	-	-	102	-	-
-	78	44	-	-	101	-	-
-	77	43	-	-	100	-	-
-	76	42	-	-	99	-	-
-	75	41	-	-	98	-	-
-	75	40	-	-	97	-	-

Brinell readings are given for both 500 kg and 3,000 kg load with 10 mm ball.
Rockwell B ratings obtained with 100 kg weight and $^1/_{16}$ inch ball.

Brinell Hardness Testing

A Swede, Johann Brinell, invented the test named for him in 1900, and it remains in use as the most often cited method of hardness testing, especially on castings and forgings. Brinell hardness is measured by indenting the surface of the metal with a hardened steel ball (10 mm in diameter) under a constant load of 500 to 3,000 kg (1100 to 6614 lbs). The lightest load is normally used for testing nonferrous alloys, and the highest load is used on harder materials such as steels and cast irons. The applied load is usually held for thirty seconds on nonferrous materials and for one-third to one-half that time on harder materials. The diameter of the indentation is then measured with a microscope having an ocular scale, and the Brinell Hardness (BHN, or HB) is calculated with the following equation.

$$BHN = P \div \text{surface area of the impression,}$$

or, $$BHN = P \div ([\pi D \div 2] \times [D - \sqrt{D^2 - d^2}])$$

where P = applied load in kilograms
D = diameter of the ball
d = diameter of the indentation, in millimeters.

For materials with BHN of 444 to 627 (which is the upper limit of BHN ratings), the steel ball must be substituted with a ball made of tungsten carbide.

When conducting Brinell tests, there are several standards that should be followed. Perhaps of primary importance is to be sure that the test is performed on a flat surface, not one that is curved or has irregularities. Indentations should not be made near an edge of the material, as there may be insufficient supporting material on the edge side of the indentation. Since indentations cause localized cold working of the material, indentations should not be made within three diameters of each other. The material's thickness at the location of the test should be a minimum of ten times the depth of the indentation.

Brinell Hardness can be converted to Rockwell C scale and Rockwell B scale hardness numbers with the following conversion formulas, courtesy of Kennametal Inc. It should be noted that even though these conversion formulas provide a reasonable degree of accuracy, the validity of converting from one measurement system to another cannot always be assured. The conversion charts on the Tables in this section may also be used for converting from one scale to another.

Rockwell C Hardness (HRC)		Equation for Converting HRC to Brinell Hardness
from	to	
21	30	BHN = 5.97 × HRC + 104.7
31	40	BHN = 8.57 × HRC + 27.6
41	50	BHN = 11.158 × HRC + 79.6
51	60	BHN = 17.515 × HRC – 401
Rockwell B Hardness (HRB)		**Equation for Converting HRB to Brinell Hardness**
from	**to**	
55	69	BHN = 1.646 × HRB + 8.7
70	79	BHN = 2.394 × HRB – 42.7
80	89	BHN = 3.279 × HRB – 114
90	100	BHN = 5.582 × HRB – 319

Rockwell Hardness Testing

The most widely used hardness test in the U.S. is the Rockwell test. The indenter in this test is specified by ASTM specifications, and it has a diamond cone point (called a Brale indenter) with a radius of 200 ± 10 μm (0.0079 in.). Unfortunately, for years the indenter size was specified as 192 μm, and older tests are not compatible with newer tests, especially in the HRC (or R_C) 63 range. In substitution for the diamond indenter, Rockwell tests may make use of hardened steel and cemented carbide balls. Standard ball diameters are $^1/_{16}$, $^1/_8$, $^1/_4$, and $^1/_2$ inch, and they are usually employed to test softer materials including soft steels and aluminum and copper alloys.

There are several different Rockwell "scales," the two most common being the B and C. The scale is determined by the indenter and load used in the test. HRB (R_B) uses a $^1/_{16}$" ball and 100 kg load and is used to measure soft steels, malleable irons, and copper and aluminum alloys as well as other softer materials. HRC uses the Brale indenter and 150 kg load and is used to measure materials harder than HRB 100 including steels, hard cast irons, pearlitic malleable iron, titanium, and deep case-hardened steel. Most of the remainder of the alphabet is used up with the other scales, which are: HRA–Brale indenter and 60 kg load, for testing shallow, case-hardened steel, cemented carbide; HRD–Brale indenter and 100 kg load for measuring medium case-hardened steel, some pearlitic malleable cast irons; HRE–uses a $^1/_8$" ball and 100 kg load to test cast iron, magnesium and some aluminum alloys, bearing metals; HRF–uses a $^1/_{16}$" ball indenter and 60 kg load to measure annealed copper alloys, thin soft sheet metals; HRG–uses a $^1/_{16}$" ball indenter and 150 kg load to test phosphorus bronze, beryllium copper, malleable irons; the upper limit for the scale is 92 HRG; HRH–Uses $^1/_8$" ball and 60 kg load for testing aluminum, zinc, lead. The following suffixes are used for bearing metals and very soft or thin materials: K ($^1/_8$" ball, 150 kg load), L ($^1/_4$" ball, 60 kg load), M ($^1/_4$" ball, 100 kg load), P ($^1/_4$" ball, 150 kg load), R ($^1/_2$" ball, 60 kg load), S ($^1/_2$" ball, 100 kg load), V ($^1/_2$" ball, 150 kg load); and there are designations for the T, W, and X suffixes which are also used for very soft materials. See **Tables 4**, **5**, and **6** for Rockwell hardness charts.

In the actual test, a minor load of 10 kg is first applied to stabilize the indenter and specimen. This is followed by the major load, which may be 60, 100, or 150 kg, and the

Table 6. Rockwell Hardness and Rockwell Hardness Superficial Hardness Comparative Scales. *(See Notes.)*

Rockwell Hardness Scales						Rockwell Superficial Hardness Scales					
A Scale	B Scale	C Scale	D Scale	E Scale	F Scale	15-N Scale	30-N Scale	45-N Scale	15-T Scale	30-T Scale	45-T Scale
85.0	-	66.0	76.0	-	-	93.0	83.0	73.0	-	-	-
84.0	-	64.0	74.0	-	-	92.0	81.0	71.0	-	-	-
83.0	-	62.0	73.0	-	-	91.0	79.0	69.0	-	-	-
81.0	-	60.0	71.0	-	-	90.0	78.0	67.0	-	-	-
80.0	-	58.0	69.0	-	-	89.0	76.0	64.0	-	-	-
79.0	-	56.0	68.0	-	-	88.0	74.0	62.0	-	-	-
78.0	-	54.0	66.0	-	-	87.0	72.0	60.0	-	-	-
77.0	-	52.0	65.0	-	-	86.0	70.0	57.0	-	-	-
75.5	-	50.0	63.0	-	-	85.5	68.0	54.5	-	-	-
74.5	-	48.0	61.5	-	-	84.5	66.5	52.5	-	-	-
73.5	-	46.0	60.0	-	-	83.5	64.5	50.0	-	-	-
72.5	-	44.0	58.5	-	-	82.5	63.0	47.5	-	-	-
71.5	-	42.0	57.0	-	-	81.5	61.0	45.5	-	-	-
70.5	-	40.0	55.5	-	-	80.5	59.5	43.0	-	-	-
69.5	-	38.0	54.0	-	-	79.5	58.0	41.0	-	-	-
68.5	-	36.0	52.5	-	-	78.5	56.0	38.5	-	-	-
67.5	-	34.0	50.5	-	-	77.5	54.5	36.0	-	-	-
66.5	106	32.0	49.5	-	116.5	76.5	52.5	34.0	94.5	85.5	77.0
64.5	104	28.5	46.5	-	115.5	75.0	49.5	30.0	94.0	84.5	75.0
63.0	102	25.5	44.5	-	114.5	73.5	47.0	26.5	93.0	83.0	73.0
61.5	100	22.5	42.0	-	113.0	72.0	44.5	23.0	92.5	81.5	71.0
60.5	98	20.0	40.0	-	112.0	70.5	42.0	20.0	92.0	80.5	69.0
59.0	96	17.0	38.0	-	111.0	69.0	39.5	17.0	91.0	79.0	67.0
57.5	94	14.5	36.0	-	110.0	68.0	37.5	14.0	90.5	77.5	65.0
56.5	92	12.0	34.0	-	108.5	66.5	35.5	11.0	89.5	76.0	63.0
55.0	90	9.0	32.0	108.5	107.5	65.0	32.5	7.5	89.0	75.0	61.0
53.5	88	6.5	30.0	107.0	106.5	64.0	30.5	5.0	88.0	73.5	59.5
52.5	86	4.0	28.0	106.0	105.0	62.5	28.5	2.0	87.5	72.0	57.5
51.5	84	2.0	26.5	104.5	104.0	61.5	26.5	0.5	87.0	70.5	55.5
50.0	80	-	24.5	103.0	103.0	-	-	-	86.0	69.5	53.5
49.0	78	-	22.5	102.0	101.5	-	-	-	85.5	68.0	51.5
47.5	76	-	21.0	100.5	100.5	-	-	-	84.5	66.5	49.5

(Continued)

Table 6. *(Continued)* **Rockwell Hardness and Rockwell Hardness Superficial Hardness Comparative Scales.** *(See Notes.)*

Rockwell Hardness Scales						Rockwell Superficial Hardness Scales					
A Scale	B Scale	C Scale	D Scale	E Scale	F Scale	15-N Scale	30-N Scale	45-N Scale	15-T Scale	30-T Scale	45-T Scale
46.5	76	-	19.0	99.5	99.5	-	-	-	84.0	65.5	47.5
45.5	74	-	17.5	98.0	98.5	-	-	-	83.0	64.0	45.5
44.0	72	-	16.0	97.0	97.0	-	-	-	82.5	62.5	43.5
43.0	70	-	14.5	95.5	96.0	-	-	-	82.0	61.0	41.5
42.0	68	-	13.0	94.5	95.0	-	-	-	81.0	60.0	39.5
41.0	66	-	11.5	93.0	93.5	-	-	-	80.5	58.5	37.5
4.0.	64	-	10.0	91.5	92.5	-	-	-	79.5	57.0	35.5
39.0	62	-	8.0	90.5	91.5	-	-	-	79.0	56.0	33.5
-	60	-	-	89.0	90.0	-	-	-	78.5	54.5	31.5
-	58	-	-	88.0	89.0	-	-	-	77.5	53.0	29.5
-	56	-	-	86.5	88.0	-	-	-	77.0	51.5	27.5
-	54	-	-	85.5	87.0	-	-	-	76.0	50.5	25.5
-	52	-	-	84.0	85.5	-	-	-	75.5	49.0	23.5
-	50	-	-	83.0	84.5	-	-	-	74.5	47.5	21.5
-	48	-	-	81.5	83.5	-	-	-	74.0	46.5	19.5
-	46	-	-	80.5	82.0	-	-	-	73.5	45.0	17.0

Notes: **Hardness Scales.** A scale readings are obtained with a Brale indenter and 60 kg load. B scale readings are obtained with a $^{1}/_{16}$ inch ball and 100 kg load. C scale readings are obtained with a Brale indenter and 150 kg load. D scale readings are obtained with a Brale indenter and 100 kg load. E scale readings are obtained with a 1/8 inch ball and 100 kg load. F scale readings are obtained with a $^{1}/_{16}$ inch ball and 60 kg load. **Superficial Hardness Scales.** 15-N scale readings are obtained with an N Brale indenter and 15 kg load. 30-N scale readings are obtained with an N Brale indenter and 30 kg load. 45-N scale readings are obtained with an N Brale indenter and 45 kg load. 15-T scale readings are obtained with a $^{1}/_{16}$ inch ball and 15 kg load. 30-T scale readings are obtained with a $^{1}/_{16}$ inch ball and 30 kg load. 45-T scale readings are obtained with a $^{1}/_{16}$ inch ball and 45 kg load.

penetration is read on a dial gage. The hardness number read from the gage varies directly with the actual material hardness, and inversely with the depth of penetration. A hard material will not allow as deep a penetration of the indenter as will a soft material. The dial gage used to measure the indentation contains 100 divisions, each of which corresponds to a penetration of 0.00008 inch.

Another version of the Rockwell test is the Superficial Rockwell Hardness Test. The test is conducted as with the standard test, but very light loads are used. The minor (initial load is 3 kg, and the major load may be 15, 30, or 45 kg. Superficial Rockwell values always give the hardness number followed by the letters N (for Brale tester), T ($^1/_{16}$" ball), W ($^1/_8$" ball), X ($^1/_4$" ball), or Y ($^1/_2$" ball) that indicate the type of tester used. A typical Superficial Rockwell designation would be 70HR45W, indicating that the material has a superficial hardness of 70HR and it was tested with a 45 kg load using a $^1/_8$" ball.

Formulas for converting HRB and HRC to BHN hardness are provided in the section immediately above on Brinell hardness.

Vickers Hardness Testing

Also known as the Diamond Pyramid Hardness Test, this test was developed in the U.K. in 1925 and uses a square base diamond indenter. The main attraction of this test is that it uses an identical indenter for all tests, thereby making comparative analysis of different materials easy because they can all be placed on the same scale. The load is applied smoothly in a Vickers test, and may vary from 1 to 120 kg. After being applied for 10 to 15 seconds, it is removed and the impression is measured. A diamond pyramid hardness (DPH) or Vickers hardness (VHN or HV) number obtained in this way is defined as the load divided by area of indentation, and is calculated with the following equation.

$$VHN = (2P \times \sin \theta/2) \div d^2$$

or,

$$VHN = 1.8544P \div d^2$$

where P = load, in kg

θ = the angle (136°) between opposite faces in the diamond

d = average length of the diagonals in the indentation, in millimeters.

A metallurgically polished surface is required for Vickers testing.

Other Hardness Tests

Scleroscope hardness testing (also known as Shore's Scleroscope Test) uses a diamond tipped hammer dropped from a fixed and standardized height. The height of the rebound of the hammer is measured and used as an indicator of the hardness of the material. It offers the ability and flexibility to perform tests rapidly, in an almost production line setting. Also, it usually leaves the specimen completely unmarked, and, like the Vickers test, all Scleroscope hardness ratings are made on a single, easily comparable scale. On the other (negative) hand, the readings are very sensitive to the surface condition of the material, and measurements of the rebound of the hammer are sometimes not as precise as a microscopic measurement of an indented material. See **Tables 4** and **5**.

Knoop indention testing is used for microhardness testing, when it is necessary to perform tests on smaller samples than required by other methods of testing. For example, to test the variation of the hardness through the thickness of a carburized case, Knoop tests could be used. A microhardness test is analogous to ordinary indention testing, the only difference being a smaller indenter and the use of much lighter loads. It should be noted that although the Knoop indenter, in conjunction with the Tukon tester, is widely used for microhardness testing in the U.S., the Vickers indenter test is prevalent in Europe. See **Tables 4** and **5**.

The Knoop indenter is a pyramidal shaped diamond, and its impression, viewed normal to the specimen surface, is rhombic in shape with the long diagonal perpendicular to, and seven times the length of, the shorter diagonal. The loads applied by the Tukon tester are from 1 to 1000 g. The length of the long diagonal of the impression is measured precisely with a microscope, and the Knoop hardness (KHN or HK) is calculated as follows.

$$KHN = P \div (L^2 \times C)$$

where P = load, in kg

L = length of longest diagonal, in millimeters

C = a constant of the indenter, relating projected area of the indentation to the square of the length of the long diagonal.

A metallurgically polished surface is required for Knoop testing.

IRHD (International Rubber Hardness Degrees) tests are carried out on rubber parts, especially O-rings. It uses a light initial force (8.46 grams for micro scales and 295.74 grams for regular scales) applied through an indenter to zero the testing apparatus scale, followed by a test force (ranging from 15.7 grams for micro scales to 597 grams for regular scales). The degree of indentation is measured to determine the hardness. Readings indicate the durometer (testing apparatus) type (either A or D), a test "value," and duration of the test. For instance, a D/50/15 reading indicated that a D type durometer was used, the test value is 50, and the test duration was 15 seconds.

Machinability ratings of materials

The machinability of a workpiece material can be defined as the comparative ease with which a material removal operation can be performed upon the material relative to the same operation on a benchmark material. A material's rating can help in the selection and adjustment of speeds and feeds for unfamiliar materials. For example, SAE-AISI resulfurized carbon steel 1212 has a machinability rating of 100%, while SAE-AISI 1030 has a machinability rating of 70%. This means that if 3 in.^{3} (49.16 cm^3) of SAE-AISI 1112 can be removed per minute with a specific tooling setup, that same setup should remove 70% of that amount, or 2.1 in.^{3} (34.41 cm^3) of 1030 carbon steel. See **Table 7** for comparative machinability ratings.

Table 7. Machinability Rating Guide. *(See text for explanation.)*

SAE-AISI Designation	Rating %	SAE-AISI Designation	Rating %	SAE-AISI Designation	Rating %	SAE-AISI Designation	Rating %	SAE-AISI Designation	Rating %
Carbon Steels									
1005	45	1019	78	1034	70	1046	57	1069A	49
1006	50	1020	72	1035	70	1049	54	1070A	49
1008	55	1021	78	1037	70	1050	54	1074A	45
1010	55	1022	78	1038	64	1050A	70	1075A	45
1011	53	1023	76	1039	64	1053	54	1078A	45
1012	55	1025	72	1040	64	1054	54	1080A	42
1013	53	1026	78	1042	64	1055A	51	1084A	45
1015	72	1029	70	1043	57	1059A	51	1085A	42
1016	78	1030	70	1044	57	1060A	51	1086A	42
1017	72	1031	70	1045	57	1064A	49	1090A	42
1018	78	1033	70	1045A	72	1065A	49	1095A	42
Resulfurized Carbon Steel									
1106	79	1117	91	1132	76	1141A	81	1151	70
1108	80	1118	91	1137	72	1144	76	1151A	81
1109	81	1119	100	1138	76	1144A	85	1211	94
1110	81	1120	81	1139	76	1145	66	1212	100
1115	81	1125	81	1140	72	1145A	78	1213	136
1116	94	1126	81	1141	70	1146	70	1215	136
High Manganese Carbon Steel									
1524	66	1536	64	1548	55	1552	49	1566A	49
1527	66	1541	57	1551	54	1561A	49	1572A	49

(Continued)

Table 7. *(Continued)* **Machinability Rating Guide.** *(See text for explanation.)*

SAE-AISI Designation	Rating %	SAE-AISI Designation	Rating %	SAE-AISI Designation	Rating %	SAE-AISI Designation	Rating %	SAE-AISI Designation	Rating %
Carbon Steel (Leaded)									
10L18	92	10L45	66	10L45A	84	10L50	60	10L50A	78
Free Machining Steel (Leaded)									
11L17	104	11L41	79	11L44	87	12L13	170	12L15	170
11L37	84	11L41A	94	11L44A	98	12L14	170	-	-
Alloy Steel (Leaded)									
41L40A	77	41L50A	70	86L20A	77	-	-	-	-
Manganese Alloy Steel									
1320A	57	1330A	60	1335A	60	1340A	57	1345A	57
Molybdenum Steel									
4012	78	4037A	72	4142A	66	4340A	57	4640	66
4017	78	4047A	66	4145A	64	4419	78	4718	60
4023	78	4118	78	4147A	64	4615	66	4720	60
4024	78	4130A	72	4150A	60	4620	66	4815A	51
4027	66	4137A	70	4161A	60	4621	66	4817A	49
4028	72	4140A	66	4320A	60	4626	60	4820	49
Nickel Chromium Molybdenum Steel									
8615	70	8625	64	8637A	70	8650A	60	8720	66
8617	66	8627	64	8640A	66	8653A	56	8822	64
8620	66	8630A	72	8645A	64	8655A	57	9255A	54
8622	66	8635A	70	8647A	60	8660A	54	9260A	51

"A" indicates annealed. Carbon Steel 1212 (100% rating) is the comparison material for ratings. The information on this Table was supplied by DoAll Company and Texaco.

(Continued)

Table 7. *(Continued)* **Machinability Rating Guide.** *(See text for explanation.)*

SAE-AISI Designation	Rating %	SAE-AISI Designation	Rating %	SAE-AISI Designation	Rating %	SAE-AISI Designation	Rating %	SAE-AISI Designation	Rating %
Chromium Steel									
5015	78	5130	57	5140A	70	5150A	64	5160A	60
5060A	60	5132A	72	5145A	66	5152A	64	E51100A	40
5120	76	5135A	72	5147A	66	5155A	60	E52100A	40
Chromium Vanadium Steel									
6102	57	6118	66	6145	66	6150A	60	6152A	60
Alloy Steel - Boron									
50B44A	70	50B50A	70	51B60A	60	94B17	66	94B30A	72
50B46A	70	50B60A	64	81B45A	66	-	-	-	-
Stainless Steel									
301	55	308	27	317	35	403	55	418	40
302	50	309	28	321	36	405	60	420	45
303	65	310	30	330	30	410	55	430F	65
304	40	314	32	347	40	416	90	440	50
Tool Steel									
A2, A3, A4	16	D5, D7	11	H24, H25	15	O1, O2, O7	16	S1, S2, S5	20
A6, A8, A9	16	H10, H11	20	H26, H42	15	O6	38	T1	14
A7	11	H13, H14	20	M2	14	P2, P3, P4	25	T4	11
A10	27	H19	20	M3	11	P5, P6	25	Ta5	8
D2, D3, D4	11	H21, H22	15	M15	8	P20, P21	22	W (All)	30

"A" indicates annealed. Carbon Steel 1212 (100% rating) is the comparison material for ratings. The information on this Table was supplied by DoAll Company and Texaco.

(Continued)

Table 7. *(Continued)* **Machinability Rating Guide.** *(See text for explanation.)*

Other Materials (Non-Steel)					
Material	**Type**	**Rating %**	**Material**	**Type**	**Rating %**
Cast and Malleable Iron	Soft Cast	81	Cast Steel	BHN 120	85
	Medium Cast	64		BHN 220	50
	Hard Cast	47		BHN 245	44
	Malleable Iron	80-106	Bronze	Al Bronze	60
Brass	Yellow	80		Mn Bronze	60
	Red	60		Ph Bronze	40
	Leaded	280		Ph, Leaded	140
	Red, Leaded	180		Si Bronze	60
Aluminum	2S	300-1500	Magnesium	Dow H	500-2000
	11S-T3	500-2000		Dow J	500-2000
	17S-T	300-1500		Dow R	1150
Nickel and Nickel Alloy	BHN 135	26	Copper and Copper Alloy	Cast	70
	Monel, Cast	45		Rolled	60
	Monel, Rolled	55		Everdur	60
	Monel "K"	35		Everdur, Leaded	120
	Inconel (B)	35		Gun Metal, Cast	60

"A" indicates annealed. Carbon Steel 1212 (100% rating) is the comparison material for ratings. The information on this Table was supplied by DoAll Company and Texaco.

Table 8. Specific Gravity of Selected Materials.

Material	Specific Gravity	Material	Specific Gravity
Aluminum	2.69	Mercury	13.58
Beryllium	1.84	Molybdenum	10.19
Babbitt	7.28	Monel	8.64
Brass (cast)	8.55	Nickel	8.79
Bronze (3-10% Al)	8.20	Platinum	21.4
Bronze (8-14% Sn)	8.20	Plutonium	19.80
Cadmium	8.65	Silver	10.52
Chromium	6.92	Stainless Steel	7.75
Cobalt	8.75	Steel	7.85
Copper	8.93	Steel (Tool)	7.72
Gold	19.29	Tin	7.28
Glass	2.59	Titanium	4.49
Iron (cast)	7.20	Tungsten	19.60
Iron (wrought)	7.79	Uranium	18.70
Lead	11.33	Vanadium	5.95
Magnesium	1.74	Water (Pure)	1.00
Manganese	7.51	Zinc	7.10

Specific gravity is the heaviness of a substance compared to an equal volume of water.

Ferrous Metals

Ferrous metals—cast iron, steel, and various alloy steels—contain iron as their base metal. By adding alloying elements, they may be altered to produce specific physical and chemical properties. Cast irons are hard, brittle iron alloys containing approximately 2 to 4% carbon and 1 to 3% silicon that possess excellent wear resistance characteristics. In the form of white cast iron, which is essentially free of graphite, it is very hard and is used where abrasion and wear resistance are required. Unalloyed white cast iron typically contains the following elements: carbon, 1.8 to 3.6%; silicon, 0.5 to 1.9%; manganese, 0.25 to 0.8%; phosphorus, 0.06 to 0.2%; and sodium, 0.06 to 0.2%.

White iron can be made into malleable iron by bringing it to an elevated temperature (960° F [1760° C]) for an extended period of time (around 60 hours) and then allowing it to cool very slowly to provide prolonged annealing. Malleable iron is stronger and tougher than white cast iron, but much more expensive to produce. Its ductility and toughness falls between that of ductile cast iron and gray cast iron. Unalloyed malleable cast iron typically contains the following elements: carbon, 2.2 to 2.9%; silicon, 0.9 to 1.9%; manganese, 0.15 to 1.2%; phosphorus, 0.02 to 0.2%; and sodium, 0.02 to 0.2%.

Another form of cast iron is gray iron, which is widely used for castings because of its excellent flow characteristics. It can be machined relatively easily, and it is widely used as a base material to dampen vibration or reduce thermal shock. Unalloyed gray cast iron typically contains the following elements: carbon, 2.5 to 4%; silicon, 1.0 to 3%; manganese, 0.2 to 1%; phosphorus, 0.002 to 1%; and sodium, 0.02 to 0.25%.

Ductile cast iron, which is also called modular iron or spheroidal graphite cast iron, is derived from the same basic materials as gray iron, but the purity requirements are more strict. Its castability is approximately equal to that of gray iron, but its ductility and strength characteristics are appreciably better. Unalloyed ductile cast iron typically contains the following elements: carbon, 3 to 4%; silicon, 1.8 to 2.8%; manganese, 0.1 to 1%; phosphorus, 0.01 to 0.1%; and sodium, 0.01 to 0.03%.

Compacted graphite cast iron is used extensively in industry as the material widely used for the heads of diesel engines and for disk brake rotors. Its properties can be changed radically by the inclusion of magnesium and titanium. Unalloyed compacted graphite cast iron typically contains the following elements: carbon, 2.5 to 4%; silicon, 1 to 3%; manganese, 0.2 to 1%; phosphorus, 0.01 to 0.1%; and sodium, 0.01 to 0.03%.

In addition to the basic forms of cast iron, the metal is commonly alloyed to increase characteristics desirable for a specific application. These alloyed varieties are commonly separated into five defining categories. The following contents relate to the individual range, and not to a particular iron alloy. 1) Abrasion resistant white irons. These alloys gain abrasion resistance from chromium contents of up to 28% (high chromium iron), as much as 7% nickel (martensitic nickel high chromium iron), and up to 3.5% molybdenum (martensitic chromium-molybdenum iron). 2) Corrosion resistant irons. Alloying elements to improve corrosion resistance include up to 4.5% manganese (nickel-chromium ductile iron), as much as 17% silicon (high silicon iron), and nickel contents as high as 36% (nickel-chromium gray iron and nickel-chromium ductile iron). 3) High resistant gray irons. Alloying elements include as much as 7% silicon (medium silicon iron), 43% nickel nickel-chromium-silicon iron), 25% aluminum (high aluminum iron), and 10% copper (nickel-chromium-silicon iron). 4) Heat resistant ductile irons. These alloys include increased silicon (up to 6%) in medium silicon ductile iron), and nickel (up to 36%) in nickel-chromium ductile iron. 5) Heat resistant white irons. These alloys contain up to 15% nickel (austentitic grade), and 35% chromium (ferritic grade).

Steel making process

Iron and steel production begins with mining iron rich mineral deposits known as iron ores. Iron is present in the ores in the form of chemical compounds, not as free iron, because the element iron is so active chemically that it does not exist in nature in the free state. The important iron compounds, as related to iron and steel production, are the oxides of iron. Important ores are hematite, magnetite, and limonite, which contain ferric oxide (Fe_2O_3), ferrosoferric oxide (Fe_3O_4), and hydrated ferric oxide ($2Fe_2O_3 . 3H_2O$), respectively. Today in the U.S., the most available iron ore is taconite, which contains less iron ore and must be upgraded in the iron making process.

Iron is extracted from the ore in a blast furnace which is essentially a tall steel shell erected on a base, often made of concrete. The blast furnace process of extracting iron from ore is a reduction smelting process. Ore and other solids (chiefly coke and limestone) are fed into the top of the furnace, and iron and waste materials are drawn off at the bottom. During the process, preheated air is blown into the furnace through tuyeres (nozzles), and within the furnace this oxygen reacts with the carbon in the coke to form carbon dioxide which, at the prevailing elevated temperature, reacts with the excess carbon present to form carbon monoxide which reacts with the iron oxides in the ore to convert them to free iron, carbon dioxide, and carbon monoxide. Reaction of the oxides with carbon monoxide accounts for a majority of the iron reduced in a blast furnace, although 20% is produced by direct reduction of the oxides reacting with carbon.

When the molten iron is drawn off, it is used either as a molten charge in steel making, or is taken to a casting machine and cast into small blocks called "pigs," each weighing approximately 100 pounds (hence the term "pig iron"). Pig iron is not pure iron. It contains considerable amounts of carbon, manganese, phosphorus, sulfur, and silicon. It must be transformed to steel using one of four different processes: the open-hearth process, the acid Bessemer process, the basic oxygen process, or the electric-arc process. Between them, these processes account for nearly all the steel tonnage produced in the U.S. Until 1969, the open-hearth furnace was the nation's major source of steel, but today the basic oxygen furnace has assumed that position. In each process, refining the raw materials is an essential operation. The raw materials used to make the steel contain various metallic and nonmetallic elements as impurities. To make steel, many of these impurities must be removed or reduced to an acceptable level. In all four major steel making processes, refining is accomplished by promoting chemical reactions—principally but not exclusively oxidation reactions. Carbon, manganese, iron, silicon, and other elements are oxidized during the refining period when many of the oxides evolve as gases and others enter the slag. Many desirable elements such as carbon and manganese are removed along with the impurities. Consequently, each process allows for the addition of selected materials to the molten metal so that the amounts of carbon, manganese, and other desirable alloying elements in the steel can be adjusted.

Steel making processes are classed as either basic or acid processes, depending on whether the furnace lining used is an acid or basic refractory material. The slag that forms during the process must be compatible with the lining: that is, a basic slag for a basic lining and an acid slag for an acid lining. Silica (SiO_2) is a common acid furnace lining, while mangensite ($MgCO_3$) and dolomite ($MgCO_3 + CaCO_3$) are basic linings. The open-hearth, Bessemer, and electric-arc processes may be operated as basic or as acid processes. The basic oxygen process is basic only. All processes conclude with the refined steel being poured into a large ladle and the slag containing impurities overflows into a slag pot. The steel is now ready to be poured into ingots, which are produced in a varicty of sizes and shapes depending on the wrought product to be manufactured.

Steel nomenclature

Annealing. A heat treat process of heating a metal to, and holding it at, a particular temperature, and then cooling it at a suitable rate to achieve a desired result such as reducing hardness, grain refinement, developing a particular microstructure, stress relieving, etc. Also see full annealing.

Austempering. A heat treat process that consists of quenching a ferrous alloy from a temperature above its transformation range in a medium having a sufficiently high rate of heat abstraction to prevent the formation of high temperature transformation products. The alloy is held at a temperature below that of pearlite formation and above that of martensite formation until transformation is complete.

Austenite. The solid solution of one or more elements in a gamma-iron (face-centered cubic iron). The solution is assumed to be carbon, unless otherwise designated; e.g., nickel austenite.

Austenitizing. The process of forming austenite in a ferrous alloy by heating it to a temperature within the transformation range (partial austenitizing), or to a temperature above the critical range (complete austenitizing).

Bainite. A transformation product of austenite. Bainite is formed by isothermal transformation of austenite at temperatures lower than that for fine pearlite and above the M_S (see the list of "transformation temperatures" in the section on Heat Treatment of Steel) temperature. Bainite is an aggregate of ferrite and cementite—when formed at temperatures in the upper portion of the transformation range it has a feathery appearance (upper bainite), while that formed at lower temperatures has an acicular or needlelike structure (lower bainite).

Basic-oxygen process. A process of steel making in which molten pig iron and scrap are refined in a converter by blowing high-purity oxygen onto the surface of the charge.

Bessemer process. A process of steel making in which air is blown through the charge of molten pig iron contained in a Bessemer converter.

Billet. A semifinished hot-rolled, forged, or extruded product, generally with a round or square cross section. For ferrous materials, the minimum diameter or thickness for a billet is a 1.5 inches, and the cross-sectional area may range from 2.25 to 36 square inches.

Blast furnace. A shaft furnace in which solid fuel is burned with an air blast to smelt ore in a continuous operation. The molten metal and molten slag are collected at the bottom of the furnace and are drawn off periodically.

Bloom. A semifinished hot-rolled product, usually with a square or rectangular cross section. If rectangular, the width is no greater than twice the thickness. The cross-sectional area of a bloom usually exceeds 36 square inches.

Blue brittleness. Brittleness that occurs in some steels after being heated to within the temperature range of 400 to 700° F (204 to 371° C), or more especially after being worked within this range. Killed steels are virtually free of blue brittleness.

Carburizing. A process in which an austenitized ferrous material is brought into contact with a carbonaceous atmosphere or medium of sufficient carbon potential as to cause absorption of carbon at the surface and, by diffusion, create a concentration gradient. Hardening by quenching follows.

Case hardening. A term used to describe one or more processes of hardening steel in which the outer portion (or case) is made substantially harder than the inner portion (or core). Most of the processes involve either enriching the surface layer with carbon and/or nitrogen, usually followed by quenching or tempering, or the selective hardening of the surface layer by means of flame or induction hardening.

Cementite. Iron carbide: a hard and brittle compound of iron and carbon (Fe_3C).

Cold working. Plastic deformation of a metal at a temperature below its recrystallization temperature. Cold working produces two effects: the surface of the steel is improved by comparison to hot worked material, and the ultimate and yield strengths of the steel are increased while ductility and toughness decrease.

Creep. A time dependent deformation of steel occurring under conditions of elevated temperature accompanied by stress intensities well within the apparent elastic limit for the temperature involved.

Critical cooling rate. The slowest rate at which steel can be cooled from above the upper critical temperature to prevent the transformation of austenite at any temperature above the M_S (see the list of "transformation temperatures" in the section on Heat Treatment of Steel) temperature.

Critical point. In an equilibrium diagram, the specific combination of composition, temperature, and pressure at which the phases of a heterogeneous system are in equilibrium.

Critical range. In ferrous alloys, the temperature ranges within which austenite forms on heating, and transforms on cooling. The heating and cooling range may overlap but they never coincide.

Critical temperature. Synonymous with critical point when pressure is constant.

Decarburization. The loss of carbon from the surface of a ferrous alloy as the result of heating in a medium that reacts with the carbon at the surface of the material.

Deoxidizer. A substance that can be added to molten metal to react with free or combined oxygen to facilitate its removal.

Ductility. The ability of a material to deform plastically without fracturing, usually measured by elongation or reduction of area in a tension test, or, for flat products such as sheet, by height of cupping in an Erichsen test.

Elastic deformation. The change in dimensions accompanying stress in the elastic region. The original dimensions are restored upon complete release of the stress.

Elastic limit. The greatest stress that a material can withstand without permanent strain remaining upon complete release of the stress.

Element. A substance that cannot be decomposed by ordinary chemical reactions: the fundamental substance from which other substances are made.

Elongation. The increase in gage length, measured after fracture of the tensile specimen within the gage length, expressed as a percentage of the original gage length. It is usually measured in two inches for rectangular tensile specimens and in 4D for round (not welded) specimens. It is a measurement of ductility.

Endothermic. A reaction attended by the absorption of heat.

Equilibrium. A dynamic condition of balance between atomic movements where the resultant change is zero and the condition appears to be static rather than dynamic.

Equilibrium diagram. A graph of the composition, temperature, and pressure limits of the phase fields in an alloy system under equilibrium conditions. In metal systems pressure is usually considered to be constant.

Etch test. An inspection procedure in which a sample is deep etched with acid and visually examined for the purpose of evaluating its structural homogeneity. Also known as a macroetch.

Exothermic. A reaction attended by the liberation of heat.

Fatigue. The phenomenon which results in the fracture of a material under repeated cyclic stresses having a maximum value lower than the tensile yield strength of the material.

Fatigue limit. The maximum or limit stress value below which a material can presumably endure an infinite number of stress cycles. Also referred to as endurance limit, and measured by the rotating beam fatigue test.

Fatigue strength. The maximum stress that can be sustained by a material for a specified number of cycles without failure. Completely reversed loading is implied unless otherwise qualified.

Ferrite. The solid solution of one or more elements (usually carbon) in body-centered cubic iron.

Flakes. In steel, internal fissures can occur in wrought products during cooling from hot forging or rolling. Their occurrence may be minimized by effective control of hydrogen, either in melting or in cooling from hot work.

Ferroalloy. An alloy of iron that contains a sufficient amount of some other element(s) to be useful as an agent for introducing the other element(s) into molten metal.

Full annealing. A thermal treatment for steel intended to decrease hardness. It is accompanied by heating above the transformation range, holding for the specified time interval, and controlled slow cooling to below that range. Subsequent cooling to ambient temperature may be accompanied either in air or in the furnace.

Gamma iron. A face-centered cubic form of pure iron, stable from 1670 to 2535° F (910 – 1390° C).

Grain. An individual crystal in a polycrystalline metal or alloy. A grain size number designation is an arbitrary value calculated from the average number of individual crystals, or grains, that appear on the etched surface of a specimen at 100 diameters magnification.

Hardenability. The property of a ferrous alloy that determines the depth and distribution of hardness that may be introduced by quenching.

Heat treatment. The operation or series of operations of heating and cooling a metal or alloy in the solid state to develop specific desired properties or characteristics.

Hooke's Law. Stress is proportional to strain. This law is valid only to the elastic point.

Hot shortness. Brittleness in metal when hot.

Hot working. The plastic deformation of a metal at a temperature above the recrystallization temperature of the metal.

Induction hardening. A quench hardening process in which heat is generated by electrical induction.

Ingot. A special kind of casting intended for rolling or forging to wrought iron.

Jominy test. A method for determining the hardenability of steel by water-quenching one end of an austentized cylindrical test specimen and measuring the resulting hardness at specified distances from the quenched end. Also called an end-quench hardenability test.

Killed steel. Steel that has been deoxidized with a strong deoxidizer such as aluminum so that little or no reaction occurs between carbon and oxygen during solidification. Generally, killed steels are used when a sound, homogeneous structure is required. Alloy steels, carburizing steels, and steels for forgings are typical applications for killed steels.

Martempering. A heat treat process that consists of quenching a ferrous alloy from an austenitizing temperature to a temperature slightly above or within the martensitic transformation range, holding the material in the quench medium until the temperature throughout the mass is essentially uniform, after which the material is removed from the quench medium and cooled slowly in air.

Martensite. A transformation product of austenite that forms below the M_S (see *Transformation temperature* below for list of abbreviations) temperature. In a metastable-phase, as formed, alpha-martensite is a supersaturated interstitial solid solution of iron carbide in ferrite having a body-centered tetragonal crystal structure characterized by an acicular or needlelike microstructural appearance. Aging or tempering alpha-martensite converts the tetragonal crystal structure to the body-centered cubic structure in which form it is called beta-martensite.

Modulus of elasticity. Within the proportional limit, the ratio of stress to corresponding strain. A measure of the rigidity of a metal. Also known as Young's modulus.

Nitriding. A surface hardening process in which certain steels are heated to, and held at, a temperature below the transformation range in contact with gaseous ammonia or other source of nascent nitrogen in order to affect a transfer of nitrogen to the surface layer of the steel. The nitrogen combines with certain alloying elements, resulting in a thin case of very high hardness.

Normalizing. A heat treat process for ferrous materials that consists of cooling the material in air from a temperature slightly above the transformation range to room temperature.

Open-hearth process. A steel making process in which pig iron and scrap are refined in a reverbatory furnace having a shallow hearth and low roof. The charge is heated by a long sweeping flame that passes over it.

Pearlite. The lamellar aggregate of ferrite and iron-carbide that results from the transformation of austenite at the lower critical point on cooling.

Phase change. The transformation of a material from one physical state to another. For example, changing from a liquid to a solid.

Poisson's ratio. The ratio of the traverse strain to the corresponding axial strain in a body subjected to uniaxial stress. It is usually applied to elastic conditions.

Proportional limit. The maximum stress at which strain remains directly proportional to stress.

Quench hardening. Hardening a ferrous alloy by austenitizing and then cooling rapidly enough so that all or some of the austenite transforms to martensite. Austenitizing temperatures for the hypoeutectoid steels are usually above the A_3 (see *Transformation temperature* below for list of abbreviations) temperature. For the hypereutectoid steels austenitizing temperatures are usually between the A_1 and A_{cm} temperatures.

Recrystallization temperature. The point at which deformed grains in heated steel break up and reform into new grains. The new grain size depends on the temperature of the metal; grain size will increase with temperature.

Reduction in area. The difference, expressed as a percentage of the original area, between the original cross section area of the tensile stress specimen and the minimum cross-section area measured after failure of the specimen. It is a measure of ductility.

Rimmed steel. A low carbon steel containing sufficient iron oxide to promote a continuous evolution of carbon monoxide during solidification of the ingot. Rimmed steel has a good surface finish and a high degree of ductility, making it useful for thin sheet, wire, tinplate, and similar products.

Semikilled steel. Steel that is incompletely deoxidized so that sufficient oxygen remains to react with carbon to form carbon monoxide to offset solidification shrinkage. Semikilled steel is used extensively for heavy structural shapes and plate.

Slab. A semifinished hot rolled product of rectangular cross section with a width greater than twice the thickness. The minimum thickness of a slab is 2 inches, and the minimal cross-sectional area is 16 square inches.

Spheroidizing. Heating and cooling to produce a spheroizoidal or globular form of carbide in steel. Also called spheroidize annealing.

Strain. The unit of change in size or shape of a body due to outside forces.

Stress. Internal forces that resist a change in the volume or shape of a material. Normal stress forces are tension, compression, and shear.

Stress relieving. A heat treatment that consists of heating a metal to a suitable temperature (normally 1000 to 1200° F [538 to 649° C]) and holding it at the temperature long enough to reduce residual stresses, and then cooling it slowly to minimize the possibility of developing new residual stresses.

Tempering. Reheating a quench hardened or normalized ferrous alloy to a temperature below the transformation range and then cooling at any suitable rate.

Tensile strength. The maximum tensile stress a material is capable of sustaining, as measured in a tension test.

Tension test. A test in which a machined or full section specimen is subjected to a measured axial load sufficient to cause fracture. The usual information derived from the test includes the elastic properties, ultimate tensile strength, elongation, and reduction in area.

Toughness. A material's ability to absorb energy and deform plastically before fracturing.

Transformation temperature. The temperature at which a phase change occurs. Often used to denote the limiting temperature of a transformation range. See the list of "transformation temperatures" in the section on Heat Treatment of Steel.

Yield point. The minimum stress at which a marked increase in strain occurs without an increase in stress.

Yield strength. The stress at which a material exhibits a specified deviation from the proportionality of stress to strain. The deviation is expressed in terms of strain, and in the offset method, a strain of 0.2% is usually specified.

Carbon and alloy steel

Carbon Steel. (AISI definition) Carbon steel is classed as such when no minimum content is specified or guaranteed for aluminum, chromium, columbium, molybdenum, nickel, titanium, tungsten, vanadium, or zirconium; when the minimum for copper does not exceed 0.40%; or when the maximum content specified or guaranteed for any of the following elements does not exceed the percentage noted: manganese 1.65%; silicon 0.60%; copper 0.60%.

The American Iron and Steel Institute (AISI) and the Society of Automotive Engineers (SAE) have developed similar designation systems by which steels are identified by chemical composition. In addition, ASTM and SAE have jointly devised a Unified Numbering System for metals and alloys in an effort to reduce confusion. **Table 1** provides numerical designations for steels by composition. In this Table, each of the carbon and alloy steels is identified by an assigned number. It will be noted that the AISI and SAE designations are essentially the same, and the UNS number is usually similar to the SAE designation in instances where AISI and SAE are different.

There are four distinct grades of carbon steels. The first two nonresulfurized grades are both referred to as "plain carbon steels." *Plain carbon steels* with 1% manganese maximum are represented by the 10*xx* designated grades. The second *plain carbon steel* group has a manganese range of 1 to 1.65% and is comprised of the 15*xx* designated grades. The chemical compositions of both of the plain carbon grades are given in **Table 2**. Plain carbon steels with low carbon contents tend to be tough and gummy in machining operations. Increases in carbon and manganese increase the hardness and provide better surface finish and chip character. For carbon contents up to 0.20/0.25%, increases in carbon result in improved machinability for both hot-rolled and cold-drawn steels. As carbon increases above this level, hardness increases to the point that tool life is affected, leading to a decrease in the machinability rating. Most carbon steels are machined in the as-rolled, or as-rolled, cold-drawn condition. Higher carbon content steels are frequently annealed

(Text continued on p. 57)

Table 1. Numerical Designations of Steels by Composition.

Classification	Groupings	Series Designation		
		AISI	SAE	UNS
Carbon and Alloy Steels				
Plain Carbon Steels (Nonresulfurized)	Plain Carbon (Mn 1.0 max.)	10xx	10xx	G10xxx
	Plain Carbon (Mn 1.0 - 1.65)	15xx	15xx	G15xxx
Carbon Steels (Free Machining)	Resulphurized	11xx	11xx	G11xxx
	Resulphurized and Rephosphorized	12xx	12xx	G12xxx
	Acid Bessemer	B11xx	11xx	-
Alloy Steels	Manganese Steel (Mn 1.75)	13xx	13xx	G13xxx
	Boron Steels	14xx	14xx	-
	Nickel Steels (Ni 3.50) (Ni 5.0)	 23xx 25xx	 23xx 25xx	 - -
	Nickel-Chromium Steels (Ni 1.25, Cr 0.65 and 0.80) (Ni 1.75, Cr 1.07) (Ni 3.50, Cr 1.50 and 1.57) (Ni 3.00, Cr 0.77)	 31xx 32xx 33xx 34xx	 31xx 32xx 33xx 34xx	 - - - -
	Molybdenum Steels (Mo 0.20 and 0.25) (Mo 0.40 and 0.52)	 40xx 44xx	 40xx 44xx	 G40xxx G44xxx
	Chromium-Molybdenum Steels (Cr 0.50, 0.80, and 0.95, Mo 0.12, 0.20, 0.25, and 0.30)	41xx	41xx	G41xxx
	Nickel-Chromium-Molybdenum Steels (Ni 1.82, Cr 0.50 and 0.80, Mo .25) (Ni 1.82, Cr 0.50, Mo 0.12 and 0.25, V 0.03 min.) (Ni 1.05, Cr 0.45, Mo 0.20 and 0.35 (Ni 0.30, Cr 0.40, Mo 0.12) (Ni 0.55, Cr 0.50, Mo 0.20) (Ni 0.55, Cr 0.50, Mo 0.25) (Ni 0.55, Cr 0.50, Mo 0.35) (Ni 3.25, Cr 1.20, Mo 0.12) (Ni 0.45, Cr 0.40, Mo 0.12) (Ni 0.55, Cr 0.20, Mo 0.20) (Ni 1.00, Cr 0.80, Mo 0.25)	 43xx 43BVxx 47xx 81xx 86xx 87xx 88xx 93xx 94xx 97xx 98xx	 43xx 43BVxx 47xx 81xx 86xx 87xx 88xx 93xx 94xx 97xx 98xx	 G43xxx - G47xxx G81xxx G86xxx G87xxx G88xxx G93xxx G94xxx G97xxx G98xxx
	Nickel-Molybdenum Steels (Ni 0.85 and1.82, Mo 0.20 and 0.25) (Ni 3.50, Mo 0.25)	 46xx 48xx	 46xx 48xx	 G46xxx G48xx
	Chromium Steels (Cr 0.27, 0.40, 0.50, and 0 .65) (Cr 0.80, 0.87, 0.92, 0.95, 1.00, and 1.05) (Cr 0.50, C 1.00 min.) (bearing steel) (Cr 1.02, C 1.00 min.) (bearing steel) (Cr 1.45, C 1.00 min.) (bearing steel)	 50xx 51xx 50xxx 51xxx 52xxx	 50xx 51xx 501xx 511xx 521xx	 G50xxx G51xxx - - -
	Chromium-Vanadium Steels	61xx	61xx	G61xxx
	Tungsten-Chromium Steels	72xx	72xx	G72xxx
	Silicon-Manganese Steels	92xx	92xx	G92xxx

(Continued)

Table 1. *(Continued)* **Numerical Designations of Steels by Composition.**

Classification	Groupings	Series Designation		
		AISI	SAE	UNS
	Boron Intensified Steels	xxBxx	xxBxx	G504xx to G506xx
	Leaded Steels	xxLxx	xxLxx	-
	Vanadium Steels	xxVxx	xxVxx	-
Stainless and Heat Resisting Steels				
Austenitic Stainless	Chromium-Nickel-Manganese Steels	2xx	202xx	S201xx, S202xx, S203xx
	Chromium-Nickel Steels	3xx	303xx	S3xxxx*
Martensitic Stainless	Hardenable Chromium	4xx	514xx	S4xxx
Ferritic Stainless	Non-hardenable Chromium	4xx	514xx	S4xxx*
Heat Resisting Steel	Low Chromium	5xx	515xx	S50xxx*

* Certain designations only.

Table 2. Composition of Nonresulfurized Plain Carbon Steel. *(Source, Bethlehem Steel.)*

AISI/SAE Number	UNS Number	Carbon %	Manganese %	Phosphorus % Max.	Sulphur %
Plain Carbon Steels (Mn 1% Maximum)					
1005*	G10050	.06 max	.35 max	.040	.050
1006*	G10060	.08 max	.25/.40	.040	.050
1008	G10080	.10 max	.30/.50	.040	.050
1010	G10100	.08/.13	.30/.60	.040	.050
1011†	G10110	.08/.13	.60/.90	.040	.050
1012	G10120	.10/.15	.30/.60	.040	.050
1013†	G10130	.11/.16	.50/.80	.040	.050
1015	G10150	.13/.18	.30/.60	.040	.050
1016	G10160	.13/.18	.60/.90	.040	.050
1017	G10170	.15/.20	.30/.60	.040	.050
1018	G10180	.15/.20	.60/.90	.040	.050
1019	G10190	.15/.20	.70/1.00	.040	.050
1020	G10200	.18/.23	.30/.60	.040	.050
1021	G10210	.18/.23	.60/.90	.040	.050
1022	G10220	.18/.23	.70/1.00	.040	.050
1023	G10230	.20/.25	.30/.60	.040	.050
1025	G10250	.22/.28	.30/.60	.040	.050
1026	G10260	.22/.28	.60/.90	.040	.050
1029	G10290	.25/.31	.60/.90	.040	.050
1030	G10300	.28/.34	.60/.90	.040	.050

(Continued)

Table 2. *(Continued)* **Composition of Nonresulfurized Plain Carbon Steel.** *(Source, Bethlehem Steel.)*

AISI/SAE Number	UNS Number	Carbon %	Manganese %	Phosphorus % Max.	Sulphur %
Plain Carbon Steels (Mn 1% Maximum)					
1033	G10330	.29/.36	.70/1.00	.040	.050
1034	G10340	.32/.38	.50/.80	.040	.050
1035	G10350	.32/.38	.60/.90	.040	.050
1037	G10370	.32/.38	.70/1.00	.040	.050
1038	G10380	.35/.42	.60/.90	.040	.050
1039	G10390	.37/.44	.70/1.00	.040	.050
1040	G10400	.37/.44	.60/.90	.040	.050
1042	G10420	.40/.47	.60/.90	.040	.050
1043	G10430	.40/.47	.70/1.00	.040	.050
1044	G10440	.43/.50	.30/.60	.040	.050
1045	G10450	.43/.50	.60/.90	.040	.050
1046	G10460	.43/.50	.70/1.00	.040	.050
1049	G10490	.46/.53	.60/.90	.040	.050
1050	G10500	.48/.55	.60/.90	.040	.050
1053	G10530	.48/.55	.70/1.00	.040	.050
1055	G10550	.50/.60	.60/.90	.040	.050
1059*	G10590	.55/.65	.50/.80	.040	.050
1060	G10600	.55/.65	.60/.90	.040	.050
1064†	G10640	.60/.70	.50/.80	.040	.050
1065†	G10650	.60/.70	.60/.90	.040	.050
1069†	G10690	.65/.75	.40/.70	.040	.050
1070	G10700	.65/.75	.60/.90	.040	.050
1074†	G10740	.70/.80	.50/.80	.040	.050
1075†	G10750	.70/.80	.40/.70	.040	.050
1078	G10780	.72/.85	.30/.60	.040	.050
1080	G10800	.75/.88	.60/.90	.040	.050
1084	G10840	.80/.93	.60/.90	.040	.050
1085†	G10850	.80/.93	.70/1.00	.040	.050
1086*	G10860	.80/.93	.30/.50	.040	.050
1090	G10900	.85/.98	.60/.90	.040	.050
1095	G10950	.90/1.03	.30/.50	.040	.050
Plain Carbon Steels (Mn 1-1.65%)					
1513	G15130	.10/.16	1.10/1.40	.040	.050
1518	G15130	.15/.21	1.10/1.40	.040	.050
1522	G15220	.18/.24	1.10/1.40	.040	.050
1524	G15240	.19/.25	1.35/1.65	.040	.050
1525	G15250	.23/.29	.80/1.10	.040	.050

(Continued)

Table 2. *(Continued)* **Composition of Nonresulfurized Plain Carbon Steel.** *(Source, Bethlehem Steel.)*

AISI/SAE Number	UNS Number	Carbon %	Manganese %	Phosphorus % Max.	Sulphur %
Plain Carbon Steels (Mn 1-1.65%)					
1526	G15260	.22/.29	1.10/1.40	.040	.050
1527	G15270	.22/.29	1.20/1.50	.040	.050
1536	G15360	.30/.37	1.20/1.50	.040	.050
1541	G15410	.36/.44	1.35/1.65	.040	.050
1547	G15470	.45/.51	1.35/1.65	.040	.050
1548	G15480	.44/.52	1.10/1.40	.040	.050
1551	G15510	.45/.56	.85/1.15	.040	.050
1552	G15520	.47/.55	1.20/1.50	.040	.050
1561	G15610	.55/.65	.75/1.05	.040	.050
1566	G15660	.60/.71	.85/1.15	.040	.050
1572	G15720	.65/.76	1.00/1.30	.040	.050

*Standard grades for wire rods and wire only. † SAE only.

Notes for Carbon Steels:
In the case of certain qualities, the foregoing standard steels are ordinarily furnished to lower phosphorus and lower sulfur maxima.

BARS AND SEMI-FINISHED:
Silicon. When silicon ranges or limits are required, the values shown in the table for Ladle Chemical Ranges and Limits apply.

RODS:
Silicon. When silicon is required, the following ranges and limits are commonly used for nonresulfurized carbon steels:

0.10 per cent maximum	0.10 to 0.20 percent	0.20 to 0.40 percent
0.07 to 0.15 percent	0.15 to 0.30 percent	0.30 to 0.60 percent

ALL PRODUCTS:
Boron. Standard killed carbon steels may be produced with a boron addition to improve hardenability. Such steels can be expected to contain 0.0005 percent minimum boron. These steels are identified by inserting the letter "B" between the second and third numerals of the AISI number, e.g., 10B46.

Lead. Standard carbon steels can be produced to a lead range of 0.15 to 0.35 percent to improve machinability. Such steels are identified by inserting the letter "L" between the second and third numerals of the AISI number, e.g., 10L45.

Copper. When copper is required, 0.20 percent minimum is generally used.

to improve machinability, particularly when they are to be cold-drawn prior to machining. The other two carbon steel groups are referred to as the free machining (or free cutting) carbon steels. Increased percentages of sulfur and/or phosphorus in these grades greatly enhances machinability. The third class of carbon steels are *resulphurized steels*, which are designated 11*xx*. The composition of these steels appears in **Table 3**. The final carbon steels group, the *resulphurized and rephosphorized steels*, is designated 12*xx* and their composition is shown in **Table 4**.

Two subcategories of the carbon steel group are "H" and "M" carbon steels. The M (for merchant) prefix steels are 10*xx* series metal with marginally altered contents of carbon and/or sulfur. Their composition is shown in **Table 5**. The H (for hardenability) suffix steels are derived from both the 10*xx* and 15*xx* series and they possess hardenability requirements in addition to composition limits. Their composition is shown in **Table 6**. Mechanical properties of selected carbon and alloy steels are given in **Table 7**.

Alloy Steel. (AISI definition) By common custom, steel is considered to be alloy steel when the maximum range given for content of alloying elements exceeds one or more of the following limits: manganese 1.65%; silicon 0.60%; copper 0.60%; or in which a definite range or a definite minimum quantity of any of the following elements is specified or required within the limits of the recognized field of constructional alloy steels; aluminum, boron, chromium up to 3.99%, cobalt, columbium, molybdenum, nickel, titanium, tungsten, vanadium, zirconium, or any other alloying element added to obtain a desired alloying effect. Small quantities of certain elements are present in alloy steels that are not specified or required. These elements are considered as incidental and may be present in the following maximum amounts: copper 0.35%; nickel 0.25%; chromium 0.20%; molybdenum 0.06%.

Low alloy steels are generally considered to obtain a total alloy content of 5% or less. All alloy steels have one or more elements, in addition to carbon, added to the iron specifically to affect a change in the properties that cannot be obtained with plain carbon steel. In alloy steels, the first two digits of the designation indicate the primary alloying agent(s) and the last two or three digits indicate the approximate middle of the carbon percentage range of the steel, in hundredths of a percent. In 4161 steel, for example, the "41" indicates that the principal alloying agent is molybdenum, and the "61" indicates that the carbon content is approximately 61% (in fact, 4160 has a carbon content range of 0.56 to 0.64%). **Table 8** provides composition information on common alloy steels. Alloy steels with specific hardenability requirements are given an 'H" suffix. Composition limits for H alloy steels are given in **Tables 9** and **10**. Mechanical properties of selected carbon and alloy steels are given in **Table 7**. Elements added to alloy steels, and their intended effect, are described below in the following section.

Carbon and alloy steels are commonly purchased as round or square bars (solid) or tubes (hollow). Approximate weights and areas for square and round bars and tubes, ranging in size or diameter from $^{1}/_{16}$ inch through 7 $^{15}/_{16}$ inches, are given in **Table 11a & b**.

Effects of alloying elements

Early civilizations grouped everything in their universe into four essential "elements" or essences identified by Aristotle: air, water, fire, and earth. A later Greek philosopher/scientist named Democritus was the first to propose that all materials are comprised of small particles that he called *atomos*, after the Greek word for invisible. It wasn't until 1868, however, that the Russian chemist Dimitri Mendeléef revealed his classification of the known elements, which became the basis for the Periodic Table of the Elements. As defined in the nomenclature section above, an element cannot be further broken down or subdivided into other substances. An element, in other words, be it gas, liquid, or solid, is a fundamental substance composed of atoms of only one kind, and they represent the simplest form of matter. There are just over 100 cataloged elements, and their selected properties are given in **Table 12**.

The essential elements of steel are iron (Fe) and carbon (C). By combining other elements, as described below, the characteristics of steel can be modified to suit an almost infinite variety of requirements.

Carbon (chemical symbol C) is the principal hardening element in steel. As carbon content increases above 0.85%, additional increases in strength and hardness through adding additional carbon are proportionally less than when below this percentage. Upon quenching, the maximum attainable hardness also increases with higher percentages of carbon content, but above a content of 0.60%, the rate of increase is very small. Ductility, toughness, and weldability generally decrease as carbon content increases.

Manganese (chemical symbol Mn) is an active deoxidizer and, in addition, combines

(Text continued on p. 78)

Table 3. Composition of Free Machining (Resulfurized) Carbon Steel. *(Source, Bethlehem Steel.)*

AISI/SAE Number	UNS Number	Carbon %	Manganese %	Phosphorus % Max.	Sulphur %
1108	G11080	.08/.13	.50/.80	.040	.08/.13
1109	-	.08/.13	.60/.90	.040	.08/.13
1110	G11100	.08/.13	.30/.60	.040	.08/.13
1116	-	.14/.20	1.10/1.40	.040	.16/.23
1117	G11170	.14/.20	1.00/1.30	.040	.08/.13
1118	G11180	.14/.20	1.30/1.60	.040	.08/.13
1119	-	.14/.20	1.00/1.30	.040	.24/.33
1132	-	.27/.34	1.35/1.65	.040	.08/.13
1137	G11370	.32/.39	1.35/1.65	.040	.08/.13
1139	G11390	.35/.43	1.35/1.65	.040	.13/.20
1140	G11400	.37/.44	.70/1.00	.040	.08/.13
1141	G11410	.37/.45	1.35/1.65	.040	.08/.13
1144	G11440	.40/.48	1.35/1.65	.040	.24/.33
1145	-	.42/.49	.70/1.00	.040	.04/.07
1146	G11460	.42/.49	.70/1.00	.040	.08/.13
1151	G11510	.48/.55	.70/1.00	.040	.08/.13

BARS AND SEMI-FINISHED:
Silicon. When silicon ranges and limits are required, the values shown in the table for Ladle Chemical Ranges and Limits apply.

RODS:
Silicon. When silicon is required, the following ranges and limits are commonly used:

Standard Steel Designations	Silicon Ranges or Limits, percent
Up to 1110 incl.	0.10 max.
1116 and over	0.10 max.; or
	0.10 to 0.20; or
	0.15 to 0.30

Table 4. Composition of Free Machining (Rephosphorized and Resulfurized) Carbon Steel. *(Source, Bethlehem Steel.)*

AISI/SAE Number	UNS Number	Carbon %	Manganese %	Phosphorus %	Sulphur %	Lead %
1211	G12110	.13 max	.60/.90	.07/.12	.10/.15	-
1212	G12120	.13 max	.70/1.00	.07/.12	.16/.23	-
1213	G12130	.13 max	.70/1.00	.07/.12	.24/.33	-
12L14	G12144	.15 max	.85/1.15	.04/.09	.26/.35	.15/.35
1215	G12150	.09 max	.75/1.05	.04/.09	.26/.35	-

Silicon. It is not common practice to produce these steels to specified limits for silicon because of its adverse effect on machinability.

Nitrogen. These grades are normally nitrogen treated unless otherwise specified.

Table 5. Composition of AISI Merchant (M) Series Carbon Steel. *(Source, Bethlehem Steel.)*

AISI/SAE Number	Carbon %	Manganese %	Phosphorus % Max.	Sulphur %
M1008	.10 max.	.25/.60	.04	0.5
M1010	.07/.14	.25/.60	.04	0.5
M1012	.09/.16	.25/.60	.04	0.5
M1015	.12/.19	.25/.60	.04	0.5
M1017	.14/.21	.25/.60	.04	0.5
M1020	.17/.24	.25/.60	.04	0.5
M1023	.19/.27	.25/.60	.04	0.5
M1025	.20/.30	.25/.60	.04	0.5
M1031	.26/.36	.25/.60	.04	0.5
M1044	.40/.50	.25/.60	.04	0.5

Table 6. Composition of Carbon and Carbon-Boron "H" Series Steel. *(Source, Bethlehem Steel.)*

AISI/SAE Number	UNS Number	Carbon %	Manganese %	Phosphorus % Max.	Sulphur % Max.	Silicon %
1038 H	H10380	.34/.43	.500/1.00	.040	.050	.15/.30
1045 H	H10450	.42/.51	.500/1.00	.040	.050	.15/.30
1522 H	H15220	.17/.25	1.00/1.50	.040	.050	.15/.30
1524 H	H15240	.18/.26	1.25/1.75*	.040	.050	.15/.30
1526 H	H15260	.21/.30	1.00/1.50	.040	.050	.15/.30
1541 H	H15410	.35/.45	1.25/1.75*	.040	.050	.15/.30
15B21 H	H15211	.17/.24	.70/1.20	.040	.050	.15/.30
15B35 H	H15351	.31/.39	.70/1.20	.040	.050	.15/.30
15B37 H	H15371	.30/.39	1.00/1.50	.040	.050	.15/.30
15B41 H	H15411	.35/.45	1.25/1.75*	.040	.050	.15/.30
15B48 H	H15481	.43/.53	1.00/1.50	.040	.050	.15/.30
15B62 H	H15621	.54/.67	1.00/1.50	.040	.050	.40/.60

* Standard H-Steels with 1.75 percent maximum manganese are classified as carbon steels.

Table 7. Mechanical Properties of Selected Carbon and Alloy Steels. See Notes. *(Source, Bethlehem Steel.)*

AISI No.	Description	Tensile Strength PSI	Yield Point PSI	Elongation %	Area Reduction %	Hardness BHN
	Carbon Steel Carburizing Grades					
1015	Annealed 1600°F	56,000	41,500	37.0	69.7	111
	Normalized 1700°F	61,500	47,000	37.0	69.6	121
	M-C 1675°F, T 350°F	75,500	44,000	30.0	69.0	156
1020	As Rolled	68,500	55,750	32.0	66.5	137
	Annealed 1600°F	57,250	42,750	36.5	66.0	111
	Normalized 1700°F	64,000	50,250	35.8	67.9	131
	M-C 1675°F, T 350°F	87,000	54,000	23.0	64.2	179

(Continued)

Table 7. *(Continued)* **Mechanical Properties of Selected Carbon and Alloy Steels.** *(Source, Bethlehem Steel.)*

AISI No.	Description	Tensile Strength PSI	Yield Point PSI	Elongation %	Area Reduction %	Hardness BHN
Carbon Steel Carburizing Grades						
1022	As Rolled	70,250	52,250	33.0	65.2	137
	Annealed 1600°F	65,250	46,000	35.0	63.6	137
	Normalized 1700°F	70,000	52,000	34.0	67.5	143
	M-C 1675°F, 350°F	89,000	55,000	25.5	57.3	179
1117	As Rolled	69,750	49,500	33.5	61.1	149
	Annealed 1575°F	62,250	40,500	32.8	58.0	121
	Normalized 1650°F	67,750	44,000	33.5	63.8	137
	M-C 1700°F, T 350°F	89,500	50,500	22.3	48.8	183
1118	As Rolled	70,500	51,500	32.3	63.0	143
	Annealed 1450°F	65,250	41,250	34.5	66.8	131
	Normalized 1700°F	69,250	46,250	33.5	65.9	143
	M-C 1700°F, T 350°F	102,500	59,250	19.0	48.9	207
Carbon Steel Water and Oil Hardening Grades—Water or Oil Quenched						
1030	Annealed 1550°F	67,250	49,500	31.2	57.9	126
	Normalized 1700°F	75,500	50,000	32.0	60.8	149
	W Q 1600°F, T 1000°F	88,000	68,500	28.0	68.6	179
	W Q 1600°F, Temp. 1100°F	85,250	63,000	29.0	70.8	170
	W Q 1600°F, Temp. 1200°F	84,500	61,500	28.5	71.4	170
1040	Annealed 1450°F	75,250	51,250	30.2	57.2	149
	Normalized 1650°F	85,500	54,250	28.0	54.9	170
	O Q 1575°F, T 1000°F	96,250	68,000	26.5	61.1	197
	O Q 1575°F, T 1100°F	91,500	64,250	28.2	63.5	187
	O Q 1575°F, T 1200°F	85,250	60,250	30.0	67.4	170
	W Q 1550°F, T 1000°F	107,750	78,500	23.2	62.6	217
	W Q 1550°F, T 1100°F	100,000	69,500	26.0	65.0	207
	W Q 1550°F, T 1200°F	93,500	68,000	27.0	67.9	197
1050	Annealed 1450°F	92,250	53,000	23.7	39.9	187
	Normalized 1650°F	108,500	62,000	20.0	39.4	217
	O Q 1550°F, T 1000°F	123,500	76,000	20.2	53.3	248
	O Q 1550°F, T 1100°F	114,000	70,500	23.5	57.6	223
	O Q 1550°F, T 1200°F	106,000	64,250	24.7	60.5	217
	W Q 1525°F, T 1000°F	131,250	92,250	20.0	55.2	262
	W Q 1525°F, T 1100°F	118,000	80,000	22.5	59.9	241
	W Q 1525°F, T 1200°F	109,000	76,500	23.7	61.2	229
1060	Annealed 1450°F	90,750	54,000	22.5	38.2	179
	Normalized 1650°F	112,500	61,000	18.0	37.2	229
	O Q 1550°F, T 900°F	145,500	93,000	16.2	44.0	293
	O Q 1550°F, T 1000°F	136,500	85,750	17.7	48.0	269
	O Q 1550°F, T 1100°F	127,750	79,000	20.0	51.7	255
1080	Annealed 1450°F	89,250	54,500	24.7	45.0	174
	Normalized 1650°F	146,500	76,000	11.0	20.6	293
	O Q 1500°F, T 900°F	181,500	112,500	13.0	35.8	352
	O Q 1500°F, T 1000°F	166,000	103,500	15.0	37.6	331
	O Q 1500°F, T 1100°F	150,000	97,000	16.5	40.3	302
1095	Annealed 1450°F	95,250	55,000	13.0	20.6	192
	Normalized 1650°F	147,000	72,500	9.5	13.5	293
	O Q 1475°°F, T 900°F	175,750	102,250	10.0	23.4	352
	O Q 1475°F, T 1000°F	159,750	95,250	13.2	32.4	321
	O Q 1475°F, T 1100°F	139,750	79,000	17.2	38.8	277
	W Q 1450°F, T 900°F	182,000	121,000	13.0	37.3	363
	W Q 1450°F, T 1000°F	165,000	102,500	16.0	41.4	311
	W Q 1450°F, T 1100°F	143,000	96,500	16.7	43.7	293

(Continued)

Table 7. *(Continued)* **Mechanical Properties of Selected Carbon and Alloy Steels.** See Notes. *(Source, Bethlehem Steel.)*

AISI No.	Description	Tensile Strength PSI	Yield Point PSI	Elongation %	Area Reduction %	Hardness BHN
Carbon Steel Water and Oil Hardening Grades—Water or Oil Quenched						
1137	Annealed 1450°F	84,750	50,000	26.8	53.9	174
	Normalized 1650°F	97,000	57,500	22.5	48.5	197
	O Q 1575°F, T 1000°F	108,000	75,750	21.3	56.0	223
	O Q 1575°F, T 1100°F	100,750	68,750	23.5	60.1	207
	O Q 1575°F, T 1200°F	97,750	68,750	23.5	60.8	201
	W Q 1550°F, T 1000°F	122,000	98,000	16.9	51.2	248
	W Q 1550°F, T 1100°F	107,750	87,750	21.0	59.2	223
	W Q 1550°F, T 1200°F	102,500	81,750	22.3	58.8	217
1141	Annealed 1500°F	86,800	51,200	25.5	49.3	163
	Normalized 1650°F	102,500	58,750	22.7	55.5	201
	O Q 1500°F, T 1000°F	110,200	75,300	23.5	58.7	229
	O Q 1500°F, T 1100°F	103,000	69,800	23.8	62.2	207
	O Q 1500°F, T 1200°F	96,300	69,600	24.8	64.1	197
1144	Annealed 1450°F	84,750	50,250	24.8	41.3	167
	Normalized 1650°F	96,750	58,000	21.0	40.4	197
	O Q 1550°F, T 1000°F	108,500	72,750	19.3	46.0	223
	O Q 1550°F, T 1100°F	102,750	68,000	21.5	51.4	212
	O Q 1550°F, T 1200°F	97,000	68,000	23.0	52.4	201
Alloy Steel Carburizing Grades						
4118	Annealed 1600°F	75,000	53,000	33.0	63.7	137
	Normalized 1670°F	84,500	56,000	32.0	71.0	156
	M-C 1700°F, T 300°F	119,000	64,500	21.0	37.5	241
	M-C 1700°F, T 450°F	115,000	64,000	22.0	49.0	235
4320	Annealed 1560°F	84,000	61,625	29.0	58.4	163
	Normalized 1640°F	115,000	67,250	20.8	50.7	235
	M-C 1700°F, T 300°F	152,500	107,250	17.0	51.0	302
	M-C 1700°F, T 450°F	148,750	105,000	17.8	55.2	285
4419	Annealed 1675°F	64,750	48,000	31.2	62.8	121
	Normalized 1750°F	75,250	51,000	32.5	69.4	143
	M-C 1700°F, T 300°F	97,250	62,750	24.2	66.4	201
	M-C 1700°F, T 450°F	94,250	58,750	25.0	68.6	197
4620	Annealed 1575°F	74,250	54,000	31.3	60.3	149
	Normalized 1650°F	83,250	53,125	29.0	66.7	174
	M-C 1700°F, T 300°F	98,000	67,000	25.8	70.0	197
	M-C 1700°F, T 450°F	98,000	66,250	27.5	68.9	192
4820	Annealed 1500°F	98,750	67,250	22.3	58.8	197
	Normalized 1580°F	109,500	70,250	24.0	59.2	229
	M-C 1700°F, T 300°F	169,500	126,500	15.0	51.0	352
	M-C 1700°F, T 450°F	163,250	120,500	15.5	53.1	331
8620	Annealed 1600°F	77,750	55,875	31.3	62.1	149
	Normalized 1675°F	91,750	51,750	26.3	59.7	183
	M-C 1700°F, T 300°F	126,750	83,750	20.8	52.7	255
	M-C 1700°F, T 450°F	124,250	80,750	19.5	54.2	248
E9310	Annealed 1550°F	119,000	63,750	17.3	42.1	241
	Normalized 1630°F	131,500	82,750	18.8	58.1	269
	M-C 1700°F, T 300°F	159,000	122,750	15.5	57.5	321
	M-C 1700°F, T 450°F	157,500	123,000	16.0	61.7	321

(Continued)

Table 7. *(Continued)* **Mechanical Properties of Selected Carbon and Alloy Steels.** See Notes. *(Source, Bethlehem Steel.)*

AISI No.	Description	Tensile Strength PSI	Yield Point PSI	Elongation %	Area Reduction %	Hardness BHN
	Alloy Steel Water Hardening Grades					
4027	Annealed 1585°F	75,000	47,250	30.0	52.9	143
	Normalized 1660°F	93,250	61,250	25.8	60.2	179
	W Q 1585°F, T 900°F	150,000	133,000	16.0	57.8	311
	W Q 1585°F, T 1000°F	139,250	122,250	18.8	60.1	285
	W Q 1585°F, T 1100°F	114,250	93,250	23.0	67.6	229
4130	Annealed 1585°F	81,250	52,250	28.2	55.6	156
	Normalized 1600°F	97,000	63,250	25.5	59.5	197
	W Q 1575°F, T 900°F	161,000	137,500	14.7	54.4	321
	W Q 1575°F, T 1000°F	144,500	129,500	18.5	61.8	293
	W Q 1575°F, T 1100°F	128,000	113,250	21.2	67.5	262
8630	Annealed 1550°F	81,750	54,000	29.0	58.9	156
	Normalized 1600°F	94,250	62,250	23.5	53.5	187
	W Q 1550°F, T 900°F	146,750	131,750	16.2	56.5	293
	W Q 1550°F, T 1000°F	134,750	123,000	18.7	59.6	269
	W Q 1550°F, T 1100°F	118,000	101,250	18.7	58.2	241
	Alloy Steel Oil Hardening Grades					
1340	Annealed 1475°F	102,000	63,250	25.5	57.3	207
	Normalized 1600°F	121,250	81,000	22.0	62.9	248
	O Q 1525°F, T 1000°F	137,750	121,000	19.2	57.4	285
	O Q 1525°F, T 1100°F	118,000	98,250	21.7	60.1	241
	O Q 1525°F, T 1200°F	112,000	96,000	23.2	62.4	229
4140	Annealed 1500°F	95,000	60,500	25.7	56.9	197
	Normalized 1600°F	148,000	95,000	17.7	46.8	302
	O Q 1550°F, T 1000°F	156,000	143,250	15.5	56.9	311
	O Q 1550°F, T 1100°F	140,250	135,000	19.5	62.3	285
	O Q 1550°F, T 1200°F	132,750	122,500	21.0	65.0	269
4150	Annealed 1525°F	105,750	55,000	20.2	40.2	197
	Normalized 1600°F	167,500	106,500	11.7	30.8	321
	O Q 1525°F, T 1000°F	175,250	159,500	14.0	46.5	352
	O Q 1525°F, T 1100°F	165,500	150,000	15.7	51.1	331
	O Q 1525°F, T 1200°F	141,000	127,500	18.7	55.7	285
4340	Annealed 1490°F	108,000	68,500	22.0	49.9	217
	Normalized 1600°F	185,500	125,000	12.2	36.3	363
	O Q 1475°F, T 1000°F	175,000	166,000	14.2	45.9	352
	O Q 1475°F, T 1100°F	164,750	159,000	16.5	54.1	331
	O Q 1475°F, T 1200°F	139,000	128,000	20.0	59.7	277
5140	Annealed 1255°F	83,000	42,500	28.6	57.3	167
	Normalized 1600°F	115,000	68,500	22.7	59.2	229
	O Q 1550°F, T 1000°F	141,000	121,500	18.5	58.9	293
	O Q 1550°F, T 1100°F	127,250	105,000	20.5	61.7	262
	O Q 1550°F, T 1200°F	117,000	94,500	22.5	63.5	235
5150	Annealed 1520°F	98,000	51,750	22.0	43.7	197
	Normalized 1600°F	126,250	76,750	20.7	58.7	255
	O Q 1525°F, T 1000°F	153,000	131,750	17.0	54.1	302
	O Q 1525°F, T 1100°F	137,000	115,250	20.2	59.5	277
	O Q 1525°F, T 1200°F	128,000	108,000	21.2	61.9	255

(Continued)

Table 7. *(Continued)* **Mechanical Properties of Selected Carbon and Alloy Steels.** See Notes. *(Source, Bethlehem Steel.)*

AISI No.	Description	Tensile Strength PSI	Yield Point PSI	Elongation %	Area Reduction %	Hardness BHN
Alloy Steel Oil Hardening Grades						
5160	Annealed 1495°F	104,750	40,000	17.2	30.6	197
	Normalized 1575°F	138,750	77,000	17.5	44.8	269
	O Q 1525°F, T 1000°F	165,500	145,500	14.5	45.7	341
	O Q 1525°F, T 1100°F	145,250	126,000	18.0	53.6	302
	O Q 1525°F, T 1200°F	128,750	110,750	20.7	55.6	262
6150	Annealed 1500°F	96,750	59,750	23.0	48.4	197
	Normalized 1600°F	136,250	89,250	21.8	61.0	269
	O Q 1550°F, T 1000°F	173,500	167,750	14.5	48.2	352
	O Q 1550°F, T 1100°F	158,250	150,500	16.0	53.2	311
	O Q 1550°F, T 1200°F	141,250	129,500	18.7	56.3	293
8650	Annealed 1465°F	103,750	56,000	22.5	46.4	212
	Normalized 1600°F	148,500	99,750	14.0	40.4	302
	O Q 1475°F, T 1000°F	172,500	159,750	14.5	49.1	352
	O Q 1475°F, T 1100°F	153,500	142,750	17.7	57.3	311
	O Q 1475°F, T 1200°F	141,000	132,000	19.5	59.8	285
8740	Annealed 1500°F	100,750	60,250	22.2	46.4	201
	Normalized 1600°F	134,750	88,000	16.0	47.9	269
	O Q 1525°F, T 1000°F	178,500	164,250	16.0	53.0	352
	O Q 1525°F, T 1100°F	149,250	134,500	18.2	59.9	302
	O Q 1525°F, T 1200°F	138,000	123,000	20.0	60.7	285
9255	Annealed 1550°F	112,750	70,500	21.7	41.1	229
	Normalized 1650°F	135,250	84,000	19.7	43.4	269
	O Q 1625°F, T 1000°F	164,250	133,750	16.7	38.3	321
	O Q 1625°F, T 1100°F	150,000	118,000	19.2	44.8	293
	O Q 1625°F, T 1200°F	138,000	106,500	21.2	48.2	277

Note: The following abbreviations are used in this Table. M-C = Mock-Carburization Temperature; T = Tempering Temperature; W Q = Water Quenching Temperature; O T = Oil Quenching Temperature. All values are for a one inch round bar, treated as indicated.

Table 8. Composition of Alloy Steels (Billets, Blooms, Slabs, Hot Rolled Bars, Cold Rolled Bars). *(Source, Bethlehem Steel.)*

AISI/SAE Number	UNS Number	Carbon %	Manganese %	Nickel %	Chromium %	Molybdenum %	Other Elements
1330	G13300	.28/.33	1.60/1.90	-	-	-	-
1335	G13350	.33/.38	1.60/1.90	-	-	-	-
1340	G13400	.38/.43	1.60/1.90	-	-	-	-
1345	G13450	.43/.48	1.60/1.90	-	-	-	-
4012††	-	.09/.14	.75/1.00	-	-	.15/.25	-
4023	G40230	.20/.25	.70/.90	-	-	.20/.30	-
4024	G40240	.20/.25	.70/.90	-	-	.20/.30	-
4027	G40270	.25/.30	.70/.90	-	-	.20/.30	S - .035/.050
4028	G40280	.25/.30	.70/.90	-	-	.20/.30	-
4032††	G40320	.30/.35	.70/.90	-	-	.20/.30	S - .035/.050
4037	G40370	.35/.40	.70/.90	-	-	.20/.30	-

(Continued)

Table 8. *(Continued)* **Composition of Alloy Steels** (Billets, Blooms, Slabs, Hot Rolled Bars, Cold Rolled Bars). *(Source, Bethlehem Steel.)*

AISI/SAE Number	UNS Number	Carbon %	Manganese %	Nickel %	Chromium %	Molybdenum %	Other Elements
4042††	G40420	.40/.45	.70/.90	-	-	.20/.30	-
4047	G40470	.45/.50	.70/.90	-	-	.20/.30	-
4118	G41180	.18/.23	.70/.90	-	.40/.60	.08/.15	-
4130	G41300	.28/.33	.40/.60	-	.80/1.10	.15/.25	-
4135††	G41350	.33/.38	.70/.90	-	.80/1.10	.15/.25	-
4137	G41370	.35/.40	.70/.90	-	.80/1.10	.15/.25	-
4140	G41400	.38/.43	.75/1.00	-	.80/1.10	.15/.25	-
4142	G41420	.40/.45	.75/1.00	-	.80/1.10	.15/.25	-
4145	G41450	.43/.48	.75/1.00	-	.80/1.10	.15/.25	-
4147	G41470	.45/.50	.75/1.00	-	.80/1.10	.15/.25	-
4150	G41500	.48/.53	.75/1.00	-	.80/1.10	.15/.25	-
4161	G41610	.56/.64	.75/1.00	-	.70/.90	.25/.35	-
4320	G43200	.17/.22	.45/.65	1.65/2.00	.40/.60	.20/.30	-
4340	G43400	.38/.43	.60/.80	1.65/2.00	.70/.90	.20/.30	-
E4340	G43406	.38/.43	.65/.85	1.65/2.00	.70/.90	.20/.30	-
4419††	-	.18/.23	.45/.65	-	-	.45/.60	-
4422††	G44220	.20/.25	.70/.90	-	-	.35/.45	-
4427††	G44270	.24/.29	.70/.90	-	-	.35/.45	-
4615	G46150	.13/.18	.45/.65	1.65/2.00	-	.20/.30	-
4617††	G47170	.15/.20	.45/.65	1.65/2.00	-	.20/.30	-
4620	G46200	.17/.22	.45/.65	1.65/2.00	-	.20/.30	-
4621††	-	.18/.23	.70/.90	1.65/2.00	-	.20/.30	-
4626	G46260	.24/.29	.45/.65	.70/1.00	-	.15/.25	-
4718††	G47180	.16/.21	.70/.90	.90/1.20	.35/.55	.30/.40	-
4720	G47200	.17/.22	.50/.70	.90/1.20	.35/.55	.15/.25	-
4815	G48150	.13/.18	.40/.60	3.25/3.75	-	.20/.30	-
4817	G48170	.15/.20	.40/.60	3.25/3.75	-	.20/.30	-
4820	G48200	.18/.23	.50/.70	3.25/3.75	-	.20/.30	-
5015††	-	.12/.17	.30/.50	-	.30/.50	-	-
5046††	G50460	.43/.48	.75/1.00	-	.20/.35	-	-
5060††	G50600	.56/.64	.75/1.00	-	.40/.60	-	-
5115††	G51150	.13/.18	.70/.90	-	.70/.90	-	-
5120	G51200	.17/.22	.70/.90	-	.70/.90	-	-
5130	G51300	.28/.33	.70/.90	-	.80/1.10	-	-
5132	G51320	.30/.35	.60/.80	-	.75/1.00	-	-
5135	G51350	.33/.38	.60/.80	-	.80/1.05	-	-
5140	G51400	.38/.43	.70/.90	-	.70/.90	-	-
5145††	-	.43/.48	.70/.90	-	.70/.90	-	-

(Continued)

Table 8. *(Continued)* **Composition of Alloy Steels** (Billets, Blooms, Slabs, Hot Rolled Bars, Cold Rolled Bars). *(Source, Bethlehem Steel.)*

AISI/SAE Number	UNS Number	Carbon %	Manganese %	Nickel %	Chromium %	Molybdenum %	Other Elements
5147††	G51470	.46/.51	.70/.95	-	.85/1.15	-	-
5150	G51500	.48/.53	.70/.90	-	.70/.90	-	-
5155	G51550	.51/.59	.70/.90	-	.70/.90	-	-
5160	G51600	.56/.64	.75/1.00	-	.70/.90	-	-
50100††	G50986	.98/1.10	.25/.45	-	.40/.60	-	-
E51100	G51986	.98/1.10	.25/.45	-	.90/1.15	-	-
E52100	G52986	.98/1.10	.25/.45	-	1.30/1.60	-	-
6118	G61180	.16/.21	.50/.70	-	.50/.70	-	V - .10/.15
6150	G61500	.48/.53	.70/.90	-	.80/1.10	-	V - .15 min
8115††	-G81150	.13/.18	.70/.90	.20/.40	.30/.50	.08/.15	-
8615	G86150	.13/.18	.70/.90	.40/.70	.40/.60	.15/.25	-
8617	G86170	.15/.20	.70/.90	.40/.70	.40/.60	.15/.25	-
8620	G86200	.18/.23	.70/.90	.40/.70	.40/.60	.15/.25	-
8622	G86220	.20/.25	.70/.90	.40/.70	.40/.60	.15/.25	-
8625	G86250	.23/.28	.70/.90	.40/.70	.40/.60	.15/.25	-
8627	G86270	.25/.30	.70/.90	.40/.70	.40/.60	.15/.25	-
8630	G86300	.28/.33	.70/.90	.40/.70	.40/.60	.15/.25	-
8637	G86370	.35/.40	.75/1.00	.40/.70	.40/.60	.15/.25	-
8640	G86400	.38/.43	.75/1.00	.40/.70	.40/.60	.15/.25	-
8642	G86420	.40/.45	.75/1.00	.40/.70	.40/.60	.15/.25	-
8645	G86450	.43/.48	.75/1.00	.40/.70	.40/.60	.15/.25	-
8650††	G86500	.48/.53	.75/1.00	.40/.70	.40/.60	.15/.25	-
8655	G86550	.51/.59	.75/1.00	.40/.70	.40/.60	.15/.25	-
8660††	G86600	.56/.64	.75/1.00	.40/.70	.40/.60	.15/.25	-
8720	G87200	.18/.23	.70/.90	.40/.70	.40/.60	.20/.30	-
8740	G87400	.38/.43	.75/1.00	.40/.70	.40/.60	.20/.30	-
8822	G88220	.20/.25	.75/1.00	.40/.70	.40/.60	.30/.40	-
9254††	G92540	.51/.59	.60/.80	-	.60/.80	-	Si - 1.20/1.60
9255††	-	.51/.59	.70/.95	-	-	-	Si - 1.80/2.20
9260	G92600	.56/.64	.75/1.00	-	-	-	Si - 1.80/2.20
9310††	-	.08/.13	.45/.65	3.00/3.50	1.00/1.40	.08/.15	-

†† SAE only

(See Alloy Steel Notes following Table 10)

Table 9. Composition of Alloy "H" Series Steels. *(Source, Bethlehem Steel.)*

AISI/SAE Number	UNS Number	Carbon %	Manganese %	Nickel %	Chromium %	Molybdenum %	Other Elements
1330 H	H13300	.27/.33	1.45/2.05	-	-	-	-
1335 H	H13350	.32/.38	1.45/2.05	-	-	-	-
1340 H	H13400	.37/.44	1.45/2.05	-	-	-	-
1345 H	H13450	.42/.49	1.45/2.05	-	-	-	-
4027 H	H40270	.24/.30	.60/1.00	-	-	.20/.30	-
4028 H	H40280	.24/.30	.60/1.00	-	-	.20/.30	S - .035/.050
4032 H ††	H40320	.29/.35	.60/1.00	-	-	.20/.30	-
4037 H	H40370	.34/.41	.60/1.00	-	-	.20/.30	-
4042 H ††	H40420	.39/.46	.60/1.00	-	-	.20/.30	-
4047 H	H40470	.44/51	.60/1.00	-	-	.20/.30	-
4118 H	H41180	.17/.23	.60/1.00	-	.30/.70	.08/.15	-
4130 H	H41300	.27/.33	.30/.70	-	.75/1.20	.15/.25	-
4135 H ††	H41350	.32/.38	.60/1.00	-	.75/1.20	.15/.25	-
4137 H	H41370	.34/.41	.60/1.00	-	.75/1.20	.15/.25	-
4140 H	H41400	.37/.44	.60/1.10	-	.75/1.20	.15/.25	-
4142 H	H41420	.39/.46	.60/1.10	-	.75/1.20	.15/.25	-
4145 H	H41450	.42/.49	.60/1.10	-	.75/1.20	.15/.25	-
4147 H	H41470	.44/.51	.60/1.10	-	.75/1.20	.15/.25	-
4150 H	H41500	.47/.54	.60/1.10	-	.75/1.20	.15/.25	-
4161 H	H41610	.55/.65	.60/1.10	-	.60/.95	.25/.35	-
4320 H	H43200	.17/.23	.40/.70	1.55/2.00	.35/.65	.20/.30	-
4340 H	H43400	.37/.44	.55/.90	1.55/2.00	.65/.95	.20/.30	-
E4340 H	H43406	.37/.44	.60/.95	1.55/2.00	.65/.95	.20/.30	-
4419 H ††	-	.17/.23	.35/.75	-	-	.45/.60	-
4620 H	H46200	.17/.23	.35/.75	1.55/2.00	-	.20/.30	-
4621 H ††	-	.17/.23	.60/1.00	1.55/2.00	-	.20/.30	-
4626 H †	-	.23/29	.40/.70	.65/1.05	-	.15/.25	-
4718 H ††	H47180	.15/.21	.60/.95	.85/1.25	.30/.60	.30/.40	-
4720 H	H47200	.17/.23	.45/.75	.85/1.25	.30/.60	.15/.25	-
4815 H	H48150	.12/.18	.30/.70	3.20/3.80	-	.20/.30	-
4817 H	H48170	.14/.20	.30/.70	3.20/3.80	-	.20/.30	-
4820 H	H48200	.17/.23	.40/.80	3.20/3.80	-	.20/.30	-
5046 H ††	H50450	.43/.50	.65/1.10	-	.13/.43	-	-
5120 H	H51200	.17/.23	.60/1.00	-	.60/1.00	-	-
5130 H	H51300	.27/.33	.60/1.00	-	.75/1.20	-	-
5132 H	H51320	.29/.35	.50/.90	-	.65/1.10	-	-
5135 H	H51350	.32/.38	.50/.90	-	.70/1.15	-	-
5140 H	H51400	.37/.44	.60/1.00	-	.60/1.00	-	-
5145 H ††	-	.42/49	.60/1.00	-	.60/1.00	-	-

(Continued)

Table 9. *(Continued)* **Composition of Alloy "H" Series Steels.** *(Source, Bethlehem Steel.)*

AISI/SAE Number	UNS Number	Carbon %	Manganese %	Nickel %	Chromium %	Molybdenum %	Other Elements
5147 H ††	H51470	.45/.52	.60/1.05	-	.80/1.25	-	-
5150 H	H51500	.47/.54	.60/1.00	-	.60/1.00	-	-
5155 H	H51550	.50/.60	.60/1.00	-	.60/1.00.	-	-
5160 H	H51600	.55/.65	.60/1.10	-	60/1.00	-	-
6118 H	H61180	.15/.21	.40/.80	-	.40/.80	-	V - .10/.15
6150 H	H61500	.47/.54	.60/1.00	-	.75/1.20	-	V - .15 min
8617 H	H86170	.14/.20	.60/.95	.35/.75	.35/.65	.15/.25	-
8620 H	H86200	.17/.23	.60/.95	.35/.75	.35/.65	.15/.25	-
8622 H	H86220	.19/.25	.60/.95	.35/.75	.35/.65	.15/.25	-
8625 H	H86250	.22/.28	.60/.95	.35/.75	.35/.65	.15/.25	-
8627 H	H86270	.24/.30	.60/.95	.35/.75	.35/.65	.15/.25	-
8630 H	H86300	.27/.33	.60/.95	.35/.75	.35/.65	.15/.25	-
8637 H	H86370	.34/.41	.70/.105	.35/.75	.35/.65	.15/.25	-
8640 H	H86400	.37/.44	.70/.105	.35/.75	.35/.65	.15/.25	-
8642 H	H86420	.39/.46	.70/.105	.35/.75	.35/.65	.15/.25	-
8645 H	H86450	.42/.49	.70/.105	.35/.75	.35/.65	.15/.25	-
8650 ††	H86500	.47/.54	.70/.105	.35/.75	.35/.65	.15/.25	-
8655 H	H86550	.50/.60	.70/.105	.35/.75	.35/.65	.15/.25	-
8660 H ††	H86600	.55/.65	.70/.105	.35/.75	.35/.65	.15/.25	-
8720 H	H87200	.17/.23	.60/.95	.35/.75	.35/.65	.20/.30	-
8740 H	H87400	.37/.44	.70/.1.05	.35/.75	.35/.65	.20/.30	-
8822 H	H88220	.19/.25	.70/.1.05	.35/.75	.35/.65	.30/.40	-
9260 H	H92600	.55/.65	.65/1.10	-	-	-	Si - 1.70/2.20
9310 H	H93100	.07/.13	.40/.70	2.95/3.55	1.00/1.45	.08/.15	-

† AISI only †† SAE only (See Alloy Steel Notes following Table 10)

Table 10. Composition of Alloy Boron and Boron "H" Series Steels. *(Source, Bethlehem Steel.)*

AISI/SAE Number	UNS Number	Carbon %	Manganese %	Nickel %	Chromium %	Molybdenum %
50B40 ††	-	.38/.43	.75/1.00	-	.40/.60	-
50B44	G50441	.43/.48	.75/1.00	-	.40/.60	-
50B46	G50461	.44/.49	.75/1.00	-	.20/.35	-
50B50	G50501	.48/.53	.75/1.00	-	.40/.60	-
50B60	G50601	.56/.64	.75/1.00	-	.40/.60	-
51B60	G51601	.56/.64	.75/1.00	-	.70/.90	-
81B45	G81451	.43/.48	.75/1.00	.20/.40	.35/.55	.08/.15
86B45 ††	-	.43/.48	.75/1.00	.40/.70	.40/.60	.15/.25
94B15 ††	-	.13/.18	.75/1.00	.30/.60	.30/.50	.08/.15
94B17	G94171	.15/.20	.75/1.00	.30/.60	.30/.50	.08/.15

(Continued)

Table 10. *(Continued)* **Composition of Alloy Boron and Boron "H" Series Steels.** *(Source, Bethlehem Steel.)*

AISI/SAE Number	UNS Number	Carbon %	Manganese %	Nickel %	Chromium %	Molybdenum %
94B30	G94301	.28/.33	.75/1.00	.30/.60	.30/.50	.08/.15
50B40 H ††	H50401	.37/.44	.65/1.10	-	.30/.70	-
50B44 H	H50441	.42/.49	.65/1.10	-	.30/.70	-
50B46 H	H50461	.43/.50	.65/1.10	-	.13/.43	-
50B50 H	H50501	.47/.54	.65/1.10	-	.30/.70	-
50B60 H	H50601	.55/.65	.65/1.10	-	.30/.70	-
51B60 H	H51601	.55/.65	.65/1.10	-	.60/1.00	-
81B45 H	H81451	.42/.49	.70/1.05	.15/.45	.30/.60	.08/.15
86B30 H	H86301	.27/.33	.60/.95	.35/.75	.35/.65	.15/.25
86B45 H ††	-	.42/.49	.70/1.05	.35/.75	.35/.65	.15/.25
94B15 H ††	H94151	.12/.18	.70/1.05	.25/.65	.25/.55	.08/.15
94B17 H	H94171	.14/.20	.70/1.05	.25/.65	.25/.55	.08/.15
94B30 H	H94301	.27/.33	.70/1.05	.25/.65	.25/.55	.08/.15

†† SAE only

Notes on Alloy Steel Tables:

1. Grades shown with prefix letter E are made only by the basic electric furnace process. All others are normally manufactured by the basic open hearth or basic oxygen processes, but may be manufactured by the basic electric furnace process with adjustments in phosphorus and sulfur.
2. The phosphorus and sulfur limitations for each process are as follows:

	Maximum (percent)	
	P	S
Basic electric	0.025	0.025
Basic open hearth or basic oxygen	0.035	0.040
Acid electric or acid open hearth	0.050	0.050

3. Minimum silicon limit for acid open hearth or acid electric furnace alloy steel is .15%.
4. Small quantities of certain elements are present in alloy steels, but are not specified or required. These elements are considered as incidental and may be present in the following maximum percentages: copper, .35; nickel, .25; chromium, .20; molybdenum, .06.
5. The listing of minimum and maximum sulfur content indicates a resulfurized steel.
6. Standard alloy steels can be produced to a lead range of .15/.35% to improve machinability.
7. Silicon range for all standard alloy steels except where noted is .15/.30%.

Table 11a. Weights and Areas of Square and Round Steel Bars $^{1}/_{16}$ Through 2 $^{1}/_{16}$ Inches. *(Source, Bethlehem Steel.)*

Size or Dia. inches	Weight (lb. per ft.)		Area (in)²		Size or Dia. inches	Weight (lb. per ft.)		Area (in)²	
	Square ■	Round ●	Square □	Round ○		Square ■	Round ●	Square □	Round ○
1/16	.013	.010	.0039	.0031	51/64	2.159	1.696	.6350	.4987
5/64	.021	.016	.0061	.0048	13/16	2.245	1.763	.6602	.5185
3/32	.030	.023	.0088	.0069	53/64	2.332	1.831	.6858	.5386
7/64	.041	.032	.0120	.0094	27/32	2.420	1.901	.7119	.5591
1/8	.053	.042	.0156	.0123	55/64	2.511	1.972	.7385	.5800
9/64	.067	.053	.0198	.0155	7/8	2.603	2.044	.7656	.6013
5/32	.083	.065	.0244	.0192	57/64	2.697	2.118	.7932	.6230
11/64	.100	.079	.0295	.0232	29/32	2.792	2.193	.8213	.6450
3/16	.120	.094	.0352	.0276	59/64	2.889	2.270	.8498	.6675
13/64	.140	.110	.0413	.0324	15/16	2.988	2.347	.8789	.6903
7/32	.163	.128	.0479	.0376	61/64	3.089	2.426	.9084	.7135
15/64	.187	.147	.0549	.0431	31/32	3.191	2.506	.9385	.7371
1/4	.212	.167	.0625	.0491	63/64	3.294	2.587	.9689	.7610
17/64	.240	.188	.0706	.0554	1	3.400	2.670	1.0000	.7854
9/32	.269	.211	.0791	.0621	1-1/32	3.616	2.840	1.0635	.8353
19/64	.300	.235	.0881	.0692	1-1/16	3.838	3.014	1.1289	.8866
5/16	.332	.261	.0977	.0767	1-3/32	4.067	3.194	1.1963	.9396
21/64	.366	.288	.1077	.0846	1-1/8	4.303	3.379	1.2656	.9940
11/32	.402	.316	.1182	.0928	1-5/32	4.545	3.570	1.3369	1.0500
23/64	.439	.345	.1292	.1014	1-3/16	4.795	3.766	1.4102	1.1075
3/8	.478	.376	.1406	.1104	1-7/32	5.050	3.966	1.4853	1.1666
25/64	.519	.407	.1526	.1198	1-1/4	5.312	4.173	1.5625	1.2272

(Continued)

Table 11a. *(Continued)* **Weights and Areas of Square and Round Steel Bars $^{1}/_{16}$ Through 2 $^{1}/_{16}$ Inches.** *(Source, Bethlehem Steel.)*

Size or Dia. inches	Weight (lb. per ft.)		Area (in)^2		Size or Dia. inches	Weight (lb. per ft.)		Area (in)^2	
	Square ■	Round ●	Square □	Round ○		Square ■	Round ●	Square □	Round ○
13/32	.561	.441	.1650	.1296	1-9/32	5.581	4.384	1.6416	1.2893
27/64	.605	.475	.1780	.1398	1-5/16	5.857	4.600	1.7227	1.3530
7/16	.651	.511	.1914	.1503	1-11/32	6.139	4.822	1.8056	1.4182
29/64	.698	.548	.2053	.1613	1-3/8	6.428	5.049	1.8906	1.4849
15/32	.747	.587	.2197	.1726	1-13/32	6.724	5.281	1.9775	1.5532
31/64	.798	.627	.2346	.1843	1-7/16	7.026	5.518	2.0664	1.6230
1/2	.850	.668	.2500	.1963	1-15/32	7.334	5.761	2.1572	1.6943
33/64	.904	.710	.2659	.2088	1-1/2	7.650	6.008	2.2500	1.7671
17/32	.960	.754	.2822	.2217	1-17/32	7.972	6.261	2.3447	1.8415
35/64	1.017	.799	.2991	.2349	1-9/16	8.301	6.520	2.4414	1.9175
9/16	1.076	.845	.3164	.2485	1-19/32	8.636	6.783	2.5400	1.9949
37/64	1.136	.893	.3342	.2625	1-5/8	8.978	7.051	2.6406	2.0739
19/32	1.199	.941	.3525	.2769	1-21/32	9.327	7.325	2.7431	2.1545
39/64	1.263	.992	.3713	.2916	1-11/16	9.682	7.604	2.8477	2.2365
5/8	1.328	1.043	.3906	.3068	1-23/32	10.044	7.889	2.9541	2.3202
41/64	1.395	1.096	.4104	.3223	1-3/4	10.413	8.178	3.0625	2.4053
21/32	1.464	1.150	.4307	.3382	1-25/32	10.788	8.473	3.1728	2.4920
43/64	1.535	1.205	.4514	.3545	1-13/16	11.170	8.773	3.2852	2.5802
11/16	1.607	1.262	.4727	.3712	1-27/32	11.558	9.078	3.3994	2.6699
45/64	1.681	1.320	.4944	.3883	1-7/8	11.953	9.388	3.5156	2.7612
23/32	1.756	1.379	.5166	.4057	1-29/32	12.355	9.704	3.6337	2.8540
47/64	1.834	1.440	.5393	.4236	1-15/16	12.763	10.024	3.7539	2.9483

(Continued)

Table 11a. *(Continued)* **Weights and Areas of Square and Round Steel Bars 1/16 Through 2 1/16 Inches.** *(Source, Bethlehem Steel.)*

Size or Dia. inches	Weight (lb. per ft.)		Area $(in)^2$		Size or Dia. inches	Weight (lb. per ft.)		Area $(in)^2$	
	Square ■	Round ●	Square □	Round ○		Square ■	Round ●	Square □	Round ○
3/4	1.913	1.502	.5625	.4418	1-31/32	13.178	10.350	3.8760	3.0442
49/64	1.993	1.565	.5862	.4604	2	13.600	10.681	4.0000	3.1416
25/32	2.075	1.630	.6103	.4794	2-1/16	14.463	11.359	4.2539	3.3410

Table 11b. Weights and Areas of Square and Round Steel Bars 2 1/8 Through 7 5/16 Inches. *(Source, Bethlehem Steel.)*

Size or Dia. inches	Weight (lb. per ft.)		Area $(in)^2$		Size or Dia. inches	Weight (lb. per ft.)		Area $(in)^2$	
	Square ■	Round ●	Square □	Round ○		Square ■	Round ●	Square □	Round ○
2-1/8	15.353	12.058	4.5156	3.5466	5-1/16	87.138	68.438	25.629	20.129
2-3/16	16.270	12.778	4.7852	3.7583	5-1/8	89.303	70.139	26.266	20.629
2-1/4	17.213	13.519	5.0625	3.9761	5-3/16	91.495	71.860	26.910	21.135
2-5/16	18.182	14.280	5.3477	4.2000	5-1/4	93.713	73.602	27.563	21.648
2-3/8	19.178	15.062	5.6406	4.4301	5-5/16	95.957	75.364	28.223	22.166
2-7/16	20.201	15.866	5.9414	4.6664	5-3/8	98.228	77.148	28.891	22.691
2-1/2	21.250	16.690	6.2500	4.9087	5-7/16	100.53	78.953	29.566	23.221
2-9/16	22.326	17.535	6.5664	5.1572	5-1/2	102.85	80.778	30.250	23.758
2-5/8	23.428	18.400	6.8906	5.4119	5-9/16	105.20	82.624	30.941	24.301
2-11/16	24.557	19.287	7.2227	5.6727	5-5/8	107.58	84.492	31.641	24.850
2-3/4	25.713	20.195	7.5625	5.9396	5-11/16	109.98	86.380	32.348	25.406
2-13/16	26.895	21.123	7.9102	6.2126	5-3/4	112.41	88.289	33.063	25.967
2-7/8	28.103	22.072	8.2656	6.4918	5-13/16	114.87	90.218	33.785	26.535
2-15/16	29.338	23.042	8.6289	6.7771	5-7/8	117.35	92.169	34.516	27.109

(Continued)

Table 11b. *(Continued)* **Weights and Areas of Square and Round Steel Bars 2 1/8 Through 7 5/16 Inches.** *(Source, Bethlehem Steel.)*

Size or Dia. inches	Weight (lb. per ft.)		Area (in)^2		Size or Dia. inches	Weight (lb. per ft.)		Area (in)^2	
	Square ■	Round ●	Square □	Round ○		Square ■	Round ●	Square □	Round ○
3	30.600	24.033	9.0000	7.0686	5-15/16	119.86	94.140	35.254	27.688
3-1/16	31.888	25.045	9.3789	7.3662	6	122.40	96.133	36.000	28.274
3-1/8	33.203	26.078	9.7656	7.6699	6-1/16	124.96	98.146	36.754	28.866
3-3/16	34.545	27.131	10.160	7.9798	6-1/8	127.55	100.18	37.516	29.465
3-1/4	35.913	28.206	10.563	8.2958	6-3/16	130.17	102.23	38.285	30.069
3-5/16	37.307	29.301	10.973	8.6179	6-1/4	132.81	104.31	39.063	30.680
3-3/8	38.728	30.417	11.391	8.9462	6-5/16	135.48	106.41	39.848	31.296
3-7/16	40.176	31.554	11.816	9.2806	6-3/8	138.18	108.52	40.641	31.919
3-1/2	41.650	32.712	12.250	9.6211	6-7/16	140.90	110.66	41.441	32.548
3-9/16	43.151	33.891	12.691	9.9678	6-1/2	143.65	112.82	42.250	33.183
3-5/8	44.678	35.090	13.141	10.321	6-9/16	146.43	115.00	43.066	33.824
3-11/16	46.232	36.311	13.598	10.680	6-5/8	149.23	117.20	43.891	34.472
3-3/4	47.813	37.552	14.063	11.045	6-11/16	152.06	119.43	44.723	35.125
3-13/16	49.420	38.814	14.535	11.416	6-3/4	154.91	121.67	45.563	35.785
3-7/8	51.053	40.097	15.016	11.793	6-13/16	157.79	123.93	46.410	36.450
3-15/16	52.713	41.401	15.504	12.177	6-7/8	160.70	126.22	47.266	37.122
4	54.400	42.726	16.000	12.566	6-15/16	163.64	128.52	48.129	37.800
4-1/16	56.113	44.071	16.504	12.962	7	166.60	130.85	49.000	38.485
4-1/8	57.853	45.438	17.016	13.364	7-1/16	169.59	133.19	49.879	39.175
4-3/16	59.620	46.825	17.535	13.772	7-1/8	172.60	135.56	50.766	39.871
4-1/4	61.413	48.233	18.063	14.186	7-3/16	175.64	137.95	51.660	40.574
4-5/16	63.232	49.662	18.598	14.607	7-1/4	178.71	140.36	52.563	41.282

(Continued)

Table 11b. *(Continued)* **Weights and Areas of Square and Round Steel Bars 2 $^1/_8$ Through 7 $^5/_{16}$ Inches.** *(Source, Bethlehem Steel.)*

Size or Dia. inches	Weight (lb. per ft.)		Area (in)^2		Size or Dia. inches	Weight (lb. per ft.)		Area (in)^2	
	Square ■	Round ●	Square □	Round ○		Square ■	Round ●	Square □	Round ○
4-3/8	65.078	51.112	19.141	15.033	7-5/16	181.81	142.79	53.473	41.997
4-7/16	66.951	52.583	19.691	15.466	7-3/8	184.93	145.24	54.391	42.718
4-1/2	68.850	54.075	20.250	15.904	7-7/16	188.08	147.71	55.316	43.445
4-9/16	70.776	55.587	20.816	16.349	7-1/2	191.25	150.21	56.250	44.179
4-5/8	72.728	57.121	21.391	16.800	7-9/16	194.45	152.72	57.191	44.918
4-11/16	74.707	58.675	21.973	17.257	7-5/8	197.68	155.26	58.141	45.664
4-3/4	76.713	60.250	22.563	17.721	7-11/16	200.93	157.81	59.098	46.415
4-13/16	78.745	61.846	23.160	18.190	7-3/4	204.21	160.39	60.063	47.173
4-7/8	80.803	63.463	23.766	18.665	7-13/16	207.52	162.99	61.035	47.937
4-15/16	82.888	65.100	24.379	19.147	7-7/8	210.85	165.60	62.016	48.707
5	85.000	66.759	25.000	19.635	7-15/16	214.21	168.24	63.004	49.483

Table 12. Selected Properties of the Elements.

Element/Symbol	Atomic Number	Relative Atomic Mass	Density		Melting Point		Boiling Point		Modulus of Elasticity	
			lb/in.^{3}	g/cm.^{3}	°F	°C	°F	°C	10^6 PSI	GPa
Aluminum/ Al	13	26.9815	0.09751	2.699	1,220.4	660	4,520	2,480	10	69
Antimony/ Sb	51	121.75	0.239	6.62	1,166.9	630.5	2,620	1610	11.3	77
Arsenic/As	33	74.9216	0.207	5.73	1,497	(817) (36 at.)	1,130	616	11	75.8
Barium/Ba	56	137.34	0.13	3.5	1,300	710	2,980	(1700)	1.8	12.4
Beryllium/Be	4	9.01218	0.0658	1.85	2,340	1284	5,020	(2400)	37	255
Bismuth/Bi	83	208.9806	0.654	9.80	520.3	271	2,590	1680	4.6	31.7
Boron/B	5	10.811	0.083	2.3	3,812	(2300)	4,620	(2550)	-	-
Cadmium/Cd	48	112.40	0.313	8.65	609.3	321	1,409	765	8	55
Calcium/Ca	20	40.88	0.056	1.55	1,560	843	2,625	2075	3	20.6
Carbon/C	6	12.01115	0.0802	2.25 (graphite) 3.51 (diamond)	6,700	(5000) Gr.	8,730	(5000)	0.7	4.8
Cerium/Ce	58	140.12	0.25	6.77	1,460	804	4,380	3500	-	-
Chromium/Cr	24	51.996	0.260	7.19	3,350	1900	4,500	2690	36	248
Cobalt/Co	27	58.9332	0.32	8.90	2,723	1492	6,420	(2900)	30	206.8
Columbium (Niobium)/Cb	41	92.9064	0.310	8.57	4,380	2468	5,970	4735	15	103.4
Copper/Cu	29	63.546	0.324	8.96	1,981.4	1083	4,700	2575	16	110.3
Gadolinium/Gd	64	157.25	0.287	7.87	-	1350	-	(3000)	-	-
Gallium/Ga	31	69.72	0.216	5.91	85.5	29.8	3,600	2420	1	6.9
Germanium/Ge	32	72.59	0.192	5.36	1,756	937	4,890	2700	11.4	79
Gold/Au	79	196.9665	0.698	19.32	1,945.4	1063	5,380	(2950)	12	82.7

(Continued)

Table 12. *(Continued)* **Selected Properties of the Elements.**

Element/Symbol	Atomic Number	Relative Atomic Mass	Density		Melting Point		Boiling Point		Modulus of Elasticity	
			lb/in.^{3}	g/cm.^{3}	°F	°C	°F	°C	10^6 PSI	GPa
Hafnium/Hf	72	178.49	0.473	11.4	3,865	2220	9,700	4450	20	137.9
Hydrogen/H	1	1.0080	3.026×10^{-6}	0.0899‡	-434.6	-259.2	-422.9	-252.5	-	-
Indium/In	49	114.82	0.264	7.31	313.5	156.4	3,630	2075	1.57	11
Iridium/Ir	77	192.22	0.813	22.5	4,449	2443	9,600	4400	75	517
Iron/Fe	26	55.847	0.284	7.87	2,802	1536	4,960	(3070)	28.5	199.4
Lanthanum/La	57	138.9055	0.223	6.16	1,535	920	8,000	(3420)	5	34.4
Lead/Pb	82	207.2	0.4097	11.34	621.3	327.4	3,160	1740	2.6	17.9
Lithium/Li	3	6.941	0.019	0.534	367	180.5	2,500	1329	1.7	11.7
Magnesium/Mg	12	24.305	0.0628	1.74	1,202	650	2,030	1103	6.5	44.8
Manganese/Mn	25	54.9380	0.268	7.43	2,273	1244	3,900	2020	23	158.5
Mercury/Hg	80	200.59	0.4896	13.55	-37.97	-38.9	675	357	-	-
Molybdenum/Mo	42	95.94	0.369	10.2	4,750	2620	8,670	4650	50	344.7
Nickel/Ni	28	58.71	0.322	8.90	2,651	1453	4,950	(3000)	30	206.8
Niobium (see Columbium)/Nb	41	92.9064	0.310	8.57	4,380	2468	5,970	4735	-	-
Nitrogen/N	7	14.0067	0.042×10^{-3}	1.2506‡	-346	-210.0	-320.4	-195.8	-	-
Osmium/Os	76	190.2	0.813	22.5	4,900	3045	9,900	>5300	80	551.5
Oxygen/O	8	15.9994	0.048×10^{-3}	1.429‡	-361.8	-218.8	-297.4		-	-
Palladium/Pd	46	106.4	0.434	12.0	2,829	1552	7,200		17	117.2
Phosphorus/P	15	30.9738	0.0658	1.82	111.4	44.1	536		-	-
Platinum/Pt	78	195.09	0.7750	21.45	3,224.3	1769	7,970		21	114.7
Plutonium/Pu	94	(244)	0.686	(19.0)	1,229	640	-		-	-
Potassium/K	19	39.102	0.031	0.86	145	63.6	1,420		0.5	3.4

(Continued)

Table 12. *(Continued)* **Selected Properties of the Elements.**

Element/Symbol	Atomic Number	Relative Atomic Mass	Density		Melting Point		Boiling Point		Modulus of Elasticity	
			lb/in.^{3}	g/cm.^{3}	°F	°C	°F	°C	10^6 PSI	GPa
Radium/Ra	88	226.0254	0.18	5.0	1,300	960	-		-	-
Rhenium/Re	75	186.2	0.765	20.0	5,733	3150	10,700		75	517
Rhodium/Rh	45	102.9055	0.4495	12.44	3,571	1960	8,100		54	372.3
Selenium/Se	34	78.96	0.174	4.81	428	220.5	1,260		8.4	57.9
Silicon/Si	14	28.086	0.084	2.4	2,605	1412	4,200		16	110.3
Silver/Ag	47	107.868	0.379	10.49	1,760.9	960.8	4,010		11	75.8
Sodium/Na	11	22.9898	0.035	0.97	207.9	97.8	1,638		1.3	8.9
Sulfer/S	16	32.06	0.0748	2.07	246.2	112.8	832.3		-	-
Tantalum/Ta	73	180.9479	0.600	16.6	5,420	3010	9,570		27	186
Tellurium/Te	52	127.60	0.225	6.24	840	449.8	2,530		6	41.3
Thallium/Tl	81	204.37	0.428	11.85	577	303	2,655		1.2	8.2
Thorium/Th	90	232.0381	0.422	11.5	3,348	1750	8,100		11.4	78
Tin/Sn	50	118.69	0.264	7.298	449.4	231.9	4,120		6	41.3
Titanium/Ti	22	47.90	0.164	4.54	3,074	1667	6,395		16.8	116
Tungsten/W	74	183.85	0.697	19.3	6,150	3380	10,700		50	344.7
Uranium/U	92	238.029	0.687	18.7	2,065	1130	7,100		29.7	205.8
Vanadium/V	23	50.9414	0.217	6.0	3,452	1920	5,430		18.4	127
Zinc/Zn	30	65.37	0.258	7.133	787	419.5	1,663		12	82.7
Zirconium/Zr	40	91.22	0.23	6.5	3,326	1860	9,030		11	75.8

‡ Density of gases are given g/dm^3 (grams/decimeter3).

with sulfur to form manganese sulfide. This enhances the machining characteristics of the free machining carbon steels. Manganese also enhances the hardenability of steel because it decreases the critical (minimum) cooling rate necessary for hardening a steel. It is present (0.3 to 0.8%) in all commercial steels. Its effectiveness depends largely upon, and is directly proportional to, the carbon content of the steel.

Phosphorus (chemical symbol P) is normally regarded as an impurity except where it is added to increase machinability and atmospheric corrosion. While it increases strength and hardness to about the same degree as carbon, it also tends to decrease ductility and toughness, particularly in steel in the quenched and tempered condition. The maximum content of most steels is up to 0.04%, but in free-machining steels the phosphorus content, which is boosted by adding phosphorus at the ladle (known as rephosphorizing), can be as high as 0.12%.

Sulfur (chemical symbol S) is not a desirable element except when machinability is an important consideration. Increasing sulfur increases machinability at all carbon levels in both alloy and plain carbon grades. Increases in sulfur up to 0.06% markedly increase machinability, but above this level machinability improves at a lower rate. Sulfur also impairs weldability, and has an adverse effect on surface quality.

Silicon (chemical symbol Si) acts as a deoxidizer in carbon and alloy steels. In the company of manganese, it produces extremely high strength steel that is combined with good ductility and shock resistance in the quenched and tempered condition. In larger quantities it adversely affects machinability and increases susceptibility to decarburization and graphitization.

Nickel (chemical symbol Ni) increases impact resistance and toughness, especially at low temperature. It also lowers the critical temperature of steel for heat treatment and enhances hardenability and it does not form carbides or other compounds that might be difficult to dissolve during heating for austenitizing. Nickel's insensitivity to variations in quenching conditions provides insurance against costly failures to attain the desired properties, particularly when the furnace is not equipped for precision control.

Chromium (chemical symbol Cr) enhances hardenability, improves wear and abrasion resistance, and promotes carburization. Of the common alloying elements, chromium is surpassed only by manganese and molybdenum in its effect on hardenability.

Molybdenum (chemical symbol Mo) enhances hardness and widens the temperature range for heat treatment. It also increases steel's high temperature tensile and creep strength and reduces susceptibility to temper brittleness.

Vanadium (chemical symbol V) is a strong deoxidizer and inhibits grain growth over a fairly large quenching range. With vanadium additions of 0.04 to 0.05%, hardenability of medium carbon steels is increased with minimum affect on grain size. For higher vanadium content, hardenability can be increased with higher austenitizing temperatures.

Copper (chemical symbol Cu) is used to improve steel's resistance to corrosion. In amounts of 0.20 to 0.50%, it does not significantly increase the mechanical properties of the steel.

Boron (chemical symbol B) increases hardenability, even when present in amounts as small as 0.0005%. Because boron is ineffective when allowed to combine with oxygen or nitrogen, its use is limited to aluminum-killed steels.

Lead (chemical symbol Pb) does not alloy with steel, but it is retained in its elemental state as a fine dispersion within the steel's structure. It is most widely used in the free-machining carbon grades where, in amounts of 0.15 to 0.35%, it effectively lubricates the cutting edge of the tool and permits an increase in cutting speed and feed and an improvement in surface finish quality without an attendant decrease in tool life.

Nitrogen (chemical symbol N) is present in almost all steels in small amounts that do not affect the characteristics of the steel. When nitrogen content reaches 0.004% or more, it combines with other elements to precipitate as a nitride, thereby increasing the steel's hardness and tensile and yield strengths while reducing its ductility and toughness.

Aluminum (chemical symbol Al) is a strong deoxidizer and inhibitor of grain growth, and it improves the nitriding characteristics of steel when present in amounts of approximately 1%.

Multiple alloying elements. A combination of two or more of the above alloying elements usually imparts some of the characteristic properties of each. Constructional chromium-nickel alloy steels, for example, develop good hardening properties, with excellent ductility, while chromium-molybdenum combinations develop excellent hardenability with satisfactory ductility and a certain amount of heat resistance. The combined effect of the two or more alloying elements on the hardenability of a steel is considerably greater than the sum of the same alloying elements if used separately. The general effectiveness of the nickel-chromium-molybdenum steels, with and without boron, can be accounted for in this way. Various combinations of alloying elements are employed by the different producers of high strength low alloy steels to achieve the mechanical properties, resistance to atmospheric corrosion, and other properties that characterize those steels. Carbon is generally maintained at a level that will insure freedom from excessive hardening after welding and will retain ductility. Manganese is used primarily as a strengthening element. Phosphorus is sometimes employed as a strengthening element and to enhance resistance to atmospheric corrosion. Copper is used to enhance resistance to atmospheric corrosion and as a strengthening element. Silicon, nickel, chromium, molybdenum, vanadium, aluminum, titanium, zirconium, and other elements are sometimes used, singly or in combination, for their beneficial effects on strength, toughness, corrosion resistance, and other desirable properties.

Manufacturing considerations—carbon and low alloy steels

Machining. The low carbon grades (0.30% carbon and less) are soft and gummy in the annealed condition and are preferably machined in the cold-worked or the normalized condition. Medium carbon grades (0.30 to 0.50% carbon) are best machined in the annealed condition, and high carbon grades (0.50 to 0.90% carbon) in the spheroidized condition. Finish machining must often be done in the fully heat treated condition for dimensional accuracy. The resulfurized grades are well known for good machinability. Nearly all carbon steels are available with 0.15 to 0.35% lead, added to improve machinability. However, resulfurized and leaded steels are not recommended for highly stressed parts. Alloy steels are generally harder than unalloyed steels of the same carbon content. Consequently, the low carbon alloy steels are somewhat easier to finish machine than their counterparts in the carbon steels. It is usually desirable to finish machine the carburizing and through-hardening grades in the final heat treated stage for better dimensional accuracy. This often requires two steps: rough machining in the annealed or hot-finished condition, then finish machining after heat treating. The latter operation, because of the relative high hardness of the material, necessitates the use of sharp, well designed, high speed steel cutting tools, proper feeds, speeds, and a generous supply of coolant. Medium and high carbon alloy grades are usually spheroidized for optimum machinability and, after heat treatment, may be finished by grinding. Many of the alloy steels are available with added sulfur or lead for improved machinability. These resulfurized and leaded alloy steels are not recommended for highly stressed applications, such as in aircraft parts.

Cold forming. The very low carbon grades have excellent cold-forming characteristics when in the annealed or normalized conditions. Medium carbon grades show progressively poorer formality with higher carbon content, and more frequent annealing is required.

The high carbon grades require special softening treatments for cold forming. Many carbon steels are embrittled by warm working or prolonged exposure in the temperature range of 300 to 700° F (150 to 370° C). The alloy steels are normally cold formed in annealed condition. Their formability depends on their carbon content. Little cold forming is done on alloy steels in the heat treated condition because of their high strength and limited ductility.

Forging. All the carbon steels exhibit excellent forgeability in the austentitic state provided the proper forging temperatures are used. As the carbon content is increased, the maximum forging temperature is decreased. At high temperatures, these steels are soft and ductile and exhibit little or no tendency to work harden. The resulfurized grades (free-machining steels) exhibit a tendency to rupture when deformed in certain high temperature ranges. Close control of forging temperatures is required. The alloy steels are only slightly more difficult to forge than carbon steels. However, maximum recommended forging temperatures are generally about 50% lower than for carbon steels of the same carbon content. Slower heating rates, shorter soaking period, and slower cooling rates are also required for alloy steels.

Welding. The low carbon grades are readily welded or brazed by all techniques. The medium carbon grades are also readily weldable but may require postwelding heat treatment. The high carbon grades are difficult to weld—preheating and postwelding treatment are usually mandatory, and special care must be taken to avoid overheating. Furnace brazing has been done successfully with all grades. Alloy steels with low carbon content are readily welded or brazed by all techniques. Alloy welding rods, comparable in strength to the base metal, are used, and moderate preheating (200–600° F *or* 98–315° C) is usually necessary. At higher carbon levels, alloy steels require higher preheating temperatures and often postwelding stress relieving. Certain alloy steels can be welded without loss of strength in the heat-affected zone provided that the welding heat is carefully controlled. If the composition and strength of the steel are such that the welded joint is reduced, the strength of the joint may be restored by heat treatment after welding. See the Welding section of this book for more detailed information.

Heat treatment. Due to the poor oxidation resistance of carbon steels, protective atmospheres must be employed during heat treatment if scaling on the surface cannot be tolerated. Also, these steels are subject to decarburization at elevated temperatures and, where surface carbon content is critical, should be heated in reducing atmospheres. For low alloy steels, various heat treatment procedures may be used to achieve any number of specific mechanical properties. In general, the annealed condition is achieved by heating to a specific condition and holding for a specified period of time. Annealing generally softens the material, producing the lowest mechanical properties. The normalized condition is achieved by holding to a slightly higher temperature than annealing, but for a shorter period of time. The purpose of normalizing varies depending on the desired properties—it can be used to increase or decrease mechanical properties. The quenched and tempered condition provides the highest mechanical properties with relative high toughness. Maximum hardness is obtained in the as-quenched condition, but alloy steels may be embrittled in this condition.

Stainless Steels

Stainless Steel. (Specialty Steel Industry of North America definition.) Stainless steels are iron-based alloys containing 10.5% or more of chromium, which is the alloying element that imparts to stainless steels their corrosion-resistant qualities. It does this by combining with oxygen to form a thin, transparent chromium-oxide protective film on the metal surface.

Austenitic stainless steels are those containing chromium, nickel, and manganese (AISI 200 series), or just chromium and nickel (AISI 300 series) as the principal alloying elements. These steels can be hardened only by cold working, heat treatment serves only to soften them. In the annealed condition, they are not attracted to a magnet, although some may become magnetic after cold working. They have excellent corrosion resistance, good formability, and the ability to develop excellent strength characteristics by cold working. When annealed, they possess maximum corrosion resistance, good yield and tensile strength, and freedom from notch effect. Some austenitic stainless steels contain titanium and niobium that are added to form more stable carbides than can be obtained with chromium. These stabilizing elements reduce susceptibility to intergranular corrosion, and prevent sensitization (the precipitation of chromium carbides, usually along grain boundaries, at temperatures ranging from 1,000 to 1,550° F [538 to 843° C]—thereby leaving the grain boundaries depleted of carbon) during welding and heat treatment. The most commonly used stabilized grades are 321 and 327.

Ferritic stainless steels (AISI 400 series) are straight chromium steels with a low carbon to chromium ratio that eliminates the effects of thermal transformation—they are not hardenable by heat treatment but can be marginally hardened by cold working. These steels are strong, attracted to a magnet, have good ductility, and are resistant to corrosion and oxidation. Their corrosion resistance can be improved by increasing the chromium or molybdenum content, while ductility can be improved by reducing carbon and nitrogen content. The ferritic steels, like the austenitic steels, are also subject to intergranular corrosion, but at temperatures in the region of 1,700° F (927° C).

Martensitic stainless steels (Also AISI 400 series) are straight chromium types having a higher carbon to chromium ratio than the ferritic group. Consequently, when cooled rapidly, they do harden (in some cases to tensile strengths of 200,000 PSI). These steels resist corrosion in mild environments, have fairly good ductility, and are always strongly magnetic. Some of the martensitic types, such as 416, 420F, and 44F, have been modified to improve machinability.

Precipitation hardened grades (AISI and UNS types S13800, S15500, S17400, and S17700) can achieve tensile and yield strengths as high as 300,000 PSI through the use of low temperature (about 900° F) aging in combination with cold working. With these steels, fabrication can be completed in an annealed condition, and then the component can be uniformly hardened without cold working or the high temperature thermal treatments that can cause distortion and heavy scaling. Prior to hardening, the machinability of these steels is approximately equal to AISI type 304 in the annealed condition.

Duplex stainless steels are dual phase materials with austenite and ferrite in close to 50-50 balance. They have excellent strength, corrosion resistance, and fabrication properties (especially those grades that contain nitrogen). Newer duplex materials with nitrogen are readily weldable and are often used in oil and gas production and chemical processing.

Basic chemical, mechanical, and physical properties of selected stainless steels are given in **Tables 13**, **14**, and **15**.

Effects of machining and fabrication on stainless steel

In all cutting operations on stainless steels, the following guidelines should be followed for maintaining corrosion resistance.

1) No contamination by ferrous (iron or steel) materials or particles should take place.

2) Mechanically cut edges will naturally form the corrosion resistant passive film that is a characteristic of stainless steel. The formation of this film on cut edges will be enhanced

(Text continued on p. 86)

Table 13. Chemical Composition of Wrought Stainless Steels.

AISI Type	UNS No.	C	Mn	P	S	Si	Cr	Ni	Mo	Other
Austenitic Grades										
201	S20100	0.15	5.50-7.50	0.060	0.030	0.75	16.00-18.00	3.50-5.50	-	0.25N
202	S20200	0.15	7.50-10.00	0.060	0.030	0.75	17.00-19.00	4.00-6.00	-	0.25N
203	S20300	0.08	5.00-6.50	0.040	0.18-0.35	1.00	16.00-19.00	5.00-6.50	0.50	1.75/2.25 Cu
205	S20500	0.12-0.25	14.00-15.50	0.060	0.030	0.75	16.50-18.00	1.00-1.75	-	0.32-0.40N
301	S30100	0.15	2.00	0.045	0.030	0.75	16.00-18.00	6.00-8.00	-	-
302	S30200	0.15	2.00	0.045	0.030	0.75	17.00-19.00	8.00-10.00	-	0.10N
303	S30300	0.15	2.00	0.20	0.15 (min.)	1.00	17.00-19.00	8.00-10.00	-	-
303Se	S30323	0.15	2.00	0.20	0.060	1.00	17.00-19.00	8.00-10.00	-	0.15(min.)Se
304	S30400	0.08	2.00	0.045	0.030	0.75	18.00-20.00	8.00-10.50	-	0.10N
305	S30500	0.12	2.00	0.045	0.030	0.75	17.00-19.00	10.50-13.00	-	-
308	S30800	0.08	2.00	0.045	0.030	1.00	19.00-21.00	10.00-12.00	-	-
309	S30900	0.20	2.00	0.045	0.030	1.00	22.00-24.00	12.00-15.00	-	-
310	S31000	0.25	2.00	0.045	0.030	1.50	24.00-26.00	19.00-22.00	-	-
314	S31400	0.25	2.00	0.045	0.030	1.50-3.00	23.00-26.00	19.00-22.00	-	-
316	S31600	0.08	2.00	0.045	0.030	0.75	16.00-18.00	10.00-14.00	2.00-3.00	0.10N
316H	S31609	0.04-0.10	2.00	0.045	0.030	0.75	16.00-18.00	10.00-14.00	2.00-3.00	-
316F	S31620	0.08	2.00	0.20	0.10 (min.)	1.00	16.00-18.00	10.00-14.00	1.75-2.50	-
317	S31700	0.08	2.0	0.045	0.030	0.75	18.00-20.00	11.00-15.00	3.00-4.00	0.10(max)N
329	S32900	0.08	2.00	0.040	0.030	0.75	23.00-28.00	2.50-5.00	1.00-2.00	-
330	S33000	0.08	2.00	0.040	0.030	0.75-1.50	17.00-20.00	34.00-37.00	-	-
384	S38400	0.08	2.00	0.045	0.030	1.00	15.00-17.00	17.00-19.00	-	-

(Continued)

Table 13. *(Continued)* **Chemical Composition of Wrought Stainless Steels.**

AISI Type	UNS No.	C	Mn	P	S	Si	Cr	Ni	Mo	Other
Ferritic Grades										
429	S42900	0.12	1.00	0.040	0.030	1.00	14.00-16.00	0.75	-	-
430	S43000	0.12	1.00	0.040	0.030	1.00	16.00-18.00	0.75	-	-
430F	S43020	0.12	1.25	0.060	0.15 (min.)	1.00	16.00-18.00	-	-	-
430FSe	S43023	0.12	1.25	0.060	0.060	1.00	16.00-18.00	-	-	0.15(min)Se
434	S43400	0.12	1.00	0.040	0.030	1.00	16.00-18.00	-	0.75-1.25	-
442	S44200	0.20	1.00	0.40	0.300	1.00	18.00-23.00	-	-	-
446	S44600	0.20	1.50	0.040	0.030	1.00	23.00-27.00	-	-	0.25N
Martensitic Grades										
403	S40300	0.15	1.00	0.040	0.030	0.50	11.50-13.00	0.60	-	-
410	S41000	0.15	1.00	0.040	0.30	1.00	11.50-13.50	0.75	-	-
414	S41400	0.15	1.00	0.040	0.030	1.00	11.50-13.50	1.25-2.50	-	-
416	S41600	0.15	1.25	0.060	0.15 (min.)	1.00	12.00-14.00	-	-	-
416Se	S41623	0.15	1.25	0.060	0.060	1.00	12.00-14.00	-	-	-
420	S42000	0.15 (min)	1.00	0.040	0.030	1.00	12.00-14.00	-	-	-
420F	S42020	Over 0.15	1.25	0.060	0.15	1.00	12.00-14.00	-	-	-
422	S42200	0.20-0.25	0.50-1.00	0.25	0.25	0.50	11.00-12.50	0.50-1.00	0.90-1.25	0.20-0.30V 0.90-1.25W
431	S43000	0.20	1.00	0.040	0.030	1.00	15.00-17.00	1.25-2.50	-	-
440A	S44002	0.60-0.75	1.00	0.040	0.030	1.00	16.00-18.00	-	0.75	-
440B	S44003	0.75-0.95	1.00	0.040	0.030	1.00	16.00-18.00	-	0.75	-
440C	S44004	0.95-1.20	1.25	0.40	0.030 (min.)	1.00	16.00/18.00	-	-	-
440F	S44020	0.95 (min.)	1.25	0.040	0.10 (min.)	1.00	16.00/18.00	-	-	-

Notes: Data are for information only and should not be used for design purposes. For design and specification, refer to appropriate ASTM specifications. Data were obtained from various sources, including AISI Steel Products Manuals, ASTM specification, and SSINA.

Table 14. Typical Mechanical Properties of Stainless Steels (Annealed Bar). *(Source, SSINA.)*

	AISI Type	UNS No.	Tensile Strength		Yield Strength (0.2% offset)		Elongation In 2" %	Hardness HBN (max)
			ksi (min.)	MPa	ksi (min.)	MPa		
Austenitic	203	S20300	75	517	30	207	40	262
	303	S30300	75	517	30	207	35	262
	303Se	S30323	75	517	30	207	35	262
	304	S30400	75 (125) *	517 (861)	30 (100) *	207 (690)	40 (10) *	-
	316	S31600	75 (125) *	517 (861)	30 (100) *	207 (690)	40 (10) *	-
	316H	S31609	75	517	30	207	40	-
	316F	S31620	75	517	30	207	40	262
Ferritic	430	S43000	70	483	40	276	20	-
	430F	S43020	70	483	40	276	20	262
Martensitic	416	S41600	75	517	40	276	22	262
	420	S42000	75	517	40	276	25	241
	420F	-	75	517	40	276	22	262
	440C	S44004	110	758	65	448	14	262
	440F	S44020	110	758	65	448	14	262

Notes: Data are for information only and should not be used for design purposes. For design and specification, refer to appropriate ASTM specifications. Data were obtained from various sources, including AISI Steel Products Manuals, ASTM specification, and individual company literature.

* Cold finished, for sizes up to $^3/_4$ inch.

Table 15. Physical Properties of Selected Stainless Steels. *(Source, SSINA.)*

Stainless Steel Type	Density lb./in.^{3}	Modulus of Elasticity PSI % 10^6	Specific Electrical Resistance at 68° F (microhm-cm)	Specific Heat Capacity Btu/lb. °F	Thermal Conductivity Btu/ft. hr. °F
Austenitic Grades					
203	0.29	28.0	74	-	9.5
303	0.29	28.0	72	0.12	9.4
303Se	0.29	28.0	72	0.12	9.4
304	0.29	28.0	72	0.12	9.4
316	0.29	28.0	74	0.12	9.4
316H	0.29	28.0	74	0.12	9.4
316F	0.29	29.0	74	0.116	8.3
Ferritic Grades					
430	0.28	29.0	60	0.11	13.8
430F	0.28	29.0	60	0.11	15.1
Martensitic Grades					
416	0.28	29.0	57	0.11	14.4
420	0.28	29.0	55	0.11	14.4
420F	0.28	29.0	55	0.11	14.5
440C	0.28	29.0	60	0.11	14.0
440F	0.28	29.0	60	0.11	14.0

Stainless Steel Type	Mean Coefficient of Thermal Expansion in./in./°F %$^{10-6}$					Magnetic Permeability (max)	Annealing Temperature °F
	32-212°F	32-600°F	32-1000°F	32-1200°F	32-1600°F		
Austenitic Grades							
203	9.4	-	-	-	11.6	1.02	1850-2050 [A]
303	9.6	9.9	10.2	10.4	-	[1.02]	1850-2050 [A]
303Se	9.6	9.9	10.2	10.4	-	[1.02]	1850-2050 [A]
304	9.6	9.9	10.2	10.4	11.0	1.02	1850-2050 [A]
316	8.9	9.0	9.7	10.3	11.1 [D]	1.02	1850-2050 [A]
316H	8.9	9.0	9.7	10.3	11.1 [D]	1.02	1850-2050 [A]
316F	9.2	9.7	10.1	-	-	-	2000 [A]
Ferritic Grades							
430	5.8	6.1	6.3	6.6	6.9 [D]	-	1250-1400
430F	5.8	6.1	6.3	6.6	6.9	-	1250-1400 [C]
Martensitic Grades							
416	5.5	6.1	6.4	6.5	-	-	1500-1650 [B] 1200-1400 [C]
420	5.7	6.0	6.5	-	-	-	1550-1650 [B] 1350-1450 [C]
420F	5.7	-	-	-	-	-	1550-1650 [B] 1350-1450 [C]
440F	5.6	-	-	-	-	-	1350-1450 [C]

Notes: Data are for information only and should not be used for design purposes. For design and specification, refer to appropriate ASTM specifications. Data were obtained from various sources, including AISI Steel Products Manuals, ASTM specification, and individual company literature.

[A] = Cool rapidly from these annealing temperatures.
[B] = Full annealing-cool slowly.
[C] = Process annealing.
[D] = Thermal range is 32-1500° F.

by a chemical (acid) passivation (see below) treatment with nitric acid.

3) Thermally cut edges will be affected in terms of chemical composition and metallurgical structure. Removal of surface layers by dressing is necessary so that impaired areas of mechanical and corrosion resistant properties are minimized.

Passivation. The thin transparent film on the surface of stainless steel is uniform, stable, and passive. This film will form spontaneously, or repair itself if damaged in air (due to the presence of oxygen) and when immersed in solutions that contain oxygen or oxidizing agents; passivity appears to be enhanced by the immersion method. Nitric acid is normally used as the oxidizing agent. It does not corrode stainless steel, does not alter critical dimensions, and will not remove heat tint, embedded iron, or other embedded surface contamination. The standard nitric acid passivation solution is made up of 10–15% (by volume) of nitric acid in water. Best results are obtained if the solution is 150° F (65° C) for the austenitic (300 series) stainless steels and 120° F (50° C) for the ferritic and martensitic (400 series) plain chromium stainless steels. The suggested immersion time is thirty minutes, followed by thorough water washing. If the component cannot be thoroughly immersed, cold swabbing with cold acid solution is normally used. Under these conditions, the following guidelines are recommended. For austentitic (300 series), use a solution of 15% nitric acid (by volume) and apply at a temperature between 65–80° F (20–25° C) for thirty to ninety minutes. For ferritics (400 series), use a solution of 12% nitric acid (by volume) and apply at 65–80° F (20–25° C) for thirty to forty five minutes. The solution should be continually swabbed on with sponges, paint brushes, or nylon pads. Thorough water rinsing must follow passivitating treatments.

Descaling. When stainless steel is heated to elevated temperatures, such as during annealing and welding, an oxide scale will form on the surface unless the material is completely surrounded by a protective atmosphere. These oxides should be removed to restore the stainless steel to its optimum corrosion resistance condition. Pickling is one of the most common methods used to remove this oxide scale. To remove scale caused by annealing austenitic stainless steel, pickling the component in a solution of 10–15% nitric acid plus 1 to 3% hydrofluoric acid (by volume), warmed to 120–140° F (50–60° C) effectively removes oxides and loosely imbedded iron and chromium depleted layers, and it leaves the surface in a clean, passivated condition. For ferritic and martensitic stainless steels, a solution of 8–12% sulfuric acid and 2% rock salt (by volume), warmed to 150–170° F (65–71° C) should be used. Pickling times are normally five to ten minutes and, upon removal, the component should be scrubbed and rinsed to remove sludge. Pickling can be done at room temperature, but exposure times must be increased.

Scale can also be removed by grit blasting, sand blasting, or glass bead blasting. Glass bead media is preferred as it is less likely to damage or contaminate the surface.

Machinability of stainless steels

Machinability of stainless steels differs substantially from carbon or low alloy steels. Most standard stainless steels are more difficult to machine, but the free-machining types are routinely machined on high production equipment. The easiest to machine are the 400 series, and the 200 and 300 series are often characterized as being the most difficult, primarily because of their gumminess and their propensity to work harden at a very rapid rate. *Figures 1* and *2* show the comparative machinability of stainless steel and other common metals.

Free-machining Stainless Steels. Certain alloying elements in stainless steels, such as sulfur, selenium, lead, copper, aluminum, calcium, or phosphorus can be added or adjusted during melting to alter the machining characteristics. These elements serve to reduce the friction created by the tool, thereby minimizing the tendency of the chip to weld to the

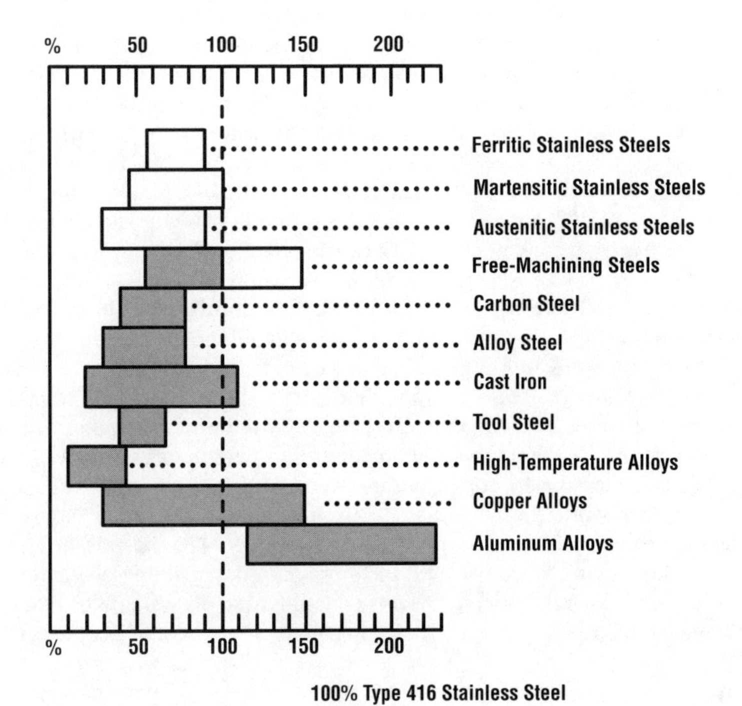

Figure 1. Comparative machinability of common metals. (Source, SSINA.)

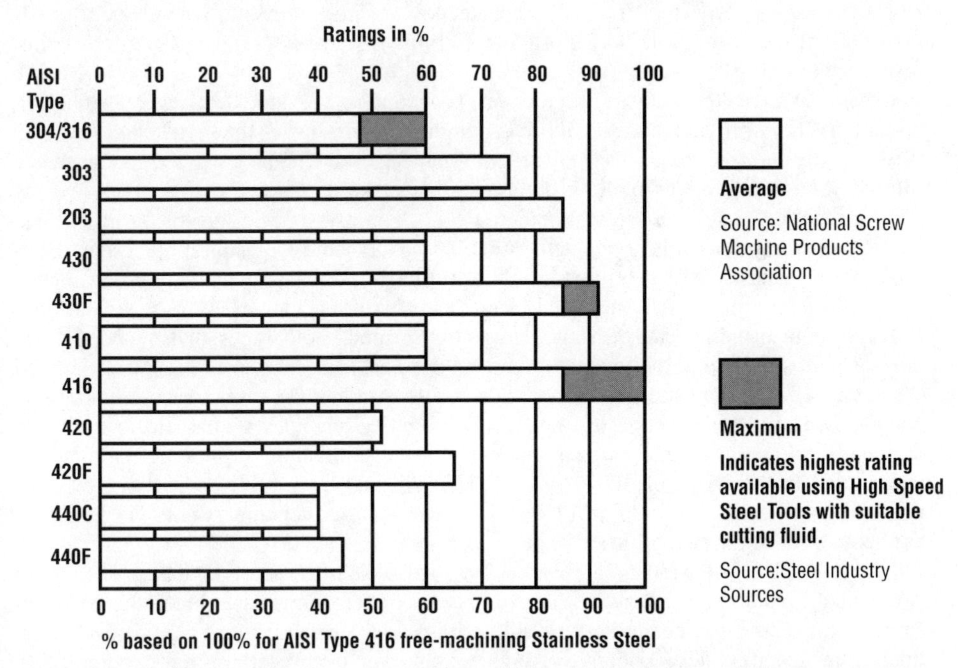

Figure 2. Comparative machinability of frequently used stainless steels and their free-machining counterparts. (Source, SSINA.)

tool. Sulfur can also be added to induce inclusions that reduce the friction forces and transverse ductility of the chips, causing them to break more readily. The improvement in machinability in the free-machining stainless steels—namely types 303, 303 Se, 203, 430F, 416, and 420F—compared to other frequently used stainless steels, can be seen in *Figure 3*. The free-machining grades can be competitive with the tougher grades and, where ease of production is an issue, should be used whenever possible. If, for example, a design demands excellent corrosion and strength of type 304, but a high machining rate is required, free-machining 303 might be a suitable substitution. While each has similar chromium, nickel, and sulfur content, type 303 can be machined at speeds 25 to 30% faster than type 304. Type 303 Se is a free-machining variation of type 304 that contains selenium rather than sulfur. Therefore, when surface finish is a primary consideration, or when cold-working may be involved (staking, swaging, spinning, or severe thread rolling in addition to machining), selecting 303 Se instead of type 304 can be advantageous. Type 203 is an austenitic free-machining grade that is a modification of the 200 series. Although, like the 200 series, it contains significant additions of manganese, it also contains nickel and copper. In the 400 series, if end-use conditions call for type 430, type 430F might be considered. The composition of type 430F is adjusted to enhance its machining characteristics while preserving as many of type 430's qualities as possible. The free-machining variation of type 410 is 416, and for type 420 the free-machining variation is 420F. The alloying elements used to improve free-machining characteristics can adversely affect corrosion resistance, transverse ductility, and other qualities including weldability. Free-machining grades are normally not used when welding, cold forming, or hot forging is specified.

Type 203. This is an austenitic grade containing increased levels of manganese in order to obtain maximum machining speeds. It is especially well suited to high production, high volume automatic screw machine work. This material is similar to type 303 but requires slightly higher machining speeds: typical turning speeds are 125–155 sfm (38–47 smm) for high-speed steel tools. Higher speeds are possible with carbide tooling. It resists a variety of organic and inorganic compounds and is comparable in corrosion resistance to type 303.

Type 303 and 303 Se. These grades are particularly well suited to screw machining operations. They provide longer tool life and higher cutting speeds than type 304. Type 303 is widely used for the following applications: shafting, valve bodies, valve trim, and valve fittings. It has desirable nongalling properties that make disassembly of parts easy. Type 303 Se is used for similar applications, and it has slightly superior corrosion properties. It also has better formability characteristics for hot- or cold-working applications. Both types machine easily with a brittle chip. In turning operations, they can be used at speeds up to 100–130 sfm (31–40 smm). Moderate cold working increases the machinability. Polishing and grinding can be satisfactorily performed. Both types resist rusting from normal atmospheric source, and they can be used in applications that expose them to sterilizing solutions, most organic chemicals and dyestuffs, and a wide variety of inorganic chemicals. They are also very nitric acid resistant, moderately resistant to sulfuric acids, and they have poor resistance to halogen acids. If, during fabrication, components are heated and cooled in the range of 800–1650° F (425–900° C), a corrective thermal treatment consisting of heating to 1900° F (1037° C) followed by quenching in water should be applied to reduce the risk of intergranular corrosion.

Type 416. This is the most readily machinable of all stainless steels, and it is particularly well suited for automatic screw machine operations (where speeds are about 165 sfm *or* 50 smm) because its use is beneficial to tool life. It is widely used for fittings, gears, housings, lead screws, shafts, valve bodies, valve stems, and valve trim, or for any part requiring considerable machining work. While not as corrosion resistant as the 300 series grades, type 416 resists atmospheric environments, fresh water, mine water, steam, carbonic acid, gasoline, crude oil, blood, alcohol, ammonia, soap, sugar solutions, etc. Maximum corrosion resistance is achieved in the heat treated condition. It resists scaling at elevated temperatures and can be used for continuous service to about 1200° F (356° C).

Type 420F. This grade is easy to machine, grind, and polish, and it has certain anti-galling and nonseizing properties. It is often used for parts made on automatic screw machines such as valve trim, pump shafts, needle valves, ball check valves, gears, cams, and pivots. It provides high hardness and good corrosion resistance. It is comparable to AISI 2315 and 2340 for automatic screw machine operations. In turning operations involving heavy-duty equipment, speeds up to 90–110 sfm (27–34 smm) and feeds up to 0.0008–0.0020 inches (0.02 to 0.05 mm) are suggested. 420F should always be used in the hardened condition for maximum corrosion resistance, and the surface should be free of scale which can be removed by grinding and polishing, or pickling (acid pickling is immersion in an aqueous solution containing about 10% nitric acid and 2% hydrofluoric acid at a maximum temperature of 140°F [60° C]). If pickled after hardening, the parts should be baked at 250–300° F (121–148° C) for at least one hour to remove acid brittleness. In the hardened condition, type 420F will resist corrosion from atmosphere, fresh water, mine water, steam, carbonic acid, crude oil, gasoline, blood, alcohol, ammonia, mercury, sterilizing solutions, soap, etc. Passivation after machining is recommended.

Type 430F. When a 16–18% straight chromium stainless steel is specified, this grade is recommended for faster cutting and reduced costs. It does not harden heat treatment. 430F is commonly used for parts requiring good corrosion resistance, such as solenoid valves, aircraft parts, and gears. It machines at approximately the same speeds (125–155 sfm *or* 38–47 smm) as AISI 1120 and 1130 steels. 430F offers superior corrosion resistance from atmospheric, fresh water, nitric acid, and dairy products, but parts must be free of scale and foreign particles. As a final treatment, after scale removal and machining, all parts should be passivated.

Fabrication of stainless steel

Stainless steels can be fabricated by methods similar to those used for carbon steels, but some modifications must be made for their yield strength and their rate of work hardening. All have work hardening rates higher than common carbon steels, but the austenitics are characterized by large increases in strength and hardness with cold work. With the exception of the resulfurized free-machining grades (303 is the common type), all stainless steels are suitable for crimping or flattening operations. The free-machining grades will withstand mild longitudinal deformation but may exhibit some tendency to splitting. In spite of their higher hardness, most martensitic and all of the ferritic grades can be successfully fabricated. **Table 16** shows the relative fabrication characteristics of three groups of stainless steels.

Tool Steels

Tool steels are used to make tools for cutting or shaping other materials. Therefore, they are usually subjected to extreme repetitive loads and their survival depends on their characteristics. Defining characteristics of tool steels are hot hardness (resistance to softening at elevated temperature), resistance to abrasive wear, resistance to breakage, resistance to deformation, resistance to adhesive wear, and toughness. There are six general classes of tool steels, plus two special purpose steels. Much of the information that follows on specific tool steels and their characteristics was extracted from *Heat Treatment, Selection, and Application of Tool Steels* by Bill Bryson (Hanser Gardner Publications)

Table 16. Relative Fabrication Characteristics of Stainless Steels. *(Source, SSINA.)*

Group	Austenitic					Ferritic		Martensitic		
Type Number	201, 202, 301, 302, 304, 304L, 305	303*	309S, 310S	316, 316L, 317, 317LMN	321, 347	430, 439	405, 442, 446	403, 410	420	440A, 440C
Characteristic										
Air Hardening	No	No	No	No	No	No	No	Yes	Yes	Yes
Blanking	F	F	F	F	F	E	E	E	E	G
Brazing, Silver	G	-	G	G	G	G	G	G	G	G
Buffing	G	-	G	G	G	G	G	G	G	G
Drawing, Deep	E	-	G	E	G	E	F	E	NR	NR
Forming, Hot	G	F	G	G	G	G	G	G	G	G
Forming, Cold	G	F	G	G	G	G	F	G	F	NR
Grinding, Ease of	F	G	F	F	F	F	F	G	E	E
Grinding, (magnetic)	No	No	No	No	No	Yes	Yes	Yes	Yes	Yes
Hardenable by Heat Treatment	No	No	No	No	No	No	No	Yes	Yes	Yes
Punching (perforating)	F	-	F	F	F	G	G	G	G	G
Polishing	G	G	G	G	F	G	G	G	G	G
Riveting, Hot	G	F	G	G	G	G	G	G	NR	NR
Riveting, Cold	G	F	G	G	G	G	G	G	NR	NR
Shearing, Cold	F	F	F	F	F	G	G	G	G	F
Soldering	G	G	G	G	G	G	G	G	G	G
Brazing	G	G	G	G	G	G	G	G	G	G

(Continued)

Table 16. *(Continued)* **Relative Fabrication Characteristics of Stainless Steels.** *(Source, SSINA.)*

Group	Austenitic					Ferritic		Martensitic		
Type Number	201, 202, 301, 302, 304, 304L, 305	303*	309S, 310S	316, 316L, 317, 317LMN	321, 347	430, 439	405, 442, 446	403, 410	420	440A, 440C
Characteristic										
Spinning	G	-	G	G	G	G	F	F	NR	NR
Welding	E	NR	E	E	E	F	F	F	F	F
Machining	F	E	F	F	F	G	F	G	F	NR

* Chemistry designed for improved machining (as are other grades, i.e., 416, 420F, 430F, 440F)

Note: Code: E = Excellent; G = Good; F = Fair; NR = Not Generally Recommended (Poor).

which contains detailed analysis of heat treating and selecting many of the popular tool steels. The machinability ratings are taken from the Properties of Materials section of this book (Table 7, page 42) and are based on AISI 1212 having a rating of 100%. Each class and subcategory is explained below, and the influencing elements in the composition of tool steels are provided in **Table 17**.

Water-hardened tool steels (W series) contain few alloying agents other than carbon. These steels are generally described as shallow hardening, and the hardened zone is thin. Their low hardenability allows for high surface hardness, and the prescribed heat treatment for these steels is relatively simple and inexpensive. Their name is derived from the fact that they are normally water quenched from austenizing temperature, but it should be remembered that water-hardening steels are the steels most subject to cracking and deformation due to the severe shock of water quenching. Also, decarb and scale are most prominent in the water hardening grades. Because of their high strength, excellent wear resistance, and resistance to splitting and galling, W series tool steels are commonly used to produce arbors, stone chisels, collets, burnishing dies, coining dies, cold extrusion dies, threading dies, dowels, gun drills, files, plug gages, taps, and most striking and woodworking tools. In fact, these were the first modern heat treatable steels and were probably first used for swords and axes. Their carbon content is in the area of 1%, and the achievable hardness range is typically between 63 and 65 R_C on the case (0.050 inch [1.27 mm] depth), making it possible to develop the sharpest edge of any steel in the marketplace. Interestingly, W1 tool steel has the same basic chemistry as AISI 1095, but the two differ in manufacturing method. Tool steels are generally manufactured under very tight rules concerning percentage of scrap versus raw material, and the grain size on the annealed material is controlled with greater accuracy. Both W1 and W2 should be preheated to 1,200° F (649° C) and hardened at 1,475° F (802° C). The optimum tempering temperature for W1 and W2 tool steels is 300° F (149° C), which will yield the following approximate results: hardness of 64 R_C, Charpy value of 80 ft lbs (108 Joules) for W1 and 77 ft lbs (104 Joules) for W2, and a machinability rating of 30 for both W1 and W2.

Cold-work tool steels are comprised of the O, A, and D series steels.

O series cold-work tool steels are oil hardening, and therefore they can be quenched with less distortion than is often experienced with the W series steels. This series is comparatively inexpensive to purchase and features high resistance to wear at normal temperatures even though softening and reduced wear resistance occur at higher temperatures. O1 is considered the work horse of general purpose tool steel, but it suffers from decarb formation during heat treatment and is subject to cracking from thermal shock when quenching. Except for machinability, O1 and O6 have very similar properties. Preheat of both steels should be done at 1,200° F (649° C), and both should be hardened at 1,475° F (802° C). When tempered at 300° F (140° C), they have a hardness of 63 R_C, a Charpy value of 33 ft lbs (45 Joules) and machinability ratings of 16 for O1 and 38 for O6). These steels are widely used for cold drawing dies, drilling bushings (O1); blanking and forming dies, threading dies, bending dies (O2); cold stamping dies, punches, and thread gages (O1 and O2). Type O7 is often used for tools that require sharp and wear resistant cutting edges.

A series cold-work tool steels are air hardening. They are primarily used for cold working, as they relinquish hardness as temperatures rise. The A7 grade, with its high percentage of carbon and chromium, has the highest wear resistance in this series. Currently, A2 is probably the most popular grade of tool steel in the U.S., with excellent wear resistance resulting from increased carbon and chrome (its wear resistance is 20 to 25% better than A6's). Its preheat temperature is 1,200°F (649° C) and its hardening

(Text continued on p. 98)

Table 17. Chemical Composition and Typical Hardness of Tool Steels. *(See Note.) (Source, Bethlehem Steel.)*

AISI Type	UNS No.	Identifying Elements %							Hardness R_c
		C	Mn	Cr	Si	W	Mo	Other	
WATER-HARDENING TOOL STEELS									
W1	T72301	0.60/1.40*	-	-	-	-	-	-	64
W2	T72302	0.60/1.40*	-	-	-	-	-	0.25V	64
W5	T72305	1.10	-	0.50	-	-	-	-	64
*Other carbon contents may be available.									
COLD-WORK TOOL STEELS									
Oil-Hardening Types									
O1	T31501	0.90	1.00	0.50	-	0.50	-	-	62
O2	T31502	0.90	1.60	-	-	-	-	-	62
O6	T31506	1.45	0.80	-	1.00	-	0.25	-	63
O7	T31507	1.20	-	0.75	-	1.75	-	-	64
Medium Alloy Air-Hardening Types									
A2	T30102	1.00	-	5.00	-	-	1.00	-	62
A3	T30103	1.25	-	5.00	-	-	1.00	1.00V	65
A4	T30104	1.00	2.00	1.00	-	-	1.00	-	62
A6	T30106	0.70	2.00	1.00	-	-	1.25	-	60
A7	T30107	2.25	-	5.25	-	1.00†	1.00	4.75	67
A8	T30108	0.55	-	5.00	-	1.25	1.25	-	60
A9	T30109	0.50	-	5.00	-	-	1.40	1.00V 1.50Ni	56
A10	T30110	1.35	1.80	-	1.25	-	1.50	1.80Ni	62
† Optional									

(Continued)

Table 17. *(Continued)* **Chemical Composition and Typical Hardness of Tool Steels.** *(See Note.) (Source, Bethlehem Steel.)*

AISI Type	UNS No.	Identifying Elements %							Hardness R_c
		C	Mn	Cr	Si	W	Mo	Other	
COLD-WORK TOOL STEELS									
High Carbon-High Chromium Types									
D2	T30402	1.50	-	12.00	-	-	1.00	1.00V	61
D3	T30403	2.25	-	12.00	-	-	-	-	61
D4	T30404	2.25	-	12.00	-	-	1.00	-	61
D5	T30405	1.50	-	12.00	-	-	1.00	3.00Co	61
D7	T30407	2.35	-	12.00	-	-	1.00	4.00V	65
SHOCK-RESISTING TOOL STEELS									
S1	T41901	0.50	-	1.50	-	2.50	-	-	58
S2	T41902	0.50	-	-	1.00	-	0.50	-	60
S5	T41905	0.55	0.80	-	2.00	-	0.40	-	60
S6	T41906	0.45	1.40	1.50	2.25	-	0.40	-	56
S7	T41907	0.50	-	3.25	-	-	1.40	-	57

AISI Type		Identifying Elements %								Hardness R_c
		C	Mn	Cr	Si	W	Mo	V	Co	
HOT-WORK TOOL STEELS										
Chromium Types										
H10	T20810	0.40	-	3.25	-	-	2.50	0.40	-	56
H11	T20811	0.35	-	5.00	-	-	1.50	0.40	-	54
H12	T20812	0.35	-	5.00	-	1.50	1.50	0.40	-	55
H13	T20813	0.35	-	5.00	-	-	1.50	1.00	-	53
H14	T20814	0.40	-	5.00	-	5.00	-	-	-	47
H19	T20819	0.40	-	4.25	-	4.25	-	2.00	4.25	57

(Continued)

Table 17. *(Continued)* **Chemical Composition and Typical Hardness of Tool Steels.** *(See Note.) (Source, Bethlehem Steel.)*

AISI Type		Identifying Elements %								Hardness R_c
		C	Mn	Cr	Si	W	Mo	V	Co	
HOT-WORK TOOL STEELS										
Tungsten Types										
H21	T20821	0.35	-	3.50	-	9.00	-	-	-	54
H22	T20822	0.35	-	2.00	-	11.00	-	-	-	52
H23	T20823	0.30	-	12.00	-	12.00	-	-	-	47
H24	T20824	0.45	-	3.00	-	15.00	-	-	-	55
H25	T20825	0.25	-	4.00	-	15.00	-	-	-	44
H26	T20826	0.50	-	4.00	-	18.00	-	1.00	-	58
Molybdenum Types										
H42	T20842	0.55	-	4.00	-	-	8.00	2.00	-	60
HIGH-SPEED TOOL STEELS										
Tungsten Types										
T1	T12001	0.75*	-	4.00	-	18.00	-	1.00	-	65
T2	T12002	0.80	-	4.00	-	18.00	-	2.00	-	66
T4	T12004	0.75	-	4.00	-	18.00	-	1.00	5.00	66
T5	T12005	0.80	-	4.00	-	18.00	-	2.00	8.00	65
T6	T12006	0.80	-	4.50	-	20.00	-	1.50	12.00	65
T8	T12008	0.75	-	4.00	-	14.00	-	2.00	5.00	65
T15	T12015	1.50	-	4.00	-	12.00	-	5.00	5.00	68
Molybdenum Types										
M1	T11301	0.85*	-	4.00	-	1.50	8.50	1.00	-	65
M2	T11302	0.85/1.00*	-	4.00	-	6.00	5.00	2.00	-	65
M3-1	T11313	1.05	-	4.00	-	6.00	5.00	2.40	-	66

(Continued)

Table 17. *(Continued)* **Chemical Composition and Typical Hardness of Tool Steels.** *(See Note.) (Source, Bethlehem Steel.)*

AISI Type		Identifying Elements %								Hardness R_c
		C	Mn	Cr	Si	W	Mo	V	Co	
HIGH-SPEED TOOL STEELS										
Molybdenum Types										
M3-2	T11323	1.20	-	4.00	-	6.00	5.00	3.00	-	66
M4	T11304	1.30	-	4.00	-	5.50	4.50	4.00	-	66
M6	T11306	0.80	-	4.00	-	4.00	5.00	1.50	12.00	66
M7	T11307	1.00	-	4.00	-	1.75	8.75	2.00	-	66
M10	T11310	0.85/1.00*	-	4.00	-	-	8.00	2.00	-	65
M30	T11330	0.80	-	4.00	-	2.00	8.00	1.25	5.00	65
M33	T11333	0.90	-	4.00	-	1.50	9.50	1.15	8.00	65
M34	T11334	0.90	-	4.00	-	2.00	8.00	2.00	8.00	65
M36	T11336	0.80	-	4.00	-	6.00	5.00	2.00	8.00	65
M41	T11341	1.10	-	4.25	-	6.75	3.75	2.00	5.00	70
M42	T11342	1.10	-	3.75	-	1.50	9.50	1.15	8.00	70
M43	T11343	1.20	-	3.75	-	2.75	8.00	1.60	8.25	70
M44	T11344	1.15	-	4.25	-	5.25	6.25	2.00	12.00	70
M46	T11346	1.25	-	4.00	-	2.00	8.25	3.20	8.25	69
M47	T11347	1.10	-	3.75	-	1.50	9.50	1.25	5.00	70
M62	T11362	1.25	-	3.75	-	6.50	10.50	2.00	-	70
* Other carbon contents may be available.										
AISI Type		**Identifying Elements %**								**Hardness R_c**
		C	**Mn**	**Cr**	**Si**	**W**	**Mo**	**Ni**	**Other**	
PLASTIC-MOLD STEELS										
P1	T51601	0.10	0.20	-	-	-	-	-	0.10V	64‡

(Continued)

Table 17. *(Continued)* **Chemical Composition and Typical Hardness of Tool Steels.** *(See Note.) (Source ,Bethlehem Steel.)*

AISI Type		Identifying Elements %								Hardness R_c
		C	Mn	Cr	Si	W	Mo	Ni	Other	
PLASTIC-MOLD STEELS										
P2	T51602	0.07	0.50	2.00	-	-	0.20	0.50	-	64 ‡
P3	T51603	0.10	0.50	0.60	-	-	-	1.25	-	64 ‡
P4	T51604	0.07	0.50	5.00	-	-	0.75	-	-	64 ‡
P5	T51605	0.10	0.30	2.25	-	-	-	-	-	64 ‡
P6	T51606	0.10	0.40	1.50	-	-	-	3.50	-	61 ‡
P20	T51620	0.35	0.75	1.70	-	-	0.40	-	-	37
P21	T51621	0.20	0.30	-	-	-	-	4.00	1.20Al	40
‡ Hardness of carburized case										
SPECIAL-PURPOSE TOOL STEELS										
Low Alloy Types										
L2	T61202	0.50/1.10*	-	1.00	-	-	-	-	0.20V	63
L6	T61206	0.70	-	0.75	-	-	0.25†	1.50	-	62
* Other carbon contents may be available. † Optional										

Note: The percentages of elements in this Table are given for identification purposes only and are not to be considered as the means of the composition ranges for the elements.

temperature is 1,775° F (968° C). When tempered at 400° F (204° C) it has a hardness of 60 R_C, a Charpy value of 17 ft lbs (23 Joules), and a machinability rating of 16. A6 has a low chrome content and is an ideal tool steel for roll forming applications in situations where galling can be a problem. It should be preheated at 1,200° F (649° C) and hardened at 1,575° F (857° C). When tempered at 400° F (204° C), it has a hardness of 59 R_C, a Charpy value of 28 ft lbs (38 Joules), and a machinability rating of 16. Because they exhibit excellent dimensional stability properties, these steels are often used for precision measuring tools, as well as cold blanking, cold coining, cold stamping, cold extrusion dies, tool shanks, engraver's tools (A2); embossing dies, powder metal dies, tool shanks, and thread roller dies (A6); and thread roller dies (A7).

D series cold-work tool steels contain high carbon and very high (12%) chromium contents. Except for D3, they are all air hardened. D3 is usually oil quenched, resulting in increased likelihood of distortion and cracking in tools made of this grade. Like the A series, they are primarily used in cold working applications, but D7, in particular has good resistance to softening at elevated temperatures, primarily due to its high vanadium content. Dimensional stability is comparable to the A series. Due to their high chrome content, these steels are difficult to work or grind. The preheat temperature for D2 is 1,200° F (649° C), and its hardening temperature is 1,850° F (1,010° C). Recommended tempering temperature is 960° F (515° C), which produces a hardness rating of 58/60 R_C, a Charpy value of 8 ft lbs (11 Joules), and a machinability rating of 11. D3 should be preheated to 1,200° F (649° C) and hardened at 1,725° C (940° C). When tempered at 400° F (204° C), its hardness is 62 R_C, its Charpy value is 6 ft lbs (8 Joules), and its machinability rating is 11. Since D3 is only available from a limited number of suppliers, substituting it with M2 should be considered. The vanadium in D7 controls and promotes a very fine grain structure in the hardened steel, and its molybdenum content (1%) provides a very slight heat resistance. D7 should be preheated to 1,400° F (760° C) and hardened at 1,975° F (1,079° C). Tempering at 400° F (204° C) results in a hardness of 62 R_C, a Charpy value of 6 ft lbs (8 Joules), and a machinability rating of 11. These steels are often used for long run dies and for precision gages, as well as for broaches, cold burnishing dies, cold cutting dies, cold drawing dies, cold extrusion dies, cold stamping dies, thread roller dies, wire drawing dies, cold mandrels, and cutters for glass and wire (D2 and D3).

Shock-resisting tool steels are the S series steels. These steels have high strength, moderate wear resistance, and a high level of toughness, making them ideal for applications where high impact loads are encountered (chisels, punches, hammers, etc.), but they are also employed in certain structural applications. They are quenched using different media: S1, S5, and S6 are usually oil quenched, S2 is water quenched, and S7, in large sections, is oil quenched while small parts are air quenched. Specific applications include screw driver bits, wood chisels, cold chisels, bolt clippers, shear blades, pipe cutters, wire cutters (S1 and S2); embossing dies (S4); swaging dies, air hammers, shear blades (S5); hot forging dies, and hot shear blades (S7). While S1 is best known for its use in chisels, it is no longer widely available. It has better compression strength in conjunction with shock resistance than S7. The preheat temperature for S1 is 1,200° F (649° C), and the hardening temperature is 1,725° F (940° C). When tempered at 400° F (204° C), its hardness is 56 R_C, its Charpy value is 185 ft lbs (250 Joules), and its machinability rating is 20. S7 has the highest shock resistance of all tool steels, and it should be preheated at 1,200° F (649° C) and hardened at 1,725° F (940° C). When tempered at 450° F (232° C), it achieves a hardness of 58 R_C, a Charpy value of 240 ft lbs, and a machinability rating of 20.

Hot-work tool steels are designated H series steels, and they are segmented into three distinct groups: chromium types, tungsten types, and molybdenum types.

Chromium hot-work tool steels, grades H10 to H19, have chromium contents ranging from 3.25 to 5%, with relatively low carbon content (0.35 to 0.40%). In thicknesses of up to 1.25 inches (30 mm), they can be air hardened. In the past, H11, H12, and H13 were the most widely used of all the hot-work tool steels (in tonnage), as they are well suited for hot die work due to their ability to endure continued exposure to high temperatures. H11, for example, provides tensile strengths of up to 250 to 300 ksi (1723 to 2068 MPa) at working temperatures of 1,000° F (540° C). Typical uses of these steels include hot extrusion dies, hot mandrels, dowel pins (H11); die casting dies (H12 and H13); and hot header dies (H13 and H19). H11 and H13 are also widely used in the construction of light metal die-casting molds. In fact, H13 is considered an excellent steel for plastic molding or die cast molds, as it can withstand temperatures up to 1,300° F (704° C) and is readily available on the market. H13 should be preheated to 1,500° F (815° C) and hardened at 1,875° F (1024° C). When tempered at 1,000° F (538° C), it has a hardness of 54 R_C, a Charpy rating of 68 ft lbs, and a machinability rating of 15. H19 has superior resistance to heat check in die casting applications, and it provides excellent wear resistance in high temperature applications. It should be preheated at 1,500° F (815° C) and hardened at 1,850° F (1,010° C). When tempered at 1,100° F (593° C), it has a hardness of 52 R_C, a Charpy value of 43 ft lbs, and a machinability rating of 15.

Tungsten hot-work tool steels (H21 to H26) contain tungsten in the range of 9 to 18%. The high alloy content makes this series slightly more brittle than other tool steels, but also significantly boosts their resistance to high temperature softening. Even though these steels can be air hardened, scaling can be minimized by quenching in oil or molten salt. Common applications for this group include forging dies for brass, hot coining dies, hot drawing dies, and hot punches (H21).

Molybdenum hot-work tool steels (H42 and H43) have similar properties to the tungsten hot-work tool steels. They do possess slightly better resistance to heat checking (parallel surface cracks caused by rapid heating and cooling), and are slightly lower priced than the tungsten grades.

High-speed tool steels are made up of two different types: the M (molybdenum) type and the T (tungsten) type. The M steels account for 95% of all high speed steel made in the U.S. The overwhelming popularity of the M series is due to the fact that they are 40% less expensive to purchase than the T series steels. As the name of this group of tool steels suggests, these steels are designed for machining at high speeds, and they have been a mainstay of the cutting tool industry since they were introduced, thanks to the addition of vanadium to tool steels in 1904. Cobalt was introduced in 1912, and the final crucial ingredient, molybdenum, was added in 1930.

Molybdenum high-speed tool steels (M1 to M62) have marginally better toughness than the tungsten steels of the same hardness, but in other respects, the M series and T series virtually share mechanical properties. Within the range, physical properties can be manipulated by choosing a steel with increased percentages of carbon and vanadium, which will result in better wear resistance, but the additional carbon will reduce the ductility; or selecting one with more cobalt, which will enhance red hardness but lower impact strength. While all M series steels, except M10, contain tungsten in amounts ranging from 3.75 to 11%, the T series steels contain only minimal amounts of molybdenum (from 0 to 1.25% maximum). In both series, achievable hardness increases proportionally to the amount of molybdenum and/or tungsten in the composition of the steel. M2 is an excellent, and extremely popular, general purpose high speed steel that has good toughness and very acceptable heat resistance. It also has the most forgiving hardening range of all of the high speed steels. M2 should be preheated at 1,500° F (816° C) and hardened at 2,200° F (1,204° C). When tempered at 1,000° F (528° C), it has a hardness of 66 R_C, a Charpy value of 10 ft lbs (13.5 Joules), and a machinability rating of 14. The M3 grade is available in two versions. Both have excellent wear and heat resistance, plus very good edge strength. M3-2 has slightly higher carbon content than M3-1, allowing it to develop a better cutting edge while losing a small amount of compressive strength. M42 can achieve a maximum hardness of 70 R_C which, along with a few other M series steels, is the highest hardness recorded for tool steels. It also holds an edge without losing hardness or significant wear resistance. M42 should be preheated at 1,500° F (816° C) and hardened at 2,175° F (1,190° C). When tempered at 950° F (510° C), it reaches a hardness of 68 R_C, achieves a Charpy value of 10 ft lbs (13.5 Joules), and has a machinability rating of 13.

Tungsten high-speed tool steels (T1 to T15) have very high red hardness and exceptional wear resistance. These steels are used primarily for making drills, form cutters, hacksaws, circular saws, slotting saws, and many other varieties of cutting tools, dies, and punches. T5 has better wear resistance, but slightly inferior shock resistance, than M42. T5 should be preheated at 1,500° F (816° C) and hardened at 2,350° F (1,288° C). Tempering at 1,000° F (528° C) will produce a hardness of 64 R_C, a Charpy value of 5 ft lbs (7 Joules), and a machinability rating of 11. Thanks to the combination of tungsten, cobalt, high carbon, and vanadium, T15 provides maximum wear resistance. Its preheat temperature is 1,500° F (816° C), and its hardening temperature is 2,250° F (1,232° C). When tempered at 1,000° F (528° C), it has a hardness of 67 R_C, a Charpy value of 3-4 ft lbs (4-5 Joules), and a machinability rating of 8.

Plastic mold steels, P2 to P21, are favored for plastic molds where heavy pressures and high temperatures (350 to 390° F [176 to 199° C]) are commonly encountered. There are several additional factors that make these steels well suited for plastic molds including good wear resistance, corrosion resistance, good thermal conductivity, and excellent toughness. The mold steel must also have certain manufacturing properties that are satisfactorily met by the P series tool steels, including acceptable machinability, dimensional stability not only during heat treatment, but also during molding operations, good polishing properties, and, in some cases, good hubbing characteristics (hubbing, which is discussed later in this section, is also called hobbing, and it is forcing a steel master "hub" or "hob"—which is an exact replica of the part to be formed—into a softer die bank to create a mold).

These properties are essential because the cost and time required to prepare a mold make it imperative that it remains in service for as long as possible. Wear resistance is important because, unless the steel has a hard, abrasion resistant surface, the repeated flow of material into the mold will rapidly erode the surface of the mold. Because many plastics are highly corrosive, and rust or other surface deterioration will destroy the surface of the mold, the steel must have good corrosion resistance. Toughness is essential because the opening and closing of the die or mold, as well as the high velocity injection of materials, expose the mold steel to repeated shocks. To avoid excessive machining, die materials should retain their size, with as little deformation as possible, during heat treating, and they must not distort from the heat generated by normal operations. The surface texture of the mold is important because the surface of the finished product will show any imperfections in the mold's surface, therefore the steel must possess structural uniformity and high surface hardness to allow it to be polished to a smooth finish. A final important characteristic of P series tool steels is that they resist losing hardness on tempering. This becomes very important to molds that are repeatedly exposed to elevated temperatures during operation, which can cause some steels to lose their hardness.

Although the P series steels remain the most popular choice for plastic molds, maraging (a combination of the words *mar*tensite and *aging*, because they are age, or precipitation, hardened at 840 to 930° F [449 to 409° C] for around three hours) steels, containing high nickel content, are also used, especially in plastic injection molding. Although these are not classified as "tool steels," their use in molds, but not in cutting operations, is responsible for their inclusion in this section. The hardness of maraging steels is between 50 and 58 R_C, and, after aging, they can obtain tensile strength ratings of 29 ksi (200 MPa). Their primary application is for use in intricate mold designs, because they shrink uniformly during heat treatment and can therefore be made to very close tolerances. Also, since carbon is not an alloying agent in these steels, decarburization does not have to be considered. The maraging steels are much more expensive than the P series steels, and they are not commonly available in large sizes, so their use remains limited.

Another non-tool steel used for molds, especially injection or compression molds with large cavities, is AISI 6F7, which has good toughness qualities and hardness between 52 and 54 R_C. GF7 is a through- (rather than case-) hardened steel. Other tool steels used for mold making include H11, H13, O1, O2, D2, and M2 and, even though these were briefly discussed, their mold steel characteristics will be briefly examined here. These through-hardened steels have an obvious advantage over surface hardened steels: modifications can be made to the cavity without requiring a new heat treatment. They also have one major disadvantage: they do not have a tough core, and therefore are more likely to experience cracking. H11 is commonly used for molds for highly abrasive plastics, usually after nitride treatment to increase surface hardness. O2, because of its high hardness and carbon content, has good wear resistance, but is not as tough as other mold steels. It is used for shallow cavities only, and especially in those with high parting line stresses. D2 is very wear resistant, but the difficulty in machining (even when annealed) and polishing D2 is widely known. Properties of the P series tool steels, plus other steels used for molds, are shown in **Table 18**.

Of the P series steels, P2 through P6, as will be discussed later, are normally used for hubbed and/or carburized cavities, while P20 and P21 are used for machined cavities. Distortion and changes in size of the latter two steels are avoided because they are supplied heat treated to hardnesses ranging from R_C 30 to R_C 60, and therefore do not require additional heat treatment after machining. In spite of this, when being used for plastics molding applications, P20 is commonly carburized to bring its surface hardness up to as high as R_C 65, but carburizing temperatures in excess of 1,600° F (871° C) can be detrimental to the polishing properties of the material. P20+Ni (3.5% Ni) can be carburized without concern about degrading its polishability. **Table 19** provides heat treating temperatures and resulting hardness values for P20 tool steel.

P21 contains nickel and aluminum as its principal alloying agents, and the aluminum content allows it to age harden in the same manner as many aluminum alloys. To obtain full hardness, it may be nitrated during aging to provide the steel with high surface hardness (to a depth of 0.006 to 0.008 inch (0.1524 to 0.2032 mm) by holding it in a gas nitriding furnace for 24 hours at 950 to 975° F (510 to 524° C). With this treatment, surface hardnesses of approximately 70 R_C can be obtained. P21 polishes to a high mirror finish, making it the preferred material for molds with critical finishing requirements.

Hubbing can, in some instances, replace machining or EDM as the preferred method for creating mold cavities. It is especially economical for making multiple cavity molds for injection molding, and it can save time and money while providing high accuracy, repeatability, and surface quality. In the process, a hubbing punch, or master hob, is slowly pressed into annealed steel under very high pressure, thereby creating the shape of the hob's face in a cavity in the annealed steel. Master hobs (or hubs) are commonly made from shock resistant S4 or S1, or cold worked die steels such as A2 or D2, and, generally, the steels most often hobbed for plastics mold cavities are P1 (the most easily hobbed mold tool steel, mainly due to its vanadium content that allows for it to have a fine grained structure after carburizing, but not widely in use), P2, P4, and P6. P5 is also hobbed without serious difficulty, but P3 is more troublesome and requires more frequent annealing. All

Table 18. Properties of Steels Commonly Used for Molds.

Type	Tensile Strength[1]		Hardness[1]	Coefficient of Heat Expansion[2]				Heat Conductivity[3]	
	ksi	MPa	BHN	68-212°F	68-752°F	20-100°C	20-400°C	660°F	350°C
P2	103	710	210	6.8	7.7	12.2	13.9	21.5	36.6
P4	52	360	120	6.6	7.3	12.0	13.2	21.4	36.5
P20	160	1100	325	6.2	7.7	11.1	13.8	19.6	33.4
P20+Ni	160	1100	325	6.2	7.7	11.1	13.8	19.7	33.5
P20+S	195	1350	400	6.2	7.7	11.1	13.9	19.6	33.4
P21	123	850	250	5.7	7.1	10.4	12.7	18.9	32.2
D2	123	850	250	5.8	6.8	10.5	12.2	12.0	20.5
H11	117	810	240	5.5	6.2	10.0	11.1	17.9	30.5
GF7	123	850	250	6.8	7.7	12.3	13.8	17.5	29.8
Maraging	157	1080	320	5.7	6.4	10.3	11.5	10.9	18.5

Notes: [1] Maximum tensile strength and hardness values are given. [2] Coefficient of heat expansion expressed in 10^{-6} in./in. °F, and 10^{-6} m/(mK) for degrees Celsius.
[3] Heat conductivity expressed in Btu/ft h °F, and W/(mK) for degrees Celsius.

Table 19. Heat Treatment Recommendations for P20 Mold Steel, Oil Quenched and Carburized.

Preheat Temperature - 1200° F (650° C) *		
Hardening Temperature - 1525° F (830° C)		
Chemistry	**Tempering Temperature**	**Hardness (R_c)**
Carbon 0.30% Manganese 0.75% Silicon 0.50% Chromium 1.65% Molybdenum 0.40%	As quenched	51
	400° F/205° C	49
	600° F/315° C	47
	800° F/425° C	44
	1000° F/450° C	39
	1100° F/595° C	33
	1200° F/650° C	26
	1250° F/675° C	21
* = if used on larger mass parts		
Gas Carburize Temperature - 1600°F (870° C)		
Furnace Cool To - 1475° F (800° C)		
Tempering Temperature	**Case Hardness (R_c)**	**Core Hardness (R_c)**
600° F/315° C	58	47
650° F/354° C	57	46
700° F/370° C	56	45
750° F/400° C	55	44
800° F/425° C	54	43
900° F/480° C	52	40
Carburizing Time (Hours)	**Average Case Depth**	
4	0.030"/0.762mm	
8	0.042"/1.0668mm	
12	0.055"/1.397mm	
16	0.062"/1.5748mm	

Reproduced from *Heat Treatment, Selection, and Application of Tool Steels* by Bill Bryson (Hanser Gardner Publications).

steels used for hubbing should be at their lowest annealed hardness, and the annealed hardnesses for these tool steels range from 110 HBN for P4, to 180 for P6. In no instance should hubbing be carried out on tool steels with hardnesses in excess of 200 HBN. Other considerations in hubbing are the diameter of the hubbing blank, which should ideally be four times the diameter of the hob, and the thickness of the blank which should be 2.5 times the depth of hubbing. Because of the extreme amounts of force required in hubbing, specialized hubbing presses are used. Smaller presses generate up to 36,000 pounds of force (1,600 kN), while the largest presses are capable of creating over 5.5 million pounds of force (25,000 kN).

Low alloy AISI tool steels, also known as the L series, were once a much larger group of tool steels, but only two remain. These steels are most commonly used for machine parts (collets, cams, forming dies, punches, etc.) as they possess good strength and toughness. Though they have similar characteristics to the W series tool steels, these steels have greater wear resistance, and L6 has nickel, which increases toughness and improves hardenability, and this steel can achieve through hardness in thicknesses up to three inches (76 mm).

L6 should be preheated to 1,500° F (815° C) and hardened at 1,525° F (830° C). When tempered at 300° F (149° C) it will obtain a hardness of 60/61 R_C, its Charpy value will be 72 ft lbs (98 Joules), and its machinability rating will be 20.

Heat Treatment of Steel - Theory

Equilibrium diagrams are graphs or maps of the temperature and composition boundaries of the phase fields that exist in an alloy system under conditions of complete equilibrium. A phase is a microscopically homogeneous body of matter. As related to alloys, a phase may be considered as a structurally homogeneous and physically distinct portion of the alloy system. For example, when an alloy solidifies, or freezes, the phases formed may be a chemical compound and a solid solution, or two solid solutions. Before an alloy solidifies completely, a solid solution may exist in equilibrium with a liquid. In each of these examples, two phases of the alloy are involved. The temperatures that appear on an equilibrium diagram are called critical temperatures or critical points. The loci of these critical temperatures are the Ae_1, Ae_3, Ae_4, and A_{cm} lines or curves. The letter A derives from the word "arrest," since these temperatures designate the point at which an arrest in the heating or cooling curve occurs. The temperature interval between the Ae_1 and Ae_3 critical points for a particular alloy is designated as the *critical range*. Ae_1 is the lower critical temperature, and Ae_3 is the upper critical temperature. Common A designations used to indicate transformation temperatures are as follows.

Ac_{cm}	For hypereutectoid steel, the temperature at which the solution of the cementite in austenite is completed on heating.
Ac_1	The temperature at which austenite begins to form on heating.
Ac_3	The temperature at which the transformation of ferrite to austenite is completed on heating.
Ac_4	The temperature at which austenite transforms to delta ferrite on heating.
Ar_{cm}	For hypereutectoid steel, the temperature at which cementite begins to precipitate from austenite on cooling.
Ar_1	The temperature at which the transformation of austenite to ferrite, or to ferrite plus cementite is completed on cooling.
Ar_3	The temperature at which austenite begins to transform to ferrite on cooling.
Ar_4	The temperature at which delta ferrite transforms to austenite on cooling.
Ae_1, Ae_3, Ae_4, Ae_{cm}	Transformation temperatures under equilibrium conditions.
M_s	(Martensite start) The temperature at which austenite starts to transform to martensite.
M_fAr	(Martensite finish) The temperature at which the formation of martensite is completed. The temperature at which austenite begins to transform to pearlite on cooling.

Critical temperatures for selected carbon and alloy steels are shown in **Table 20**.

The rate at which an iron-carbon alloy is heated or cools affects the critical temperature of the alloy. In practice, during manufacture, strict equilibrium conditions are not maintained during heating and cooling because equilibrium could be achieved only if the heating or cooling rate was extremely slow. In commercial enterprises, strictly adhering to equilibrium rates is not feasible, and generally, the greater the rate of heating or cooling, the greater the deviation from the equilibrium critical temperatures. As can be seen from the display above, the critical points on heating are identified by the symbol Ac followed

Table 20. Critical Temperatures for Selected Carbon and Alloy Steels. See Notes. *(Source, Bethlehem Steel.)*

AISI No.	A_{c1} Temperature		A_{c3} Temperature		A_{r1} Temperature		A_{r3} Temperature	
	°F	°C	°F	°C	°F	°C	°F	°C
1015	1390	754	1560	849	1390	754	1510	821
1020	1350	732	1540	838	1340	727	1470	799
1022	1360	738	1530	832	1300	704	1440	782
1117	1345	729	1540	838	1340	727	1450	788
1118	1330	721	1515	824	1175	635	1385	752
1030	1350	732	1485	807	1250	677	1395	757
1040	1340	727	1445	785	1250	677	1350	732
1050	1340	727	1420	771	1250	677	1320	716
1060	1355	735	1400	760	1250	677	1300	704
1080	1350	732	1370	743	1250	677	1280	693
1095	1350	732	1365	741	1265	685	1320	716
1337	1330	721	1450	788	1180	638	1310	710
1141	1330	721	1435	779	1190	643	1230	666
1144	1335	724	1400	760	1200	649	1285	696
4118	1380	749	1520	827	1260	682	1430	777
4320	1350	732	1485	807	840	449	1330	721
4419	1380	749	1600	871	1420	771	1510	821
4620	1300	704	1490	810	1220	660	1335	724
4820	1310	710	1440	782	780	416	1215	657
8620	1380	749	1520	827	1200	649	1400	760
E9310	1350	732	1480	804	810	432	1210	654
4027	1370	743	1510	821	1320	716	1410	766
4130	1400	760	1510	821	1305	707	1400	760
8630	1365	741	1465	796	1205	652	1335	724
1340	1340	727	1420	771	1160	627	1195	646
4140	1395	757	1450	788	1280	693	1330	721
4150	1390	754	1450	788	1245	674	1290	699
4340	1350	732	1415	768	720	382	890	477
5140	1370	743	1440	782	1260	682	1320	716
5150	1345	729	1445	785	1240	671	1310	710
5160	1380	749	1420	771	1280	693	1310	710
6150	1395	757	1445	785	1290	699	1315	713
8650	1325	718	1390	754	910	488	1230	666
8740	1370	743	1435	779	1160	627	1265	685
9255	1410	766	1480	804	1270	688	1330	721

A_{c1} is the temperature at which austenite begins to form on heating.
A_{c3} is the temperature at which the transformation of ferrite to austenite is completed on heating.
A_{r1} is the temperature at which the transformation of austenite to ferrite, or to ferrite plus cementite, is completed on cooling.
A_{r3} is the temperature at which austenite begins to transform to ferrite on cooling.

by a number subscript. The letter "c" is from the French word for heating, "chauffage." The symbol Ar, plus a number subscript, indicates critical points on cooling, and the letter "r" is taken from "refroidissement," the French word for cooling. When it is necessary to identify critical points for equilibrium, the symbol Ae is normally used used, with the "e" inserted to designate equilibrium. However, equilibrium temperatures are sometimes given as A_1, etc., without the "e."

A typical iron-carbon equilibrium diagram is shown in *Figure 3*. Point "C" on the diagram indicates the eutectic point where a liquid solution is converted into two or more intimately mixed solids on cooling; the number of solids formed is the same as the number of components in the solution. A eutectic is defined as an intimate mechanical mixture of two or more phases having a definite melting point and a definite combustion. Point "S" is the eutectoid point where an isothermal reversible reaction in which a solid solution is converted into two or more intimately mixed solids on cooling. The number of solids formed is the same as the number of components in the solution. Therefore, a eutectoid is a mechanical mixture of two or more phases having a definite composition and a definite temperature of transformation within the solid state.

The heat treatment of steels is essentially a process of controlled departure from equilibrium heating and cooling. When a steel is cooled, under equilibrium conditions, from above the Ae_3 critical temperature, the austenite remaining when the Ae_1 temperature

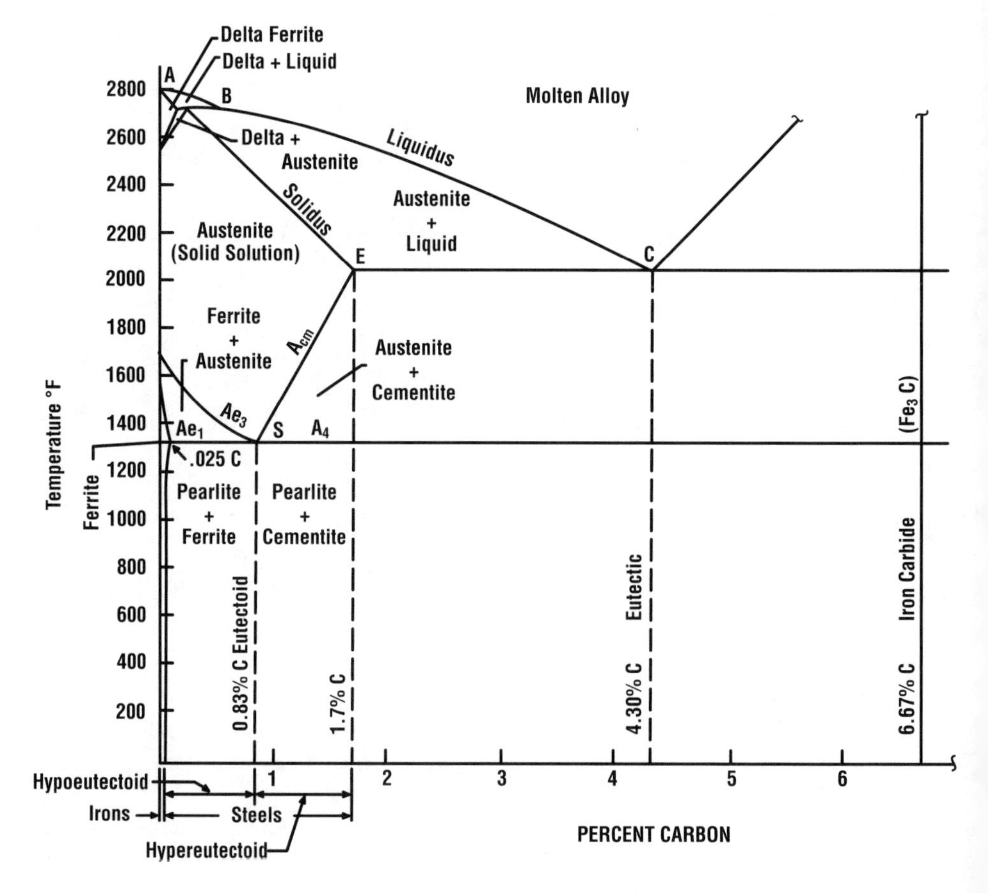

Figure 3. An iron-carbon equilibrium diagram.

is reached transforms into pearlite. With more rapid cooling, the faster the cooling rate, the further the transformation temperature Ar_1 is depressed below the Ae_1 temperature. The eutectoid, pearlite, that forms when austenite is cooled below the Ar_1 temperature, is a lamellar product made up of alternate plates of ferrite and cementite. The pearlite formed under equilibrium cooling is a coarse lamellar product. The process of transformation of pearlite from austenite involves diffusion of carbon, and since the diffusion rate is a function of temperature, faster cooling rates with attendant lower transformation temperatures produce pearlite with a finer lamellar structure. A finer dispersion of phases tends to produce greater strength and hardness and reduces the ductility of the steel. With very rapid cooling, the austenite transformation occurs at a low temperature and the resulting structure is not pearlite, but a structure known as martensite. Therefore, by controlling the cooling rate, the temperature of the austenite transformation and the structure and properties of the steel are controlled.

Fine grain size is usually specified for applications involving hardening by thermal treatment. As carbon and alloy steels are heated to a temperature just above their upper critical temperature, they transform to austenite of uniformly fine grain size. If heated to higher temperatures, coarsening of the grains will eventually occur. The temperature at which this transformation occurs is to some extent dependent on the composition of the steel, but the primary influence is the type and degree of deoxidation used in the steel making process. Time at temperature also influences the degree of coarsening. Deoxidizers such as aluminum, and alloying elements such as vanadium, titanium, and columbium, inhibit grain growth and thereby increase the temperature at which coarsening of the austenite grains occurs. Aluminum is most commonly used to control grain size because of its low cost and dependability. For steels used in the quenched and tempered condition, a fine grain size at the quenching temperature is almost always preferred, because fine austenitic grain size is conducive to good ductility and toughness. Coarse grain size enhances hardenability, but also increases the tendency of the steel to crack during thermal treatment. Grain size is observed by examining a polished and etched specimen under 100 diameters magnification, and comparing it with a standard. It is impossible to produce steel with a single grain size, so a range of numbers is usually recorded. For identification purposes, a steel is considered to be fine grained if it is "5 to 8 inclusive," and coarse grained if it is "1 to 5 inclusive." These requirements are usually considered fulfilled if 70% of the grains observed fall within these ranges.

The time-temperature relationships for austenite transformations are graphically presented with isothermal transformation diagrams, commonly known as Time-Temperature-Transformation (T-T-T) diagrams. These diagrams, which are a plot of the times of the beginning and end of austenite transformations as a function of temperature, have been developed for various plain carbon and alloy steels. The data required for plotting a T-T-T diagram are developed through a series of isothermal transformation studies that determine the austenite transformation(s) that occur in a steel held at a constant temperature for increasing periods of time. These studies consist of heating small, thin pieces of specimen steel above the upper critical temperature, and holding it at the temperature long enough for homogeneous austenite to form. Each specimen is then removed and quenched in a lead or salt bath that is held at the desired study temperature. Thin specimens are used so that cooling to the bath temperature can be assumed to be instantaneous, and each specimen is held in the isothermal bath for a different period of time, ranging from seconds to hours. Each is then quenched in brine or water to transform any remaining austenite to martensite. The structure of each specimen is then examined to determine which transformation product(s), if any, formed during the isothermal cycle, and what percentage of the total structure transformed into each product. The collected data are

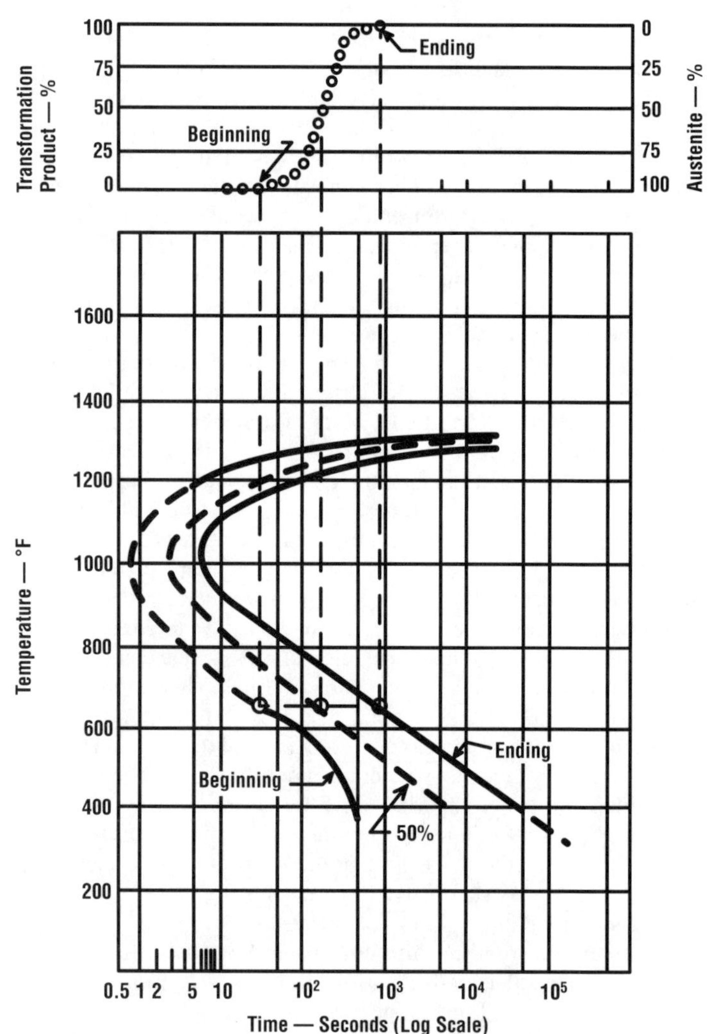

Figure 4. The use of transformational curves to develop T-T-T diagrams.

used to construct T-T-T diagrams in which time, from the beginning to the end of the transformation process, is plotted as a function of temperature. The basic methodology for creating a T-T-T diagram is shown in *Figure 4*.

Depending on the transformation temperatures, and the composition of the steel, austenite will transform into one or more of the final constituents: proeutectoid ferrite, proeutectoid cementite, pearlite, upper bainite, lower bainite, or martensite. Generally, for carbon and low alloy steels, the temperature ranges over which austenite transforms into the various constituents, under isothermal cooling, are as indicated on the transformation diagram shown in *Figure 5*. As can be seen, pearlitic microstructures are formed from about 1,300 to 1,000° F (705 to 538° C). Pearlite, as we have seen, is a constituent with a lamellar structure of alternating plates of ferrite and cementite. When the transformation occurs at

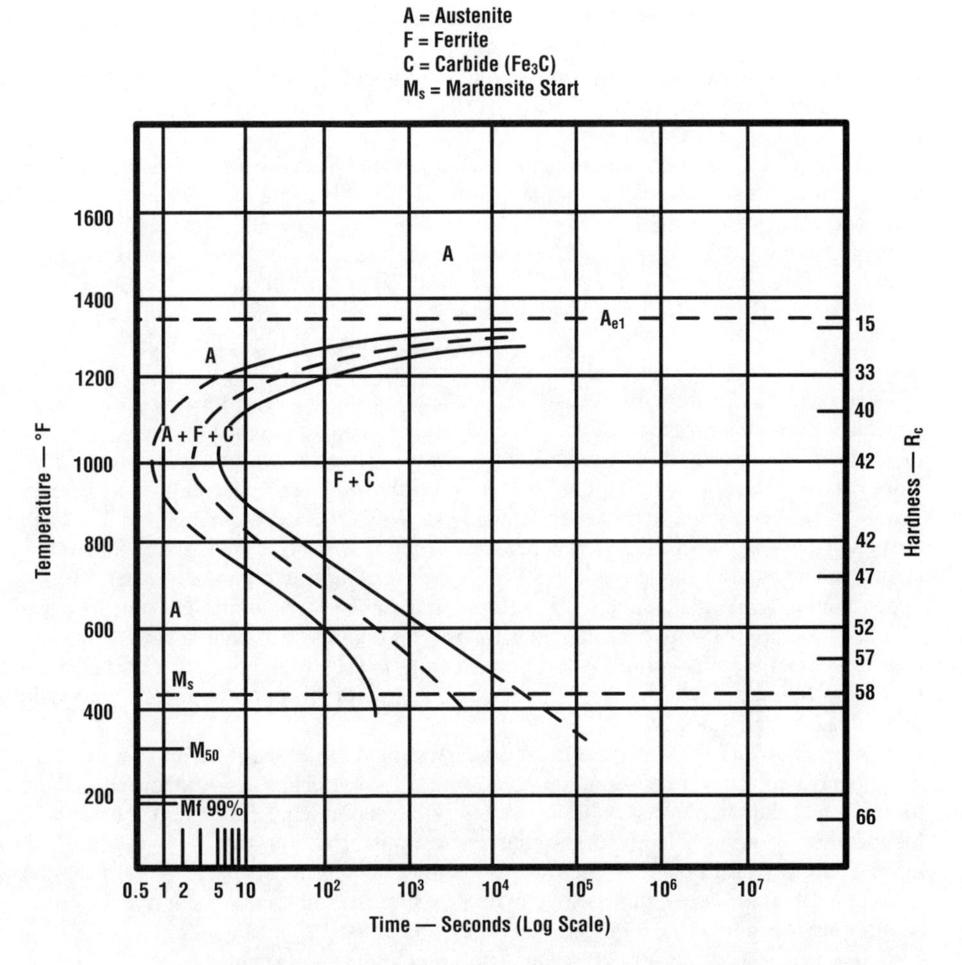

Figure 5. The isothermal transformation diagram of a eutectoid carbon steel.

about 1,000° F, the lamellar structure is very fine and difficult to resolve with an optical microscope. The transformation of austenite to pearlite is a process of nucleation and growth. As the nucleus grows into the grain, it absorbs carbon atoms from the surrounding austenite. When the carbon concentration of the surrounding austenite has been sufficiently reduced, ferrite nucleates and grows along the surface of the cementite plate. Because ferrite can dissolve only about 0.02% carbon, carbon atoms are rejected by the ferrite as it forms, resulting in a buildup of carbon at the ferrite-austenite interface that continues to increase until the concentration level is sufficiently high, and at that point a new cementite nucleus forms. The sideways nucleation is repeated while, simultaneously, growth occurs at the edges of the ferrite and cementite plates. The alternating lamellae of cementite and ferrite originate from a single cementite nucleus called a pearlite colony.

The rate of nucleation and the rate of growth of pearlite colonies increase as the temperature decreases from Ae_1 down to about 1,100° F (593° C). As a result, at lower transformation temperatures the spacing between the pearlite lamellae becomes smaller and the metal becomes harder. Pearlite formed just below the Ae_1 temperature (about

1,300° F) is coarse pearlite with a lamellar spacing in the order of 10^{-3} mm, and a hardness of about R_C 15. The pearlite formed at about 1,100° F, however, is fine pearlite with spacing in the order of 10^{-4} mm and a hardness of about R_C 40. The fine pearlite formed at lower transformation temperatures is harder, tougher, and more ductile than pearlite formed at the higher temperatures.

As temperatures near the low end of the pearlite transformation range, another constituent forms from austenite. In plain carbon steel, this constituent, known as bainite, is formed only by isothermal transformation treatments and does not form when the steel is cooled continuously from above the critical temperature. Bainite forms from about 1,050° F (566° C) down to about 400° F (205° C). Steels that are transformed in the range where bainite and pearlite transformation temperatures overlap have structures containing both pearlite and bainite but, as the transformation temperatures are lowered, bainite becomes the predominant constituent and pearlite disappears. Like pearlite, bainite is a mixture of ferrite and cementite, but the two are not arranged in lamellar form as in a pearlite structure. The transformation of austenite to bainite is also thought to involve a process of nucleation and growth accompanied by carbon diffusion. Bainite apparently grows from a ferrite nucleus into a plate like structure with each plate composed of ferrite matrix in which carbide particles are embedded. When viewed in section, bainite has a characteristic acicular (needle like) appearance. Bainite formed at the higher temperatures in the transformation range has a feathery appearance and is commonly referred to as upper-bainite, while that formed at lower temperatures (lower-bainite) has a pronounced acicular structure. As with pearlite, the hardness and toughness of bainite both increase as the transformation temperature is lowered. Pearlite is usually tougher than upper-bainite of similar hardness, while lower-bainite compares favorably with tempered martensite on the basis of toughness.

The transformations of austenite to pearlite or bainite are time and temperature dependent, while the transformation of austenite to martensite is an athermal change, meaning that the transformation is dependent primarily on temperature and is essentially independent of time. Therefore, the martensite transformation process is considerably different from pearlite and bainite transformations. As can be seen in *Figure 5*, neither pearlite nor bainite forms immediately upon reaching an isothermal reaction temperature. An incubation period is required, and the steel must be held at temperature for a sufficient period of time for the reaction to be completed. By contrast, when austenite reaches the M_s (martensite start) temperature, the austenite immediately transforms and, if the steel is held at that temperature, the small amount of martensite that has formed is all that will form. Unless the steel is cooled to a still lower temperature, the transformation is arrested.

At some temperature below the M_s temperature, the transformation of austenite to martensite will essentially be complete. This temperature is designated M_f (martensite finish). At temperatures between M_s and M_f, fractional transformation will occur. This makes it possible to quench a steel to the temperature at which 50% of the austenite will transform to martensite, and then isothermally transform the remaining austenite to lower-bainite.

The transformation of austenite to martensite does not involve the diffusion of carbon. The instantaneous transformation of a small volume of austenite to form a martensite needle involves a shear displacement of the iron atoms in the austenite space lattice. (A space lattice is a geometric construction formed by a regular array of points—called lattice points—that represent the three-dimensional arrangement of atoms in space for a specific crystal structure.) The martensite formed in this way has a body-centered tetragonal (having four angles and four sides) crystal structure. In this form, it is called alpha-martensite and consists of ferrite and carbon, or finely divided iron-carbide ($F_{e3}C$) in a metastable (unstable) structure that is considered to be in transition between the face-centered cubic structure of austenite, and the body-centered cubic structure of ferrite. The instantaneous athermal transformation of austenite to martensite at relatively low temperatures does not allow the carbon atoms to diffuse out of the lattice, so they remain in solution in the highly stressed transition lattice. The tetragonal crystal structure of alpha martensite transforms to the body-centered cubic structure known as beta-martensite upon slight heating or long standing. Essentially, the trapped carbon escapes or is thrown out of the crystal lattice, allowing the stressed lattice to shrink down to the body-centered cubic structure.

Alpha-martensite has an acicular structure similar to that of lower-bainite. It is the hardest and most brittle of the microstructures that can be obtained in a given steel. Its hardness is a function of carbon content, ranging from a theoretical maximum of R_C 65 in eutectoid alloys to R_C 40 or less in low carbon steels. Alpha-martensite is usually tempered to increase its ductility and toughness, although these changes are usually at the expense of hardness and strength. In tempering, the steel is heated to some temperature below the critical temperatures and cooled at a suitable rate. The microstructure and mechanical properties of tempered steel depend on the temperature and duration of the tempering cycle. As temper time and temperature increase, the carbide particles agglomerate and become progressively larger. Therefore, an increase in temperature and/or time at temperature usually results in lowering the hardness and strength of a steel while increasing its ductility and toughness.

To develop a martensitic structure, a steel must be austenitized at a rate sufficient to prevent the formation of ferrite, pearlite, or bainite. Most steels must be cooled very rapidly if pearlite formation is to be avoided. This is evident in *Figure 4*, which shows that the transformation of austenite to pearlite begins in one second or less between the temperatures of 1,100 and 950° F (593 and 510° C). In this range, often referred to as the pearlite nose of the transformation curve, the transformation is complete in less than ten seconds. An isothermal transformation diagram shows the changes in microstructure that occur when a steel is cooled instantly to some reaction temperature, and held at that temperature long enough for the reaction to go to completion. The diagram for a given steel shows what structure or structures are formed by isothermal transformation at any selected temperature, the time required for the steel to be at temperature before a reaction will begin, and the time needed to complete the reaction.

Fine grained austenite transforms to pearlite more rapidly than does coarser grained austenite. This can be explained by the fact that, as grain size decreases, the proportion of grain boundary material to total mass increases. Because pearlite nucleates at the grain boundaries in homogeneous austenite, transformation to pearlite begins more quickly and progresses faster in fine grained austenite than coarse grained austenite.

Figures 5 and *6* demonstrate the effect of carbon on the isothermal transformation reactions in plain carbon steels. For the steel with the higher carbon content (*Figure 5*), the transformations are shown to start later and to progress more slowly than do comparable reactions in the lower carbon steel (*Figure 6*). In effect, higher carbon contents shift the isothermal transformation curves to the right. In other words, transformations start later and proceed more slowly as carbon content is increased. Most of the common metallic alloying elements also tend to retard the start of isothermal transformations and increase the length of time required to complete them.

Compared to isothermal transformations, transformations under continuous cooling take longer to start and begin at lower temperatures. In effect, the isothermal transformation curve is shifted downward and to the right. This can be seen by the reactions of a plain carbon steel of eutectoid composition. In actual practice, the steel can be fully hardened

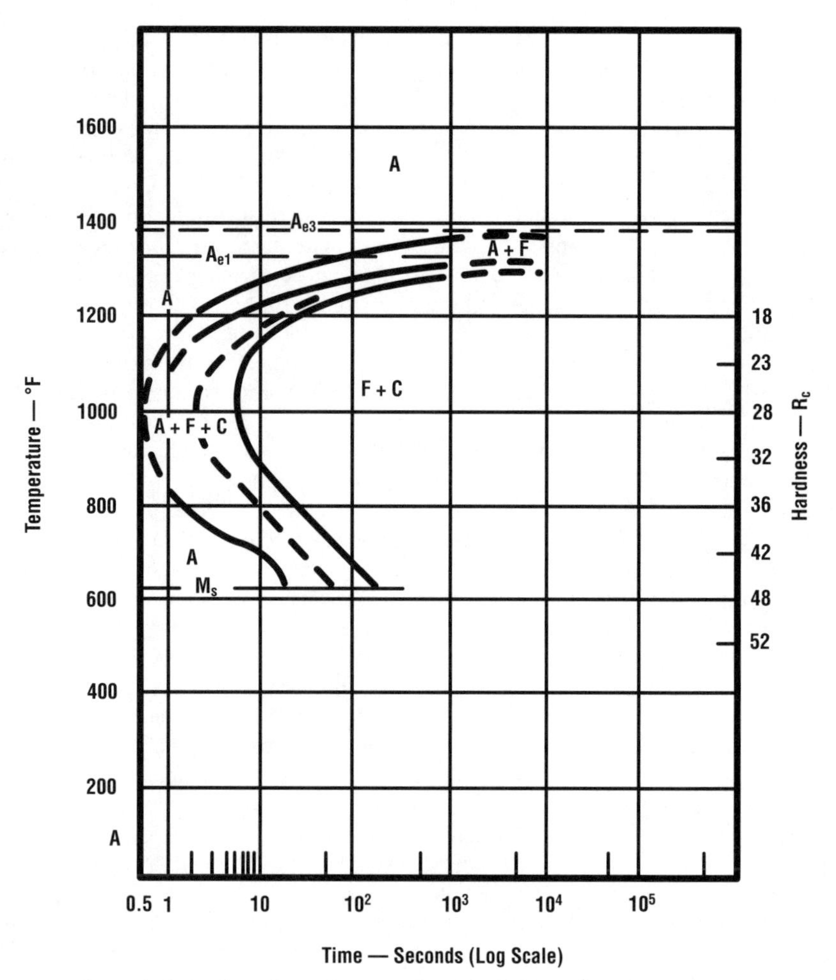

Figure 6. The isothermal transformation diagram of a hypoeutectoid carbon steel.

by cooling at a rate of 250° F per second (140° C/s). However, the isothermal diagram for such steel, as seen in *Figure 7*, indicates that a cooling rate of at least 400° F per second (220° C/s) is required to prevent the cooling rate curve from intersecting the transformation start curve at the pearlite knee. Theoretically, when the two curves intersect, transformation to pearlite should begin and should continue until the cooling curve passes out of the pearlite field.

If a steel is to be hardened to a martensitic structure, it must be austenitized and then cooled at a rate that is fast enough to prevent the formation of any of the other transformation products such as pearlite, ferrite, or bainite. The slowest rate that avoids all other transformation is called the critical cooling rate for martensite. As would be expected, different steels have different critical cooling rates, and the variations can be significant. Steels with relatively low critical cooling rates are high hardenability steels, and steels that must be cooled rapidly in order to obtain a martensitic structure are low hardenability steels. Hardenability is essentially a measure of the ability of a steel to transform to martensite—that is, to harden. It is also a measure of the depth to which a

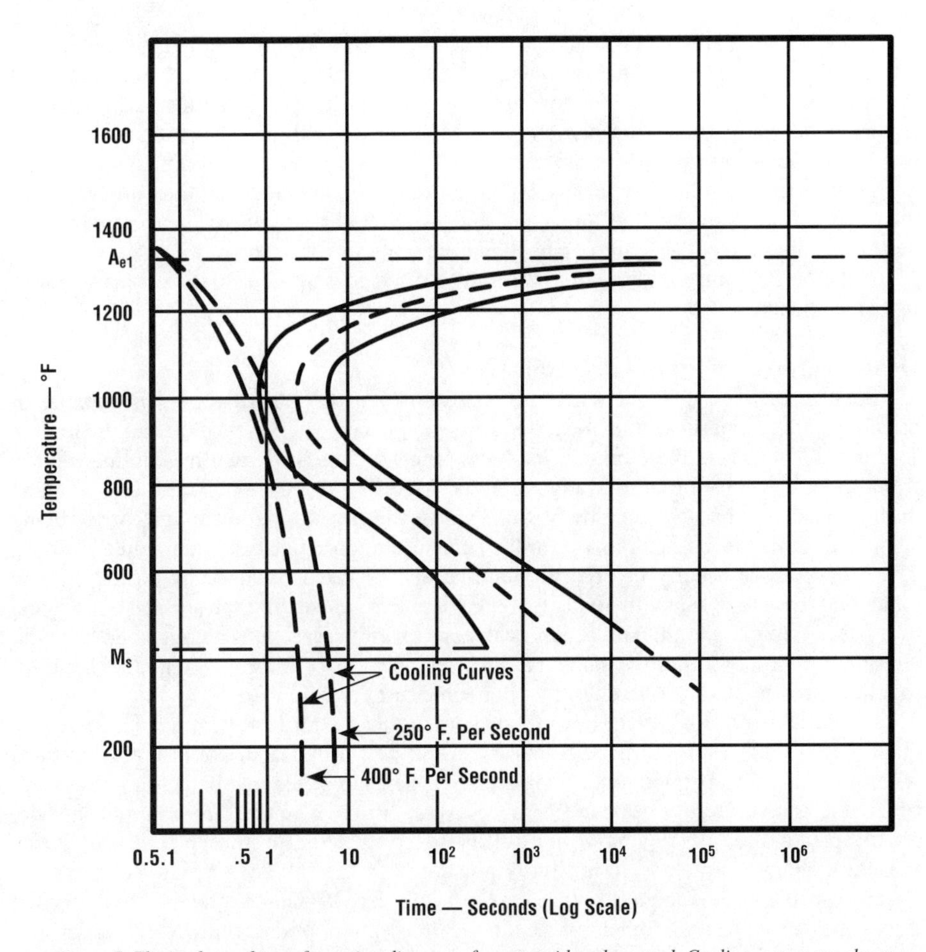

Figure 7. The isothermal transformation diagram of a eutectoid carbon steel. Cooling curves are shown.

steel will transform to martensite under given cooling conditions. Low hardenability steels respond to only a limited depth, which classes them as shallow hardening steels. With high hardenability steels, it is possible to harden thicker sections to greater depths.

The most common method of cooling a steel to obtain a martensitic structure is by quenching, which is rapid cooling of the steel from above the critical temperature by immersion in a medium that is capable of cooling it at the required rate. The cooling rate is determined by the hardenability of the steel and the size of the piece being quenched. Brine, water, and oils, listed in order of decreasing severity of quench, are the most common quenching media. In a piece of steel of any appreciable size, the cooling rates at the surface and at the center are different, and the difference increases with the severity of the quench.

Jominy end-quench test. The most widely used hardenability test is the Jominy end-quench test. Its popularity is attributable to its convenience, as only a single specimen is required, and in one operation the specimen is exposed to a wide range of cooling rates varying from a rapid water quench on one end of the specimen to a slow air quench on the other end. The test specimen is a bar four inches long and one inch in diameter that is heated to an austenitizing temperature and held there long enough to develop a uniform austenite structure, and then placed in a fixture and quenched. Quenching is accomplished by a gentle stream of water that is directed at, and only allowed to contact, one flat end of the specimen. The test bar is thereby subjected to a full range of cooling rates that vary from a rapid water quench on one end to a slow air cool on the other. After quenching, two flat surfaces are ground on the specimen. These surfaces are positioned 180° apart, are ground to a depth of at least 0.015 inch (0.381 mm), and run the full length of the specimen. Starting at the quenched end, Rockwell C hardness measurements are performed at $^1/_{16}$ inch (1.6 mm) apart along the length of the bar over a span of at least 2.5 inches (63.5 mm). The results are then plotted to show hardness versus distance from the quenched end.

Heat Treatment of Steel – Procedures

Most steel is used in the as-rolled condition, but thermal treatment is capable of broadening the properties attainable. There are two general categories of thermal treatment: 1) those that increase the strength, hardness, and toughness of steel by cooling rapidly from above the transformation range; and 2) those that decrease hardness and promote uniformity either by cooling slowly from above the transformation range, or by being subjected to prolonged heating within the transformation range, and then cooled slowly. The first category usually involves through hardening that is induced by quenching and tempering, or one of many methods that are capable of enhancing the surface of the steel to a predetermined depth. The second category encompasses normalizing together with annealing in order to improve machinability, toughness or cold forming characteristics, or to relieve stresses and restore ductility after some form of cold working.

Furnaces of various descriptions (and ages) are in use for heat treating. The essential quality of a heat treating furnace is that it must maintain indicated uniform temperatures within a relatively narrow thermal range, and the sensing gage must be able to accurately report the temperature within the furnace. Most furnaces in use are nonatmospheric, meaning that they do not have the ability to control the precise mixture of gases present during treatment. While this is not normally a problem, precautions must be taken—especially at temperatures in excess of 1,200° F (649° C)—to protect the surfaces of steel parts during heating, otherwise scaling, carburization, or decarburization may result. The most modern furnaces are capable of creating a vacuum environment, which allows customizing the pressure so that it is compatible with the steel being treated.

As reviewed earlier, the predictable change to the microstructure that can be affected by heating and cooling solid steel at selected rates under controlled rates is the basis for the heat treatment of steel. The practical aspects of steel heat treatment are considered in the following discussion of annealing, normalizing, quench and tempering, martempering, and austempering. Consult **Tables 21** through **26** for detailed information on temperatures required for thermally treating carbon and low alloy steels, and see **Tables 26** through **29** for details for thermally treating stainless steels.

Annealing

Steel is annealed to reduce hardness, relieve stresses, to develop a particular microstructure with specific physical and mechanical properties, to improve machinability and formability. Basically, it consists of a heating cycle, a holding period, and a controlled cooling cycle. Various annealing procedures are in use, each of which is identified by a descriptive term.

Full annealing is a relatively straightforward heat treatment in which the steel is heated to a temperature above the A_3 critical temperature and held at the temperature long enough to allow the solution of carbon and other alloying elements in the austenite. The

(Text continued on p. 120)

Table 21. Thermal Treatments for Carbon Steels-Water and Oil Hardening Grades. *(Source, Bethlehem Steel.)*

Type[1]	Normalizing Temp.		Annealing Temp.		Hardening Temp.		Quenching Medium
	°F	°C	°F	°C	°F	°C	
1030	-	-	-	-	1575-1600	857-871	Water or Caustic
1035	-	-	-	-	1550-1600	829-871	Water or Caustic
1037	-	-	-	-	1525-1575	802-857	Water or Caustic
1038 [2]	-	-	-	-	1525-1575	802-857	Water or Caustic
1039 [2]	-	-	-	-	1525-1575	802-857	Water or Caustic
1040 [2]	-	-	-	-	1525-1575	802-857	Water or Caustic
1042	-	-	-	-	1500-1550	816-829	Water or Caustic
1043 [2]	-	-	-	-	1500-1550	816-829	Water or Caustic
1045 [2]	-	-	-	-	1500-1550	816-829	Water or Caustic
1046 [2]	-	-	-	-	1500-1550	816-829	Water or Caustic
1050 [2]	1600-1700	871-927	-	-	1500-1550	816-829	Water or Caustic
1053	1600-1700	871-927	-	-	1500-1550	816-829	Water or Caustic
1060	1600-1700	871-927	1400-1500	760-816	1575-1625	857-885	Oil
1074 †	1550-1650	829-899	1400-1500	760-816	1575-1625	857-885	Oil
1080	1550-1650	829-899	1400-1500 3	760-816 [3]	1575-1625	857-885	Oil [4]
1084	1550-1650	829-899	1400-1500 3	760-816 [3]	1575-1625	857-885	Oil [4]
1085 †	1550-1650	829-899	1400-1500 3	760-816 [3]	1575-1625	857-885	Oil [4]
1090	1550-1650	829-899	1400-1500 3	760-816 [3]	1575-1625	857-885	Oil [4]
1095	1550-1650	829-899	1400-1500 3	760-816 [3]	1575-1625	857-885	Water and Oil
1137	-	-	-	-	1550-1600	829-871	Oil
1141	1600-1700	871-927	1400-1500	760-816	1500-1550	816-829	Oil
1144	1600-1700	871-927	-	-	1500-1550	816-829	Oil
1145	-	-	-	-	1475-1500	802-816	Water or Oil
1146	-	-	-	-	1475-1500	802-816	Water or Oil
1151	1600-1700	871-927	-	-	1475-1500	802-816	Water or Oil
1536	1600-1700	871-927	-	-	1500-1550	816-829	Water or Oil
1541	1600-1700	871-927	1400-1500	760-816	1500-1550	816-829	Oil
1548	1600-1700	871-927	-	-	1500-1550	816-829	Oil
1552	1600-1700	871-927	-	-	1500-1550	816-829	Oil
1566	1600-1700	871-927	-	-	1575-1625	857-885	Oil

Notes: When tempering is required, temperature should be selected to effect desired hardness.

† Indicates SAE designation only. No AISI number assigned.

1 AISI/SAE designation numbers.

2 These grades are commonly used for parts where induction hardening is employed, although all steels from 1030 up can be induction hardened.

3 Spheroidal structures are often required for machining purposes and should be cooled very slowly or be isothermally transformed to produce the desired structure.

4 May be water or brine quenched by special techniques such as partial immersion, or time quenched; otherwise subject to quench cracking.

Table 22. Thermal Treatments for Carbon Steels-Carburizing Grades. *(Source, Bethlehem Steel.)*

Type[1]	Carburizing Temp.		Cooling Medium	Reheat Temp.		Cooling Medium	Carbonitriding Temp.		Cooling Medium
	°F	°C		°F	°C		°F	°C	
1010	-	-	-	-	-	-	1450-1650	788-899	Oil
1015	-	-	-	-	-	-	1450-1650	788-899	Oil
1016	1650-1700	899-927	Water or Caustic [2]	-	-	-	1450-1650	788-899	Oil
1018	1650-1700	899-927	Water or Caustic [2]	1450	788	Water or Caustic [2]	1450-1650	788-899	Oil
1019	1650-1700	899-927	Water or Caustic [2]	1450	788	Water or Caustic [2]	1450-1650	788-899	Oil
1020	1650-1700	899-927	Water or Caustic [2]	1450	788	Water or Caustic [2]	1450-1650	788-899	Oil
1022	1650-1700	899-927	Water or Caustic [2]	1450	788	Water or Caustic [2]	1450-1650	788-899	Oil
1026	1650-1700	899-927	Water or Caustic [2]	1450	788	Water or Caustic [2]	1450-1650	788-899	Oil
1030	1650-1700	899-927	Water or Caustic [2]	1450	788	Water or Caustic [2]	1450-1650	788-899	Oil
1109	1650-1700	899-927	Water or Oil	1400-1450	760-788	Water or Caustic [2]	-	-	-
1117	1650-1700	899-927	Water or Oil	1450-1600	788-871	Water or Caustic [2]	1450-1650	788-899	Oil
1118	1650-1700	899-927	Oil	1450-1600	788-871	Oil	-[3]	-	-
1513	1650-1700	899-927	Oil	1450	788	Oil	-[3]	-	-
1518	1650-1700	899-927	Oil	1450	788	Oil	-[3]	-	-
1522	1650-1700	899-927	Oil	1450	788	Oil	-[3]	-	-
1524	1650-1700	899-927	Oil	1450	788	Oil	-[3]	-	-
1525	1650-1700	899-927	Oil	1450	788	Oil	-[3]	-	-
1526	1650-1700	899-927	Oil	1450	788	Oil	-[3]	-	-
1527	1650-1700	899-927	Oil	1450	788	Oil	-[3]	-	-

Notes: Normalizing is generally unnecessary for fulfilling either dimensional or machinability requirements of parts made from the above grades. Where dimensioning is of vital importance, normalizing temperatures of at least 50° F (28° C) above the carburizing temperatures are sometimes required to minimize distortion.

Tempering temperatures are usually 250-400° F (121-204° C), but higher temperatures may be used when permitted by the hardness specification for the finished parts.

[1] AISI/SAE designation numbers.

[2] 3% sodium hydroxide.

[3] Higher manganese steels, such as 1118 and the 1500 series are not usually carbonitrided. If carbonitriding is performed, care must be taken to limit the nitrogen content because high nitrogen will increase their tendency to retain austenite.

Table 23. Representative Heat Treatments for Selected Low-Carbon and Low-Alloy Steels.
(Source, MIL-HDBK-723.)

Stress Relief Temperatures			
Material	**°F**	**°C**	**Soak Time at Temp.**
Low-Alloy Steels (after heat treat at 150 to 180,000 psi)	700) 25	371) 14	1 hour per inch (25.4 mm) of cross section

Annealing Cycle							
Material	**Number**	**Annealing Temp.**		**Furnace Cooling Cycle***		**Soak Time (hr.)**	**Heatup Time**
		°F	**°C**	**°F**	**°C**		
Low-Carbon	1018	1575 to 1650	857 to 899	From 1575 to 1300	857 to 704	1 hr. for sections to 1 in. thick. Add 1/2 hr. for each additional 1 in. (25.4 mm) of thickness.	1 hr. per in. (25.4 mm) of material thickness.
	1020	1575 to 1650	857 to 899	From 1575 to 1290	857 to 699		
	1025	1575 to 1650	857 to 899	From 1575 to 1290	857 to 699		
	1030	1550 to 1625	829 to 885	From 1550 to 1200	829 to 649		
	1035	1550 to 1625	829 to 885	From 1550 to 1200	829 to 649		
Low-Alloy	4130	1450 to 1550	788 to 829	From 1450 to 900	788 to 482		
	4140	1450 to 1550	788 to 829	From 1450 to 900	788 to 482		
	4340	1450 to 1550	788 to 829	From 1450 to 900	788 to 482		
	5150	1500 to 1600	816 to 871	From 1500 to 900	816 to 482		
	6150	1550 to 1650	829 to 899	From 1550 to 1000	829 to 538		

* Cooling Rate is 50° F (28° C) per hour. After reaching lower temperature, rate of cooling is unimportant.

Normalizing Cycle					
Material	**Number**	**Normalizing Temp.**		**Soak Time (hr.)**	**Heatup Time**
		°F	**°C**		
Low-Carbon	1015	1650 to 1700	899 to 927	1 hr. for sections to 1 in. thick. Add 1/2 hr. for each additional 1 in. of thickness	1 hr. per in. of material thickness.
	1020	1650 to 1700	899 to 927		
	1025	1625 to 1675	885 to 913		
	1030	1625 to 1675	885 to 913		
	1035	1600 to 1650	871 to 899		
Low-Alloy	4130	1600 to 1750	871 to 954		
	4140	1550 to 1700	829 to 927		
	4340	1550 to 1700	829 to 927		
	5150	1550 to 1700	829 to 927		
	6150	1600 to 1750	871 to 954		

Austenizing Cycle							
Material	**Number**	**Austenizing Temp.**		**Soak Time**			**Heatup Time**
		°F	**°C**	**Thickness inch**	**Thickness mm**	**Time (Hours)**	
Low-Carbon	1025	1575 to 1650	857 to 899	1/2 or less	12.7 or less	1/4	1 hr. per in. of material thickness.
	1030	1550 to 1600	829 to 871	1	25.4	1/3	
	1035	1525 to 1575	802 to 857	2	50.8	1/2	
Low-Alloy	4130	1500 to 1600	816 to 871	3	76.2	3/4	
	4140	1550 to 1600	829 to 871	4	101.6	1-1/4	

(Continued)

Table 23. *(Continued)* **Representative Heat Treatments for Selected Low-Carbon and Low-Alloy Steels.** *(Source, MIL-HDBK-723.)*

Austenizing Cycle							
Material	Number	Austenizing Temp.		Soak Time			Heatup Time
		°F	°C	Thickness inch	Thickness mm	Time (Hours)	
Low-Alloy	4340	1500 to 1550	816 to 829	5	127	1-1/2	1 hr. per in. of material thickness.
	5150	1475 to 1550	802 to 829	-	-	-	
	6150	1550 to 1625	829 to 885	-	-	-	

Tempering Cycle for Low-Alloy Steels								
Type	Tempering Temperature for Indicated Tensile Strength						Soak Time	Heatup Time
	125 ksi - °F	860 MPa - °C	150 ksi - °F	1034 MPa - °C	180 ksi - °F	1240 MPa - °C		
4130	950 to 1150	510 to 621	800 to 1000	427 to 538	700 to 900	371 to 482	1 hr. per in. of material thickness.	1 hr. per in. of material thickness.
4140	1050 to 1250	566 to 677	950 to 1150	510 to 621	800 to 1000	527 to 538		
4340	1075 to 1225	579 to 663	975 to 1075	524 to 579	850 to 975	454 to 524		

The information presented in this table is intended for general cases. Temperatures and times should be adjusted to compensate for differences in equipment, chemical composition, size and shape of parts being treated, and other variables.

Table 24. Thermal Treatments for Alloy Steels-Directly Hardenable Grades. *(Source, Bethlehem Steel.)*

Type[1]	Normalizing Temp.		Annealing Temp.[4]		Hardening Temp.[5]		Quenching Medium
	°F	°C	°F	°C	°F	°C	
1330	1600-1700 [2]	871-927 [2]	1550-1650	843-899	1525-1575	802-857	Water or Oil
1335	1600-1700 [2]	871-927 [2]	1550-1650	843-899	1500-1550	829-829	Oil
1340	1600-1700 [2]	871-927 [2]	1550-1650	843-899	1500-1550	829-829	Oil
1345	1600-1700 [2]	871-927 [2]	1550-1650	843-899	1500-1550	829-829	Oil
4037	-	-	1550-1575	843-857	1525-1575	802-857	Oil
4042 †	-	-	1550-1575	843-857	1525-1575	802-857	Oil
4047	-	-	1450-1550	788-843	1500-1575	829-857	Oil
4130	1600-1700 [2]	871-927 [2]	1450-1550	788-843	1500-1600	829-871	Water or Oil
4135 †	-	-	1450-1550	788-843	1500-1600	829-871	Oil
4137	-	-	1450-1550	788-843	1500-1600	829-871	Oil
4140	-	-	1450-1550	788-843	1500-1600	829-871	Oil
4142	-	-	1450-1550	788-843	1500-1600	829-871	Oil
4145	-	-	1450-1550	788-843	1500-1550	829-829	Oil
4147	-	-	1450-1550	788-843	1500-1550	829-829	Oil
4150	-	-	1450-1550	788-843	1500-1550	829-829	Oil
4161	-	-	1450-1550	788-843	1500-1550	829-829	Oil [6]
4340	1600-1700 [2]	871-927 [2]	1450-1550	788-843	1500-1550	829-829	Oil [3]
50B40 †	1600-1700 [2]	871-927 [2]	1500-1600	829-871	1500-1550	829-829	Oil
50B44	1600-1700 2	871-927 2	1500-1600	829-871	1500-1550	829-829	Oil
5046 †	1600-1700 2	871-927 2	1500-1600	829-871	1500-1550	829-829	Oil
50B46	1600-1700 2	871-927 2	1500-1600	829-871	1500-1550	829-829	Oil

(Continued)

Table 24. *(Continued)* **Thermal Treatments for Alloy Steels-Directly Hardenable Grades.**
(Source, Bethlehem Steel.)

Type[1]	Normalizing Temp.		Annealing Temp.[4]		Hardening Temp.[5]		Quenching Medium
	°F	°C	°F	°C	°F	°C	
50B50	1600-1700 [2]	871-927 [2]	1500-1600	829-871	1475-1550	802-829	Oil
5060 †	1600-1700 [2]	871-927 [2]	1500-1600	829-871	1475-1550	802-829	Oil
50B60	1600-1700 [2]	871-927 [2]	1500-1600	829-871	1475-1550	802-829	Oil
5130	1600-1700 [2]	871-927 [2]	1450-1550	788-843	1525-1575	802-857	Water, Caustic, or Oil
5132	1600-1700 [2]	871-927 [2]	1450-1550	788-843	1525-1575	802-857	Oil
5135	1600-1700 [2]	871-927 [2]	1500-1600	829-871	1500-1550	829-829	Oil
5140	1600-1700 [2]	871-927 [2]	1500-1600	829-871	1500-1550	829-829	Oil
5145 †	1600-1700 [2]	871-927 [2]	1500-1600	829-871	1500-1550	829-829	Oil
5147 †	1600-1700 [2]	871-927 [2]	1500-1600	829-871	1475-1550	802-829	Oil
5150	1600-1700 [2]	871-927 [2]	1500-1600	829-871	1475-1550	802-829	Oil
5155	1600-1700 [2]	871-927 [2]	1500-1600	829-871	1475-1550	802-829	Oil
5160	1600-1700 [2]	871-927 [2]	1500-1600	829-871	1475-1550	802-829	Oil
51B60	1600-1700 [2]	871-927 [2]	1500-1600	829-871	1475-1550	802-829	Oil
50100 †	-	-	1350-1450	732-788	1425-1475	774-802	Water
					1500-1600	829-871	Oil
51100	-	-	1350-1450	732-788	1425-1475	774-802	Water
					1500-1600	829-871	Oil
52100	-	-	1350-1450	732-788	1425-1475	774-802	Water
					1500-1600	829-871	Oil
6150	-	-	1550-1650	843-899	1500-1625	829-885	Oil
81B45	1600-1700 [2]	871-927 [2]	1550-1650	843-899	1500-1575	829-857	Oil
8630	1600-1700 [2]	871-927 [2]	1450-1550	788-843	1525-1600	802-871	Water or Oil
8637	-	-	1500-1600	829-871	1525-1575	802-857	Oil
8640	-	-	1500-1600	829-871	1525-1575	802-857	Oil
8642	-	-	1500-1600	829-871	1500-1575	829-857	Oil
8645	-	-	1500-1600	829-871	1500-1575	829-857	Oil
86B45	-	-	1500-1600	829-871	1500-1575	829-857	Oil
8650 †	-	-	1500-1600	829-871	1500-1575	829-857	Oil
8655	-	-	1500-1600	829-871	1475-1550	802-829	Oil
8660 †	-	-	1500-1600	829-871	1475-1550	802-829	Oil
8740	-	-	1500-1600	829-871	1525-1575	802-857	Oil
8254 †	-	-	-	-	1500-1650	829-899	Oil
9255 †	-	-	-	-	1500-1650	829-899	Oil
9260	-	-	-	-	1500-1650	829-899	Oil
94B30	1600-1700 [2]	871-927 [2]	1450-1550	788-843	1500-1625	829-885	Oil

Note: When tempering is required, temperature should be selected to effect desired hardness (see footnotes 3 and 6).

† Indicates SAE designation only. No AISI number assigned.

(Continued)

Table 24. *(Continued)* **Thermal Treatments for Alloy Steels-Directly Hardenable Grades.** *(Source, Bethlehem Steel.)*

1 All steels are fine grain.
2 These steels should be either normalized or annealed for optimum machinability.
3 Temper at 1,100-1,225° F (593-663° C).
4 The specific annealing cycle is dependent on the alloy content of the steel, the type of subsequent machining, and the desired surface finish.
5 With the exception of 4340, 50100, 51100, and 52100, these steels are frequently hardened and tempered to a final machinable hardness without preliminary thermal treatment.
6 Temper above 700° F (371° C).

steel is cooled from the annealing temperature at a slow rate so that the transformation is completed in the high temperature region of the pearlite range. Full annealing produces a structure of relatively soft coarse pearlite and, depending on the composition of the steel, ferrite or carbide may also be present. This process is also known as *Solution Annealing*. It softens the steel, but it is primarily used to improve the machinability of medium carbon steels. While it is simple, it is also a very slow process, as the steel must be cooled very slowly from the annealing (austenitizing) temperature to a temperature below that at which the transformation is completed.

Isothermal annealing is effective in obtaining either a lamellar or spheroidized structure. If a lamellar pearlitic structure is desired, the steel is austenitized above the upper transformation temperature and then cooled to and held at the proper temperature for austenitization to occur. To obtain a spheroidized structure, a lower austenitizing temperature is used so that some carbide remains undissolved. Cooling and transformation as for the pearlitic structure above will result in a spheroidized structure. An advantage of isothermal annealing is that it accelerates the cooling and transformation temperature, as well as the cooling subsequent to transformation, allowing for appreciable time savings compared with some other processes.

Spheroidize annealing is used to achieve a spheroidal (globular) carbide in a ferrite matrix , primarily to obtain optimum cold forming characteristics. One method consists of holding the steel at a temperature just below Ae_1 (holding between Ac_1 and Ac_3 for at least part of the time is normally involved). In order to achieve full spheroidization of the carbides by this method, the steel usually must be held at the temperature for long periods of time. Heating and cooling the steel to temperatures slightly above and slightly below the Ae_1 temperature is another method that will produce a spheroidized structure. Also, if the carbide is not completely dissolved during austenitizing, and the steel is slowly cooled in a manner similar to full annealing, a spheroidized structure can be developed. A spheroidized structure is sometimes desirable to develop minimum hardness and maximum ductility to facilitate cold forming, or to improve the machinability of high carbon steels.

Process annealing is also known as *stress relief annealing* and it is carried out at subcritical temperatures (below Ac_1) on cold worked, normalized, welded, or straightened steel to restore its ductility and reduce its hardness through recrystallization. It consists of heating steel to some temperature below, and usually near, Ac_1 and holding it at that temperature for an appropriate period, then air cooling it to ambient temperature.

Normalizing

Normalizing is performed by reheating steel to about 150 to 200° F (83 to 111° C) above its critical temperature, and then cooling it in air. The resulting fine grained pearlitic structure enhances the uniformity of mechanical properties and, for some grades, improves machinability. Notch toughness, in particular, is much better than in the as-rolled condition. For large sections, and when freedom from residual stresses or lower hardness are

(Text continued on p. 126)

Table 25. Thermal Treatments for Alloy Steels-Carburizing Grades. *(Source, Bethlehem Steel.)*

Type[1]	Pretreatments	Carburizing Temp.[5]		Cooling Method	Reheat temp.		Quenching Medium	Reheat temp.	
		°F	°C		°F	°C		°F	°C
4012 †	Normalize [2]	1650-1700	899-927	Quench in Oil [7]	-	-	-	250-350	121-177
4023	Normalize [2]	1650-1700	899-927	Quench in Oil [7]	-	-	-	250-350	121-177
4024	Normalize [2]	1650-1700	899-927	Quench in Oil [7]	-	-	-	250-350	121-177
4027	Normalize [2]	1650-1700	899-927	Quench in Oil [7]	-	-	-	250-350	121-177
4028	Normalize [2]	1650-1700	899-927	Quench in Oil [7]	-	-	-	250-350	121-177
4032	Normalize [2]	1650-1700	899-927	Quench in Oil [7]	-	-	-	250-350	121-177
4118	Normalize [2]	1650-1700	899-927	Quench in Oil [7]	-	-	-	250-350	121-177
4320	Normalize [2] and Cycle Anneal [4]	1650-1700	899-927	Quench in Oil [7]	-	-	-	250-350	121-177
				Cool Slowly	1525-1550 [9]	829-843 [9]	Oil		
4419 †	Normalize [2] and Cycle Anneal [4]	1650-1700	899-927	Quench in Oil [7]	-	-	-	250-350	121-177
4422 †	Normalize [2] and Cycle Anneal [4]	1650-1700	899-927	Quench in Oil [7]	-	-	-	250-350	121-177
4427 †	Normalize [2] and Cycle Anneal [4]	1650-1700	899-927	Quench in Oil [7]	-	-	-	250-350	121-177
4615	Normalize [2] and Cycle Anneal [4]	1650-1700	899-927	Quench in Oil [7]	-	-	-	250-300	121-149
				Cool Slowly	1525-1550 [9]	829-843 [9]	Oil		
				Quench in Oil	1525-1550 [8]	829-843 [9]	Oil		
4617 †	Normalize [2] and Cycle Anneal [4]	1650-1700	899-927	Quench in Oil [7]	-	-	-	250-300	121-149
				Cool Slowly	1525-1550 [9]	829-843 [9]	Oil		
				Quench in Oil	1525-1550 [8]	829-843 9	Oil		
4620	Normalize [2] and Cycle Anneal [4]	1650-1700	899-927	Quench in Oil [7]	-	-	-	250-300	121-149
				Cool Slowly	1525-1550 [9]	829-843 [9]	Oil		
				Quench in Oil	1525-1550 [8]	829-843 [9]	Oil		

(Continued)

Table 25. *(Continued)* **Thermal Treatments for Alloy Steels-Carburizing Grades.** *(Source, Bethlehem Steel.)*

Type[1]	Pretreatments	Carburizing Temp.[5]		Cooling Method	Reheat temp.		Quenching Medium	Reheat temp.	
		°F	°C		°F	°C		°F	°C
4621 †	Normalize [2] and Cycle Anneal [4]	1650-1700	899-927	Quench in Oil [7]	-	-	-	250-300	121-149
				Cool Slowly	1525-1550 [9]	829-843 [9]	Oil		
				Quench in Oil	1525-1550 [8]	829-843 [9]	Oil		
4626	Normalize [2] and Cycle Anneal [4]	1650-1700	899-927	Quench in Oil [7]	-	-	-	250-300	121-149
				Cool Slowly	1525-1550 [9]	829-843 [9]	Oil		
				Quench in Oil	1525-1550 [8]	829-843 [9]	Oil		
4718 †	Normalize [2] and Cycle Anneal [4]	1650-1700	899-927	Quench in Oil [7]	-	-	-	250-300	121-149
				Cool Slowly	1525-1550 [9]	829-843 [9]	Oil		
				Quench in Oil	1525-1550 [8]	829-843 [9]	Oil		
4720	Normalize [2] and Cycle Anneal [4]	1650-1700	899-927	Quench in Oil [7]	1525-1550 [8]	829-843 [9]	Oil	250-350	121-177
4815	Normalize, [2] Temper, [3] and Cycle Anneal [4]	1650-1700	899-927	Quench in Oil [7]	-	-	-	250-325	121-1
				Cool Slowly	1475-1525 [9]	802-829	Oil		
				Quench in Oil	1475-1525 [8]	802-829	Oil		
4817	Normalize, [2] Temper, [3] and Cycle Anneal [4]	1650-1700	899-927	Quench in Oil [7]	-	-	-	250-325	121-163
				Cool Slowly	1475-1525 [9]	802-829	Oil		
				Quench in Oil	1475-1525 [8]	802-829	Oil		
4820	Normalize, [2] Temper, [3] and Cycle Anneal [4]	1650-1700	899-927	Quench in Oil [7]	-	-	-	250-325	121-163
				Cool Slowly	1475-1525 [9]	802-829	Oil		
				Quench in Oil	1475-1525 [8]	802-829	Oil		
5015 †	Normalize [2]	1650-1700	899-927	Quench in Oil [7]	-	-	-	250-325	121-163
5115 †	Normalize [2]	1650-1700	899-927	Quench in Oil [7]	-	-	-	250-325	121-163
5120	Normalize [2]	1650-1700	899-927	Quench in Oil [7]	-	-	-	250-325	121-163

(Continued)

Table 25. *(Continued)* **Thermal Treatments for Alloy Steels-Carburizing Grades.** *(Source, Bethlehem Steel.)*

Type[1]	Pretreatments	Carburizing Temp.[5]		Cooling Method	Reheat temp.		Quenching Medium	Reheat temp.	
		°F	°C		°F	°C		°F	°C
6118	Normalize [2]	1650	899	Quench in Oil [7]	-	-	-	325	163
8115 †	Normalize [2] and Cycle Anneal [4]	1650-1700	899-927	Quench in Oil [7]	-	-	-	250-350	121-177
				Cool Slowly	1550-1600 [9]	843-871	Oil		
				Quench in Oil	1550-1600 [8]	843-871	Oil		
8615	Normalize [2] and Cycle Anneal [4]	1650-1700	899-927	Quench in Oil [7]	-	-	-	250-350	121-177
				Cool Slowly	1550-1600 [9]	843-871	Oil		
				Quench in Oil	1550-1600 [8]	843-871	Oil		
8617	Normalize [2] and Cycle Anneal [4]	1650-1700	899-927	Quench in Oil [7]	-	-	-	250-350	121-177
				Cool Slowly	1550-1600 [9]	843-871	Oil		
				Quench in Oil	1550-1600 [8]	843-871	Oil		
8620	Normalize [2]	1650-1700	899-927	Quench in Oil [7]	-	-	-	250-350	121-177
				Cool Slowly	1550-1600 [9]	843-871	Oil		
				Quench in Oil	1550-1600 [8]	843-871	Oil		
8622	Normalize [2]	1650-1700	899-927	Quench in Oil [7]	-	-	-	250-350	121-177
				Cool Slowly	1550-1600 [9]	843-871	Oil		
				Quench in Oil	1550-1600 [8]	843-871	Oil		
8625	Normalize [2]	1650-1700	899-927	Quench in Oil [7]	-	-	-	250-350	121-177
				Cool Slowly	1550-1600 [9]	843-871	Oil		
				Quench in Oil	1550-1600 [8]	843-871	Oil		
8627	Normalize [2]	1650-1700	899-927	Quench in Oil [7]	-	-	-	250-350	121-177
				Cool Slowly	1550-1600 [9]	843-871	Oil		
				Quench in Oil	1550-1600 [8]	843-871	Oil		

(Continued)

Table 25. *(Continued)* **Thermal Treatments for Alloy Steels-Carburizing Grades.** *(Source, Bethlehem Steel.)*

Type[1]	Pretreatments	Carburizing Temp.[5]		Cooling Method	Reheat temp.		Quenching Medium	Reheat temp.	
		°F	°C		°F	°C		°F	°C
8720	Normalize [2]	1650-1700	899-927	Quench in Oil [7]	-	-	-	250-350	121-177
				Cool Slowly	1550-1600 [9]	843-871	Oil		
				Quench in Oil	1550-1600 [8]	843-871	Oil		
8822	Normalize [2]	1650-1700	899-927	Quench in Oil [7]	-	-	-	250-350	121-177
				Cool Slowly	1550-1600 [9]	843-871	Oil		
				Quench in Oil	1550-1600 [8]	843-871	Oil		
9310	Normalize [2] and Temper [3]	1600-1700	871-927	Quench in Oil	1450-1525 [8]	788-829	Oil	250-325	121-177
				Cool Slowly	1450-1525 [9]	788-829	Oil		
94B15	Normalize [2]	1650-1700	899-927	Quench in Oil [7]	-	-	-	250-350	121-177
94B17	Normalize [2]	1650-1700	899-927	Quench in Oil [7]	-	-	-	250-350	121-177

† Indicates SAE designation only. No AISI number assigned.
1 All steels are fine grain. Indicates heat treatments may not be appropriate for coarse grain steels.
2 Normalizing temperatures should be at least equal to carburizing temperature, followed by air cooling.
3 After normalizing, reheat to temperature of 1,100-1,200° F (593-649° C) and hold at temperature for approximately one hour per inch (25.4 mm) of maximum section, or for four hours minimum.
4 Where cycle annealing is desired, heat to at least as high as the carburizing temperature, hold for uniformity, cool rapidly to 1,000-1,250° F (538-677° C), hold one to three hours, then air cool or furnace cool to obtain a structure suitable for machining and finishing.
5 It is general practice to reduce carburizing temperatures to approximately 1,550° F (843° C) before quenching to minimize distortion and retained austenite. For 4800 series steels, the carburizing temperature is reduced to approximately 1,500° F (816° C) before quenching.
6 Temperatures higher than those shown are used in some instances where application applies.
7 This treatment is most commonly used and generally produces a minimum of distortion.
8 This treatment is used where maximum grain refinement is required and/or where parts are subsequently ground on critical dimensions. A combination of good case and core properties is secured with somewhat greater distortion than is obtained by a single quench from the carburizing treatment.
9 In this treatment, the parts are slowly cooled—preferably under a protective atmosphere. They are then reheated and oil quenched. A tempering operation follows as required. This treatment is used when machining must be done between carburizing and hardening, or when facilities for quenching from the carburizing cycle are not available. Distortion is at least equal to that obtained by a single quench from the carburizing cycle, as described in note 5.

Table 26. Recommended Heating and Holding Time for Annealing, Normalizing, Austenitizing, and Stress Relieving Carbon Steel, Low Alloy Steel, and Martensitic Stainless Steel. *(Source, MIL-H-6875G.)*

Thickness[5]		Heat-up Time (Minutes)[4]		Minimum Holding Time [2,3]
Inches	Millimeters	Furnace[1]	Salt Bath	Minutes
0.250 and under	6.35 and under	20	10	15
0.251-0.500	6.3754 to 12.7	30	10	25
0.501-1.000	12.7254 to 25.4	45	10	30
1.001-1.500	25.4254 to 38.1	60	15	30
1.501-2.000	38.1254 to 50.8	75	20	30
2.001-2.500	50.8245 to 63.5	90	25	40
2.501-3.000	63.5254 to 76.2	105	30	45
3.001-3.500	76.2254 to 88.9	120	35	55
3.501-4.000	88.9254 to 101.6	135	40	60
4.001-5.000	101.6254 to 127	165	50	75
5.001-6.000	127.0254 to 152.4	195	60	90
6.001-7.000	152.4254 to 177.8	225	75	105
7.001-8.000	177.8254 to 203.2	255	90	120

[1] For unplated parts only. Copper plated parts require at least fifty percent longer heat-up time and the heat-treating facility should: (a) determine the appropriate heat-up time as a function of maximum part thickness and (b) establish suitable process controls for ensuring that the parts reach the required heat-treat temperature prior to start of holding time.

[2] Maximum holding time should not exceed twice the recommended minimum time. In all cases, holding time shall not start until parts or material have reached specified heat-treat temperature.

[3] Minimum stress relieving time shall be one hour for stress relieving temperatures up to 850° (F 454° C), inclusive, and 2 hours for higher stress relieving temperatures.

[4] Heat-up time starts when all temperature indicators rise to within 10° F (5.5° C) of set temperature. These times are suitable for simple solid shapes heated from all surfaces. Longer times are necessary for complex shapes and/or parts not uniformly heated.

[5] Thickness is minimum dimension of heaviest section.

Table 27. Recommended Holding Time for Annealing Austenitic Stainless Steel. *(Source, MIL-H-6875G.)*

Thickness[1]		Minimum Holding Time for Full Annealing (Minutes)
Inches	Millimeters	Atmosphere Furnace[2]
Up to 0.100	Up to 2.54	20
0.101 to 0.250	2.5654 to 6.35	25
0.251 to 0.500	6.3754 to 12.7	45
0.501 to 1.00	12.7254 to 25.4	60
1.01 to 1.50	25.5654 to 38.1	75
1.51 to 2.00	38.354 to 50.8	90
2.01 to 2.50	51.054 to 63.5	105
2.51 to 3.00	63.754 to 76.2	120

[1] Thickness is the minimum dimension of the heaviest section of a part or the minimum dimension of the heaviest section of a multi-layer load.

[2] Holding time starts when all temperature indicators rise to within 10° F (5.5° C) of set temperature. For continuous and repetitive batch heat treatment, the holding time may be lowered provided the solution of carbides is assured per ASTM A 262.

Table 28. Annealing Procedures for Austenitic Stainless Steels. *(Source, MIL-H-6875H.)*

Type	Annealing Temperature		Quenching Medium[1]
	°F	°C	
201, 202, 301, 302, 303, 304, 304L, and 308 [2]	1850-2050	1010-1121	Water
309, 310, 316, 316L [2]	1900-2050	1038-1121	Water
321 [3]	1750-2050	954-1121	Air or Water
347, 348 [3]	1800-2050	982-1121	Air or Water

[1] Other means of cooling are permitted provided that the cooling rate is rapid enough to prevent carbide precipitation.

[2] Do not stress relieve the unstabilized grades, except 304L and 316L, between 875° ± 25° F and 1,500° F (468° ± 14° C and 815° C). Stress relieving of stabilized grains should be at 1,650° F (899° C) for one hour.

[3] When stress relieving after welding is specified, hold for one-half hour minimum at specified temperature or hold for two hours at 1,650° ± 25° F (899° ± 14° C).

desired, it can be followed by a stress relief treatment. As a preliminary treatment to quenching and tempering, it is used to develop a more uniform microstructure and facilitate the solution of carbides and alloying elements. When necessary, normalizing can be followed by tempering, usually at 1,000 to 1,300° F (538 to 704° C), to reduce hardness and improve toughness.

Quenching and Tempering

Hardening by quenching and tempering is the heat treatment commonly used to develop martensitic structures with the desired combination of toughness and strength. The process is divided into separate operations: heating, quenching, and tempering.

Austenitizing is performed prior to hardening by heating steel to a temperature above its critical range. The steel should be held at the temperature long enough for the carbon and other alloying elements to dissolve, but not long enough for excessive grain growth to occur.

Quenching is the rapid cooling of steel from the austenitizing temperature to a temperature below M_s (martensite start temperature). The cooling rate must be rapid enough to prevent the formation of other transformation products such as pearlite, bainite, ferrite, or cementite. The choice of cooling medium is dependent on the desired cooling rate, which is determined by the composition (hardenability) of the steel, and the size and shape of the section being quenched. Quenching sets up high thermal and transformation stresses that can cause distortion or cracking of the part, therefore it is usually desirable to keep those stresses at a minimum by cooling at a rate that is just slightly faster than the critical cooling rate as determined by the hardenability of the steel and the size and shape of the part being quenched. The most common quenching media are water, mineral oil, and, of course, forced air.

Quenching water is normally held at about 65° F (18° C). As the water temperature increases during the quenching process, the treated steel tends to develop a surrounding envelope of steam that inhibits the water's cooling capacity. The tendency to form a steam envelope can be reduced by adding salt to the water to create brine (5 to 10% sodium chloride). Sodium hydroxide is even more efficient, and both solutions are generally used on very shallow hardening steels to attain high surface hardness while retaining a ductile core. Brine solutions are commonly used for carbon and low alloy steel as well as martensitic stainless steel. In both brine and water quenching, agitation of the quenching medium is especially important because it produces more uniform cooling as well as accelerates the rate of cooling.

Table 29. Thermal Treatment Procedures for Martensitic Stainless Steels. *(Source, MIL-H-6875H.)*

Type	Annealing °F		Annealing °C		Transformation Temp.		Quenching Medium		Subcritical Anneal Temp. Air Cool	
	Temp.	Cool Rate[1]	Temp.	Cool Rate[1]	°F	°C			°F	°C
403	1500-1600	25 to 50°/hr to 1100°	829-871	14 to 28°/hr to 593°	1750-1850	954-1010	Oil or Air		1200-1450	649-788
410	1500-1600	25 to 50°/hr to 1100°	829-871	14 to 28°/hr to 593°	1750-1850	954-1010	Oil or Air		1200-1450	649-788
416	1500-1650	25 to 50°/hr to 1100°	829-899	14 to 28°/hr to 593°	1750-1850	954-1010	Oil or Air		1200-1450	649-788
420	1550-1650 for 6 hours	25 to 50°/hr to 1100° then water quench	829-899 for 6 hours	14 to 28°/hr to 593° then water quench	1750-1850	954-1010	Oil or Air		1350-1450	732-788
440C	1550-1600 for 6 hours + 1300 for 4 hours	25 to 50°/hr to 1100°	829-899 for 6 hours + 704 for 4 hours	14 to 28°/hr to 593°	1900-1950	1037-1065	Oil or Air [2]		1250-1350	677-732
Tempering Temperature to Achieve Indicated Tensile Strength										
Type	100 ksi	690 MPa	120 ksi	830 MPa	Avoid Tempering or Holding Within Range[3]		180 ksi	1240 MPa	200 ksi	1380 MPa
	°F	°C	°F	°C			°F	°C	°F	°C
403	1300	704	1100	593	700-1100	371-593	500	260	-	-
410	1300	704	1100	593	700-1100	371-593	500	260	-	-
416	1300	704	1075	579	700-1075	371-579	500	260	-	-
420	1300 4	704 4	1075	579	700-1075	371-579	-	-	600	315
440C	Temper at 325° F (163° C) for R_C 58 minimum. Temper at 375° F (190° C) for R_C 57 minimum. Temper at 450° F (232° C) for R_C 55 minimum.									

[1] Cooling rate performed in furnace.
[2] Quench for 440C shall be followed by refrigeration to -100° F (-73° C) or lower for two hours. Double temper to remove retained austenite.
[3] Tempering these alloys in the range listed results in decreased impact strength and reduced corrosion resistance. However, tempering in this range is sometimes necessary to obtain the strength and ductility required.
[4] Temper 420 steel 300° F (149° C) for RC 52 minimum; 400° F (204° C) for RC 50 minimum; 600° F (315° C) for RC 48 minimum.

Quenching oils are mineral oils that usually have a viscosity in the range of 100 to 110 SUS at 40° C. They can be used to provide a wide variety of cooling rates dependent on viscosity and their mineral base. Normally, quenching oils have a temperature of between 60 and 160° F (15 and 71° C) and are generally maintained at temperatures below 200° F (93° C) during the quench. Due to the quenching properties of oils, which are less effective than water, they are not usually used for low-hardenability steels.

Polymer solutions are also used for quenching, but only three have achieved wide acceptance in the U.S.: PAG (polyalkylene glycol), PVP (polyvinyl pyrrolidone), and PVA (polyvinyl alcohol). Proponents of polymer quenching fluids attest that they are superior to either water or oil.

Tempering is the process of heating quench hardened or normalized steel to a temperature below the transformation range, holding it at that temperature for a suitable time, and cooling it at an appropriate rate. The martensite formed during quenching is very hard, brittle, and highly stressed, and tempering is used to remove these stresses and to improve the ductility of the steel, usually to the detriment of strength and hardness. The stress relief and the recovery of ductility are achieved by the precipitation of carbide from the supersaturated unstable alpha martensite and through diffusion and coalescence of the carbide as tempering proceeds.

Tempering is usually carried out at a temperature range between 350 and 1,300° F (177 to 704° C) and the usual time at temperature ranges from thirty minutes to four hours. For many carbon and low alloy steels, ductility and toughness increase when tempered at temperatures up to 400° F (204° C). In the approximate temperature range of 450 to 700° F (232 to 371° C), notch toughness decreases as tempering temperatures are increased—consequently, quench hardened steels are rarely hardened in this range. Tempering in the range of 200 to 400° F (93 to 204° C) is used when it is important to retain hardness and strength and effect a modest improvement in toughness. In the highest temperature range of 700 to 1,250° F (371 to 677° C), tempering causes an appreciable increase in toughness and ductility of the steel as well as a simultaneous decrease in hardness and strength.

Martempering is a process used to reduce high stresses, that can lead to distortion and cracking, caused by the transformation of austenite to martensite during rapid cooling of a steel through the martensite transformation range. It is an interrupted quenching process in which steel is quenched from the austenitizing temperature into hot oil or brine at a temperature near, but slightly above, the M_s temperature of the alloy. The steel is held in the quenching medium long enough for its temperature to stabilize, after which time it is removed and allowed to cool slowly in air. While the steel cools slowly, the transformation of austenite to martensite takes place. Due to the slow cooling rate, a relative uniform temperature is maintained throughout the mass of the steel, and the severe thermal gradients that are characteristic of conventional quenching are thereby avoided. The martensite forms at a uniform rate throughout the piece, and the stresses developed during transformation are much lower than for conventional quenching. The lower stresses induced by martempering, in turn, lessen the distortion of the treated part.

Martempering is normally used on steels of medium hardenability, such as those that are conventionally hardened by oil quenching. A modified martempering process consists of quenching from the austenitizing temperature to a temperature slightly below M_S. The higher cooling rates obtained by the lower temperature quench, in effect, allows the treatment of steels of lower hardenability that can be treated by the standard tempering process, and alloy steels are generally more suitable than carbon steels for martempering. An isothermal transformation diagram illustrating the martempering cycle is shown in *Figure 8*.

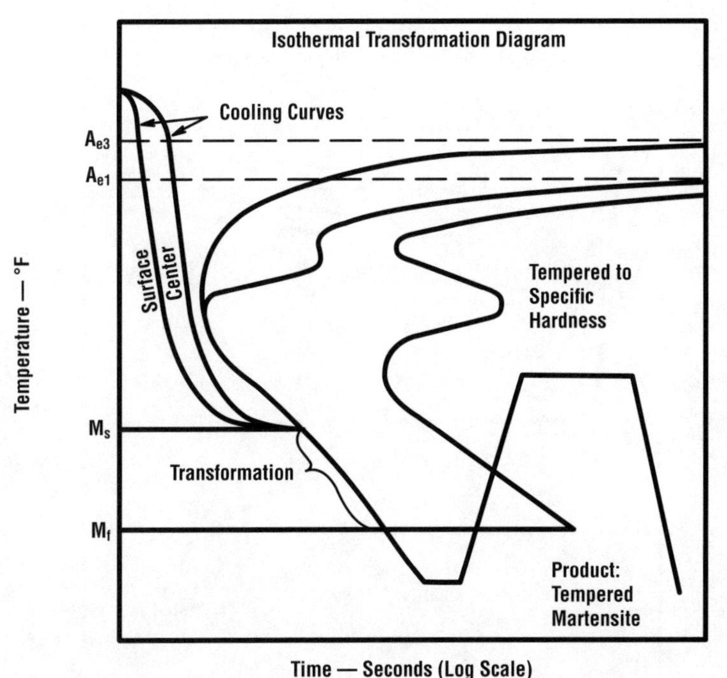

Figure 8. Schematic representation of the martempering cycle.

Austempering is used to isothermally transform austenite to lower bainite. Lower bainite compares favorably to tempered martensite with respect to strength and hardness and, at comparable hardness, it exhibits superior ductility. Austempering is, therefore, an alternate method of heat treating steels to develop high strength and hardness in combination with good ductility and toughness. The process consists of the following steps.

Heating the steel to a temperature within its austenitizing range, which usually occurs between 1,450 and 1,600° F (788 and 871° C).

Quenching the steel in a constant temperature bath at the desired transformation temperature in the lower bainite region, normally in the range of 500 to 750° F (260 to 399° C).

Holding the steel in the bath for a sufficient period of time to allow for the austenite to transform isothermally to bainite.

Cooling the steel to room temperature in still air.

Since austempering is an isothermal transformation process carried out at relatively high temperatures, transformation stresses induced by normal quenching are reduced, and distortion is minimized. It is usually substituted for conventional quenching and tempering in order to achieve higher ductility or notch toughness at a given hardness, and/or to decrease the distortion and cracking associated with normal quenching. *Figure 9* shows an isothermal transformation diagram for the austempering cycle.

Surface Hardening Steel

In many industrial applications it is necessary to develop a high surface hardness on a steel part so that it can resist wear and abrasion. This can be achieved by increasing the hardness of a high carbon steel, but high hardness is then accompanied by low ductility and toughness. In many applications, the poor ductility and toughness cannot be tolerated throughout the entire part, and another solution to the problem must be found.

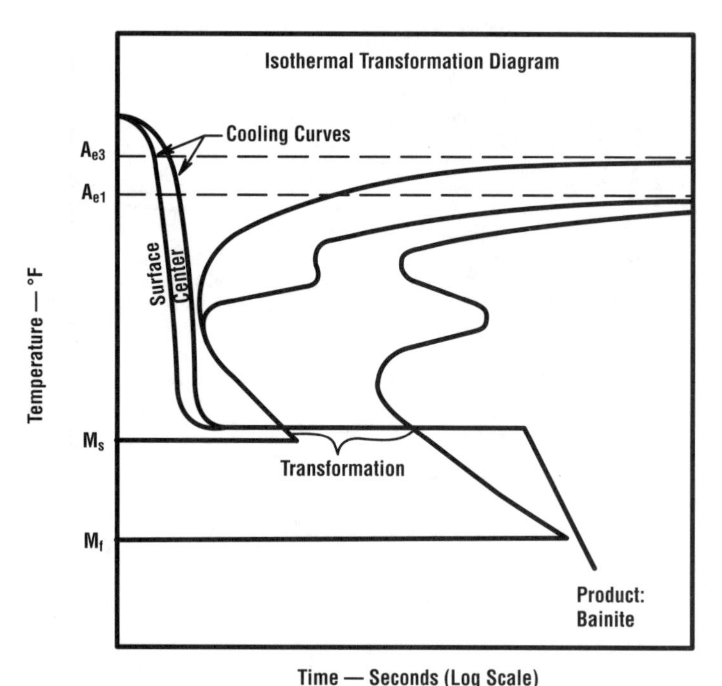

Figure 9. Schematic representation of the austempering cycle.

Low carbon steels can be treated to develop a hard surface, or "case," while the interior of the steel, or core, is unaffected and retains its normal ductility and toughness. Surface hardening (or case hardening) processes may be divided into two classifications, 1) those in which the composition of the surface materials must be changed, and 2) those in which the composition of the surface materials is not changed. Carburizing, nitriding, carbonitriding, and cyaniding processes are used to satisfy the first classification, and flame hardening and induction hardening can be used to create the characteristics of the second classification.

Carburizing is usually applied to plain carbon or low alloy steels with less than 0.20% carbon. Low alloy steels are heated in contact with a carbonaceous material to develop a surface layer on the steel that has a high carbon content. Upon quenching, the high carbon core remains relatively soft, resulting in a steel with a wear resistant exterior surface and a soft, tough core. There are three basic methods of carburizing steel: pack carburizing, gas carburizing, and liquid carburizing.

Pack carburizing is a process in which steel parts are placed in containers and carbonaceous solids (usually charcoal, or coke mixed with a suitable energizing catalyst such as barium carbonate, potassium carbonate carbonate, or sodium carbonate) are packed around them. The carbon monoxide gas that forms is actually the carburizing medium. The chemical reaction is $3F_e + 2CO \rightarrow F_{e3}C + CO_2$.

The carbon atoms diffuse into the steel, which, for carburizing, is heated above the critical range, and goes into solid solution in the austenite. The temperature range for pack carburizing is from 1,500 to 1,750° F (815 to 954° C), with the temperature selected being about 100° F (56° C) above the A_{c3} point of the alloy being treated. The depth of core and the carbon concentration gradient of carburized parts are both governed by the carburizing temperature, time, and the original composition of the steel, but depths are usually at least 0.04 inch (1.0160 mm), and up to 0.060 inch

(1.5240 mm) or more can be obtained.

Gas carburizing exposes the steel parts to carburizing gases such as methane, propane, butane, or from vaporized hydrocarbon liquids. The process can be performed in either batch or continuous furnaces at temperatures from 1,550 to 1,750° F (843 to 954° C), with 1,700° F (927° C) being a popular choice as it allows for rapid carburizing with minimum wear to furnace components. Case depths of between 0.10 and 0.40 inch (0.254 and 1.0160 mm) are common, with depth being dependent on time and temperature. Gas carburizing is generally more versatile and efficient than pack carburizing.

Liquid carburizing immerses the steel in a molten salt bath containing barium cyanide or sodium cyanide at temperatures ranging from 1,550 to 1,700° F (843 to 927° C) for depths up to 0.030 inch (0.7620 mm), but higher temperature baths can penetrate depths up to 0.250 inch (6.35 mm) in some circumstances. The primary application of liquid carburizing is to rapidly induce case depths of between 0.040 and 0.080 inch (1.0160 and 2.0320 mm). The case usually absorbs some nitrogen in addition to carbon, which enhances surface hardness.

Vacuum carburizing can provide good uniformity and reliability, but it requires a vacuum furnace. In vacuum carburizing, austenitization takes place in a vacuum, carburization is induced by introducing hydrocarbon gas under partial pressure, and quenching is done in either oil or gas. It is performed as follows. The first step is to bring the steel to carburization temperature, typically 1,550 to 1900° F (843 to 1038° C), and hold it until it has reached a uniform temperature throughout. This is performed in a vacuum of 0.30 to 0.50 torr (40 to 67 Pa). Next, hydrocarbon gas (usually methane or propane, or a mixture) is added to the atmosphere under partial pressure (usually between 10 and 50 torr (1.3 and 6.6 kPa) for graphite furnaces, and 100 to 200 torr (13 to 25 kPa) in ceramic furnaces for a period of approximately one-half hour. A partial vacuum of 0.50 to 1.0 torr (67 to 135 Pa) is then reintroduced, and the steel is held at carburiding temperature for another half-hour or longer while the carbon diffuses inward from the surface. Finally, the steel is quenched in oil, which normally takes place under a partial pressure of nitrogen.

Plasma or *Ion carburizing* requires an oxygen-free vacuum and an electric flow between the part (the cathode) and a counter electrode (the anode). It normally takes place in a methane gas atmosphere of 1 to 20 torr (0.20 and 2.7 kPa), and at temperatures 1,650 to 1,830° F (899 to 1000° C). The process is essentially insensitive to the composition of the steel, provides uniform treatment results, and, because the glow discharge from the anode accelerates ions bearing carbon directly toward the steel surface, it is fast.

Many carbon and low alloy steels are carburized, although steels containing between 0.15 and 0.20% carbon are usually selected. The heat treatment selected to carburize the steel will depend on the carburizing temperature, composition of the core and case, and the properties that the steel must obtain. In some cases, the steel may be quenched directly from the carburizing temperature and then retempered, but in some instances it is desirable to slowly cool the steel from the carburizing temperature and then reheat it to a temperature slightly above Ae_4, and then quench and retemper. Double reheat and quench operations are also employed for some steels. In some circumstances it is necessary to develop a case locally, and not over the entire surface of a part. In such cases, carburization can be preserved in local surface areas by protecting those areas with copper plating, or covering them with a copper bearing commercial paste, thereby allowing the carbon to penetrate only the exposed areas. Another effective method is to machine the part after carburization to remove the case from those areas where a soft surface is required.

Gas nitriding is a surface hardening process that employs nitrogen to induce the formation of nitrides on the case. For successful nitriding, it is necessary to use alloy steels that contain aluminum, chromium, molybdenum, tungsten, and/or vanadium, which are elements that combine with nitrogen to form hard nitrides. The nitriding medium is ammonia gas, and the process temperature is 900 to 1,100° F (482 to 593° C). Case depth is dependent on time and temperature, but is usually in the range of 0.015 to 0.020 inch (0.3810 to 0.5080 mm). A significant advantage is that parts may be hardened, tempered, *and* machined prior to nitriding, because the process does not require post tempering or quenching that can impair dimensional instability or cause distortion. Nitrided parts exhibit exceptional wear resistance qualities and have very little tendency to gall and seize, making them especially well suited for applications involving metal-to-metal wear. Additional positive effects are high fatigue resistance and improved corrosion resistance.

Liquid nitriding uses a molten salt bath containing either cyanides or cyanates, heated to 900 to 1,100° F (same range as for gas nitriding). The process is very similar to gas nitriding, and treated parts exhibit nearly identical qualities. Parts to be treated must have the elements necessary for nitriding.

Plasma or *Ion nitriding* uses the same principles as plasma carburizing, which is discussed above. It is performed at relatively low temperatures (700° F [371° C]) because the glow discharge of the anode to the grounded workpiece introduces nitrogen directly to its surface.

Carbonitriding exposes carbon and alloy steels to a gaseous atmosphere from which they simultaneously absorb carbon and nitrogen. The process is a modified gas carburizing process in which ammonia is introduced into the gas carburizing atmosphere. Operating temperatures range from 1,425 to 1,650° F (774 to 899° C) for parts that are to be quenched, to 1,200 to 1,450° F (649 to 788° C) when a liquid quench is not required. Because nitrogen increases the hardenability of the case, full hardness can be achieved by less severe quenching, and distortion is minimized. Case depths of 0.003 to 0.0.0 inch (0.0762 to 0.7620 mm) are developed.

Cyaniding, sometimes called *liquid carbonitriding*, is similar to liquid carburizing except that the molten salt bath contains higher percentages of sodium cyanide, ranging from 30 to 97%. The steel absorbs both carbon and nitrogen from the molten bath and, by controlling the bath composition and operating temperature, it is possible to regulate, within limits, the relative amount of carbon and nitrogen in the case. Operating temperatures range from 1,400 to 1,600° F (760 to 871° C), just above Ac_1 for the steel being treated. The case is hard and somewhat brittle (due to the presence of nitrides), and rather thin, usually less than 0.010 inch (0.2540 mm) deep. Cyanided parts are extremely wear resistant.

Flame hardening consists of rapidly heating the surface area to be hardened with a flame to a temperature above the upper critical temperature of the alloy. With oxy-gas fuels hardness can be induced to a depth of less than 0.125 inch (3.18 mm). The hardened surface is then quenched in a suitable manner. Heating and cooling cycles are precisely controlled to attain the desired depth of hardening consistency. Steels suitable for flame hardening are usually in the range of 0.30 to 0.60% carbon, with hardenability appropriate for the depth to be hardened and the quenchant used. Various quenching media can be used, and are usually sprayed directly on the surface a short distance behind the flame. Immediate tempering is required to avoid cracking caused by residual stresses, and may be done by conventional furnace tempering or flame tempering.

Induction hardening uses an induced electrical current to heat steel above the upper critical temperature. When the high frequency alternating current is sent through a coil (or inductor) surrounding a steel part (or conductor) a magnetic field is created that heats the part by induced energy. Heating results primarily from the resistance of the part to the flow of currents created by the induced voltage, and from hysteresis losses caused by the rapidly alternating magnetic field if the part is magnetic. Therefore, most plain carbon and alloy steels heat most rapidly below the Curie point (approximately the upper critical temperature–the Curie point is the temperature that marks the transition between ferromagnetism and paramagnetism, or between the ferroelectric phase and the paraelectric phase), where they are ferromagnetic, and less rapidly above that point.

With conventional induction generators, heat is developed primarily on the surface of the part, and the total depth of heating is dependent on the frequency of the alternating current passing through the coil, the rate at which heat is conducted from the surface to the interior, and the length of the heating cycle. For surface hardening, frequencies of 10,000 to 50,000 cycles per second, using high power and short heating cycles, are used, while lower frequencies and long heating cycles are preferred for through hardening by induction. Quenching may be done with a water spray, but oil quenching is also possible by immersing the part in an oil bath after reaching hardening temperature.

The hardness in induction heating is, as in conventional heating, a function of the carbon content of the steel, and higher hardness values for a given carbon content can be obtained for induction surface hardened parts. The increment of added hardness may be as much as 5 R_C points for steels containing 0.30% carbon, and less for those with lower carbon content. Conventional hardening temperatures can generally be used when induction heating plain carbon grades, and alloy grades containing carbide forming elements such as chromium, molybdenum, and vanadium. The hardening temperature must be increased, however, if the normal influence of the alloying element is desired. Increased hardening temperatures do not increase the austenitic grain size, because grain growth is inhibited by the undissolved carbides. Generally, steels heated to conventional hardening temperatures by induction heating show a similar or somewhat finer grain size than steels heated in the furnace for hardening.

Nonferrous Metals

Metals that contain no iron are known as nonferrous. Seven of these nonferrous elements are produced as pure metals or alloys in significant quantities: aluminum, copper, lead, magnesium, nickel, tin, and zinc.

Aluminum

Aluminum alloys have found wide acceptance in engineering design primarily because they are relative lightweight, have a high strength to weight ratio, have superior corrosion resistance, and they are comparatively inexpensive. For some applications they are favored because of their high thermal and electrical conductivity, ease of fabrication, and ready availability. In fact, aluminum is the fourth most widely distributed of the elements, following oxygen, nitrogen, and silicon. Aluminum alloys weigh approximately 0.1 pound per cubic inch, which is one-third the weight of iron (0.28 lb/in.^{3}), and copper (0.32 lb/in.^{3}). It is just slightly heavier than magnesium (0.066 lb/in.^{3}) and somewhat lighter than titanium (0.163 lb/in.^{3}). In its commercially pure state, aluminum is a relatively weak metal, having a tensile strength of approximately 12,000 PSI. However, with the addition of such alloying elements as manganese, silicon, copper, magnesium, and/or zinc, and with proper heat treatment or cold working, the tensile strength of aluminum can approach 100,000 PSI. Physical property ranges for aluminum are given in **Table 1**.

Generally, the strengths of aluminum alloys decrease and toughness increases with increases in temperature, and with time at temperature above room temperature. The effect is usually greatest over the temperature range of 212 to 400° F (100 to 204° C). Exceptions to the general trends are tempers developed by solution heat treating without subsequent aging, for which the initial elevated temperature exposure results in some age hardening and reduction in toughness. Further time at temperature beyond that required to achieve peak hardness results in the aforementioned decrease in strength and increase in toughness.

Corrosion resistance of aluminum is attributable to its self-healing nature; a thin, invisible skin of aluminum oxide forms when the metal is exposed to the atmosphere. Pure aluminum will form a continuous protective oxide film, while high strength alloyed forms will sometimes become pitted as a result of localized galvanic corrosion at sites of alloying-constituent concentration. As a conductor of electricity, aluminum compares favorably with copper. Although the conductivity of the electroconductor grade of aluminum is only 62% that of the International Annealed Copper Standard (IACS), on a pound for pound basis the power loss for aluminum is less than half that of copper—an advantage where weight and space are primary requirements. As a heat conductor, aluminum ranks high among the metals. It is especially useful in heat exchangers and other applications requiring rapid heat dissipation. As a reflector of radiant energy, aluminum is excellent throughout the entire range of wavelengths—from the ultraviolet end of the spectrum through the visible and infrared bands, to the electromagnetic wave frequencies of radio and radar.

Alclad sheet (sometimes called clad sheet) is a product consisting of an aluminum alloy sheet having on one or both surfaces a layer of aluminum or aluminum alloy integrally bonded to the surface of the base metal. In general, alclad sheets have mechanical properties slightly lower than those of the bare alloy sheets of the same thickness. However, the corrosion resistant qualities of the aluminum alloy sheet are improved by the cladding.

Aluminum is easily fabricated. It can be cast by any method, rolled to any reasonable thickness, stamped, hammered, forged, or extruded. It is readily turned, milled, bored, or machined. It can be joined by several welding processes (see the Welding section

Table 1. Physical Property Ranges for Wrought and Cast Aluminum Alloys. *(Source, MIL-HDBK-694A.)*

<table>
<tr><th rowspan="2">Property</th><th colspan="2">Range</th><th rowspan="2" colspan="2">Notes</th></tr>
<tr><th>Wrought Alloys</th><th>Cast Alloys</th></tr>
<tr><td>Specific Gravity</td><td>2.70 to 2.82</td><td>2.57 to 2.95</td><td colspan="2">About one-third that of steel.</td></tr>
<tr><td>Weight (lb/cubic in.)</td><td>0.095 to 0.102</td><td>0.093 to 0.107</td><td colspan="2">Approximately 173 lb./cubic ft</td></tr>
<tr><td>Electrical Conductivity (Int'l. Annealed Copper Std.)</td><td>30% to 60%</td><td>21% to 47%</td><td>About 59% for 99.9% Al</td><td rowspan="2">Values for electrical and thermal conductivity depend on the composition and condition of the alloys. Both are increased by annealing, and decreased by adding alloying elements to pure aluminum. Both are also decreased by heat treatment, cold work, and aging.</td></tr>
<tr><td>Thermal Conductivity (cgs units at 77° F)</td><td>0.29 to 0.56</td><td>0.21 to 0.40</td><td>About 0.53 for 99% Al</td></tr>
<tr><td>Coefficient of Thermal Expansion</td><td>10.8 to 13.2</td><td>11.0 to 14.0</td><td colspan="2">Roughly double that of ordinary steels and cast irons; substantially greater than copper-alloy materials. Alloying elements other than silicon have little effect on the expansion of aluminum. Considerable amounts of silicon (12%) appreciably decrease the dimensional changes induced by varying temperatures. Where a low coefficient of thermal expansion is desirable (as in engine pistons) an aluminum alloy containing a relatively high percentage of silicon may be specified.</td></tr>
<tr><td>Reflectivity</td><td colspan="2">See Notes</td><td colspan="2">Greater than for any other metal. Suitably treated, aluminum sheet of high purity may yield a reflectivity for light greater than 80%.</td></tr>
<tr><td>Modulus of Elasticity</td><td colspan="4">In the range of 10.0 to 10.6 PSI $\times 10^6$ (69 to 73 GPa) for wrought alloys.
In the range of 10.0 to 11.9 PSI $\times 10^6$ (69 to 82 GPa) for casting alloys.</td></tr>
<tr><td>Modulus of Rigidity</td><td colspan="2">3.9×10^6 (average)</td><td colspan="2">Average, all alloys.</td></tr>
<tr><td>Poisson's Ratio</td><td colspan="2">0.33</td><td colspan="2">Average, all alloys.</td></tr>
<tr><td>Ultimate Torsional Strength (% of Ultimate Tensile Strength)</td><td colspan="2">65</td><td colspan="2">Average all alloys.</td></tr>
</table>

of this book for guidelines on welding, brazing, and soldering aluminum). Aluminum can also be coated with a wide variety of surface finishes for decorative and protective purposes.

Types of Aluminum

The Aluminum Association Alloy and Temper Designation System is used to identify wrought and casting alloys. With few exceptions, aluminum alloys are designed either for casting or for use in wrought products, but not for both—some general purpose alloys are available but, on the whole, compositions are formulated to satisfy specific requirements. The first digit in the alloy designation system identifies the principal alloying constituent of the metal. For the wrought alloys, the second digit notes variations of the initial alloy. For the casting alloys, the second digit, together with the third, are used to set the designation off from others—even though they have no reference to a specific alloy, these numbers make the designation unique. For wrought alloys, the third and fourth digits serve the same purpose as the second and third for the casting alloys—they are chosen to make the designation unique. The fourth digit on the casting alloys designation indicates casting (0) or ingot (1.2). Variations of similar casting alloys are preceded by A, B, or C. Tempers use a letter plus number designation system. The four-digit designation system, plus the letters of the temper designation system, is detailed in **Table 2**. The letter-number combination designations used for tempering are detailed later in this section. See **Table 3** for chemical compositions of wrought alloys, **Table 4** for their mechanical properties, and **Table 5** for their physical properties. Chemical compositions for casting alloys are given in **Table 6**. Mechanical properties of sand casting alloys are provided in **Table 7**, and the same information for permanent mold casting alloys appears in **Table 8**. Physical properties of casting alloys are shown in **Table 9**.

Table 2. Aluminum Association Alloy and Temper Designation System.

Wrought Alloy System	Cast Alloy System	Temper System
1xxx – Pure Al (99.00% or greater)	1xx.x – Pure Al (99.00% or greater)	F – As fabricated
2xxx – Al-Cu Alloys	2xx.x – Al-Cu Alloys	O – Annealed
3xxx – Al-Mn Alloys	3xx.x – Al-Si + Cu and/or Mg	H – Strain hardened (wrought alloys only)
4xxx – Al-Si Alloys	4xx.x – Al-Si	W – Solution heat treated
5xxx – Al-Mg Alloys	5xx.x – Al-Mg	T – Thermally treated to produce tempers other than F, O, H (usually solution heat treated, quenched, and precipitation hardened)
6xxx – Al-Mg-Si Alloys	7xx.x – Al-Zn	
7xxx – Al-Zn Alloys	8xx.x – Al-Sn	
8xxx– Al + Other Elements	9xx.x – Al + Other Elements	
9xxx – Unused	6xx.x – Unused	

Characteristics of Wrought Alloys by Series Designations

1xxx Series – Pure Aluminum. This series is made up of commercially pure aluminum, ranging from the baseline 1100 (99.00% minimum Al) to relatively purer 1050/1350 (99.50% min.) and 1175 (99.75 min.). Some alloys in this series, such as 1350, which is used especially for electrical applications, have relatively tight controls on those impurities that might impair performance of the alloy for its intended use. The 1xxx series are strain hardenable, but would not be used where strength is a prime consideration. Instead, the emphasis would be on those applications where extremely high corrosive resistance, high formability, and/or good electrical conductivity are required. Examples are foil and strip for packaging, chemical equipment, tank, car, or truck bodies, spun hollowware, and elaborate sheet metal work.

2xxx Series – Al-Cu Alloys. The alloys in this series are heat treatable and possess, in individual alloys, good combinations of high strength (especially at elevated temperatures), toughness, and, in specific cases, weldability. They are not resistant to atmospheric corrosion, so painting or cladding is suggested to resist environmental exposure. The higher strength 2xxx series are used primarily for aircraft (2024 is an example) and truck bodies (2014), usually in bolted or riveted construction. Specific members of this series (2219 and 2048) are readily welded and are often used in aerospace applications where welding is the preferred joining method. Several of these alloys have been developed for the aircraft industry and have high toughness (e.g., 2124, 2324, 2419), and higher control on any impurities that may diminish their resistance to unstable fracture. Alloys 2011, 2017, and 2117 are widely used for fasteners and screw machine stock. A recent addition to the 2xxx series is 2195, a "weldalite" alloy that, with 2219 and 2419, has been used for fuel tanks and booster rockets on the space shuttle. As a group, these alloys are noteworthy for their excellent strengths at elevated and cryogenic temperatures, and creep resistance at elevated temperatures. Typically, 2xxx alloys have a tensile strength range of 27,000 to 62,000 PSI (186 to 427 MPa).

(Text continued on p. 156)

Table 3. Chemical Composition of Wrought Aluminum and Aluminum Alloys. *(Source, Aluminum Association, Inc.)*

Alloy	Si	Fe	Cu	Mn	Mg	Cr	Zn	Ti	V	Other (Specified)	Other (Unspecified)		Al
											Each	Total	
1050	0.25	0.40	0.05	0.05	0.05	-	0.05	0.03	0.05	-	0.03	-	99.50 min.
1060	0.25	0.35	0.05	0.03	0.03	-	0.05	0.03	0.05	-	0.03	-	99.60 min.
1100	0.95 Si + Fe		0.05- 0.20	0.05	-	-	0.10	-	-	-	0.05	0.15	99.0 min.
1200	1.00 Si + Fe		0.05	0.05	-	-	0.10	0.05	-	-	0.05	0.15	99.0 min.
1230	0.70 Si + Fe		0.10	0.05	0.05	-	0.10	0.03	0.05	-	0.03	-	99.30 min
1235	0.65 Si + Fe		0.05	0.05	0.05	-	0.10	0.06	0.05	-	0.03	-	99.35 min.
1145	0.55 Si + Fe		0.05	0.05	0.05	-	0.05	0.03	0.05	-	0.03	-	99.45 min.
1345	0.30	0.40	0.10	0.05	0.05	-	0.05	0.03	0.05	-	0.03	-	99.45 min.
1350	0.10	0.40	0.05	0.01	-	0.01	0.05	0.02 V + Ti		0.03 Ga	0.03	0.10	99.50 min.
1175	0.15 Si + Fe		0.10	0.02	0.02	-	0.04	0.02	0.05	0.03 Ga	0.02	-	99.75 min.
2011	0.40	0.7	5.0-6.0	-	-	-	0.30	-	-	-	0.05	0.15	Balance
2014	0.50-1.2	0.7	3.9-5.0	0.40-1.2	0.20-0.8	0.10	0.25	0.15	-	-	0.05	0.15	Balance
2017	0.20-0.8	0.7	3.5-4.5	0.40-1.0	0.40-0.8	0.10	0.25	0.15	-	-	0.05	0.15	Balance
2117	0.8	0.7	2.2-3.0	0.20	0.20-0.50	0.10	0.25	-	-	-	0.05	0.15	Balance
2018	0.9	1.0	3.5-4.5	0.20	0.45-0.9	0.10	0.25	-	-	1.7-2.3 Ni	0.05	0.15	Balance
2218	0.9	1.0	3.5-4.5	0.20	1.2-1.8	0.10	0.25	-	-	1.7-2.3 Ni	0.05	0.15	Balance
2219	0.20	0.30	5.8-6.8	0.20-0.40	0.02	-	0.10	0.02-0.10	0.05-0.15	-	0.05	0.15	Balance
2419	0.15	0.18	5.8-6.8	0.20-0.40	0.02	-	0.10	0.02-0.10	0.05-0.15	-	0.05	0.15	Balance
2024	0.50	0.50	3.8-4.9	0.30-0.9	1.2-1.8	0.10	0.25	0.15	-	-	0.05	0.15	Balance
2124	0.20	0.30	3.8-4.9	0.30-0.9	1.2-1.8	0.10	0.25	0.15	-	-	0.05	0.15	Balance
2025	0.50-1.2	1.0	3.9-5.0	0.40-1.2	0.05	0.10	0.25	0.15	-	-	0.05	0.15	Balance
2036	.050	0.50	2.2-3.0	0.10-0.4	0.30-0.6	0.10	0.25	0.15	-	-	0.05	0.15	Balance
2048	0.15	0.20	2.8-3.8	0.20-0.6	1.2-1.8	-	0.25	0.10	-	-	0.05	0.15	Balance

(Continued)

Table 3. *(Continued)* **Chemical Composition of Wrought Aluminum and Aluminum Alloys.** *(Source, Aluminum Association, Inc.)*

Alloy	Si	Fe	Cu	Mn	Mg	Cr	Zn	Ti	V	Other (Specified)	Other (Unspecified)		Al
											Each	Total	
2195	0.12	0.15	3.7-4.3	0.25	0.25-0.8	-	0.25	0.10	-	-	0.05	0.15	Balance
3003	0.6	0.7	0.05-0.20	1.0-1.5	-	-	0.10	-	-	-	0.05	0.15	Balance
3004	0.30	0.7	0.25	1.0-1.5	0.8-1.3	-	0.25	-	-	-	0.05	0.15	Balance
3104	0.6	0.8	0.05-0.25	0.8-1.4	0.8-1.3	-	0.25	0.10	0.05	0.05 Ga	0.05	0.15	Balance
3005	0.6	0.7	0.30	1.0-1.5	0.20-0.6	0.10	0.25	0.10	-	-	0.05	0.15	Balance
3105	0.6	0.7	0.30	0.30-0.8	0.20-0.8	0.20	0.40	0.10	-	-	0.05	0.15	Balance
4032	11.0-13.5	1.0	0.50-1.3	-	0.8-1.3	0.10	0.25	-	-	0.50-1.3 Ni	0.05	0.15	Balance
4043	4.5-6.0	0.8	0.30	0.05	0.05	-	0.10	0.20	-	-	0.05	0.15	Balance
4045	9.0-11.0	0.8	0.30	0.05	0.05	-	0.10	0.20	-	-	0.05	0.15	Balance
4047	11.0-13.0	0.8	0.30	0.15	0.10	-	0.20	-	-	-	0.05	0.15	Balance
5005	0.30	0.7	0.20	0.20	0.50-1.1	0.10	0.25	-	-	-	0.05	0.15	Balance
5050	0.40	0.7	0.20	0.10	1.1-1.8	0.10	0.25	-	-	-	0.05	0.15	Balance
5052	0.25	0.40	0.10	0.10	2.2-2.8	0.15-0.35	0.10	-	-	-	0.05	0.15	Balance
5252	0.08	0.10	0.10	0.10	2.2-2.8	-	0.05	-	0.05	-	0.03	0.10	Balance
5154	0.25	0.40	0.10	0.10	3.1-3.9	0.15-0.35	0.20	0.20	-	-	0.05	0.15	Balance
5254	0.45 Si + Fe		0.05	0.01	3.1-3.9	0.15-0.35	0.20	0.05	-	-	0.05	0.15	Balance
5454	0.25	0.40	0.10	0.50-1.0	2.4-3.0	0.05-0.20	0.25	0.20	-	-	0.05	0.15	Balance
5754	0.40	0.40	0.10	0.50	2.6-3.6	0.30	0.20	0.15	-	-	0.05	0.15	Balance
5056	0.30	0.40	0.10	0.05-0.20	4.5-5.6	0.05-0.20	0.10	-	-	-	0.05	0.15	Balance
5356	0.25	0.40	0.10	0.05-0.20	4.5-5.5	0.05-0.20	0.10	0.06-0.20	-	-	0.05	0.15	Balance
5456	0.25	0.40	0.10	0.50-1.0	4.7-5.5	0.05-0.20	0.25	0.20	-	-	0.05	0.15	Balance
5556	0.25	0.40	0.10	0.50-1.0	4.7-5.5	0.05-0.20	0.25	0.05-0.20	-	-	0.05	0.15	Balance
5457	0.08	0.10	0.20	0.15-0.45	0.8-1.2	-	0.05	-	0.05	-	0.03	0.10	Balance
5657	0.08	0.10	0.10	0.03	0.6-1.0	-	0.05	-	0.05	0.03 Ga	0.02	0.05	Balance

(Continued)

Table 3. *(Continued)* **Chemical Composition of Wrought Aluminum and Aluminum Alloys.** *(Source, Aluminum Association, Inc.)*

Alloy	Si	Fe	Cu	Mn	Mg	Cr	Zn	Ti	V	Other (Specified)	Other (Unspecified)		Al
											Each	Total	
5182	0.20	0.35	0.15	0.20-0.50	4.0-5.0	0.10	0.25	0.10	-	-	0.05	0.15	Balance
5083	0.40	0.40	0.10	0.40-1.0	4.0-4.9	0.05-0.25	0.25	0.15	-	-	0.05	0.15	Balance
5086	0.40	0.50	0.10	0.20-0.7	3.5-4.5	0.05-0.25	0.25	0.15	-	-	0.05	0.15	Balance
6101	0.30-0.7	0.50	0.10	0.03	0.35-0.8	0.03	0.10	-	-	-	0.03	0.10	Balance
6201	0.50-0.9	0.50	0.10	0.03	0.6-0.9	0.03	0.10	-	-	-	0.03	0.10	Balance
6111	0.6-1.1	0.40	0.50-0.9	0.10-0.45	0.50-1.0	0.10	0.15	0.10	-	-	0.10	0.05	Balance
6351	0.7-1.3	0.50	0.10	0.40-0.8	0.40-0.8	-	0.20	0.20	-	-	0.20	0.05	Balance
6061	0.40-0.8	0.7	0.15-0.40	0.15	0.8-1.2	0.04-0.35	0.25	0.15	-	-	0.05	0.15	Balance
6262	0.40-0.8	0.7	0.15-0.40	0.15	0.8-1.2	0.04-0.14	0.25	0.15	-	-	0.05	0.15	Balance
6063	0.20-0.6	0.35	0.10	0.10	0.45-0.9	0.10	0.10	0.10	-	-	0.05	0.15	Balance
6463	0.20-0.6	0.15	0.20	0.05	0.45-0.9	-	0.05	-	-	-	0.05	0.15	Balance
6066	0.9-1.8	0.50	0.7-1.2	0.6-1.1	0.8-1.4	0.40	0.25	0.20	-	-	0.05	0.15	Balance
6070	1.0-1.7	0.50	0.15-0.40	0.40-1.0	0.50-1.2	0.10	0.25	0.15	-	-	0.05	0.15	Balance
7049	0.25	0.35	1.2-1.9	0.20	2.0-2.9	0.10-0.22	7.2-8.2	0.10	-	-	0.05	0.15	Balance
7050	0.12	0.15	2.0-2.6	0.10	1.9-2.6	0.04	5.7-6.7	0.06	-	-	0.05	0.15	Balance
7150	0.12	0.15	1.9-2.5	0.10	2.0-2.7	0.04	5.9-6.9	0.06	-	-	0.05	0.15	Balance
7075	0.40	0.50	1.2-2.0	0.30	2.1-2.9	0.18-0.28	5.1-6.1	0.20	-	-	0.05	0.15	Balance
7175	0.15	0.20	1.2-2.0	0.10	2.1-2.9	0.18-0.28	5.1-6.1	0.10	-	-	0.05	0.15	Balance
7475	0.10	0.12	1.2-1.9	0.06	1.9-2.6	0.18-0.25	5.2-6.2	0.06	-	-	0.05	0.15	Balance
7178	0.40	0.50	1.6-2.4	0.30	2.4-3.1	0.18-0.28	6.3-7.3	0.20	-	-	0.05	0.15	Balance
8017	0.10	0.55-0.8	0.10-0.20	-	0.01-0.05	-	0.05	-	-	-	0.03	0.10	Balance
8030	0.10	0.30-0.8	0.15-0.30	-	0.05	-	0.05	-	-	-	0.03	0.10	Balance
8176	0.03-0.15	0.40-1.0	-	-	-	-	0.10	-	-	0.03 Ga	0.05	0.15	Balance
8177	0.10	0.25-0.45	0.04	-	0.04-0.12	-	0.05	-	-	-	0.03	0.10	Balance

Table 4. Mechanical Properties of Wrought Aluminum Alloys. *(Source, Aluminum Association, Inc.)*

Alloy	Temper	Ultimate Tensile Strength		Tensile Yield Strength		Ultimate Shearing Strength		Fatigue Endurance Limit		Elongation %[1]		Brinell Hardness
										0.0625" Thick	0.50" Dia.	
		ksi	MPa	ksi	MPa	ksi	MPa	ksi	MPa			BHN
1060	O	10	69	4	28	7	48	3	21	43	-	19
	H18	19	131	18	124	11	76	6.5	45	6	-	35
1100	O	13	90	5	34	9	62	5	34	35	45	23
	H18	24	165	22	152	13	90	9	62	5	15	44
1350	O	12	83	4	28	8	55	-	-	-	(d)	-
2011	T3	55	379	43	296	32	221	18	124	-	15	95
2014	O	27	186	14	97	18	124	13	90	-	18	45
	T4, T451	62	427	42	290	38	262	20	138	-	20	105
	T6, T651	70	483	60	414	42	290	18	124	-	13	135
2017	O	26	179	10	69	18	124	13	90	-	22	45
	T4, T451	61	421	37	255	37	255	-	-	22	-	-
2018	T61	61	421	46	317	39	269	17	117	-	12	120
2024	O	27	186	11	76	18	124	13	90	20	22	47
	T4, T351	68	469	47	324	41	283	20	138	20	19	120
2025	T6	58	400	37	255	35	241	18	124	-	19	110
2117	T4	43	296	24	165	28	193	14	97	-	27	70
2218	T72	48	331	37	255	30	207	-	-	-	11	95
2219	O	25	172	11	76	-	-	-	-	18	-	-
	T31, T351	52	359	36	248	-	-	-	-	17	-	-
	T37	57	393	46	317	-	-	-	-	11	-	-
	T62	60	414	42	290	-	-	15	103	10	-	-

(Continued)

Table 4. *(Continued)* **Mechanical Properties of Wrought Aluminum Alloys.** *(Source, Aluminum Association, Inc.)*

Alloy	Temper	Ultimate Tensile Strength		Tensile Yield Strength		Ultimate Shearing Strength		Fatigue Endurance Limit		Elongation %[1]		Brinell Hardness
										0.0625" Thick	0.50" Dia.	
		ksi	MPa	ksi	MPa	ksi	MPa	ksi	MPa			BHN
2219	T81, T851	66	455	51	352	-	-	15	103	10	-	-
	T87	69	476	57	393	-	-	15	103	10	-	-
3003	O	16	110	6	41	11	76	7	48	30	40	28
	H12	19	131	18	124	12	83	8	55	10	20	35
	H18	29	200	27	186	16	110	19	131	4	10	55
3004	O	26	179	10	69	16	110	14	97	20	25	45
	H32	31	214	25	172	17	117	15	103	10	17	52
	H34	35	241	29	200	18	124	15	103	9	12	63
	H36	38	262	33	228	20	138	16	110	5	9	70
	H38	41	283	36	248	21	145	16	110	5	6	77
4032	T6	55	379	46	317	38	262	16	110	-	9	120
5005	O	18	124	6	41	11	76	-	-	25	-	28
	H12	20	138	19	131	14	97	-	-	10	-	-
	H14	23	159	22	152	14	97	-	-	6	-	-
	H16	26	179	25	172	15	103	-	-	5	-	-
	H18	29	200	28	193	16	110	-	-	4	-	-
	H32	20	138	17	117	14	97	-	-	11	-	36
	H34	23	159	20	138	14	97	-	-	8	-	41
	H36	26	179	24	165	15	103	-	-	6	-	46
	H38	29	200	27	186	16	110	-	-	5	-	51
5050	O	21	145	8	55	15	103	12	83	24	-	36

(Continued)

Table 4. *(Continued)* **Mechanical Properties of Wrought Aluminum Alloys.** *(Source, Aluminum Association, Inc.)*

Alloy	Temper	Ultimate Tensile Strength		Tensile Yield Strength		Ultimate Shearing Strength		Fatigue Endurance Limit		Elongation %[1]		Brinell Hardness
										0.0625" Thick	0.50" Dia.	
		ksi	MPa	ksi	MPa	ksi	MPa	ksi	MPa			BHN
5050	H32	25	172	21	145	17	117	13	90	9	-	46
	H34	28	193	24	165	18	124	13	90	8	-	53
	H36	30	207	26	179	19	131	14	97	7	-	58
	H38	32	221	29	200	20	138	14	97	6	-	63
5052	O	28	193	13	90	18	124	16	110	25	30	47
	H32	33	228	28	193	20	138	17	117	12	18	60
	H34	38	262	31	214	21	145	18	124	10	14	68
	H36	40	276	35	241	23	159	19	131	8	10	73
	H38	42	290	37	255	24	165	20	138	7	8	77
5056	O	42	290	22	152	26	179	20	138	-	35	65
	H18	63	434	59	407	34	234	22	152	-	10	105
5083	O	42	290	21	145	25	172	-	-	-	22	-
5086	O	38	262	17	117	23	159	-	-	22	-	-
	H32, H116	42	290	12	83	-	-	-	-	12	-	-
	H34	47	324	37	255	27	186	-	-	10	-	-
	H112	39	269	19	131	-	-	-	-	14	-	-
5154	O	35	241	17	117	22	152	17	117	27	-	58
	H32	39	269	30	207	22	152	18	124	15	-	67
	H34	42	290	33	228	24	165	19	131	13	-	73
	H36	45	310	36	248	26	179	20	138	12	-	78
	H38	48	331	39	269	28	193	21	145	10	-	80

(Continued)

Table 4. *(Continued)* **Mechanical Properties of Wrought Aluminum Alloys.** *(Source, Aluminum Association, Inc.)*

Alloy	Temper	Ultimate Tensile Strength		Tensile Yield Strength		Ultimate Shearing Strength		Fatigue Endurance Limit		Elongation %[1]		Brinell Hardness
										0.0625" Thick	0.50" Dia.	
		ksi	MPa	ksi	MPa	ksi	MPa	ksi	MPa			BHN
5154	H112	35	241	17	117	-	-	17	117	25	-	63
5252	H25	34	234	25	172	21	145	-	-	11	-	68
	H38, H28	41	283	35	241	23	159	-	-	5	-	75
5254	O	35	241	17	117	22	152	17	117	27	-	58
	H32	39	269	30	207	22	152	18	124	15	-	67
	H34	42	290	33	228	24	165	19	131	13	-	73
	H36	45	310	36	248	26	179	20	138	12	-	78
	H38	48	331	39	269	28	193	21	145	10	-	80
	H112	35	241	17	117	-	-	17	117	25	-	63
5454	O	36	248	17	117	23	159	-	-	22	-	62
	H34	44	303	35	241	26	179	-	-	10	-	81
5456	O	45	310	23	159	-	-	-	-	-	24	-
5457	O	19	131	7	48	12	83	-	-	22	-	32
	H25	26	179	23	159	16	110	-	-	12	-	48
	H38, H28	30	207	27	186	18	124	-	-	6	-	55
5657	H25	23	159	20	138	14	97	-	-	12	-	40
	H38, H28	28	193	24	165	15	103	-	-	7	-	50
6061	O	18	124	8	55	12	83	9	62	25	30	30
	T4, T451	35	241	21	145	24	165	14	97	22	25	65
6063	O	13	90	7	48	10	69	8	55	-	-	25
	T4	25	172	13	90	-	-	-	-	22	-	-

(Continued)

Table 4. *(Continued)* **Mechanical Properties of Wrought Aluminum Alloys.** *(Source, Aluminum Association, Inc.)*

Alloy	Temper	Ultimate Tensile Strength		Tensile Yield Strength		Ultimate Shearing Strength		Fatigue Endurance Limit		Elongation %[1]		Brinell Hardness
										0.0625" Thick	0.50" Dia.	
		ksi	MPa	ksi	MPa	ksi	MPa	ksi	MPa			BHN
6063	T6	35	241	31	214	22	152	10	69	12	-	73
6066	O	22	152	12	83	14	97	-	-	-	18	43
	T6, T651	57	393	52	359	34	234	16	110	-	12	120
6101	T6	32	221	28	193	20	138	-	-	15	-	71
7049	T73	75	517	65	448	44	303	-	-	-	12	135
7075	T6, T651	83	572	73	503	48	331	23	159	11	11	150
7178	T6, T651	88	607	78	538	-	-	-	-	10	11	-

[1] Elongation is the percent elongation in two inch (50.8 mm) specimens—one $^{1}/_{16}$" thick (1.6 mm), the other $^{1}/_{2}$" thick (12.7 mm).

Table 5. Physical Properties of Wrought Aluminum Alloys. *(Source, MIL-HDBK-694A.)*

Alloy	Temper	Density lb/cu in.	Thermal Conductivity		Coef. Thermal Expan.		Electrical Conductivity[3]	Melting Point °F/°C	
			CGS[1]	Eng.[2]	μin °F	μm °C		Solidus	Liquidus[4]
1060	O	0.098	0.56	1625	13.1	23.6	62	1195/645	1215/655
	H18	0.098	0.53	1540			61	1195/645	1215/655
1100	O	0.098	0.53	1540	13.1	23.6	59	1190/643	1215/655
	H18	0.098	0.52	1510			57	1190/643	1215/655
2011	T3	0.102	0.34	990	12.7	22.9	39	1005/540	1190(d)/643(d)
2014	O	0.101	0.46	1340	12.8	23.0	50	945/507	1180(e)/638(e)
	T4	0.101	0.29	840			34	945/507	1180(e)/638(e)
	T6	0.101	0.37	1070			40	945/507	1180(e)/638(e)
2017	O	0.099	0.41	1190	13.2	23.6	50	955/513	1185(e)/640(e)
	T4	0.099	0.29	840			34	955/513	1185(e)/640(e)
2018	T61	0.101	0.37	1070	12.4	22.3	40	945/507	1180(d)/638(d)
2024	O	0.100	0.45	1310	12.9	23.2	50	935/500	1180(e)/638(e)
	T4	0.100	0.29	840			30	935/500	1180(e)/638(e)
2025	T6	0.101	0.37	1070	12.6	22.7	40	970/520	1185(e)/640(e)
2117	T4	0.099	0.37	1070	13.2	23.75	40	1030/555	1200(d)/650(d)
2218	T72	0.102	0.37	1070	12.4	22.3	40	940/505	1175(e)/635(e)
2219	O	0.103	0.41	1190	12.4	22.3	44	1010/543	1190(e)/643(e)
	T31, T37	0.103	0.27	780			28	1010/543	1190(e)/643(e)
	T62, T81, T87	0.103	0.30	870			30	1010/543	1190(e)/643(e)
3003	O	0.099	0.46	1340	12.9	23.2	50	1190/643	1210/655
	H12	0.099	0.39	1130			42	1190/643	1210/655
	H18	0.099	0.37	1070			41	1190/643	1210/655
3004	A11	0.098	0.39	1130	13.3	23.9	42	1165/630	1210/655

(Continued)

Table 5. *(Continued)* **Physical Properties of Wrought Aluminum Alloys.** *(Source, MIL-HDBK-694A.)*

Alloy	Temper	Density lb/cu in.	Thermal Conductivity		Coef. Thermal Expan.		Electrical Conductivity[3]	Melting Point °F/°C	
			CGS[1]	Eng.[2]	μin °F	μm °C		Solidus	Liquidus[4]
4032	T6	0.097	0.33	960	10.8	19.4	40	990/532	1060(e)/570(e)
5005	A11	0.098	0.48	1390	13.2	23.75	52	1170/632	1210/655
5050	A11	0.097	0.46	1340	13.2	23.75	50	1155/625	1205/650
5052	A11	0.097	0.33	960	13.2	23.75	35	1125/607	1200/650
5056	O	0.095	0.28	810	13.4	24.1	29	1055/568	1180/638
	H38	0.095	0.26	750			27	1055/568	1180/638
5083	O	0.096	0.28	810	13.2	23.75	29	1095/590	1180/638
5086	A11	0.096	0.30	870	13.2	23.75	31	1085/585	1185/640
5154	A11	0.096	0.30	870	13.3	23.9	32	1100/593	1190/643
5252	A11	0.097	0.33	960	13.2	23.75	35	1125/607	1200/650
5254	A11	0.096	0.30	870	13.3	23.9	32	1100/593	1190/643
5357	A11	0.098	0.40	1160	13.2	-	-	-	-
5454	O	0.097	0.32	930	13.1	23.6	34	1115/600	1195/645
	H38	0.097	0.32	930			34	1115/600	1195/645
5456	O	0.096	0.28	810	13.3	23.9	29	1055/568	1180/638
5457	A11	0.098	0.45	1310	13.1	23.75	46	1165/630	1210/655
5657	A11	0.098	0.33	960	13.1	23.75	54	1180/638	1215/657
6061	O	0.098	0.41	1190	13.1	23.6	47	1080/580	1205(d)/650(d)
	T4	0.098	0.37	1070			40	1080/580	1205(d)/650(d)
6063	T6	0.098	0.48	1390	13.0	23.4	58	1140/615	1210/655
	T42	0.098	0.46	1340			50	1140/615	1210/655
6066	O	0.098	0.37	1070	12.9	23.2	40	1045/565	1195(e)/645(e)
	T6	0.098	0.35	1010			37	1045/565	1195(e)/645(e)

(Continued)

Table 5. *(Continued)* **Physical Properties of Wrought Aluminum Alloys.** *(Source, MIL-HDBK-694A.)*

Alloy	Temper	Density lb/cu in.	Thermal Conductivity		Coef. Thermal Expan.		Electrical Conductivity[3]	Melting Point °F/°C	
			CGS[1]	Eng.[2]	μin °F	μm °C		Solidus	Liquidus[4]
6101	T6	0.098	0.52	1510	13.0	23.4	57	1150/620	1210/655
	T61	0.098	0.53	1540			59	1150/620	1210/655
	T62	0.098	0.52	1510			58	1150/620	1210/655
	T64	0.098	0.54	1570			60	1150/620	1210/655
7001	T6	0.102	0.29	840	13.0	-	-	-	-
7072	O	0.098	0.53	1540	13.1	23.6	59	1185/640	1215/655
7075	T6	0.101	0.29	840	13.1	23.6	3	890/475	1175(g)/635(g)
7178	T6	0.102	0.30	870	13.0	23.4	31	890/475	1165(g)/630(g)
7079	T6	0.099	0.29	840	13.1	-	-	-	-

Notes: [1] CGS = cal/cm/cm2/°C/sec at 25° C (77° F).
[2] Eng. = English Units = btu/in./ft^2/°F/hour at 77° F.
[3] At 68° F (20° C). International Annealed Copper Standard (IACS). Coefficient of Thermal Expansion is based on change in length per °F or °C, from 68 to 212° F (20 to 100° C).
[4] Letter symbols represent points on phase equilibrium diagram for two component systems.

Table 6. Chemical Composition Limits of Aluminum Casting Alloys. *(Source, Aluminum Association, Inc.)*

Alloy	Use*	Si	Cu	Zn	Fe	Mg	Ni	Mn	Ti	Other (specified)	Others	
											Each	Total
201.0	S	0.10	4.0-5.2	-	0.15	0.15-0.55	-	0.20-0.50	0.15-0.35	-	0.05	0.10
204.0	S&P	0.20	4.2-5.0	0.10	0.35	0.15-0.35	0.05	0.10	0.15-0.30	-	0.05	0.15
208.0	S&P	2.5-3.5	3.5-4.5	1.0	1.2	0.10	0.35	0.50	0.25	-	-	0.50
222.0	S&P	2.0	9.2-10.7	0.8	1.5	0.15-0.35	0.50	0.50	0.25	-	-	0.35
242.0	S&P	0.7	3.5-4.5	0.35	1.0	1.2-1.8	1.7-2.3	0.35	0.25	0.25 Cr	0.05	0.15
295.0	S	0.7-1.5	4.0-5.0	0.35	1.0	0.03	-	0.35	0.25	-	0.05	0.15
296.0	P	2.0-3.0	4.0-5.0	0.50	1.2	0.05	0.35	0.35	0.25	-	-	0.35
308.0	P	5.0-6.0	4.0-5.0	1.0	1.0	0.10	-	0.50	0.25	-	-	0.50
319.0	S&P	5.5-6.5	3.0-4.0	1.0	1.0	0.10	0.35	0.50	0.25	-	-	0.50
328.0	S	7.5-8.5	1.0-2.0	1.5	1.0	0.20-0.6	0.25	0.20-0.6	0.25	0.35 Cr	-	0.50
332.0	P	8.5-10.5	2.0-4.0	1.0	1.2	0.50-1.5	0.50	0.50	0.25	-	-	0.50
333.0	P	8.0-10.0	3.0-4.0	1.0	1.0	0.05-0.50	0.50	0.50	0.25	-	-	0.50
336.0	P	11.0-13.0	0.50-1.5	0.35	1.2	0.7-1.3	2.0-3.0	0.35	0.25	-	0.05	-
354.0	S&P	8.6-9.4	1.6-2.0	0.10	0.20	0.40-0.6	-	0.10	0.20	-	0.05	0.15
355.0	S&P	4.5-5.5	1.0-1.5	0.35	0.6	0.40-0.6	-	0.50	0.25	0.25 Cr	0.05	0.15
C355.0	S&P	4.5-5.5	1.0-1.5	0.10	0.20	0.40-0.6	-	0.10	0.20	-	0.05	0.15
356.0	S&P	6.5-7.5	0.25	0.35	0.6	0.20-0.45	-	0.35	0.25	-	0.05	0.15
A356.0	S&P	6.5-7.5	0.20	0.10	0.20	0.25-0.45	-	0.10	0.20	-	0.05	0.15
357.0	S&P	6.5-7.5	0.05	0.05	0.15	0.45-0.6	-	0.03	0.20	-	0.05	0.15
A357.0	S&P	6.5-7.5	0.20	0.10	0.20	0.40-0.7	-	0.10	0.04-0.20	-	0.05	0.15
359.0	S&P	8.5-9.5	0.20	0.10	0.20	0.50-0.7	-	0.10	0.20	-	0.05	0.15
360.0	D	9.0-10.0	0.6	0.50	2.0	0.40-0.6	0.50	0.35	-	0.15 Sn	-	0.25
A360.0	D	9.0-10.0	0.6	0.50	1.3	0.40-0.6	0.50	0.35	-	0.15 Sn	-	0.25

(Continued)

Table 6. *(Continued)* **Chemical Composition Limits of Aluminum Casting Alloys.** *(Source, Aluminum Association, Inc.)*

Alloy	Use*	Si	Cu	Zn	Fe	Mg	Ni	Mn	Ti	Other (specified)	Others	
											Each	Total
380.0	D	7.5-9.5	3.0-4.0	3.0	2.0	0.10	0.50	0.50	-	0.35 Sn	-	0.50
A380.0	D	7.5-9.5	3.0-4.0	3.0	1.3	0.10	0.50	0.50	-	0.35 Sn	-	0.50
383.0	D	9.5-11.5	2.0-3.0	3.0	1.3	0.10	0.30	0.50	-	0.15 Sn	-	0.50
384.0	D	10.5-12.0	3.0-4.5	3.0	1.3	0.10	0.50	0.50	-	0.35 Sn	-	0.50
390.0	D	16.0-18.0	4.0-5.0	0.10	1.3	0.45-0.65	-	0.10	0.20	-	-	0.20
B390.0	D	16.0-18.0	4.0-5.0	1.5	1.3	0.45-0.65	0.10	0.50	0.10	-	-	0.20
392.0	D	18.0-20.0	0.40-0.80	0.50	1.5	0.80-1.20	0.50	0.20-0.60	0.20	0.30 Sn	-	0.50
413.0	D	11.0-13.0	1.0	0.50	2.0	0.10	0.50	0.35	-	0.15 Sn	-	0.25
A413.0	D	11.0-13.0	1.0	0.50	1.3	0.10	0.50	0.35	-	0.15 Sn	-	0.25
C433.0	D	4.5-6.0	0.6	0.50	2.0	0.10	0.50	0.35	-	0.15 Sn	-	0.25
443.0	S&P	4.5-6.0	0.6	0.50	0.8	0.05	-	0.50	0.25	0.25 Cr	-	0.35
B443.0	S&P	4.5-6.0	0.15	0.35	0.8	0.05	-	0.35	0.25	-	0.05	0.15
A444.0	P	6.5-7.5	0.10	0.10	0.20	0.05	-	0.10	0.20	-	0.05	0.15
512.0	S	1.4-2.2	0.35	0.35	0.6	3.5-4.5	-	0.8	0.25	0.25 Cr	0.05	0.15
513.0	P	0.30	0.10	1.4-2.2	0.40	3.5-4.5	-	0.30	0.20	-	0.05	0.15
514.0	S	0.35	0.15	0.15	0.50	3.5-4.5	-	0.35	0.25	-	0.05	0.15
518.0	D	0.35	0.25	0.15	1.8	7.5-8.5	0.15	0.35	-	0.15 Sn	-	0.25
520.0	S	0.25	0.25	0.15	0.30	9.5-10.6	-	0.15	0.25	-	0.05	0.15
535.0	S&P	0.15	0.05	-	0.15	6.2-7.5	-	0.10-0.25	0.10-0.25	-	0.05	0.15
705.0	S&P	0.20	0.20	2.7-3.3	0.8	1.4-1.8	-	0.40-0.6	0.25	0.20-0.40 Cr	0.05	0.15
707.0	S&P	0.20	0.20	4.0-4.5	0.8	1.8-2.4	-	0.40-0.6	0.25	0.20-0.40 Cr	0.05	0.15
710.0	S	0.15	0.35-0.65	6.0-7.0	0.50	0.6-0.8	-	0.05	0.25	-	0.05	0.15
711.0	P	0.30	0.35-0.65	6.0-7.0	0.7-1.4	0.25-0.45	-	0.05	0.20	-	0.05	0.15

(Continued)

Table 6. *(Continued)* **Chemical Composition Limits of Aluminum Casting Alloys.** *(Source, Aluminum Association, Inc.)*

Alloy	Use*	Si	Cu	Zn	Fe	Mg	Ni	Mn	Ti	Other (specified)	Others	
											Each	Total
712.0	S	0.30	0.25	5.0-6.5	0.50	0.50-0.65	-	0.10	0.15-0.25	0.40-0.6 Cr	0.05	0.20
713.0	S&P	0.25	0.40-1.0	7.0-8.0	1.1	0.20-0.50	0.15	0.6	0.25	0.35 Cr	0.10	0.25
771.0	S	0.15	0.10	6.5-7.5	0.15	0.8-1.0	-	0.10	0.10-0.20	0.06-0.20 Cr	0.05	0.15
850.0	S&P	0.7	0.7-1.3	-	0.7	0.10	0.7-1.3	0.10	0.20	-	-	0.30
851.0	S&P	2.0-3.0	0.7-1.3	-	0.7	0.10	0.30-0.7	0.10	0.20	-	-	0.30
852.0	S&P	0.40	1.7-2.3	-	0.7	0.6-0.9	0.9-1.5	0.10	0.20	-	-	0.30

* S = sand casting, P = permanent mold casting, D = die casting.

Table 7. Mechanical Properties for Aluminum Sand Casting Alloys.
(Source, Aluminum Association, Inc. and QQ-A-601F.)

Alloy	Temper	Ultimate Tensile Strength		Yield Tensile Strength		% Elongation in Two Inches, or 4 × Dia.	Hardness BHN
		ksi	MPa	ksi	MPa		
201.0	T7	60	414	50	345	3.0	110-140
204.0	T4	45	310	28	193	6.0	-
208.0	F	19	131	12	83	1.5	40-70
208.0	T55	21	145	-	-	-	75
222.0	0	23	159	20	138	-	65-95
222.0	T61	30	207	40	276	-	100-130
242.0	0	23	159	18	124	-	55-85
242.0	T571	29	200	30	207	-	70-100
242.0	T61	32	221	20	138	-	90-120
242.0	T77	24	165	13	90	1.0	60-90
295.0	T4	29	200	13	90	6.0	45-75
295.0	T6	32	221	20	138	3.0	60-90
295.0	T62	36	248	28	193	-	80-110
295.0	T7	29	200	16	110	3.0	55-85
319.0	F	23	159	13	90	1.5	55-85
319.0	T5	25	172	26	179-	-	65-95
319.0	T6	31	214	20	138	1.5	65-95
328.0	F	25	172	14	97	1.0	45-75
328.0	T6	34	234	21	145	1.0	65-95
355.0	T51	25	172	18	124	-	50-80
355.0	T6	32	221	20	138	2.0	70-105
355.0	T7	35	241	36	248	-	70-100
355.0	T71	30	207	22	152	-	60-95
C355.0	T6	36	248	25	172	2.5	75-105
356.0	F	19	131	-	-	2.0	40-70
356.0	T51	23	159	16	110	-	45-75
356.0	T6	30	207	20	138	3.0	55-90
356.0	T7	31	214	29	200	-	60-90
356.0	T71	25	172	18	124	3.0	45-75
A356.0	T6	34	234	24	165	3.5	70-105
443.0	F	17	117	7	49	3.0	25-55
B433.0	F	17	117	6	41	3.0	25-55
512.0	F	17	117	10	69	-	35-65
514.0	F	22	152	9	62	6.0	35-65
520.0	T4	42	290	22	152	12.0	60-90
535.0	0	35	241	18	124	9.0	-
535.0	F or T5	35	241	18	124	9.0	60-90
705.0	F or T5	30	207	17	117	5.0	50-80

(Continued)

Table 7. *(Continued)* **Mechanical Properties for Aluminum Sand Casting Alloys.**
(Source, Aluminum Association, Inc. and QQ-A-601F.)

Alloy	Temper	Ultimate Tensile Strength		Yield Tensile Strength		% Elongation in Two Inches, or 4 × Dia.	Hardness BHN
		ksi	MPa	ksi	MPa		
707.0	T5	33	228	22	152	2.0	70-100
707.0	T7	37	255	30	207	1.0	65-95
710.0	F or T5	32	221	20	138	2.0	60-90
712.0	F or T5	34	234	25	172	4.0	60-90
713.0	F or T5	32	221	22	152	3.0	60-90
771.0	0	36	248	27	186	1.5	-
771.0	T5	42	290	38	262	1.5	85-115
771.0	T51	32	221	27	186	3.0	70-100
771.0	T52	36	248	30	207	1.5	70-100
771.0	T53	36	248	27	186	1.5	-
771.0	T6	42	290	35	241	5.0	75-105
771.0	T71	48	331	45	310	2.0	105-135
850.0	T5	16	110	11	76	5.0	30-60
851.0	T5	17	117	11	76	3.0	30-60
852.0	T5	24	165	18	124	-	45-75

Table 8. Mechanical Properties for Aluminum Permanent Mold Casting Alloys.
(Source, Aluminum Association, Inc. and QQ-A-596E.)

Alloy	Temper	Ultimate Tensile Strength		Yield Tensile Strength		% Elongation in Two Inches, or 4 × Dia.	Hardness BHN
		ksi	MPa	ksi	MPa		
204.0	T4	48	331	29	200	8.0	-
208.0	T4	33	228	15	103	4.5	60-90
208.0	T6	35	241	22	152	2.0	75-105
208.0	T7	33	228	16	110	3.0	65-95
213.0	F	23	159	-	-	-	-
222.0	T551	30	207	-	-	-	100-130
222.0	T65	40	276	-	-	-	125-155
242.0	T571	34	234	-	-	-	90-120
242.0	T61	40	276	-	-	-	95-125
296.0	T4	33	228	-	-	4.5	-
296.0	T6	35	241	-	-	2.0	75-105
296.0	T7	33	228	-	-	3.0	-
308.0	F	24	165	-	-	-	55-85
319.0	F	28	193	14	97	1.5	70-100
319.0	T6	34	234	-	-	2.0	75-105
332.0	T5	31	214	-	-	-	90-120
333.0	F	28	193	-	-	-	65-100
333.0	T5	30	207	-	-	-	70-105

(Continued)

Table 8. *(Continued)* **Mechanical Properties for Aluminum Permanent Mold Casting Alloys.** *(Source, Aluminum Association, Inc. and QQ-A-596E.)*

Alloy	Temper	Ultimate Tensile Strength		Yield Tensile Strength		% Elongation in Two Inches, or 4 × Dia.	Hardness BHN
		ksi	MPa	ksi	MPa		
333.0	T6	35	241	-	-	-	85-115
333.0	T7	31	214	-	-	-	75-105
336.0	T551	31	214	-	-	-	90-120
336.0	T65	40	276	-	-	-	110-140
354.0	T61	48	331	37	255	3.0	-
354.0	T62	52	359	42	290	2.0	-
355.0	T51	27	186	-	-	-	60-90
355.0	T6	37	255	-	-	1.5	75-105
355.0	T62	42	290	-	-	-	90-120
355.0	T7	36	248	-	-	-	70-100
355.0	T71	34	234	27	186	-	65-95
C355.0	T61	40	276	30	207	3.0	75-105
356.0	F	21	145	-	-	3.0	40-70
356.0	T51	25	172	-	-	-	55-85
356.0	T6	33	228	22	152	3.0	65-95
356.0	T7	25	172	-	-	3.0	60-90
356.0	T71	25	172	-	-	3.0	60-90
A356.0	T61	37	255	26	179	5.0	70-100
357.0	T6	45	310	-	-	3.0	75-105
A357.0	T61	45	310	36	248	3.0	85-115
359.0	T61	45.0	310	34	234	4.0	75-105
359.0	T62	47	324	38	262	3.0	85-115
443.0	F	21	145	7	49	2.0	30-60
B443.0	F	21	145	6	41	2.5	30-60
A444.0	T4	20	138	-	-	20.0	-
513.0	F	22	152	12	83	2.5	45-75
535.0	F	35	241	18	124	8.0	60-90
705.0	T5	37	255	17	117	10.0	55-85
707.0	T5	42	290	-	-	4.0	-
707.0	T7	45	310	35	241	3.0	80-110
711.0	T1	28	193	18	124	7.0	55-85
713.0	T5	32	221	22	152	4.0	60-90
850.0	T5	18	124	-	-	8.0	30-60
851.0	T5	17	117	-	-	3.0	30-60
851.0	T6	18	124	-	-	8.0	-
852.0	T5	27	186	-	-	3.0	55-85

Note: The values on this Table represent properties obtained from separately cast test bars. Average properties of specimens cut from castings shall not be less than 75% of tensile and yield strength values and shall not be less than 25% of elongation values given above.

Table 9. Physical Properties of Aluminum Casting Alloys.

Alloy	Temper	Use[5]	Density lb/cu in.	Thermal Conductivity		Coef. Thermal Expan.[4]		Electrical Conductivity[3]	Melting Range (Approx.)	
				CGS[1]	Eng.[2]	µin °F	µm °C		°F	°C
201.0	T7	P	0.101	0.29	840	19.3	34.7	32-34	1060-1200	570-650
208.0	F	S	0.101	0.29	840	12.2	22.0	31	970-1170	520-630
222.0	O	S	0.107	0.38	1100	-	-	41	970-1160	520-625
	T61	S	0.107	0.31	900	12.3	22.1	33	970-1160	520-625
242.0	O	S	0.102	0.40	1160	-	-	44	990-1180	530-635
	T77	S	0.102	0.36	1040	12.3	22.1	38	980-1180	525-635
	T571	P	0.102	0.32	930	12.5	22.5	34	980-1180	525-635
	T61	P	0.102	0.32	930	12.5	22.5	33	980-1180	525-635
295.0	T4	S	0.102	0.33	960	12.7	22.9	35	970-1190	520-645
	T62	S	0.102	0.34	990	12.7	22.9	35	970-1190	520-645
296.0	T4	P	0.101	0.32	930	12.2	22.0	33	970-1170	520-630
	T6	P	0.101	0.32	930	12.2	22.0	33	970-1170	520-630
308.0	F	P	0.101	0.34	990	11.9	21.4	37	970-1140	520-615
319.0	F	S	0.101	0.27	780	12.0	21.6	27	970-1120	520-605
	F	P	0.101	0.28	810	12.0	21.6	28	970-1120	520-605
332.0	T5	P	0.100	0.25	720	11.5	20.7	26	970-1080	520-580
333.0	F	P	0.100	0.25	720	11.5	20.7	26	970-1090	520-585
	T5	P	0.100	0.29	840	11.5	20.7	29	970-1090	520-585
	T6	P	0.100	0.28	810	11.5	20.7	29	970-1090	520-585
	T7	P	0.100	0.34	990	11.5	20.7	35	970-1090	520-585
355.0	T51	S	0.098	0.40	1160	12.4	22.3	43	1020-1150	550-620
	T6	S	0.098	0.34	990	12.4	22.3	36	1020-1150	550-620
	T7	S	0.098	0.39	1130	12.4	22.3	42	1020-1150	550-620
	T6	P	0.098	0.36	1040	12.4	22.3	39	1020-1150	550-620

(Continued)

Table 9. *(Continued)* **Physical Properties of Aluminum Casting Alloys.**

Alloy	Temper	Use[5]	Density lb/cu in.	Thermal Conductivity		Coef. Thermal Expan.[4]		Electrical Conductivity[3]	Melting Range (Approx.)	
				CGS[1]	Eng.[2]	µin °F	µm °C		°F	°C
C355.0	T61	S	0.098	0.35	1010	12.4	22.3	39	1020-1150	550-620
356.0	T51	S	0.097	0.40	1160	11.9	21.4	43	1040-1140	560-615
	T6	S	0.097	0.36	1040	11.9	21.4	39	1040-1140	560-615
	T7	S	0.097	0.37	1070	11.9	21.4	40	1040-1140	560-615
A356.0	T6	S	0.098	0.36	1040	11.9	21.4	40	1040-1130	560-610
357.0	T6	S	0.098	0.36	1040	11.9	21.4	39	1040-1140	560-615
443.0	F	S	0.097	0.35	1010	12.3	22.1	37	1070-1170	575-630
512.0	F	S	0.096	0.35	1010	12.7	22.9	38	1090-1170	590-630
513.0	F	P	0.097	0.32	930	13.3	23.9	34	1080-1180	580-640
514.0	F	S	0.096	0.33	960	13.3	23.9	35	1110-1180	600-640
520.0	T4	S	0.093	0.21	600	14.0	25.2	21	840-1110	450-600
535.0	F	S	0.091	0.24	690	13.1	23.6	23	1020-1170	550-630
705.0	F	S	0.100	0.25	720	13.1	23.6	25	1110-1180	600-640
710.0	F	S	0.102	0.33	960	13.4	24.1	35	1110-1200	600-650
712.0	F	S	0.102	0.38	1100	13.1	23.6	40	1110-1180	600-640
713.0	F	S	0.104	0.37	1070	13.3	23.9	37	1100-1170	595-630
850.0	T5	S	0.103	0.44	1280	-	-	47	440-1200	225-650
851.0	T5	S	0.102	0.40	1160	12.6	22.7	43	450-1170	230-630
852.0	T5	S	0.104	0.42	1210	12.9	23.2	45	410-1180	210-635

Notes: [1] CGS = cal/cm/cm^2/°C/sec at 25° C (77° F).
[2] Eng. = English Units = btu/in./ft^2/°F/hour at 77° F.
[3] At 68° F (20° C). International Annealed Copper Standard (IACS).
[4] Coefficient of Thermal Expansion is based on change in length per °F or °C, from 68 to 212° F (20 to 100° C).
[5] S = sand casting, P = permanent mold casting, D = die casting.

3xxx Series – Al-Mn Alloys. These medium strength alloys are strain hardenable, have excellent corrosion resistance, and are readily welded, brazed, or soldered. Because of its superiority in handling many foods and chemicals, alloy 3003 is widely used in cooking utensils as well as in builder's hardware. Alloy 3105 is widely used for roofing and siding, and variations of the 3xxx series are used in sheet and tubular form for heat exchangers in vehicles and power plants. Because of its use in the bodies of beverage cans, alloy 3004 (and its modification 3104) is among the most widely used of all aluminum alloys. Typically, 3xxx alloys have a tensile strength range of 16,000 to 41,000 PSI (110 to 283 MPa).

4xxx Series – Al-Si Alloys. These are medium high strength heat treatable alloys that are widely used in complex shaped forgings. Alloy 4032 is used principally for forgings such as aircraft pistons. 4043, on the other hand, is one of the most widely used filler alloys for gas metal arc and gas tungsten arc welding of 6xxx series alloys for use in structural and automotive applications. These alloys have good flow characteristics as a result of their high silicon content which, in the case of forgings, ensures the filling of complex dies, and in welding allows complete filling of crevices and seams in welded joints. Typically, 4xxx alloys have a tensile strength range of 25,000 to 55,000 PSI (172 to 379 MPa).

5xxx Series – Al-Mg Alloys. These strain hardenable alloys have moderately high strength, excellent corrosion resistance (even in salt water), and very high toughness even at cryogenic temperatures to near absolute zero. They can be welded with a variety of techniques, even at thicknesses up to 7 $^{7}/_{8}$ inches (20 cm), which has contributed to their popularity in bridge construction and other building applications, marine use, and for use in storage tanks and pressure vessels where temperatures as low as –270° F (–168° C) may be encountered. Alloys 5052, 5086, and 5083 are usually chosen for structural use. Specialty alloys include 5182, used for beverage can ends, 5754 which is used for automotive panels and frames, and 5252, 5454, and 5657 which are all used for bright trim applications. In this series, strength increases with higher magnesium content, but care must be taken to avoid the use of 5xxx alloys with more than 3% Mg content (such as 5454 and 5754) in applications where they receive continuous exposure to temperatures above 212° F (100° C) where they become sensitized and susceptible to stress corrosion cracking. Typically, 5xxx alloys have a tensile strength of 18,000 to 51,000 PSI (124 to 352 MPa).

6xxx Series – Al-Mg-Si Alloys. The alloys in this series are heat treatable, have moderately high strength, excellent extrudability properties, and excellent corrosion resistance. Because they are easily welded and extruded, they are the first choice for architectural and structural members where particular strength or stiffness criticality is important. Alloy 6061 is often used in welded structures such as vehicle frames, railroad cars, and pipelines. 6063 is widely used in extruded form for bridge construction and automobile frames. Specialty alloys in the series include 6066-T6 (used for high strength forgings), 6111 (used for automotive body panels), and 6101 and 6201, which are used for high strength electrical bus and electrical conductor wire, respectively. Typically, 6xxx alloys have a tensile strength range of 18,000 to 58,000 PSI (124 to 400 MPa).

7xxx Series – Al-Zn Alloys. These are, among the Al-Zn-Mg-Cu aluminum alloy versions, the highest strength of all aluminum alloys. They are heat treatable. Alloys 7150 and 7475, especially, are noted for their combination of strength and fracture toughness. Historically, the widest application of the 7xxx alloys has been in the aircraft industry where fracture critical design concepts have led to the development of high-toughness alloys. Although this series is not weldable by routine commercial processes, it is regularly used in riveted construction. Because their atmospheric corrosion resistance is not as high as the 5xxx or 6xxx series, 7xxx series alloys are usually coated or, for sheet and plate, used in an alclad version. Special tempers such as the T73 type are required whenever stress corrosion cracking may be a problem. Typically, 7xxx alloys have a tensile strength range of 32,000 to 88,000 PSI (220 to 607 MPa).

8xxx Series – Alloys with Al + other elements not covered by other series. These alloys contain less frequently used alloying elements such as lead, nickel, and lithium. They are heat treatable and have high conductivity, strength, and hardness. Applications include electrical conductors, bearings, and the aerospace industry. Typically, 8xxx alloys have a tensile strength in the range of 17,000 to 35,000 PSI (117 to 241 MPa).

Characteristics of Casting Alloys by Series Designation

The casting aluminum alloys contain larger percentages of alloying agents such as silicon and copper than are found in the wrought alloys. Their elongation and strength (especially fatigue) properties are relatively lower than the wrought products, mainly because current casting practice is unable to reliably prevent casting defects, even though recent innovations such as squeeze casting and thixocasting have resulted in considerable improvements.

2xx.x Series – Al-Cu Alloys. Both sand and permanent mold castings are produced from this series, which is heat treatable. These alloys possess high strength at room and elevated temperatures, and many of the metals in this group are high toughness alloys. Heat treated alloy 201.0 is the strongest of the casting alloys. However, since its castability is somewhat limited by a tendency to microporosity and hot tearing, it is best suited to investment casting. 201.0 is commonly used in machine tool construction and aircraft construction. Typically, 2xx.x alloys have a tensile strength range of 19,000 to 65,000 PSI (131 to 448 MPa).

3xx.x Series – Al-Si+Cu or Mg Alloys. These are among the most widely used casting alloys. Their high silicon content contributes to fluidity, and their heat treatability provides high strength options. They may be cast by sand and permanent mold techniques (319.0 and 356.0/A356.0), die casting (360.0, 380.0/A380.0, and 390.0) and other methods such as squeeze and forge casing (357.0/A357.0). Typically, 3xx.x alloys have a tensile strength range of 19,000 to 40,000 PSI (131 to 276 MPa).

4xx.x Series – Al-Si Alloys. Although not heat treatable, this series features very good castability and excellent weldability. They have a low melting point (570° F [299° C]), moderate strength, good corrosion resistance, and high elongation before rupture. This series is widely used when intricate, thin walled, leak-proof, fatigue resistant castings are required. 4xx.x alloys have excellent fluidity and can be used in sand, permanent mold, and die casting operations. Typically, 4xx.x alloys have a tensile strength in the range of 17,000 to 25,000 PSI (117 to 172 MPa).

5xx.x Series – Al-Mg alloys. This series is relatively tough to cast, but offers good finishing characteristics and machinability, and excellent corrosion resistance—they are suitable for salt water and other similar corrosive environments. The 512.0 and 514.0 is commonly used for door and window frames where they can be anodized or color coated. Other alloys in this series are used for cooking utensils and aircraft fittings. These alloys are not heat treatable and are sand, permanent mold, and die castable. Typically, 5xx.x alloys have a tensile strength range of 17,000 to 25,000 PSI (117 to 172 MPa).

7xx.x Series – Al-Zn Alloys. Because these alloys are more difficult to cast, they tend to be used only when excellent finishing and machinability characteristics are important (such as in furniture, office machinery, etc.). They are heat treatable, and may be cast with both sand and permanent mold techniques. Typically, 7xx.x alloys have a tensile strength range of 10,000 to 55,000 PSI (69 to 379 MPa).

8xx.x Series – Al-Sn Alloys. These alloys are also relatively difficult to cast, but they are uniquely suited for bushings and have excellent machinability characteristics. They are heat treatable, and can cast with sand or permanent molds. Bearings and bushings are the most common applications for these alloys. Typically, 8xx.x alloys have a tensile strength in the range of 15,000 to 30,000 PSI (103 to 207 MPa).

Temper designation system for aluminum alloys

Temper designations indicate mechanical or thermal treatment of the alloy. It follows the alloy number and is always preceded by a dash, as in 2014–T6. Basic designations consist of letters. Subdivisions of the basic designations, where required, are indicated by one or more digits following the letter. These designate specific sequences of basic treatments, but only operations recognized as significantly influencing the characteristics of the product are indicated. Should some other variation of the same sequence of basic operations be applied to the same alloy, resulting in different characteristics, additional digits are added to the designation.

The basic temper designations and subdivisions are as follows.

–F *As Fabricated*. Applies to products that acquire some temper from shaping processes not having special control over the amount of strain hardening or thermal treatment. For wrought products, there are no mechanical property limits.

–O *Annealed*. Applies to wrought products that are annealed to obtain the lowest strength temper, and to cast products that are annealed to improve ductility and dimensional stability. The O may be followed by a digit other than zero, indicating a product in the annealed condition that has specific characteristics. It should be noted that variations of the –O temper shall not apply to products that are strain hardened after annealing and in which the effect of strain hardening is recognized in the mechanical properties or other characteristics. The following temper designation has been assigned for wrought products that are high temperature annealed to accentuate ultrasonic response and provide dimensional stability.

- –O1 *Thermally Treated at Approximately the Same Time and Temperature Required for Solution Heat Treatment and Slow Cooled to Room Temperature*. Applicable to products that are to be machined prior to solution heat treatment by the user. Mechanical property limits are not applicable.

–H *Strain Hardened*. Wrought products only. Applies to products that have their strength increased by strain hardening with or without supplemental thermal treatments to produce partial softening. The –H is always followed by *two* or more digits. The first digit indicates the specific combination of basic operations as follows.

- –H1 *Strain Hardened Only*. Applies to products that are strain hardened to obtain the desired mechanical properties without supplementary thermal treatment. The number that follows this designation indicates the degree of strain hardening (see below).
- –H2 *Strain Hardened and then Partially Annealed*. Applies to products that are strain hardened more than the desired final amount and then reduced in strength to the desired level by partial annealing. For alloys that age-soften at room temperature, the –H2 tempers have approximately the same ultimate strength as the corresponding –H3 tempers. For other alloys, the –H2 tempers have approximately the same ultimate strength and slightly longer elongations as the corresponding –H1 tempers. The number following this designation indicates the degree of strain hardening

remaining after the product has been partially annealed (see below).

–H3 *Strain Hardened and then Stabilized.* Applies to products that are strain hardened and then stabilized by low temperature heating in order to slightly lower their strength and increase ductility. This designation applies only to the alloys that contain magnesium: unless stabilized gradually, they will age soften at room temperature. The number following this designation indicates the degree of strain hardening remaining after the product has been strain hardened a specified amount and then stabilized (see below).

–H4 *Strain Hardened and Lacquered or Painted.* Applied to products that are strain hardened and subsequently subjected to a thermal operation during lacquering or painting. The number following this designation indicates the degree of strain hardening remaining after the product has been thermally treated as part of the lacquering or painting cure operation (see below).

The digit following the –H1, –H2, –H3, and –H4 designation indicates the final degree of strain hardening. The hardest commercially practical temper is designated by the numeral 8 (full hard). Tempers between –O (annealed) and 8 (full hard) are designated by numerals 1 through 7. Materials having an ultimate strength about midway between that of the –O temper and that of the 8 temper are designated by the numeral 4 (half hard); between –O and 4 by the numeral 2 (quarter hard); between 4 and 8 by the numeral 6 (three-quarter hard); etc. Numeral 9 indicates extra hard tempers with tensile strength exceeding the –H*x*8 temper by 2 ksi (13.79 MPa) or more.

A third digit, when used, indicates that the degree of control of temper, or the mechanical properties, are different from, but within the range of, those for the two digit –H temper designation to which it is added. Numerals 1 through 9 may be arbitrarily assigned and registered with the Aluminum Association for an alloy and product to indicate a specific degree of control of temper or specific mechanical property limits. Zero has been assigned to indicate degrees of control of temper, or mechanical property limits negotiated between the manufacturer and purchaser that are not used widely enough to justify registration.

The following three-digit –H temper designations have been assigned for wrought products in all alloys.

–H*x*11 Applies to products that are strain hardened less than the amount required for a controlled –H*x*1 temper.

–H112 Applies to products that acquire some temper from shaping processes not having special control over the amount of strain hardening or thermal treatment, but for which there are mechanical property limits or mechanical property testing is required.

–H311 Applies to products that are strain hardened less than the amount required for a controlled –H31 temper.

The following three-digit –H temper designations have been assigned.

Patterned or Embossed Sheet	*Fabricated From*	*Patterned or Embossed Sheet*	*Fabricated From*
–H114	*–O*	*–H264*	*–H25*
–H124	*–H11*	*–H364*	*–H35*
–H224	*–H21*	*–H174*	*–H16*
–H324	*–H31*	*–H274*	*–H26*
–H134	*–H12*	*–H374*	*–H36*
–H234	*–H22*	*–H184*	*–H17*

(Continued)

Patterned or Embossed Sheet	*Fabricated From*	*Patterned or Embossed Sheet*	*Fabricated From*
–H334	*–H32*	*–H284*	*–H27*
–H144	*–H13*	*–H384*	*–H37*
–H244	*–H23*	*–H194*	*–H18*
–H344	*–H33*	*–H 294*	*–H28*
–H154	*–H14*	*–H394*	*–H38*
–H254	*–H24*	*–H195*	*–H19*
–H354	*–H34*	*–H295*	*–H29*
–H164	*–H15*	*–H395*	*–H39*

–W *Solution Heat Treated.* Applies to products whose strength naturally changes at room temperature after solution heat treatment. The change might take place over weeks, or years, and the designation is specific only when the period of natural aging is determined: for example,W 2.5 hours.

–T *Thermally Treated to Produce Tempers Other than – F, – O, or – H.* Applies to products that are thermally treated with or without supplementary strain hardening to produce stable tempers. The –T is always followed by one or more digits. Numerals 1 to 10 have been assigned to indicate specific sequences of basic treatment, as follows.

–T1 *Cooled from an Elevated Temperature Shaping Process and then Naturally Aged to a Substantially Stable Condition.* Applies to castings, extrusions, etc., that are not cold worked after an elevated temperature shaping process. Also applies to products that are cold worked by flattened or straightened after cooling, but any effects incurred by these processes are not accounted for in the property limits.

–T2 *Cooled from an Elevated Temperature Shaping Process and then Cold Worked and Naturally Aged to a Substantially Stable Condition.* Applies to products that are cold worked to improve strength after cooling from a hot working process. Also applies to products that are cold worked by flattening or straightening after cooling, and any effects incurred in these processes are accounted for in the property limits.

–T3 *Solution Heat Treated and then Cold Worked and Naturally Aged to a Substantially Stable Condition.* Applies to products that are cold worked to improve strength, or in which the effects of cold work in flattening or straightening is recognized in applicable specifications.

–T4 *Solution Heat Treated and Naturally Aged to a Substantially Stable Condition.* Applies to products that are not cold worked after solution heat treatment, but in which the effect of cold work in flattening or straightening may be recognized in applicable specifications.

–T5 *Cooled from an Elevated Temperature Shaping Process and then Artificially Aged.* Applies to products that are artificially aged after an elevated temperature rapid cool fabrication process, such as casting or extrusion, to improve mechanical properties and/or dimensional stability.

–T6 *Solution Heat Treated and then Artificially Aged.* Applies to products that are not cold worked after solution heat treatment, but in which the effect of cold work in flattening or straightening may be recognized in applicable specifications.

–T7 *Solution Heat Treated and then Stabilized or Overaged.* Applies to products that are stabilized to carry them beyond the point of maximum hardness, providing control of growth and/or residual stress. Cast products that are artificially aged after solution treatment to provide dimensional and strength stability are covered by this designation.

–T8 *Solution Heat Treated, Cold Worked, and then Artificially Aged.* Applies to products that are cold worked to improve strength, or in which the effect of cold work in flattening or straightening is recognized in applicable specifications.

–T9 *Solution Heat Treated, Artificially Aged, and then Cold Worked.* Applies to products that are cold worked specifically for the purpose of gaining strength.

–T10 *Cooled from an Elevated Temperature Shaping Process, Cold Worked, and then Artificially Aged.* Applies to products that are artificially aged after an elevated temperature rapid cool fabrication process, such as casting or extrusion, and then cold worked to improve strength.

A period of natural aging at room temperature may occur between or after the operations listed for tempers –T3 through –T10. Control of this period is exercised when it is metallurgically important. Additional digits may be added to designations –T1 through –T10 to indicate a variation in treatment that significantly alters the characteristics of the product. These may be assigned for an alloy and product to indicate a specific treatment or specific mechanical property limits. The following additional digits have been assigned for wrought products in all alloys.

–T*x*51 *Stress Relieving by Stretching.* Applies to products that are stress relieved by stretching the following amounts after solution heat treatment or after cooling from an elevated temperature shaping process. The products receive no further straightening after stretching.

Plate: 1.5 to 3% permanent set.

Rolled or cold finished rod, and bar: 1 to 3% permanent set.

Die or ring forgings and rolled rings: 1 to 5% permanent set.

Applies directly to plate and rolled or cold finished rod and bar. These products receive no further straightening after stretching. Applies to extruded rod, bar, and shapes when designated as follows.

–T*x*510 Applies to extruded rod, bar, shapes, and tube, and to drawn tube, when stretched the indicated amounts amounts after solution heat treatment or after cooling from an elevated temperature shaping process. These products receive no further straightening after stretching.

Extruded rod, bar, shapes, and tube: 1 to 3% permanent set.

Drawn tube: 0.50 to 3% permanent set.

–T*x*511 Applies to extruded rod, bar, shapes, and tube, and tube to be drawn, that receive minor straightening after stretching to comply with standard tolerances.

–T*x*52 *Stress Relieved by Compressing.* Applies to products that are stress relieved by compression by the following amounts after solution heat treatment.

Extrusions: 1 to 5% permanent set.

Forgings: 1 to 5% permanent set.

–T*x*53 *Stress Relieved by Thermal Treatment.*

–T*x*54 *Stress Relieved by Combined Stretching and Compression.* Applies to die forgings that are stress relieved by restriking cold in the finish die.

The following two-numeral temper designations have been assigned for wrought products in all alloys.

–T42 Applies to products solution heat treated and naturally aged by the user that attain mechanical properties different from the –T4 temper.

–T62 Applies to products solution heat treated and artificially aged by the user that attain mechanical properties different from the –T6 temper.

Unregistered Tempers. The letter P has been assigned to denote –H, –T, and –O temper variations that are negotiated between manufacturer and purchaser. The letter P immediately follows the temper designation that most nearly pertains.

Heat treating aluminum alloys

The heat treatment processes commonly used to improve the properties of aluminum alloys are solution heat treatment, precipitation (age) hardening, and annealing. Dimensional changes in aluminum alloys during thermal treatment are generally negligible, but in some cases the changes may have to be considered in manufacturing. Because of the many variables involved, there are no tabular values for these dimensional changes, but some guidelines can be followed. For instance, in the artificial aging of alloy 2219 from the "–T42," "–T351," and "–T37" tempers to the "–T62," "–T851," and "–T87" tempers, respectively, a net dimensional growth of 0.00010 to 0.0015 inch per inch (0.00254 to 0.0381 mm per 25.4 mm) may be anticipated. Additional growth of as much as 0.0010 in./in. (0.0254 mm/25.4 mm) may occur during subsequent service of a year or more at 300° F (149° C), or equivalent shorter exposure at higher temperatures. The dimensional changes that occur during the artificial aging of other wrought heat treatable alloys are less than one-half that for alloy 2219 under the same conditions.

Solution heat treatment. Solution heat treatment (–W temper) is used to redistribute the alloying constituents that segregate from the aluminum during cooling from the molten state. It consists of 1) heating the alloy to a temperature at which the soluble constituents will form a homogeneous mass by solid diffusion, 2) holding the mass at that temperature until diffusion takes place, and then 3) quenching the alloy rapidly to retain the homogeneous condition. **Table 10** shows suitable temperatures for solution heat treating wrought alloys and casting alloys. If the specified maximum temperature is exceeded, there is the danger of localized melting and lowering the mechanical properties of the alloy. Excessive overheating will cause severe blistering. If the temperature is below the minimum specified, solution will not be complete, resulting in underdevelopment of the product's physical properties and lowered corrosion resistance. After solution heat treating, the product is soaked the required time to bring about the necessary degree of solid solution. Soaking times vary with thickness, and **Tables 11** and **12** provide minimal soaking times for wrought alloys and casting alloys. Alclad products should be soaked at the minimal temperature necessary to develop the required mechanical properties. Longer soaking may allow the alloying constituents of the base metal to diffuse through the alclad coating. When this occurs, corrosion resistance is adversely affected.

Precipitation (or "age") hardening. In the quenched condition, heat treated alloys are supersaturated solid solutions that are comparatively soft and workable, but unstable, depending on composition. At room temperature, the alloying constituents of some alloys (–W temper) tend to precipitate from the solution spontaneously, causing the metal to harden in about four days. This is called "natural aging." It can be retarded or even arrested to facilitate fabrication by holding the alloy at subzero temperatures until ready for forming. Other alloys age more slowly at room temperature, and may take years to reach maximum strength and hardness. These alloys can be aged artificially (called "artificial aging" or "elevated precipitation heat treatment"), to stabilize them and improve their qualities, by heating them to moderately elevated temperatures for specified lengths of time. A small amount of cold working, after solution heat treatment, produces a substantial increase in yield strength, some increase in tensile strength, and some loss of ductility. The effect on the properties developed will vary with different compositions.

(Text continued on p. 168)

Table 10. Recommended Temperatures for Solution Heat Treating of Aluminum Alloys. *(Source, MIL-H-6088G.)*

Alloy	Products/ Limitations[1]	Metal Temperature[5]		Temper Designation		
		°F	°C	Immediately After Quenching[2]	After Natural Aging[3]	After Stress Relief[4]
Wrought Products (Excluding Forgings)						
2011	Wire, rod, bar	945-995	507-535	-W	-T3[6], -T4	-T451
2014	Flat sheet	925-945	496-507	-W	-T3[6], -T42	-
	Coiled sheet	925-945	496-507	-W	-T4, -T42	-
	Plate	925-945	496-507	-W	-T4, -T42	-T451
	Wire, rod, bar	925-945	496-507	-W	-T4	-T451
	Extrusions	925-945	496-507	-W	-T4, -T42	-T4510, -T4511
	Drawn tube	925-945	496-507	-W	-T4	-
2017	Wire, rod, bar	925-950	496-510	-W	-T4	-T451
	Rivets	925-950	496-510	-W	-T4	-
2024	Flat sheet	910-930	488-499	-W	-T3[6], -T361[6], -T42	-
	Coiled sheet	910-930	488-499	-W	-T4, -T42, -T3[6]	-
	Rivets	910-930	488-499	-W	-T4	-
	Plate	910-930	488-499	-W	-T4, -T42, -T361[6]	-T351
	Wire, rod, bar	910-930[7]	488-499	-W	-T4, -T36[6], -T42	-T351
	Extrusions	910-930	488-499	-W	-T3[6], -T42	-T3510, T3511
	Drawn tube	910-930	488-499	-W	-T3[6], -T42	-
2048	Sheet, plate	910-930	488-499	-W	-T4, -T42	-T351
2117	Wire, rod, bar	925-950	496-510	-W	-T4	-
	Rivets	890-950	477-510	-W	-T4	-
2124	Plate	910-930	488-499	-W	-T4[6], -T42	-T351
2219	Sheet	985-1005	529-540	-W	-T31[6], -T37[6], -T42	-
	Plate	985-1005	529-540	-W	-T31[6], -T37[6], -T42	-T351
	Rivets	985-1005	529-540	-W	-T4	-
	Wire, rod, bar	985-1005	529-540	-W	-T31[6], -T42	-T351
	Extrusions	985-1005	529-540	-W	-T31[6], -T42	-T3510, -T3511
6010	Sheet	1045-1065	563-574	-W	-T4	-
6013	Sheet	1045-1065	563-574	-W	-T4	-
6061	Sheet	960-1075[8]	516-579	-W	-T4, -T42	-
	Plate	960-1075	516-579	-W	-T4, -T42	-T451
	Wire, rod, bar	960-1075	516-579	-W	-T4, -T42	-T451
	Extrusions	960-1075	516-579	-W	-T4, -T42	-T4510, -T4511
	Drawn tube	960-1075	516-579	-W	-T4, -T42	-
6063	Extrusions	960-985	516-529	-W	-T4, -T42	-T4510, -T4511
	Drawn tube	960-980	516-527	-W	-T4, -T42	N/A
6066	Extrusions	960-1010	516-543	-W	-T4, -T42	-T4510, -T4511
	Drawn tube	960-1010	516-543	-W	-T4, -T42	-

(Continued)

Table 10. *(Continued)* **Recommended Temperatures for Solution Heat Treating of Aluminum Alloys.** *(Source, MIL-H-6088G.)*

Alloy	Products/ Limitations[1]	Metal Temperature[5]		Temper Designation		
		°F	°C	Immediately After Quenching[2]	After Natural Aging[3]	After Stress Relief[4]
Wrought Products (Excluding Forgings)						
6262	Wire, rod, bar	960-1050	516-543	-W	-T4	-T451
	Extrusions	960-1050	516-543	-W	-T4	-T4510, -T4511
	Drawn tube	960-1050	516-543	-W	-T4	-
6951	Sheet	975-995	524-535	-W	-T4, -T42	-
7001	Extrusions	860-880	460-471	-W	-	-W510[2], -W511[2]
7010	Plate	880-900	471-482	-W	-	-W51[2]
7039	Sheet	840-860[9]	449-460	-W	-	-
	Plate	840-860[9]	449-460	-W	-	-W51[2]
7049/ 7149	Extrusions	860-885	460-474	-W	-	-W510[2] -W511[2]
7050	Sheet	880-900	471-482	-W	-	-
	Plate	880-900	471-482	-W	-	-W51[2]
	Extrusions	880-900	471-482	-W	-	-W510[2] -W511[2]
	Wire, rod, rivets	880-900	471-482	-W	-	-
7075	Sheet	860-930[10]	460-499	-W	-	-
	Plate[11]	860-930	460-499	-W	-	-W51[2]
	Wire, rod, bar[11]	860-930	460-930	-W	-	-W51[2]
	Extrusions	860-880	460-471	-W	-	-W510[2] -W511[2]
	Drawn tube	860-880	460-471	-W	-	-
7150	Extrusions	880-900	471-482	-W	-	-W510[2] -W511[2]
	Plate	880-895	471-479	-W	-	-W51[2]
7178	Sheet[13]	860-930	460-499	-W	-	-
	Plate[13]	860-910	460-488	-W	-	-W51[2]
	Extrusions	860-880	460-471	-W	-	-W510[2] -W511[2]
7475	Sheet	880-970	471-521	-W	-	-
	Plate	880-970	471-521	-W	-	-
7475 Alclad	Sheet	880-945	471-507	-W	-	-
Forgings[14]						
2014	Die forgings	925-945	496-507	-W	-T4, -T41	-
	Hand forgings	925-945	496-507	-W	-T4, -T41	-T452
2018	Die forgings	940-970	504-521	-W	-T4, -T41	-

(Continued)

Table 10. *(Continued)* **Recommended Temperatures for Solution Heat Treating of Aluminum Alloys.** *(Source, MIL-H-6088G.)*

Alloy	Products/ Limitations[1]	Metal Temperature[5]		Temper Designation		
		°F	°C	Immediately After Quenching[2]	After Natural Aging[3]	After Stress Relief[4]
Forgings[14]						
2024	Die & hand forgings	910-930	488-499	-W	-T4	-T352
2025	Die forgings	950-970	510-521	-W	-T4	-
2218	Die forgings	940-960	504-516	-W	-T4, -T41	-
2219	Die & hand forgings	985-1005	529-540	-W	-T4	-T352
2618	Die & hand forgings	975-995	524-535	-W	-T4, -T41	-
4032	Die forgings	940-970	504-521	-W	-T4	-
6053	Die forgings	960-980	516-527	-W	-T4	-
6061	Die & hand forgings	960-1075	516-579	-W	-T4, -T41	-T452
	Rolled rings	960-1025	516-552	-W	-T4, -T41	-T452
6066	Die forgings	960-1010	516-543	-W	-T4	-
6151	Die forgings	950-980	510-527	-W	-T4	-
	Rolled rings	950-980	510-527	-W	-T4	-T452
7049/ 7149	Die & hand forgings	860-885	460-474	-W	-	-W52[2]
7050	Die & hand forgings	880-900	471-482	-W	-	-W52[2]
7075	Die & hand forgings	860-890[9]	460-477	-W	-	-W52[2]
	Rolled rings	860-890[9]	460-477	-W	-	-W52[2]
7076	Die & hand forgings	850-910	454-488	-W	-	-
7175	Die forgings	See Note 15		-W	-	-
	Hand forgings	See Note 15		-W	-	-
Castings (all mold practices)[16]						
A201.0[18]	-	945-965 followed by 970-995	496-518 521-535	-	-T4	-
A206.0 (206)[18]	-	945-965 followed by 970-995	496-518 521-535	-	-T4	-
222.0 (122)	-	930-960	499-516	-	-T4	-
242.0 (142)	-	950-980	510-527	-	-T4, -T41	-
295.0 (195)	-	940-970	504-521	-	-T4	-

(Continued)

Table 10. *(Continued)* **Recommended Temperatures for Solution Heat Treating of Aluminum Alloys.** *(Source, MIL-H-6088G.)*

Alloy	Products/ Limitations[1]	Metal Temperature[5]		Temper Designation		
		°F	°C	Immediately After Quenching[2]	After Natural Aging[3]	After Stress Relief[4]
Castings (all mold practices)[16]						
296.0 (B295.0)	-	935-965	502-518	-	-T4	-
319.0 (319)	-	920-950	493-510	-	-T4	-
328.0 (Red X-8)	-	950-970	510-521	-	-T4	-
333.0 (333)	-	930-950	499-510	-	-T4	-
336.0 (A332.0)	-	950-970	510-521	-	-T45	-
A336.0 (A332.0)	-	940-970	504-521	-	-T45	-
354.0 (354)	-	980-995	527-535	-	-T4	-
355.0 (355), C355.0	-	960-995	516-535	-	-T4	-
356.0 (356), A356.0 (A356)	-	980-1025[12]	527-552	- -	-T4 -T4	- -
357.0 (357), A357.0 (A357)	-	980-1025[12]	527-552	- -	-T4 -T4	- -
359.0 (359)	-	980-1010	527-543	-	-T4	-
520.0 (220)	-	800-820	427-438	-	-T4	-
705.0[17]	-	-		-	T1 T5	-
707.0[17]	-	-	-		T1	-
712.0[17]	-	990 -	532	-	T4 T1	-
713.0[17]	-	-		-	T1	-
850.0[17]	-	-		-	T1	-
851.0[17]	-	-		-	T1	-
852.0[17]	-	-		-	T1	-

[1] The term "wire, rod, and bar" as used herein refers to rolled or cold finished wire, rod, and bar. The term "extrusions" refers to extruded wire, rod, bar, shapes, and tube.
[2] This temper is unstable and generally not available.
[3] Applied only to those alloys that will naturally age to a substantially stable condition.

(Continued)

Table 10. *(Continued)* **Recommended Temperatures for Solution Heat Treating of Aluminum Alloys.** *(Source, MIL-H-6088G.)*

[4] For rolled or extruded products, metal is stress relieved by stretching after quenching; and for forgings, metal is stress relieved by stretching or compression after quenching.

[5] When a difference between the maximum and minimum temperatures of a range listed herein exceeds 20° F (11° C), any 20° F temperature range (or 30° F [17° C] range for 6061) within the entire range may be utilized, provided that no exclusions or qualifying criteria are cited herein or in the applicable material specification.

[6] Cold working subsequent to solution heat treatment and prior to any precipitation heat treatment is necessary.

[7] Temperatures as low as 900° F may be used, provided that every heat treat lot is tested to show that the requirements of the applicable material specification are met, and analysis of test data to show statistic conformance to the specification limits is available for review.

[8] Maximum temperature for alclad 6061 sheet should not exceed 1000° F (538° C).

[9] Other temperatures may be necessary for certain sections, conditions, and requirements.

[10] It must be recognized that under some conditions melting can occur when heating 7075 alloy above 900° F (482° C) and that caution should be exercised to avoid this problem. In order to minimize diffusion between the cladding and the core, alclad 7075 sheet in thicknesses of 0.020 inch (0.508 mm) or less may be solution heat-treated at 850° F to 930° F (454 to 499° C).

[11] For plate thicknesses over 4 inches and for rod diameters or bar thicknesses over 4 inches (101.6 mm), a maximum temperature of 910° (488° C) F is recommended to avoid melting.

[12] Heat treatment above 1010° F (593° C) may require an intermediate solution heat treatment of one hour at 1000-1010° F (538-543° C) to prevent eutectic melting of magnesium rich phases.

[13] Under some conditions melting can occur when heating this alloy above 900° F (482° C).

[14] Unless otherwise indicated, hand forgings include rolled rings, and die forgings include rolled rings, and die forgings include impacts.

[15] Heat-treating procedures are at present proprietary among producers. At least one such procedure is patented (U.S. Patent Number 3,791,876).

[16] Former commercial designation is shown in parentheses.

[17] Unless otherwise specified, solution heat treatment is not required. Castings should be quickly cooled after shake-out or stripping from molds, so as to obtain a fine tin distribution.

[18] In general, product should be soaked for two hours in the range 910-930° F (488-499° C) prior to heating into the solution heat-treating range. Other presolution heat-treating temperature ranges may be necessary for some configurations and sizes.

Table 11. Recommended Soaking Time for Solution Heat Treatment of Wrought Products. *(Source, MIL-H-6088G.)*

Thickness[2]		Soaking Time (Minutes)[1]			
		Salt Bath[3]		Air Furnance[4]	
Inches	Millimeters	Min.	Max. (Alclad)[5]	Min.	Max. (Alclad)[5]
0.016 and under	0.4064 and under	10	15	20	25
0.017 to 0.020 incl.	0.4381 to 0.5080 incl.	10	20	20	30
0.021 to 0.032 incl.	0.5334 to 0.8128 incl.	15	25	25	35
0.033 to 0.063 incl.	0.8382 to 1.6002 incl.	20	30	30	40
0.064 to 0.090 incl.	1.6256 to 2.2860 incl.	25	35	35	45
0.091 to 0.124 incl.	2.3114 to 3.1496 incl.	30	40	40	50
0.125 to 0.250 incl.	3.1750 to 6.3500 incl.	35	45	50	60
0.251 to 0.500 incl.	6.3754 to 12.700 incl.	45	55	60	70
0.501 to 1.000 incl.	12.7254 to 25.400 incl.	60	70	90	100
1.001 to 1.500 incl.	25.425 to 38.100 incl.	90	100	120	130
1.501 to 2.000 incl.	38.125 to 50.800 incl.	105	115	150	160
2.001 to 2.500 incl.	50.825 to 63.500 incl.	120	130	180	190

(Continued)

Table 11. *(Continued)* **Recommended Soaking Time for Solution Heat Treatment of Wrought Products.** *(Source, MIL-H-6088G.)*

Thickness[2]		Soaking Time (Minutes)[1]			
		Salt Bath[3]		Air Furnance[4]	
Inches	Millimeters	Min.	Max. (Alclad)[5]	Min.	Max. (Alclad)[5]
2.501 to 3.000 incl.	63.525 to 76.200 incl.	135	160	210	220
3.001 to 3.500 incl.	76.225 to 88.900 incl.	150	175	240	250
3.501 to 4.000 incl.	88.925 to 101.600 incl.	165	190	270	280

[1] Longer soaking times may be necessary for specific forgings. Shorter soaking times are satisfactory when the soak time is accurately determined by thermocouples attached to the load or when other metal temperature-measuring devices are used.
[2] The thickness is the minimum dimension of the heaviest section.
[3] Soaking time in salt-bath furnaces begins at time of immersion, except when, owing to a heavy charge, the temperature of the bath drops below the specified minimum; in such cases, soaking time begins when the bath reaches the specified minimum.
[4] Soaking time in air furnaces begins when all furnace control instruments indicate recovery to the minimum of the process range.
[5] For alclad metals, the maximum recovery time (time between charging furnace and recovery of furnace instruments) should not exceed 30 minutes for thicknesses up to 0.050 inch (12.700 mm), 60 minutes for 0.050 or greater but less than 0.102 inch (2.591 mm), and 120 minutes for 0.102 or greater.

Table 12. Recommended Soaking Time for Solution Heat Treatment of Cast Alloys. *(Source, MIL-H-6088G.)*

Alloy	Soaking Time (Hours)
A201.0 and A206.0	2 at 910-930° F (488-409° C) followed by 2-8 at 945-965° F (340-518° C) followed by 8-24 at 970-995° F (521-535° C)
222.0	6 to 18 incl.
242.0	2 to 10 incl.
295.0	6 to 18 incl.
296.0	4 to 12 incl.
319.0	6 to 18 incl.
328.0	12
336.0, A336.0	8 hr. then water quench to 150-212° F (66-100° C)
354.0	10 to 12 incl.
355.0 and C355.0	6 to 24 incl.
356.0 and A356.0	6 to 24 incl.
357.0 and A357.0	8 to 24 incl.
359.0	10 to 14 incl.
520.0	18

Table 13 shows recommended precipitation hardening treatments for wrought products, forgings, and castings.

Annealing. Annealing is used to effect recrystallization, essentially complete precipitation, or to remove internal stresses (annealing for the purpose of obliterating the hardening effects of cold working will remove the effects of heat treating). For most alloys annealing consists of heating to approximately 650° F (343° C) at a controlled rate. The rate is dependent on such factors as thickness, type of anneal desired, and method employed. Cooling rate is unimportant, but drastic quenching is not recommended because of the strains produced. **Table 14** lists recommended annealing conditions for work hardened wrought aluminum alloys. In order to avoid excessive oxidation and grain growth, annealing temperatures should not exceed 775° F (412° C). Soaking castings for two hours at 650 to 750° F (343 to 393° C) and then cooling them for two hours will relieve them of residual stress and provide dimensional stability.

Salt baths versus air chamber furnaces

Salt baths. The time required to bring the product to temperature is shortened, and uniform temperature is more easily maintained in salt baths than in air chamber furnaces. Also, when solution heat treating in molten salt, the danger of generating porosity is greatly diminished. After prolonged use, however, there is some decomposition of the sodium nitrate to form compounds that, when dissolved in the quenching water, attack the aluminum alloys. The addition to the salt bath of about 0.50 ounce of sodium or potassium dichromate per 100 pounds of nitrate tends to inhibit this attack. It should be noted that nitrate salt baths can present an explosive hazard when heat treating 5xx.x series casting alloys.

Air chamber furnaces. Air chamber furnaces are more flexible and economical for handling large volumes of work. When solution heat treating certain aluminum alloys, it is necessary to control the atmosphere in order to avoid the generation of porosity. Such porosity lowers the mechanical properties of aluminum alloys and may be manifested as large numbers of minute blisters over the surface of the product. In severe cases, the product may crack when it is quenched. Furnace products of combustion contain water vapor and may contain gaseous compounds of sulfur, both of which tend to cause porosity during solution heat treatment. For this reason, furnaces that permit their products of combustion to come into contact with the load are not recommended for the solution heat treatment of alloys that may become porous during such treatment. Anodic oxide films or the metal coating on alclad products protect the underlying material from this effect. To some degree, certain fluoroborates will also protect against or minimize porosity.

Effects of quenching

Quenching is the sudden chilling of the metal in oil or water. It increases the strength and corrosion resistance of the alloy and "freezes" the structure and distribution of the alloying constituents that existed at the temperature just prior to cooling. The properties of the alloy are governed by its composition and characteristics, its thickness of cross section, and the rate at which it is cooled. This rate is controlled by proper choice of both type and temperature of cooling medium. Rapid quenching, as in cold water, will provide maximum corrosion resistance and is used for items made from sheet, tube, extrusions, and small forgings. Rapid quenching is preferred over a less dramatic quench that would increase the mechanical properties. A slower quench, done in boiling or hot water, is used for heavy sections and large forgings. It tends to minimize distortion and cracking that can result from uneven cooling. The corrosion resistance of forging alloys is not affected by the temperature of the quench water. Also, corrosion resistance of thicker sections is generally less critical than that of thinner sections. **Table 15** provides maximum time delays, from furnace to immersion, for various product thicknesses.

Formability and machinability of aluminum and aluminum alloys

Aluminum alloys are readily formed hot or cold by common fabrication processes. Generally, pure aluminum is more easily worked than the alloys, and annealed tempers

(Text continued on p. 180)

Table 13. Recommended Precipitation Hardening Heat Treating Condition. *(Source, MIL-H-6088G.)*

Alloy	Temper Before Aging	Limitations	Age Hardening Heat Treatment[1] Metal Temperature[4] °F	°C	Aging Time [2, 13]	Temper Designation After Indicated Treatment
Wrought Products (Excluding Forgings)						
2011	-W	-	Room Temp.	Room Temp.	96 (minimum)	-T4, -T42
	-T3	-	310-330	154-166	14	-T8
	-T4	-	-	-	-	-
	-T451	-	-	-	-	-
2014	-W	-	Room Temp.	Room Temp.	96 (minimum)	-T4, -T42
	-T3	Flat sheet	310-330	154-166	18	-T6
	-T4, -T42 [3]	-	340-360	171-182	10	-T6, -T62
	-T451 [3]	-	340-360	171-182	10	-T651
	-T4510	Extrusions	340-360	171-182	10	-T6510
	-T4511	Extrusions	340-360	171-182	10	-T6511
2017	-W	-	Room Temp.	Room Temp.	96 (minimum)	-T4
	-T4	-	-	-	-	-
	-T451	-	-	-	-	-
2024	-W	-	Room Temp.	Room Temp.	96 (minimum)	-T4, -T42
	-T3	Sheet and drawn tube	365-385	185-196	12	-T81
	-T4	Wire, rod, bar	365-385	185-196	12	-T6
	-T3	Extrusions	365-385	185-196	12	-T81
	-T36	Wire	365-385	185-196	8	-T86
	-T42	Sheet and plate	365-385	185-196	9	-T62
	-T42	Sheet only	365-385	185-196	16	-T72
	-T42	Other than sheet and plate	365-385	185-196	16	-T62
	-T351	Sheet and plate	365-385	185-196	12	-T851
	-T361		365-385	185-196	8	-T861
	-T3510	Extrusions	365-385	185-196	12	-T8510
	-T3511		365-385	185-196	12	-T8511
2048	-W	-	Room Temp.	Room Temp.	96 (minimum)	-T4, -T42
	-T42	Sheet and plate	365-385	185-196	9	-T62
	-T351	-	365-385	185-196	12	-T851
2117	-W	Wire, rod, bar and rivets	Room Temp.	Room Temp.	96 (minimum)	-T4
2124	-W	Plate	Room Temp.	Room Temp.	96 (minimum)	-T4, -T42
	-T4		365-385	185-196	9	-T6
	-T42		365-385	185-196	9	-T62
	-T351		365-385	185-196	12	-T851
2219	-W	-	Room Temp.	Room Temp.	96 (minimum)	-T4, -T42
	-T31	Sheet	340-360	171-182	18	-T81

(Continued)

Table 13. *(Continued)* **Recommended Precipitation Hardening Heat Treating Condition.** *(Source, MIL-H-6088G.)*

Alloy	Temper Before Aging	Limitations	Age Hardening Heat Treatment[1]			Temper Designation After Indicated Treatment
			Metal Temperature[4]		Aging Time [2, 13]	
			°F	°C		
Wrought Products (Excluding Forgings)						
2219	-T31	Extrusions	365-385	185-196	18	-T81
	-T31	Rivets	340-360	171-182	18	-T81
	-T37	Sheet	315-335	157-168	24	-T87
	-T37	Plate	340-360	171-182	18	-T87
	-T42	-	365-385	185-196	36	-T62
	-T351	-	340-360	171-182	18	-T851
	-T351	Rod and bar	365-385	185-196	18	-T851
	-T3510	Extrusions	365-385	185-196	18	-T8510
	-T3511		365-385	185-196	18	-T8511
6010	-W	Sheet	340-360	171-182	8	-T6
6013	-W	Sheet	Room Temp.	Room Temp.	336	-T4
	-T4 [22]	-	365-385	185-196	4	-T6
6061	-W	-	Room Temp.	Room Temp.	96 (minimum)	-T4, -T42
	-T1	Rods, bar, shapes, and tube, extruded	340-360	171-182	8	-T5
	-T4 [14]	Except extrusions	310-330	154-166	18	-T6
	-T451		310-330	154-166	18	-T651
	-T42		310-330	154-166	18	-T62
	-T4	Extrusions	340-360	171-182	8	-T6
	-T42		340-360	171-182	8	-T62
	-T4510		340-360	171-182	8	-T6510
	-T4511		340-360	171-182	8	-T6511
6063	-W	Extrusions	Room Temp.	Room Temp.	96 (minimum)	-T4, -T42
	-T1	-	350-370	177-188	3	-T5, -T52
	-T1	-	415-435	213-224	1-2	-T5, -T52
	-T4	-	340-360	171-182	8	-T6
	-T4	-	350-370	166-166	6	-T6
	-T42	-	340-360	171-182	8	-T62
	-T42	-	350-370	154-166	6	-T62
	-T4510	-	340-360	171-182	8	-T6510
	-T4511	-	340-360	171-182	8	-T6511
6066	-W	Extrusions	Room Temp.	Room Temp.	96 (minimum)	-T4, -T42
	-T4	-	340-360	171-182	8	-T6
	-T42	-	340-360	171-182	8	-T62
	-T4510	-	340-360	171-182	8	-T6510
	-T4511	-	340-360	171-182	8	-T6511
6262	-W	-	Room Temp.	Room Temp.	96 (minimum)	-T4

(Continued)

Table 13. *(Continued)* **Recommended Precipitation Hardening Heat Treating Condition.** *(Source, MIL-H-6088G.)*

Alloy	Temper Before Aging	Limitations	Age Hardening Heat Treatment[1]			Temper Designation After Indicated Treatment
			Metal Temperature[4]		Aging Time [2, 13]	
			°F	°C		
Wrought Products (Excluding Forgings)						
6262	-T4	Wire, rod, bar and drawn tube	330-350	166-177	8	-T6
	-T451	-	330-350	166-177	8	-T651
	-T4	Extrusions	340-360	171-182	12	-T6
	-T4510		340-360	171-182	12	-T6510
	-T4511		340-360	171-182	12	-T6511
6951	-W	-	Room Temp.	Room Temp.	96 (minimum)	-T4, -T42
	-T4	Sheet	310-330	154-166	18	-T6
	-T42	-	310-330	154-166	18	-T62
7001	-W	Extrusions	240-260	116-127	24	-T6
	-T510		240-260	116-127	24	-T6510
	-T511		240-260	116-127	24	-T6511
7010	-W51 [21]	Plate	240-260 plus 330-350	116-127 plus 166-177	6-24 6-15	- -T7651
			240-260 plus 330-350	116-127 plus 166-177	6-24 9-18	 -T7451 [17]
			240-260 plus 330-350	116-127 plus 166-177	6-24 15-24	 T7351
7039	-W [15]	Sheet	165-185 plus 310-330	74-85 plus 154-166	16 14	 -T61
	-W51 [15]	Plate	165-185 plus 310-330	74-85 plus 154-166	16 14	 -T64
7049, 7149	-W511	Extrusions	Room Temp. followed by 240-260 followed by 320-330	Room Temp. followed by 116-127 followed by 160-166	48 24 12-14	-T76510, -T76511
			Room Temp. followed by 240-260 followed by 325-335	Room Temp. followed by 116-127 followed by 163-168	48 24-25 12-21	-T73510, -T73511 [19]
7050	-W51 [8]	Plate	240-260 plus 315-335	116-127 plus 157-168	3-6 12-15	 -T7651
			240-260 plus 315-335	116-127 plus 157-168	3-6 24-30	 -T7451 [17]

(Continued)

Table 13. *(Continued)* **Recommended Precipitation Hardening Heat Treating Condition.** *(Source, MIL-H-6088G.)*

Alloy	Temper Before Aging	Limitations	Age Hardening Heat Treatment[1]			Temper Designation After Indicated Treatment
			Metal Temperature[4]		Aging Time [2,13]	
			°F	°C		
Wrought Products (Excluding Forgings)						
7050	-W510 [8]	Extrusions	240-260 plus 315-335	116-127 plus 157-168	3-8 15-18	-T76510
	-W511 [8]		240-260 plus 315-335	116-127 plus 157-168	3-8 15-18	-T76511
	-W [8]	Wire, rod, rivets	245-255 plus 350-360	118-124 plus 177-182	4 min. 8 min.	-T73
7075	-W [7]	-	240-260	116-127	24	-T6, -T62
	-W [5, 8, 11]	Sheet and plate	215-235 plus 315-335	101-113 plus 157-168	6-8 24-30	-T73
	-W [8, 11]		240-260 plus 315-335	116-127 plus 157-168	3-5 15-18	-T76
	-W [6, 8, 11]	Wire, rod, bar	215-235 plus 340-360	101-113 plus 171-182	6-8 8-10	-T73
	-W [5, 8, 11]	Extrusions	215-235 plus 340-360	101-113 plus 171-182	6-8 6-8	-T73
	-W [8, 11]		240-260 plus 310-330	116-127 plus 154-166	3-5 18-21	-T76
	-W51 [5, 8, 11]	Plate	215-235 plus 315-335	101-235 plus 157-168	6-8 24-30	-T7351
	-W51 [8, 11]		240-260 plus 315-335	116-127 plus 157-168	3-5 15-18	-T7651
	-W51 [10]	-	240-260	116-127	24	-T651
	-W51 [6, 8, 11]	Wire, rod, bar	215-235 plus 340-360	101-113 plus 171-182	6-8 8-10	-T7351
	-W510 [7]	Extrusions	240-260	116-127	24	-T6510
	-W511 [7]		240-260	116-127	24	-T6511
	-W50 [5, 8, 11]		215-235 plus 340-360	101-113 plus 171-182	6-8 6-8	-T73510
	-W511 [5, 8, 11]		215-235 plus 340-360	101-113 plus 171-182	6-8 6-8	-T73511

(Continued)

Table 13. *(Continued)* **Recommended Precipitation Hardening Heat Treating Condition.** *(Source, MIL-H-6088G.)*

Alloy	Temper Before Aging	Limitations	Age Hardening Heat Treatment[1]			Temper Designation After Indicated Treatment
			Metal Temperature[4]		Aging Time [2, 13]	
			°F	°C		
Wrought Products (Excluding Forgings)						
7075	-W510 [5, 8, 11]	Extrusions	240-260 plus 310-330	116-127 plus 154-166	3-5 18-21	-T76510
	-W511 [8, 11]		240-260 plus 310-330	116-127 plus 154-166	3-5 18-21	-T76511
	-T6 [8]	Sheet	315-335	157-168	24-30	-T73
	-T6 [8]	Wire, rod, bar	340-360	171-182	8-10	-T73
	-T6 [8]	Extrusions	340-360 310-330	171-182 154-166	6-8 18-21	-T73 -T76
	-T651 [8]	Plate	315-335 315-335	157-168 157-168	24-30 15-18	-T7351 -T7651
	-T651 [8]	Wire, rod, bar	340-360	171-182	8-10	-T7351
	-T6510 [8]	Extrusions	340-360 310-330	171-182 154-166	6-8 18-21	-T73510 -T76510
	-T6511 [8]		340-360 310-330	171-182 154-166	6-8 18-21	-T73511 -T76511
7150	-W510, -W511	Extrusions	240-260 plus 310-330	116-127 plus 154-166	8 4-6 [20]	-T6510, -T6511
	-W51	Plate	240-260 plus 300-320	116-127 plus 149-160	24 12	-T651
7178	-W	-	240-260	116-127	24	-T6, -T62
	-W [8, 11]	Sheet	240-260 plus 315-335	116-127 plus 157-168	3-5 15-18	-T76
	-W [8, 11]	Extrusions	240-260 plus 310-330	116-127 plus 154-166	3-5 18-21	-T76
	-W51	Plate	240-260	116-127	24	-T651
	-W51 [8, 11]		240-260 plus 315-335	116-127 plus 157-168	3-5 15-18	-T7651
	-W510	Extrusions	240-260	116-127	24	-T6510
	-W510 [8, 11]	-	240-260 plus 310-330	116-127 plus 154-166	3-5 18-21	-T76510
	-W511	Extrusions	240-260	116-127	24	-T6511
	-W511 [8, 11]		240-260 followed by 310-330	116-127 followed by 154-166	3-5 18-21	-T76511

(Continued)

Table 13. *(Continued)* **Recommended Precipitation Hardening Heat Treating Condition.** *(Source, MIL-H-6088G.)*

Alloy	Temper Before Aging	Limitations	Age Hardening Heat Treatment[1]			Temper Designation After Indicated Treatment
			Metal Temperature[4]		Aging Time [2, 13]	
			°F	°C		
Wrought Products (Excluding Forgings)						
7475	-W	Sheet	240-260 followed by 315-325	116-127 followed by 157-163	3 8-10	-T761
	-W51	Plate	240-260	116-127	24	-T651
7475 Alclad	-W	Sheet	250-315	121-157	3	-T61
Forgings						
2014	-W	-	Room Temp.	Room Temp.	96 (minimum)	-T4
	-T4	-	330-350	166-177	10	-T6
	-T41	-	340-360	171-182	5-14	-T61
	-T452	Hand forgings	330-350	166-177	10	-T652
2018	-W	Die forgings	Room Temp.	Room Temp.	96 (minimum)	-T4
	-T41	Die forgings	330-350	166-177	10	-T61
2024	-W	Die and hand forgings	Room Temp.	Room Temp.	96 (minimum)	-T4
	-W52	Hand forgings	Room Temp.	Room Temp.	96 (minimum)	-T352
	-T4	Die and hand forgings	365-385	171-182	12	-T6
	-T352	Hand forgings	365-385	171-182	12	-T852
2025	-W	Die forgings	Room Temp.	Room Temp.	96 (minimum)	-T4
	-T4	Die forgings	330-350	166-177	10	-T6
2218	-W	Die forgings	Room Temp.	Room Temp.	96 (minimum)	-T4, -T41
	-T4	Die forgings	330-350	166-177	10	-T61
	-T41	Die forgings	450-470	232-243	6	-T72
2219	-W	-	Room Temp.	Room Temp.	96 (minimum)	-T4
	-T4	-	365-385	171-182	26	-T6
	-T352	Hand forgings	340-360	171-182	18	-T852
2618	-W	-	Room Temp.	Room Temp.	96 (minimum)	-T4
	-T41	Die forgings	380-400	193-204	20	-T61
4032	-W	Die forgings	Room Temp.	Room Temp.	96 (minimum)	-T4
	-T4	Die forgings	330-350	166-177	10	-T6
6053	-W	Die forgings	Room Temp.	Room Temp.	96 (minimum)	-T4
	-T4	Die forgings	330-350	166-177	10	-T6
6061	-W	Die and hand forgings	Room Temp.	Room Temp.	96 (minimum)	-T4
	-T41	Die and hand forgings	340-360	171-182	8	-T61
	-T452	Rolled rings and hand forgings	340-360	171-182	8	-T652
6066	-W	Die forgings	Room Temp.	Room Temp.	96 (minimum)	-T4
	-T4	Die forgings	340-360	171-182	8	-T6

(Continued)

Table 13. *(Continued)* **Recommended Precipitation Hardening Heat Treating Condition.** *(Source, MIL-H-6088G.)*

Alloy	Temper Before Aging	Limitations	Age Hardening Heat Treatment[1]			Temper Designation After Indicated Treatment
			Metal Temperature[4]		Aging Time[2, 13]	
			°F	°C		
Forgings						
6151	-W	Die forgings	Room Temp.	Room Temp.	96 (minimum)	-T4
	-T4	Die forgings	330-350	166-177	10	-T6
	-T452	Rolled Rings	330-350	166-177	10	-T652
7049	-W -W52	Die and hand forgings	Room Temp. followed by 240-260 followed by 320-330	Room Temp. followed by 116-127 followed by 160-166	48 24 10-16	-T73 -T7352
7050	-W	Die forgings	240-260 plus 340-360	116-127 plus 171-182	3-6 6-12	-T74[16]
	-W52	Hand forgings	240-260 plus 340-360	116-127 plus 171-182	3-6 6-8	-T7452[18]
7075	-W	-	240-260	116-127	24	-T6
	-W[8, 11]	- - -	215-235 plus 340-360	101-113 plus 171-182	6-8 8-10	-T73
	-W52	Hand forgings	240-260	116-127	24	-T652
	-W52[8, 11]	- - -	215-235 plus 340-360	101-113 plus 171-182	6-8 6-8	-T7352
	-W51	Rolled Rings	215-235 plus 340-360	101-113 plus 171-182	6-8 6-8	-T7351
	-W	Die and hand forgings	215-235 plus 340-360	101-113 plus 171-182	6-8 6-8	-T74[16]
7076	-W	Die and hand forgings	265-285	129-140	14	-T6
7149	-W	Die and hand forgings	Room Temp.	Room Temp.	48	
	-W52	-	240-260 plus 320-340	116-127 plus 160-171	24 10-16	-T73, -T7352
7175	-W52	Hand forgings	240-260	116-127		
	-W	Die and hand forgings	215-235 plus 340-360	101-113 plus 171-182	6-8 6-8	-T74[16]
Castings (all mold practices)						
201.0	-T4	-	300-320	149-160	10-24	-T6
A201.0	-T4	-	360-380	177-193	5 (minimum)	-T7
A206.0	-T4	-	380-400	193-204	5 (minimum)	-T7
222.0	-F	-	330-350	166-177	16-22	-T551
	-T4	-	380-400	193-204	10-12	-T61
	-T4	-	330-350	166-177	7-9	-T65

(Continued)

Table 13. *(Continued)* **Recommended Precipitation Hardening Heat Treating Condition.** *(Source, MIL-H-6088G.)*

Alloy	Temper Before Aging	Limitations	Age Hardening Heat Treatment[1]			Temper Designation After Indicated Treatment
			Metal Temperature[4]		Aging Time [2, 13]	
			°F	°C		
Castings (all mold practices)						
242.0	-F	-	320-350	160-177	22-26	-T571
	-T41	-	400-450	204-232	1-3	-T61
295.0	-T4	-	300-320	149-160	12-20	-T62
296.0	-T4	-	300-320	149-160	1-8	-T6
	-T4	-	490-510	254-266	4-6	-T7
319.0	-T4	-	300-320	149-160	1-6	-T6
328.0	-T4	-	300-320	149-160	2-5	-T6
333.0	-F	-	390-410	199-210	7-9	-T5
	-T4	-	300-320	149-160	2-5	-T6
	-T4	-	490-510	254-266	4-6	-T7
336.0	-T45	-	300-350	149-177	14-18	-T65
354.0	-T41	-	300-320	149-160	10-12	-T61
	-T41	-	330-350	166-177	6-10	-T62
355.0 and C355.0	-F	-	430-450	221-232	7-9	-T51
	-T4	-	300-320	149-160	1-6	-T6
	-T4	-	300-320	149-160	10-12	-T61
	-T4	-	330-350	166-177	14-18	-T62
	-T4	-	430-450	221-232	3-5	-T7
	-T4	-	465-485	240-252	4-6	-T71
356.0 and A356.0	-T4	-	-	-	-	-
	-F	-	430-450	221-232	6-12	-T51
	-T4	-	300-320	149-160	1-6	-T6
	-T4	-	300-320	149-160	6-10	-T61
357.0 and A357.0	-T4	-	300-340	149-171	2-12	-T6
359.0	-T4	-	300-320	149-160	8-12	-T61
	-T41	-	330-350	166-177	6-10	-T62
520.0	-T4	-	300-320	149-160	20-12	-T61
	-T41	-	330-350	166-177	6-10	-T62
705.0	-W	-	200-220 or Room Temp.	93-104 or Room Temp.	10 21 days	-T5
707.0	-F	-	300-320 or Room Temp.	149-160 or Room Temp.	3-5 21 days	-T5
712.0	-F	-	300-320 or Room Temp.	149-160 or Room Temp.	9-11 21 days	-T5
	-F	-	Room Temp.	Room Temp.	96 (minimum)	-T1
713.0	-F	-	240-260 or Room Temp.	116-127 or Room Temp.	16 21 days	-T5

(Continued)

Table 13. *(Continued)* **Recommended Precipitation Hardening Heat Treating Condition.** *(Source, MIL-H-6088G.)*

Alloy	Temper Before Aging	Limitations	Age Hardening Heat Treatment[1]			Temper Designation After Indicated Treatment
			Metal Temperature[4]		Aging Time [2, 13]	
			°F	°C		
Castings (all mold practices)						
850.0	-F	-	420-440	216-226	7-9	-T5
851.0	-F	-	420-440	216-226	7-9	-T5
852.0	-F	-	420-440	216-226	7-9	-T5

[1] To produce the stress-relieved tempers, metal that has been solution heat-treated in accordance with -W temper, material must be stretched or compressed as required before aging. In instances where a multiple stage aging treatment is used, the metal may be, but need not be, removed from the furnace and cooled between aging steps.

[2] The time at temperature will depend on time required for load to reach temperature. The times shown are based on rapid heating with soaking time measured from the time the load reached the minimum temperature shown.

[3] Alternate treatment of 18 hours at 305°-330° F (151.6°-165.6° C) may be used for sheet and plate.

[4] When the interval of the specified temperature range exceeds 20° F (11.1- C), any 20° temperature range (or 30° [16.6° C] range for 6061) within the entire range may be utilized provided that no exclusions or qualifying criteria are cited herein or in the applicable material specification.

[5] Alternate treatment of 6 to 8 hours at 215° to 235° F (101.6° to 112.7° C) followed by a second stage of 14 to 18 hours at 325° to 345° F (162.7° to 173.8° C) may be used providing a heating-up rate of 25° F (13.8° C) per hour is used.

[6] Alternate treatment of l0 to 14 hours at 340° to 360° F (171.1° to 182.2° C) may be used providing a heating-up rate of 25° F (13.8° C) per hour is used.

[7] For extrusions an alternate three-stage treatment comprised of 5 hours at 200° to 220° F (93.3° to 104.4° C) followed by 4 hours at 240° to 260° F (115.6° to 126.7° C) followed by 4 hours at 290° to 310° F (143.3° to 154.4° C) may be used.

[8] The aging of aluminum alloys 7049, 7050, 7075 and 7178 from any temper to the -T7 type tempers requires closer control on aging practice variables such as time, temperature, heating-up rates, etc., for any given item. In addition to the above, when re-aging material in the -T6 temper series to the -T7 type temper series, the specific condition of the T6 temper material (such as its property level and other effects of processing variables) is extremely important and will affect the capability of the re-aged material to conform to the requirements specified for the applicable -T7 type tempers.

[9] Old or former commercial designation is shown in parentheses.

[10] For plate, an alternate treatment of 4 hours at 195°-215° F (90.6°-101.6° C) followed by a second stage of 8 hours at 305°-325° F (151.6°-162.7° C) may be used.

[11] With respect to -T73, -7351, -T73510, -T73511, -T7352, -T76, -T76510 and -T76511 tempers, a license has been granted to the public under U.S. Patent 3,198,676 and these times and temperatures are those generally recommended by the patent holder. Counterpart patents exist in several countries other than the United States. Licenses to operate under these counterpart patents should be obtained from the patent holder.

[12] A heating-up rate of 50°-75° F (27.7°-47.6° C) per hour is recommended.

[13] The 96 hour minimum aging time required for each alloy listed with temper designation -W is not necessary if artificial aging is to be employed to obtain tempers other than that derived from room temperature aging. (For example, natural aging—96 hours—to achieve the -T4 or -T42 temper for 2014 alloy is not necessary prior to artificial aging to obtain a -T6 or -T62 temper.)

[14] An alternate treatment comprised of 8 hours at 350° F (176.7° C) also may be used.

[15] A heating-up rate of 35° F (19.4° C) per hour from 135° F (57.2° C) is recommended.

[16] Formerly designated as -T736 temper.

[17] Formerly designated as -T73651 temper.

[18] Formerly designated as -T73652 temper.

[19] Longer times are to be used with section thicknesses less than 2 inches (50.5 mm).

[20] Soak time of 4 hours for extrusions with leg thickness less than 0.8 inch (20.3 mm) and 6 hours for extrusions having thicker legs.

[21] An alternative treatment is to omit the first stage and heat at a rate no greater than 36° F (20° C) per hour.

[22] Does not require the 14-day room temperature age.

Table 14. Recommended Annealing Conditions for Wrought Aluminum and Aluminum Alloys.[1]
(Source, MIL-H-6088G.)

-0 Temper is Obtained After These Annealing Conditions							
Alloy	Metal Temperature[3]		Time at Temperature	Alloy	Metal Temperature		Time at Temperature
	°F	°C			°F	°C	
1060	650	349	See Note 2	5086	650	349	See Note 2
1100	650	349	See Note 2	5154	650	349	See Note 2
1350	650	349	See Note 2	5254	650	349	See Note 2
2014	760	404	See Note 4, 2-3 hours	5454	650	349	See Note 2
2017	760	404	See Note 4, 2-3 hours	5456	650	349	See Note 2
2024	760	404	See Note 4, 2-3 hours	5457	650	349	See Note 2
2036	725	385	See Note 4, 2-3 hours	5652	650	349	See Note 2
2117	760	404	See Note 4, 2-3 hours	6005	760	404	See Note 4, 2-3 hours
2219	760	404	See Note 4, 2-3 hours	6013	775	413	See Note 4, 2-3 hours
3003	775	413	See Note 2	6053	760	404	See Note 4, 2-3 hours
3004	650	349	See Note 2	6061	760	404	See Note 4, 2-3 hours
3105	650	349	See Note 2	6063	760	404	See Note 4, 2-3 hours
5005	650	349	See Note 2	6066	760	404	See Note 4, 2-3 hours
5050	650	349	See Note 2	7001	760	404	See Note 5, 2-3 hours
5052	650	349	See Note 2	7075	760	404	See Note 5, 2-3 hours
5056	650	349	See Note 2	7175	760	404	See Note 5, 2-3 hours
5083	650	349	See Note 2	7178	760	404	See Note 5, 2-3 hours

1 This table should be used for information and guidance purposes only. It is desirable to test specimen before establishing final temperatures.
2 Time in furnace should be no longer than necessary to get center of load to the desired temperature, taking into consideration the thickness or diameter of metal. Rate of cooling is unimportant.
3 Metal temperature variation in the annealing furnace should not be greater than +10° F, -15° F (+5.5° C, -8.3° C).
4 This annealing removes the effects of the solution heat treatment. Cooling rate must be 50° F (28° C) per hour from annealing temperature to 500° F (260° C). The rate of subsequent cooling is unimportant.
5 This annealing removes the effects of the solution heat treatment by cooling at an uncontrolled rate in the air to 400° F (204° C) or less followed by a reheating to 450° F (232° C) for 4 hours and cooling at room atmosphere conditions.

Table 15. Maximum Quench Delay for Immersion Quenching. (See Note.) *(Source, MIL-H-6088G.)*

Nominal Thickness		Maximum Time (Seconds)[1]
Up to 0.016 inch incl.	Up to 0.4064 mm incl.	5
0.017 to 0.031 inch incl.	0.4318 to 0.7874 mm incl.	7
0.032 to 0.090 inch incl.	0.8128 to 2.2860 mm incl.	10
0.091 inch and over	2.3114 mm and over	15

Note: Quench delay time begins when the furnace door starts to open or when the first corner of the load emerges from a salt bath, and ends when the last corner of the load is immersed in the quenchant. With the exception of alloy 2219, the maximum quench delay times may be exceeded (for example, with extremely large loads or long lengths) if performance tests indicate that all portions of the load will be above 775° F (413° C) when quenched. For alloy 2219, the maximum quench delay times may be exceeded if performance tests indicate all parts will be above 900° F when quenched.

[1] Shorter times than shown may be necessary to ensure that the minimum temperature of 7178 alloy is above 775° F (412° C) when quenched.

are more easily worked than the hard tempers. Also, the naturally aged tempers afford better formability than the artificially aged tempers. For example, the 99% metal (alloy 1100), in the annealed temper "–O," has the best forming characteristics, and alloy 7075, in the full heat treated temper "–T6," is among the most difficult to form because of its hardness.

In the process of forming, the metal hardens and strengthens due to the working effect. In cold drawing, the changes in tensile strength and other properties can become quite large depending on the amount of work and the alloy's composition. In bending (a form of cold working), the bend radius and the thickness of the metal must also be considered (see **Table 16** for appropriate bend radii for specified thicknesses). Most forming is done when cold, and choosing the proper temper usually permits the completion of the fabrication without the necessity of intermediate anneal. In some difficult drawing operations, however, intermediate annealing may be required between successive draws. Hot forming is usually done at temperatures of 300–400° F (149–204° C), which allows the metal to be easily worked without appreciably lowering its strength provided that the heating periods do not exceed thirty minutes. In general, a combination of the shortest possible time with the lowest temperature that will provide the desired results is recommended. Forming is also done in the as-quenched condition on those alloys that age spontaneously at room temperature after solution heat treatment (–W temper). In these instances, the quenched material should be refrigerated to retard hardening until forming is complete.

Selecting the proper temper is important when specifying aluminum for forming operations. When nonheat treatable alloys are to be formed, the temper chosen should be sufficiently soft to permit the desired bend radius or draw depth. In more difficult forming operations, material in the annealed temper "–O" should be used. For less severe forming requirements, material in one of the harder tempers, such as "–H14," may be satisfactorily formed. When heat treatable alloys are to be used for forming, the shape should govern the selection of the alloy and its temper. Maximum formability of the heat treatable alloys is obtained in the annealed temper. However, limited formability can be effected in the fully heat treated temper, provided the bend radii are large enough.

Clues to the formability of an alloy are its percent of elongation, and the difference between the yield and ultimate tensile strengths. As a rule, the higher the elongation value and/or the wider the range between the yield and tensile strengths, the better the forming characteristics.

Machinability—the ease with which a material can be finished by cutting—is characterized by fast cutting speeds, small chip size, smoothness of the machined surface,

Table 16. Approximate Radii for 90° Cold Bend of Wrought Aluminum Alloys. *(Source, MIL-HDBK-694A.)*

Designation		Radius Required (in terms of sheet thickness)							
Alloy	**Temper**	**1/64 inch .397 mm**	**1/32 inch .794 mm**	**1/16 inch 1.588 mm**	**1/8 inch 3.175 mm**	**3/16 inch 4.762 mm**	**1/4 inch 6.35 mm**	**3/8 inch 9.525 mm**	**1/2 inch 12.7 mm**
1100	-O	0	0	0	0	0	0	0	1-2
	-H14	0	0	0	0	0-1	0-1	0-1	2-3
	-H18	0-1	0.5-1.5	1-2	1.5-3	2-4	2-4	3-5	3-6
3003	-O	0	0	0	0	0	0	0	1-2
	-H14	0	0	0	0-1	0-1	1-1.5	1-2.5	1.5-3
	-H18	0.5-1.5	1-2	1.5-3	2-4	3-5	4-6	4-7	5-8
5052	-O	0	0	0	0	0-1	0-1	0.5-1.5	1-2
	-H34	0	0	0-1	0.5-1.5	1-2	1.5-3	2-3	2.5-3.5
	-H38	0.5-1.5	1-2	1.5-3	2-4	3-5	4-6	4-7	5-8
5083	-O	-	-	0-0.5	0-1	0-1	0.5-1.5	1.5-2	1.5-2.5
2014 Clad	-O	0	0	0	0	0-1	0-1	1.5-3	3-5
	-T3	1-2	1.5-3	2-4	3-5	4-6	4-6	5-7	5.5-8
	-T4	1-2	1.5-3	2-4	3-5	4-6	4-6	5-7	5.5-8
	-T6	2-4	3-5	3-5	4-6	5-7	6-10	7-10	8-11
2024	-O	0	0	0	0	0-1	0-1	1.5-3	3-5
	-T3	1.5-3	2-4	3-5	4-6	4-6	5-7	6-8	6-9
	-T4	1.5-3	2-4	3-5	4-6	4-6	5-7	6-8	6-9
	-T81	3.5-5	4.5-6	5-7	6.5-8	7-9	8-10	9-11	9-12
5456	-O	-	-	-	0-1	0.5-1	0.5-1	0.5-1.5	0.5-2
	-H321	-	-	-	2-3	2-3	3-4	3-4	3-4
6061	-O	0	0	0	0	0-1	0-1	0.5-2	1-2.5
	-T4	0-1	0-1	0.5-1.5	1-2	1.5-3	2-4	2.5-4	3-5
	-T6	0-1	0.5-1.5	1-2	1.5-3	2-4	3-4	3.5-5.5	4-6
7075	-O	0	0	0-1	0.5-1.5	1-2	1.5-3	2.5-4	3-5
	-T6	2-4	3-5	4-6	5-7	5-7	6-10	7-11	7-12

and good tool life. While some aluminum alloys are excellent for machining, others are very troublesome. The troublesome alloys are soft and gummy, produce long and stringy chips, and must be machined at slow cutting rates. The harder alloys and tempers afford better machinability. In general, alloys containing copper, zinc, or magnesium as the principal alloying constituents are the most readily machined. Other compositions, containing bismuth and lead, are also usually machinable as they were specially designed for high speed screw machine work. Compositions containing more than 10% silicon are usually the most difficult to machine, and even content of 5% silicon alloys exhibit a gray surface rather than machining to a bright finish.

Wrought alloys that have been heat treated have fair to good machining characteristics. These alloys are easier to machine to a good finish when they are in the full hard temper condition, rather than annealed. Wrought alloys that are not heat treated, regardless of temper, tend to gumminess. Finally, wrought compositions having copper as the principal alloying element are more easily machined than those that have been hardened mainly by magnesium silicide.

Corrosion Resistance and Protective Finishes

Aluminum and aluminum alloys owe their corrosion resistance to the oxide film that forms on the surface upon exposure to oxygen. This coating prevents further oxidation, but, in some environments, supplementary protection is required. The degree of inherent corrosion resistance of the alloy depends on the composition and the thermal history of the metal. Magnesium, silicon, or magnesium silicide enhance the corrosion resistance properties of aluminum alloys.

Cladding, chemical treatment, electrolytic oxide finishing, electroplating, and applications of organic or inorganic coatings are used for supplementary protection of aluminum alloys. Cladding is perhaps the most effective means of protection, and the process consists of applying layers (approximately 2 to 15% of the total thickness) of pure aluminum or a corrosion resistant aluminum alloy to the surface of the ingot, and then hot working the ingot to cause the cladding metal to weld to the core. In subsequent hot working and fabricating, the cladding becomes alloyed with the core and is reduced in thickness proportionally. The cladding not only serves to protect the core metal, it also affords protection by electrolytic action, even when the surface is scratched, because the cladding is anodic to the base metal and therefore corrodes sacrificially.

Some chemical treatments result in the formation of oxide films, while others etch the metal and lower the corrosion resistance by removing the oxide film. Chemical finishes, though widely used as paint bases because they are slightly porous, are not as successful as electrolytic finishes, and electrolytic oxide finishing is perhaps the most widely used method for protecting aluminum. It consists of treating the metal in an electrolyte capable of giving off oxygen, using the metal as an anode. The film formed by the process is an aluminum oxide that is thin, hard, inert, and minutely porous, and can be left as is, painted, or dyed. Electroplating plating aluminum alloys requires great care in preparing the surface of the metal. It must be buffed to remove any scratches or defects, cleaned thoroughly to remove all grease, dirt, or other foreign material, and it must be given a coating of pure zinc (by immersion in a zincate solution). After plating, the surface is buffed and finished like other metals.

Organic and inorganic coatings range from paints and lacquers to vitreous enamels. When providing a protective (rather than decorative) painted coating, the surface should first be etched with a cleaner containing phosphoritic acid (or similar) to remove contaminates and deposit a thin phosphate film. A prime coat such as zinc chromate can then be applied, followed by paint, varnish, or lacquer.

Stress-Corrosion Cracking

The high strength heat treatable wrought alloys in certain tempers are susceptible to stress-corrosion cracking, depending on product, section size, and the direction and magnitude of stress. These alloys include 2014, 2025, 2618, 7075, 7150, 7175, and 7475 in the "–T6" temper, and 2014, 2024, 2124, and 2219 in "–T3" and "–T4" tempers. Other alloy-temper combinations—notably 2024, 2124, 2219, and 2519 in the "–T6" or "–T8" type tempers, and 7010, 7049, 7050, 7075, 7149, 7175, and 7475 in the "–T3" type temper—are decidedly more resistant, and sustained tensile stresses of 50 to 75% of the minimum yield strength may be permitted without concern about stress corrosion cracking. The "–T74" and "–T76" tempers of 7010, 7075, 7475, 7049, 7149, and 7050 provide an intermediate degree of resistance to stress-corrosion cracking (superior to the "–T6" temper but not as good as the "–T73" temper of 7075). To assist in the selection of materials, letter ratings indicating the relative resistance to stress-corrosion cracking of various mill products are presented in **Table 17**. This table is based on ASTM G64, which contains more detailed information regarding the rating system and the procedures used to determine the ratings.

Where short periods at elevated temperatures of 150 to 500° F (66 to 260° C) may be encountered, the precipitation heat treated tempers of 2024 and 2219 are recommended over the naturally aged tempers. Alloys 5083, 5086, and 5456 should not be used under high constant applied stress for continuous service at temperatures exceeding 150° F (66° C), because of the hazard of developing susceptibility to stress-corrosion cracking. In general, the "–H34" through "–H38" tempers of 5086, and the "–H32" through "–H38" tempers of 5083 and 5456 are not recommended, because these tempers can become susceptible to stress-corrosion cracking.

To avoid stress-corrosion cracking, practices such as the use of press or shrink fits, taper pins, clevis joints (in which tightening of the bolt imposes a bending load on female lugs), and straightening or assembly operations that result in sustained surface stresses (especially when acting in the short transverse grain orientation) should be avoided in these high strength alloys: 2014–T451, –T4, –T6, –T651, –T652; 2024–T3, –T351, –T4; 7075–T6, –T651, –T652; 7150–T6151; and 7475–T6, –T651. When straightening or forming is necessary, it should be performed when the material is in the freshly quenched condition or at an elevated temperature to minimize the residual stresses induced. Where elevated temperature forming is performed on 2014–T4 and –T451, or 2024–T3 and –T351, a subsequent precipitation heat treatment for the "–T6" or "–T651," or "–T81" or "–T851," temper is recommended.

It is good engineering practice to control short transverse tensile stress at the surface of structural parts at the lowest practicable level. Careful attention should be given in all stages of manufacturing, beginning with design of the part configuration, to choose practices in the heat treatment, fabrication, and assembly that avoid unfavorable combinations of end-grain microstructure and sustained tensile stress. The greatest danger arises when residual, assembly, and service stress combine to produce high sustained tensile stress at the metal surface. It is imperative that, for materials with low resistance to stress-corrosion cracking in the short transverse grain orientation, every effort be taken to keep the level of sustained tensile stress close to zero.

Cryogenic effects

In general, the strengths (including fatigue strength) of aluminum alloys increase with decreases in temperature below room temperature. The increase is greatest over the range of minus 100 to minus 423° F (minus 73 to minus 253° C), with the upper range being the temperature of liquid hydrogen. The strengths at minus 452° F (minus 269°

(Text continued on p. 187)

Table 17. Resistance to Stress-Corrosion Ratings for High Strength Aluminum Alloy Products. See Note. *(Source, MIL-HDBK-5H.)*

Alloy and Temper[1]	Test Direction[2]	Rolled Plate	Rod and Bar[3]	Extruded Shapes	Forging
2014-T6	L	A	A	A	B
	LT	B[4]	D	B[4]	B[4]
	ST	D	D	D	D
2024-T3, T4	L	A	A	A	[5]
	LT	B[4]	D	B[4]	[5]
	ST	D	D	D	[5]
2024-T6	L	[5]	A	[5]	A
	LT	[5]	B	[5]	A[4]
	ST	[5]	B	[5]	D
2024-T8	L	A	A	A	A
	LT	A	A	A	A
	ST	B	A	B	C
2124-T8	L	A	[5]	[5]	[5]
	LT	A	[5]	[5]	[5]
	ST	B	[5]	[5]	[5]
2219-T351X, T37	L	A	[5]	A	[5]
	LT	B	[5]	B	[5]
	ST	D	[5]	D	[5]
2219-T6	L	A	A	A	A
	LT	A	A	A	A
	ST	A	A	A	A
2219-T85XX, T87	L	A	[5]	A	A
	LT	A	[5]	A	A
	ST	A	[5]	A	A
6061-T6	L	A	A	A	A
	LT	A	A	A	A
	ST	A	A	A	A
7049-T73	L	A	[5]	A	A
	LT	A	[5]	A	A
	ST	A	[5]	B	A
7049-T76	L	[5]	[5]	A	[5]
	LT	[5]	[5]	A	[5]
	ST	[5]	[5]	C	[5]
7050-T74	L	A	[5]	A	A
	LT	A	[5]	A	A
	ST	B	[5]	B	B
7050-T76	L	A	A	A	[5]
	LT	A	B	A	[5]
	ST	C	B	C	[5]
7075-T76	L	A	A	A	A
	LT	B[4]	D	B[4]	B[4]
	ST	D	D	D	D
7075-T73	L	A	A	A	A
	LT	A	A	A	A
	ST	A	A	A	A
7075-T74	L	[5]	[5]	[5]	A
	LT	[5]	[5]	[5]	A
	ST	[5]	[5]	[5]	B

(Continued)

Table 17. *(Continued)* **Resistance to Stress-Corrosion Ratings for High Strength Aluminum Alloy Products.** See Note. *(Source, MIL-HDBK-5H.)*

Alloy and Temper[1]	Test Direction[2]	Rolled Plate	Rod and Bar[3]	Extruded Shapes	Forging
7075-T76	L	A	5	A	5
	LT	A	5	A	5
	ST	C	5	C	5
7149-T73	L	5	5	A	A
	LT	5	5	A	A
	ST	5	5	B	A
7175-T74	L	5	5	5	A
	LT	5	5	5	A
	ST	5	5	5	B
7475-T6	L	A	5	5	5
	LT	B[4]	5	5	5
	ST	D	5	5	5
7475-T73	L	A	5	5	5
	LT	A	5	5	5
	ST	A	5	5	5
7475-T76	L	A	5	5	5
	LT	A	5	5	5
	ST	C	5	5	5

Note: Ratings were determined from stress corrosion tests performed on at least ten random lots for which test results showed 95% conformance at the 95% confidence level when tested at the stresses indicated below. A practical interpretation of these ratings follows the rating definition.

A - Equal or greater than 75% of the specified minimum yield strength. Very high. No record of service problems and SCC not anticipated in general applications.

B - Equal or greater than 50% of the specified minimum yield strength. High. No record of service problems and SCC not anticipated at stresses of the magnitude caused by solution heat treatment. Precautions must be taken to avoid high sustained tensile stress exceeding 50% of the minimum specified yield strength produced by any combination of sources including heat treatment, straightening, forming, fit-up, and sustained service loads.

C - Equal or greater than 25% of the specified minimum yield strength. Intermediate. SCC not anticipated if the total sustained tensile strength is less than 25% of the minimum specified yield strength. This rating is designated for the short transverse direction in improved products used primarily for high resistance to exfoliation corrosion in relatively thin structures where applicable short transverse stresses are unlikely.

D - Fails to meet the criterion for the rating C. Low. SCC failures have occurred in service or would be anticipated if there is any sustained tensile stress in the designated test direction. This rating currently is designated only for the short transverse direction in certain materials.

The above stress levels are not to be interpreted as "threshold" stresses, and are not recommended for design. Other documents, such as MIL-STD-1568, NAS SD-24, and MSFC-SPEC-522A, should be consulted for design recommendations.

[1] The ratings apply to standard mill products in the types of tempers indicated, including stress-relieved tempers, and could be invalidated in some cases by application of nonstandard thermal treatments of mechanical deformation at room temperature by the user.

[2] Test direction refers to orientation of the stressing direction relative to the directional grain structure typical of wrought materials, which in the case of extrusions and forgings may not be predictable from the geometrical cross section of the product.

L-Longitudinal: parallel to the direction of principal metal extension during manufacture of the product.

LT-Long Transverse: perpendicular to direction of principal metal extension. In products whose grain structure clearly shows directionality (width to thickness ratio greater than two) it is that perpendicular direction parallel to the major grain dimension.

ST-Short Transverse: perpendicular to direction of principal metal extension and parallel to minor dimension of grains in products with significant grain directionality.

[3] Sections with width-to-thickness ratio equal to or less than two for which there is no distinction between LT and ST.

[4] Rating is one class lower for thicker sections: extrusion, 1 inch and over; plate and forgings, 1.5 inches and over.

[5] Ratings not established because the product is not offered commercially.

This table is based upon ASTM G 64.

Table 18. Typical Fatigue Strengths of Wrought Aluminum Alloys. *(Source, MIL-HDBK-694.)*

Alloy	Temper	Repeated Flexure Fatigue Strength				
		0.1 ksi/0.7 MPa	1.0 ksi/6.9 MPa	10 ksi/69 MPa	100 ksi/689 MPa	500 ksi/3450 MPa
		Million Cycles to Failure				
1100	-O	-	6.5	5.5	5	5
	-H16	14	11.5	10	9	8
3003	-O	10.5	9	8	7.5	7
	-H14	17	12	10	9	9
	-H18	19	14	11.5	10.5	10
5052	-O	23.5	19.5	17.5	16.5	16
	-H34	26	20.5	19	18	18
	-H38	29.5	24	22.5	21	20
2011	-T3	35	26.5	22.5	19.5	18
2014	-T6	39	30	24	19	18
2017	-T4	42	34	27	22	20
2018	-T61	42	29	23	19.5	17
2024	-T4	43	31	24	21	20
4032	-T6	37	30	23.5	18	16
6061	-T6	31	23	17	14.5	13.5
6063	-T42	19.5	16	13.5	11	9.5
	-T5	20.5	15.5	12	10.5	9.5
	-T6	23.5	16.5	13.5	11	9.5
6151	-T6	30	22	17	13	12
7075	-T6	40	29	24	22	22

Alloy	Temper	75°F/24°C	300°F/149°C	400°F/204°C	500°F/260°C
		Fatigue Strength ksi/MPa			
3003	-H18	10	7.5	5	3.5
2014	-T6	18	12	8	5
2024	-T4	20	14	9	6
5052	-H36	18.5	12.5	9.5	6
6061	-T6	16	11	7.5	4.5
7075	-T6	22	12	8.5	7

Table 19. Effect of Temperature on Thermal Coefficient of Linear Expansion for Wrought Aluminum Alloys. *(Source, MIL-HDBL-694.)*

Alloy	Average Coefficient, 10^{-6} in./in./°F			
	-58 to +68° F -49 to 20° C	68 to 212° F 20 to 100° C	68 to 392° F 20 to 200° C	68 to 572° F 20 to 300° C
1100	12.2	13.1	13.7	14.2
2011	11.9	12.8	13.4	13.9
2014	12.0	12.3	13.1	13.6
2017	12.1	12.7	13.3	13.9
2018	11.7	12.4	12.9	13.4

(Continued)

Table 19. *(Continued)* **Effect of Temperature on Thermal Coefficient of Linear Expansion for Wrought Aluminum Alloys.** *(Source, MIL-HDBL-694.)*

Alloy	Average Coefficient, 10^{-6} in./in./°F			
	-58 to +68° F -49 to 20° C	68 to 212° F 20 to 100° C	68 to 392° F 20 to 200° C	68 to 572° F 20 to 300° C
2024	11.9	12.6	13.2	13.7
2025	12.1	12.6	13.1	13.6
2117	12.1	13.0	13.6	14.0
2218	11.7	12.4	13.0	13.5
3003	12.0	12.9	13.5	13.9
4032	10.3	10.8	11.3	11.7
5052	12.3	13.2	13.8	14.3
5056	12.5	13.4	14.0	14.5
6053	12.1	12.8	13.4	14.0
6061	12.1	13.0	13.5	14.1
6063	12.1	13.0	13.6	14.2
6151	12.1	12.8	13.4	13.9
7075	12.1	12.9	13.5	14.4

C)—the temperature of liquid helium—are nearly the same as at minus 423° F. For most alloys, elongation and various indices of toughness remain nearly constant or increase with decrease in temperature, but the 7000 series experiences modest reductions. None of the alloys exhibits a marked transition in fracture resistance over a narrow range of temperature indicative of embrittlement. The tensile and shear moduli of aluminum alloys also increase with decreasing temperature, so that at minus100, minus 320, and minus 423° F (minus 73, minus 196, and minus 253° C), they are approximately 5, 12, and 16% (respectively) above the room temperature values.

Selecting aluminum alloys

As in selecting any other material used in engineering design, several factors must be considered when choosing an aluminum alloy for maximum value and optimum performance. The factors include service conditions, production run, and the relative costs of suitable machining and fabricating processes. Within limits, the selection of a specific composition for a particular application may be simplified by determining the requirements for mechanical or physical properties, determining which alloys meet the selected criteria, and from the chosen alloys, pick the one best suited for the fabrication and finish machining requirements. One critical characteristic when considering wrought alloys is fatigue strength, and typical fatigue strengths of selected alloys are given in **Table 18**. The effects of temperature on aluminum's characteristics are shown in **Table 19** (coefficient of linear expansion), **Table 20** (ultimate tensile strength), **Table 21** (yield strength), and **Table 22** (elongation).

The choice of an alloy for casting is, to a great extent, governed by the type of mold that is to be used. The type of mold sets guidelines for such factors as intricacy of design, size, cross section, tolerance, surface finish, and number of castings to be produced. Sand molds are particularly suited to large castings, large tolerances, and small runs. They are not acceptable for the production of thin sections (less than $^{3}/_{16}$ inch [4.76 mm]), or smooth finishes. Permanent molds, which are particularly well suited to cast iron, provide better

(Text continued on p. 191)

Table 20. Typical Effect of Temperature on Ultimate Tensile Strength for Wrought Aluminum Alloys. *(Source, MIL-HDBK-694.)*

Alloy	Temper	Percent of Ultimate Strength at 75°F/24°C								
		-320°F -196°C	-110°F -79°C	-20°F -29°C	+212°F +100°C	300°F 149°C	400°F 204°C	500°F 260°C	600°F 316°C	700°F 371°C
1100	-O	189	115	104	77	65	46	31	19	16
	-H18	144	109	104	92	75	25	17	10	8
3003	-O	206	122	107	81	69	53	38	25	19
	-H14	164	109	103	95	82	64	34	18	14
	-H18	143	110	104	90	79	48	26	14	10
5052	-O	158	106	101	100	86	64	43	27	18
	-H34	144	106	101	100	82	60	32	20	13
	-H38	141	105	101	98	81	55	29	18	12
5083	-O	141	105	101	100	68	52	36	25	14
2011	-T3	-	-	-	85	51	29	12	6	4
2014	-T6	120	104	103	89	57	23	13	9	6
2017	-T4	128	104	103	90	64	36	19	10	7
2018	-T61	118	104	103	92	74	31	16	9	7
2024	-T3	-	-	-	94	79	41	17	11	8
	-T4	127	106	104	94	66	40	18	12	8
	-T81	-	-	-	94	79	41	17	11	8
2117	-T4	-	-	-	84	70	37	17	10	7
2218	-T61	-	-	-	95	70	37	17	9	7
4032	-T6	119	105	102	91	67	24	14	9	6
6053	-T6	-	-	-	86	68	35	15	11	8
6061	-T6	133	110	105	93	76	42	17	10	7
6063	-T42	153	120	108	100	95	41	20	14	11
	-T5	138	108	105	89	74	33	17	11	9
	-T6	135	109	103	89	60	26	13	9	7
6151	-T6	125	107	104	88	56	25	14	10	8
7075	-T6	123	105	103	80	30	17	13	10	8
7079	-T6	-	-	-	86	44	20	14	10	6

Table 21. Typical Effect of Temperature on Yield Strength for Wrought Aluminum Alloys. *(Source, MIL-STD-694.)*

Alloy	Temper	Percent of Yield Strength at 75°F/24°C								
		-320°F -196°C	-110°F -79°C	-20°F -29°C	+212°F +100°C	300°F 149°C	400°F 204°C	500°F 260°C	600°F 316°C	700°F 371°C
1100	-O	123	107	103	100	90	70	40	30	20
	-H18	-	-	-	82	64	18	9	7	4
3003	-O	145	116	105	92	83	75	58	42	33
	-H14	120	104	101	90	76	43	19	12	10
	-H18	122	107	103	78	59	33	15	9	7
5052	-O	121	100	100	100	100	85	62	38	23
	-H34	118	103	100	97	87	48	26	16	10
	-H38	116	101	100	100	78	40	22	14	8
5083	-O	-	-	-	100	82	77	50	34	20
2011	-T3	-	-	-	79	44	26	9	5	4
2014	-T6	113	103	102	93	58	22	12	8	6
2017	-T4	132	104	101	92	75	42	24	12	9
2018	-T61	110	103	101	94	87	28	14	6	5
2024	-T3	133	106	101	96	92	44	18	12	8
	-T4	-	-	-	96	77	43	19	13	8
	-T81	-	-	-	95	78	34	14	9	6
2117	-T4	-	-	-	88	71	50	23	15	8
2218	-T61	-	-	-	95	80	36	14	7	6
4032	-T6	103	100	100	96	72	20	12	6	4
6053	-T6	-	-	-	88	75	38	12	8	6
6061	-T6	116	105	103	95	78	38	12	6	5
6063	-T42	126	119	110	108	115	50	26	19	16
	-T5	116	105	104	95	86	31	17	12	10
	-T6	116	106	102	90	64	21	11	8	6
6151	-T6	115	106	104	91	58	22	13	10	8
7075	-T6	124	105	102	85	29	16	12	9	6
7079	-T6	-	-	-	88	44	19	12	9	6

Table 22. Typical Effect of Temperature on Elongation for Wrought Aluminum Alloys. *(Source, MIL-HDBK-694.)*

Alloy	Temper	Percent Elongation									
		-320°F -196°C	-110°F -79°C	-20°F -29°C	+75°F +24°C	+212°F +100°C	300°F 149°C	400°F 204°C	500°F 260°C	600°F 316°C	700°F 371°C
1100	-O	-	-	-	45	45	55	65	75	80	85
	-H18	-	-	-	15	15	20	65	75	80	85
3003	-O	49	45	44	43	40	47	60	65	70	70
	-H14	32	18	16	16	16	16	20	60	70	70
	-H18	-	-	-	10	10	11	18	60	70	70
5052	-O	-	-	-	30	35	45	65	80	100	120
	-H32	30	21	18	14	16	25	40	80	100	120
	-H38	-	-	-	8	9	20	40	80	100	120
5083	-O	-	-	-	25	35	45	60	70	95	120
2011	-T3	-	-	-	15	16	25	35	45	90	125
2014	-T6	-	-	-	13	14	15	35	45	65	70
2017	-T4	-	-	-	22	18	16	28	45	95	100
2018	-T61	-	-	-	12	12	12	25	40	60	100
2024	-T3	-	-	-	17	16	11	23	55	75	100
	-T4	-	-	-	19	19	17	27	55	75	100
	-T81	-	-	-	7	8	11	23	55	75	100
2117	-T4	-	-	-	27	16	20	35	55	80	110
2218	-T61	-	-	-	13	14	17	30	70	85	100
4032	-T6	-	-	-	9	9	9	30	50	70	90
6053	-T6	-	-	-	13	13	13	25	70	80	90
6061	-T6	25	20	19	17	18	20	28	60	8585	90
6063	-T42	-	-	-	33	18	20	40	75	80	105
	-T5	-	-	-	22	18	20	40	75	80	105
	-T6	-	-	-	18	15	20	40	75	80	105
6151	-T6	-	-	-	17	19	22	40	50	50	50
7075	-T6	-	-	-	11	15	30	60	65	80	65
7079	-T6	-	-	-	13	18	37	60	100	175	175

surface finishes and closer tolerances than sand molds, but the minimum thicknesses that can be produced are about the same. Permanent molds are better suited for longer runs because they do not require the pattern equipment or molding operations needed for sand castings. Dies are especially well suited for long production runs and, even though they are relatively expensive, their initial cost can be justified by savings in finishing costs, plus their high production rate. Other advantages are the ability to produce thinner cross sections, closer tolerances, smoother surfaces, and more intricate designs. **Table 23** gives casting properties of several alloys, plus their principal use.

Choosing a wrought alloy is influenced by the proposed method of fabrication and the design requirements of the finished part. Although a variety of compositions and tempers will generally produce the desired mechanical and physical properties, the number of compositions and tempers amenable to the various fabrication techniques is, in some instances, limited. The fabrication technique that will provide the greatest economy is

Table 23. Casting Properties and Principal Uses of Selected Aluminum Casting Alloys. *(Source, QQ-A-610F.)*

Alloy	Specific Gravity*	Casting Properties	Response to Heat Treatment	Principal Use	Machinability
208.0	2.79	Good	Yes	Miscellaneous castings. Composition is such as to make it more available during periods of shortages of primary aluminum.	Good
213.0	2.87	Good	No	General	Excellent
222.0	2.91	Good	Yes	Same as alloy 242.0. Automotive pistons.	Excellent
242.0	2.78	Fair	Yes	Castings requiring strength and hardness at elevated temperatures, such as air-cooled cylinder heads and pistons. Aircraft and diesel engine pistons.	Good
295.0	2.81	Good	Yes	General, where high strength ductility and resistance to shock are required.	Very good
296.0	2.78	Good	Yes	General, where high strength and high ductility are required.	Good
308.0	2.77	Excellent	No	General	Good
319.0	2.79	Excellent	Yes	High strength and general purpose alloy for intricate castings.	Good
328.0	2.73	Excellent	Yes	General, where high strength is required.	Good
332.0	2.76	Fair	Yes	Automotive and diesel pistons.	Fair
333.0	2.77	Good	Yes	General, high strength castings.	Good
336.0	2.70	Fair	Yes	Automotive and diesel pistons.	Fair
355.0	2.69	Good to Excellent	Yes	General, where high strength and corrosion resistance are required. Also retains strength at elevated temperatures. General, where pressure tightness such as in pump bodies, liquid cooled cylinder heads, is required.	Good
C355.0	2.69	Good	Yes	Same as alloy 355.0.	Good
356.0	2.66	Excellent	Yes	General, where high strength and corrosion resistance are required.	Good
A356.0	2.66	Excellent	Yes	Same as alloy 356.0.	Good
357.0	2.66	Excellent	Yes	Same	Good

(Continued)

Table 23. *(Continued)* **Casting Properties and Principal Uses of Selected Aluminum Casting Alloys.** *(Source, QQ-A-596A.)*

Alloy	Specific Gravity*	Casting Properties	Response to Heat Treatment	Principal Use	Machinability
B443.0	2.64	Excellent	No	General, with maximum corrosion resistance and for leak proof castings of intricate design.	Fair
512.0	2.65	Good	No	Same as alloy 514.0.	Good
513.0	2.68	Fair	No	General	Excellent
514.0	2.65	Fair	No	Castings requiring superior resistance to corrosion.	Very good
520.0	2.57	Fair	Yes	Castings requiring maximum strength, elongation and resistance to shock, require special founding practice.	Excellent
535.0	2.62	Good	Yes	Applications requiring excellent shock resistance and corrosion resistance.	Excellent
705.0	2.76	Good	Note [1]	Same as alloy 712.0. General, where high strength and ductility are required, particularly without heat treatment, in addition to excellent corrosion resistance.	Excellent
707.0	2.77	Good	Yes, but not for normal application [1]	Same as alloy 712.0.	Excellent
710.0	2.76	Fair	Note [1]	Same as alloy 712.0.	Excellent
712.0	2.81	Good	Note [1]	General, where high strength and ductility are required, particularly without heat treatment, in addition to excellent corrosion resistance.	Excellent
713.0	2.81	Good	Note [1]	Same as alloy 712.0.	Excellent
771.0	2.81	Good	Yes	Applications requiring excellent dimensional stability or shock resistance combined with very high yield strength.	Excellent
850.0	2.88	Fair	Yes	Bearings	Excellent
851.0	2.83	Fair	Yes	Bearings	Excellent
852.0	2.88	Fair	Yes	Bearings	Excellent

* Specific gravity value is approximate.

[1] These alloys, when properly cast, develop their highest strength after aging at room temperature for several weeks, or after short artificial aging at slightly elevated temperatures.

Table 24. Principal Characteristics and Uses of Wrought Aluminum Alloys. *(Source, MIL-HDBK-694.)*

Alloy	Outstanding Characteristics	Recommended Uses
Nonheat Treatable Alloys		
1100	Very good formability, weldability, and resistance to corrosion. Relatively low strength but high ductility.	General purpose material for drawing and stamping, and for a miscellany of parts where high strength is not required.
3003	Good formability and weldability, very good resistance to corrosion. Appreciably higher strength than 1100.	General purpose material for drawing and stamping. Miscellaneous parts where higher strength is needed than that provided by 1100.

(Continued)

Table 24. *(Continued)* **Principal Characteristics and Uses of Wrought Aluminum Alloys.** *(Source, MIL-HDBK-694.)*

Alloy	Outstanding Characteristics	Recommended Uses
Nonheat Treatable Alloys		
5052	Moderate mechanical properties, stronger and harder than 1100 and 3003. Fairly good formability. Readily weldable. Excellent resistance to corrosion by salt water.	General purpose alloy where fairly high strength is required. For marine and outside applications, fuel and hydraulic lines, and tanks.
Heat Treatable Alloys		
2011	Excellent free-machining qualities. Fairly high mechanical properties.	Stock for screw-machine products. Bolts, nuts, screws, and a great diversity of parts made on automatic screw machines.
2014	High mechanical properties including yield and tensile strength, fatigue, and hardness. Fair formability and forging qualities. Readily machinable.	Most commonly used alloy where high strength is required. General structural applications, heavy duty forgings, and strong fittings.
Clad 2014	A sheet product that combines the high mechanical properties of 2014 with the good corrosion resistance of 6053.	For structures requiring high unit strength together with good resistance to various corrosive environments.
2017	A bar, rod, and wire alloy having relatively high strength, and good machining qualities.	Screw-machine products, fittings, and structural applications where relatively high strength is required. Now largely superseded by newer alloys.
2018 2218	Both retain strength well at elevated temperatures.	Forged pistons and cylinder head for internal-combustion engines. Suitable for various types of high temperature services. Forged cylinder heads and pistons.
2024	A high strength alloy with mechanical properties intermediate between 2014 and 6061.	General purpose material for various structural applications where good strength is required. Fittings and screw-machine products.
Clad 2024	A sheet product that combines the mechanical properties of 2024 with the corrosion resistance of 1230 aluminum alloy.	For structural applications requiring good strength together with resistance to corrosion.
2025	Fairly high mechanical properties. Good forging qualities.	Specialty forging alloy. Applications mostly confined to propellers for superchargers and engines.
4032	Retains strength well at elevated temperatures.	Forged pistons for internal-combustion engines.
6151	Fairly good mechanical properties. Excellent forging qualities. Good resistance to corrosion.	General purpose material for ordinary forgings. Small press forgings and intricate pieces which are difficult to forge in the harder alloys.
6061	Good mechanical properties. Superior brazing and welding qualities. Good forging characteristics, workability, and resistance to corrosion.	General structural purposes. Marine and outside work. Transportation equipment. Many small forged parts. Various extrusion applications.
7075	Affords maximum strength and endurance limit. Not readily formed. Poorest forging qualities.	Structural applications requiring maximum yield and tensile strength. Section thickness limited to 3 inches.
Clad 7075	A sheet product that combines the mechanical properties of 7075 with improved corrosion resistance.	Structural applications where the highest strength together with maximum corrosion resistance is necessary.

to some degree governed by the quantity to be manufactured, making it necessary to take into account the processes and tooling that must be used for each method. Although aluminum can be formed by any of the conventional methods, it is especially well suited to extrusion, drawing, and forging. Principal characteristics of selected wrought alloys are given in **Table 24**. When choosing an aluminum alloy for any wrought product, it should be remembered that for corresponding tempers, the ease of fabricating *decreases* as the strength increases. Also, as strength increases, the cost increases, so strength is related to cost.

Aluminum extrusions have numerous applications, and are especially useful for producing architectural shapes. Extrusion is less expensive than roll forming, but it cannot produce sections as thin, and the die design requirements for extrusion require careful thought so that the metal flows uniformly in both thick and thin sections. Extrusion alloys are specifically designed for their intended use. Alloy 7075–T6 is often chosen when high strength is required. Alloy 2014–T6 has similar qualities, but is not as strong. Alloy 2024 –T6 is useful for thinner sections, and 6061 has superior forming qualities, resistance to corrosion, and high yield strength. Alloy 6063, either as extruded (–T42), or artificially aged (–T5), provides adequate strength for many applications and does not discolor from anodic oxide finishing. When high resistance to corrosion is required, extruded shapes of alloy 1100 and 3003 are commonly used.

For aluminum alloys, drawing is much the same as with other metals. While more expensive than extrusion, it yields products with much closer tolerances. In drawing aluminum, tool radii are important, and a tool thickness of four to eight times metal thickness is usually satisfactory. Too small a radius can cause tensile fracture, and too large a radius may induce wrinkling. Alloys of the nonhardenable variety, such as 1100, 3003, 5050, and 5052, are often used as they may be deformed to a greater extent before rupturing.

Forgings are used when higher strength is required, or where the forging process is especially suited for manufacturing the part. Aluminum may be either press forged or drop forged, using special forging stock produced in the form of an extruded bar or shape. Press forging, while slower than drop forging, affords greater flexibility in design, higher accuracy, and lower die cost. Ratings of alloys for forging are shown in **Table 25**.

Table 25. Relative Ratings of Aluminum Alloys for Forging. *(Source, MIL-HDBK-694R.)*

Alloy	Strength	Cold Weldability	Corrosion Resistance	Machinability	Electrical Conductance	Hardness	Forgability
1100	4 - 3	1 - 3	1	4 - 3	2	4 - 3	1
2011	2	3 - 4	3 - 4	1	3	2	-
2014	1	3 - 4	3 - 4	2	3	1 - 2	3
Alclad 2014	1	3 - 4	1	2	3	-	-
2017	1	3	3 - 4	2	4	2	-
2018	1	-	3	2	3	2	3
2024	1	3 - 4	3 - 4	2	4	1	-
Alclad 2024	1	4	1	2	4	-	-
2117	3	2	3	3	-	-	-
2218	2	-	3	2	3	2	4
3003	4 - 3	1 - 3	1	4 - 3	3	4 - 3	1

Table 25. *(Continued)* **Relative Ratings of Aluminum Alloys for Forging.** *(Source, MIL-HDBK-694R.)*

Alloy	Strength	Cold Weldability	Corrosion Resistance	Machinability	Electrical Conductance	Hardness	Forgability
4032	2	-	3	3	4	2	3
5052	3	1 - 3	1	4 - 3	4	3 - 2	-
5056	2	1 - 3	1 - 3	4 - 3	4	-	-
5083	2	3	1	4 - 3	4	2	-
5456	2	3	1 - 2	4 - 3	4	2	-
6061	3 - 2	3 - 4	1	3	3	3 - 2	-
6063	3 - 2	2 - 3	1	3	2	3 - 2	-
6151	2	-	2	3	3	2	1
7075	1	4	3	2	4	1	4
Alclad 7075	1	4	1	2	4	-	-
7079	1	4	3	2	4	1	-

Note: Relative ratings are in decreasing order of merit. The first number in numbered pairs is a rating of the softest temper, and the second number is a rating of the hardest temper.

Copper

Properties of copper and copper alloys

While copper ranks third behind iron/steel and aluminum in production, almost 50% of the total is consumed in the manufacture of copper wire. Commercially available copper and its alloys are identified by Unified Numbering System Designations that use the former Copper Alloy Number as the base of the new number. The old Copper Alloy Number 377, for example, now has the UNS designation C37700. New copper alloy designations are assigned by the Copper Development Association, and are awarded to new alloys on the following basis: 1) the full chemical composition must be disclosed; 2) the alloy must be in commercial use, or intended for commercial use; and 3) the composition must not fall within the limits of any designated alloy already available. In the designation system, numbers from C10000 through C79999 denote wrought alloys, and numbers from C80000 through C99999 denote cast alloys.

In Europe, CEN and ISO designations are used for copper alloys. The CEN system uses a six-character system, with the first character being C for copper, the second character being W (wrought), B (ingots), C (castings), or M (master alloys). The last character is a letter used to designate the material group. The ISO system is based on the element symbols of the material, in descending order of magnitude. **Table 1** explains the CEN system, and **Table 2** shows selected ISO/CEN conversions.

Description of wrought copper alloys

Characteristics and popular uses of wrought copper rod and bar alloys are given below. Composition of wrought copper alloys are given in **Tables 3** through **8**; **Table 9** provides mechanical properties; and **Table 10** provides physical properties.

Coppers. C10100 – C15815 are wrought coppers, with a minimum copper content of 99.3%.

C10100, C10200, C10300, C10800: Conductivity 101% IACS (International Annealed Copper Standard), the highest among coppers. Used for electric and electronic conductors, wave guides, cavity resonators, superconductor matrixes, vacuum tube and solid-state devices, glass-to-metal seals.

C10400, C10500, C10700: Conductivity 100% IACS. Very ductile, good hot strength and annealing resistance. Used for electrical conductors operating at elevated temperatures where pure coppers would soften too readily.

C11000, C11300, C11400, C11500, C11600: Conductivity 10% IACS. Ductile, anneal resistant. C11000 is primarily used for plumbing fittings and some electrical components not requiring extensive machining. Silver bearing grades have good annealing resistance and are used for electrical components and commutator components.

C12000, C12100, C12200, C12900: Deoxidizing provides improved embrittlement resistance during welding. Conductivity 95% IACS (C12000), and 85% IACS (12200). Used for electromechanical hardware conductors, bus bars, etc., where welding is required.

C14500, C14520, C14700: Conductivity 90–96% IACS. Free machining coppers. Used for welding and cutting torch tips, products requiring high conductivity and high machinability. C14700 has less directionality in ductility than C14500, and is better for crimped conductors, cold-formed parts, but is somewhat more costly than C14500.

C15000: High conductivity combined with good elevated temperature strength and deformation resistance. Used primarily for resistance welding caps RWMA Class I, and sometimes used for Class II applications.

C15715, C15725, C15760: Good resistance to softening after high temperature

(Text continued on p. 231)

Table 1. CEN European Numbering System for Copper and Copper Alloys.

Material Groups	Number Ranges Available for Positions 3, 4, and 5	Final Letter Designating Material Group	Number Range Allocated to Materials
Copper	001-999 001-999	A B	001-049A 050-099B
Miscellaneous Copper Alloys	001-999 001-999	C D	100-149C 150-199D
Miscellaneous Copper Alloys	001-999 001-999	E F	200-249E 250-299F
Copper-aluminum Alloys	001-999	G	300-349G
Copper-nickel Alloys	001-999	H	350-399H
Copper-nickel-zinc Alloys	001-999	J	400-449J
Copper-tin Alloys	001-999	K	459-499K
Copper-zinc Alloys, binary	001-999	L M	500-549L 550-599M
Copper-zinc-lead Alloys	001-999	N P	600-649P 650-699P
Copper-zinc Alloys, complex	001-999	R	700-749R 750-799S
Copper material not standardized by CEN/TC 133	800-999	A-S*	800-999*
* Letter as appropriate for the Material Group			

The first character in CEN designations is C (copper), followed by one of four second letters: W (wrought), B (ingot), C (casting), or M (master alloy).

Table 2. Equivalent ISO/CEN Copper and Copper Alloy Materials Designations.

Materials	Material Designations	
	ISO Symbols	CEN Numbers
Coppers	Cu-ETP Cu-OF	CW 004A CW 008A
Wrought Brasses	CuZn37 CuZn39Pb3 CuZn20Al2As CuZn40Mn1Pb1AlFeSn	CW 508L CW 614N CW 702R CW 721R
Other Wrought Alloys	CuNi2Si CuAl10Fe1 CuNi30Mn1Fe	CW 111C CW 305G CW 354H
Cast Alloys	CuZn33Pb2-GB CuZn33Pb2-GS CuSn12-GB CuSn12-GS	CB 750S CC 750S CB 483K CC 483K
Master Alloys	CuAl50 (A)-M CuCr10-M CuS20-M	CM 344G CM 204E CM 220E

Table 3. Chemical Composition of Wrought Coppers. *(Source, Copper Development Association.)*

Copper Number	Designation	Designation	Cu + Ag % (min.)	Ag (min.)		As	Sb	P	Te	Other (Named)
				%	Troy Oz.					
C10100 [1]	OFE	Oxygen Free Electronic	99.99 [2]	-	-	0.0005	0.0004	0.0003	0.0002	note [3]
C10200 [1]	OF	Oxygen Free	99.95 [2]	-	-	-	-	-	-	0.0010 Oxygen
C10300	OFXLP	-	99.95 [4]	-	-	-	-	0.001-0.005	-	-
C10400 [1]	OFS	Oxygen Free with Ag	99.95 [2]	0.027	8	-	-	-	-	0.0010 Oxygen
C10500 [1]	OFS	Oxygen Free with Ag	99.95 [2]	0.034	10	-	-	-	-	0.0010 Oxygen
C10700 [1]	OFS	Oxygen Free with Ag	99.95 [2]	0.085	25	-	-	-	-	-
C10800 [1]	OFLP	-	99.95 [4]	-	-	-	-	0.005-0.012	-	0.005 Oxygen
C10910 [1]	-	-	99.95 [2]	-	-	-	-	-	-	0.02 Oxygen
C10920	-	-	99.90	-	-	-	-	-	-	0.02 Oxygen
C10930	-	-	99.90	0.044	13	-	-	-	-	0.02 Oxygen
C10940	-	-	99.90	0.085	25	-	-	-	-	note [5]
C11000 [1]	ETP	Electrolytic Tough Pitch	99.90	-	-	-	-	-	-	note [5]
C11010 [1]	RHC	Remelted High Conductivity	99.90	-	-	-	-	-	-	note [5]

(Continued)

Table 3. *(Continued)* **Chemical Composition of Wrought Coppers.** *(Source, Copper Development Association.)*

Copper Number	Designation	Designation	Cu + Ag % (min.)	Ag (min.)		As	Sb	P	Te	Other (Named)
				%	Troy Oz.					
C11020 [1]	FRHC	Fire-Refined High Conductivity	99.90	-	-	-	-	-	-	note [5]
C11030 [1]	CRTP	Chemically Refined Tough Pitch	99.90	-	-	-	-	-	-	note [5]
C11040 [1]	-	-	99.90	-	-	0.0005	0.0004	-	0.0002	note [6]
C11100 [1]	-	Electrolytic Tough Pitch, Anneal Resistant	99.90	-	-	-	-	-	-	note [7]
C11300 [1]	STP	Tough Pitch With Ag	99.90	0.027	8	-	-	-	-	note [5]
C11400 [1]	STP	Tough Pitch With Ag	99.90	0.034	10	-	-	-	-	note [5]
C11500 [1]	STP	Tough Pitch With Ag	99.90	0.054	16	-	-	-	-	note [5]
C11600 [1]	STP	Tough Pitch With Ag	99.90	0.085	25	-	-	-	-	note [5]
C11700	-	-	99.90 [8]	-	-	-	-	0.04	-	0.004-0.02 B
C12000	DLP	Phosphorus-Deoxidized, Low Residual Phosphorus	99.90	-	-	-	-	0.004-0.012	-	-
C12100	-	-	99.90	0.014	4	-	-	0.005-0.012	-	-
C12200 [9]	DHP	Phosphorus-Deoxidized, High Residual Phosphorus	99.90	-	-	-	-	0.015-0.040	-	-

(Continued)

Table 3. *(Continued)* **Chemical Composition of Wrought Coppers.** *(Source, Copper Development Association.)*

Copper Number	Designation	Designation	Cu + Ag % (min.)	Ag (min.)		As	Sb	P	Te	Other (Named)
				%	Troy Oz.					
C12210	-	-	99.90	-	-	-	-	0.015-0.025	-	-
C12220	-	-	99.90	-	-	-	-	0.040-0.065	-	-
C12300	-	-	99.90	-	-	-	-	0.015-0.040	-	-
C12500	FRTP	-	99.88	-	-	0.012	0.003	-	-	0.025 Te + Se 0.003 Bi 0.004 Pb 0.050 Ni
C12510	-	-	99.90	-	-	-	0.003	0.03	-	0.025 Te + Se 0.005 Bi 0.020 Pb 0.050 Ni 0.05 Fe 0.05 Sn 0.080 Zn
C12900	FRSTP	Fire-Refined Tough Pitch with Ag	99.88	0.054	16	0.012	0.003	-	0.025 [10]	0.050 Ni 0.003 Bi 0.004 Pb
C14180	-	-	99.90	-	-	-	-	0.075	-	0.02 Pb 0.01 Al
C14181	-	-	99.90	-	-	-	-	0.002	-	0.002 Cd 0.005 C 0.002 Pb 0.002 Zn
C14200	DPA	Phosphorus-Deoxidized, Arsenical	99.40	-	-	0.15-0.50	-	0.015-0.040	-	-
C14300	-	Cadmium Copper, Deoxidized	99.90 [11]	-	-	-	-	-	-	0.05-0.15 Cd

(Continued)

Table 3. *(Continued)* **Chemical Composition of Wrought Coppers.** *(Source, Copper Development Association.)*

Copper Number	Designation	Designation	Cu + Ag % (min.)	Ag (min.)		As	Sb	P	Te	Other (Named)
				%	Troy Oz.					
C14410	-	-	99.90 [12]	-	-	-	-	0.005-0.020	-	0.05 Fe 0.05 Pb 0.10-0.20 Sn
C14415	-	-	99.96 [12]	-	-	-	-	-	-	0.10-0.15 Sn
C14420	-	-	99.90 [13]	-	-	-	-	-	0.005-0.05	0.04-0.15 Sn
C14500 [14]	-	Tellurium-Bearing	99.90 [62]	-	-	-	-	0.004-0.012	0.40-0.7	-
C14510	-	Tellurium-Bearing	99.85 [62]	-	-	-	-	0.010-0.030	0.30-0.7	0.05 Pb
C14520	DPTE	Phosphorus-Deoxidized, Tellurium-Bearing	99.90 [62]	-	-	-	-	0.004-0.020	0.40-0.7	-
C14530	-	-	99.90 [10, 12]	-	-	-	-	0.001-0.010	0.003-0.023 [10]	0.0030.023 Sn
C14700	-	Sulfur-Bearing	99.90 [14, 65]	-	-	-	-	0.002-0.005	-	0.20-0.50 S [65]
C15000	-	Zirconium Copper	99.80 [19]	-	-	-	-	-	-	0.10-0.20 Zr
C15100	-	-	99.85 [19]	-	-	-	-	-	-	0.05-0.15 Zr
C15150	-	-	99.9 min.	-	-	-	-	-	-	0.15-0.030 Zr
C15500	-	-	99.75	0.027-0.10	8-30	-	-	0.040-0.080	-	0.08-0.13 Mg
-	**Cu + Ag % (min.)**			**Al [15]**	**Fe**	**Pb**	**Oxygen**		**Boron**	
C15715	99.62			0.13-0.17	0.01	0.01	0.12-0.19		-	
C15720	99.52			0.18-.22	0.01	0.01	0.16-0.24		-	
C15725	99.43			0.23-0.27	0.01	0.01	0.20-0.28		-	
C15760	98.77			0.58-0.62	0.01	0.01	0.52-0.59		-	
C15815	97.82			0.13-0.17	0.01	0.01	0.19		1.2-1.8	

(Continued)

Table 3. *(Continued)* **Chemical Composition of Wrought Coppers.** *(Source, Copper Development Association.)*

Notes:

[1] These are high conductivity coppers which have in the annealed condition a minimum conductivity of 100% IACS, except for Alloy C10100 which has a minimum conductivity of 101% IACS.

[2] Copper is determined by the difference between the impurity total and 100%.

[3] The following additional maximum limits shall apply: Bi, 1 ppm (0.0001%); Cd, 1.0 ppm (0.0001%); Fe, 10 ppm (0.0010%); Pb, 5 ppm (0.0005%); Mn, 0.5 ppm (0.00005%); Ni, 10 ppm (0.0010%); Oxygen, 5 ppm (0.0005%); Se, 3 ppm (0.0003%); Ag, 25 ppm (0.0025%); S, 15 ppm (0.0015%); Sn, 2 ppm (0.0002%); Zn, 1 ppm (0.0001%).

[4] Includes P.

[5] Oxygen and trace elements may vary depending on the process.

[6] The following additional maximum limits shall apply: Se, 2 ppm (0.0002%); Bi, 1.0 ppm (0.00010%); Group Total, Te + Se + Bi, 3 ppm (0.0003%). Sn, 5 ppm (0.0005%); Pb, 5 ppm (0.0005%); Fe, 10 ppm (0.0010%); Ni, 10 ppm (0.0010%); S, 15 ppm (0.0015%); Ag, 25 ppm (0.0025%); Oxygen, 100-650 ppm (0.010-0.065%). The total maximum allowable of 65 ppm (0.0065%) does not include oxygen.

[7] Small amounts of Cd or other elements may be added by agreement to improve the resistance to softening at elevated temperatures.

[8] Includes B + P.

[9] This includes oxygen-free copper which contains P in an amount agreed upon.

[10] Includes Te + Se.

[11] Includes Cd. Deoxidized with lithium or other suitable elements as agreed upon.

[12] Includes Cu + Ag + Sn.

[13] Includes Te + Sn.

[14] Includes oxygen-free or deoxidized grades with deoxidizers (such as phosphorus, boron, lithium or others) in an amount agreed upon.

[15] All aluminum present as Al_2O_3; 0.04% oxygen present as Cu_2O with a negligible amount in solid solution with copper.

[19] Cu + Sum of Named Elements, 99.9% min.

[62] Includes Te.

[65] Includes Cu + Ag + S + P.

Table 4. Chemical Composition of Wrought High Copper Alloys. *(Source, Copper Development Association.)*

Copper Alloy Number	Cu + Ag	Fe	Sn	Ni	Co	Cr	Si	Be	Other (named)
C16200	Rem. [16]	0.02	-	-	-	-	-	-	0.7-1.2 Cd
C16500	Rem. [16]	0.02	0.50-0.7	-	-	-	-	-	0.6-1.0 Cd
C17000	Rem. [16]	[17]	-	[17]	[17]	-	0.20	1.60-1.79	0.20 Al
C17200	Rem. [16]	[17]	-	[17]	[17]	-	0.20	1.80-2.00	0.20 Al
C17300	Rem. [16]	[17]	-	[17]	[17]	-	0.20	1.80-2.00	0.20 Al 0.20-0.60 Pb
C17410	Rem. [16]	0.20	-	-	0.35-0.6	-	0.20	0.15-0.50	0.20 Al
C17450	Rem. [16]	0.20	0.25	0.50-1.0	-	-	0.20	0.15-0.50	0.20 Al 0.50 Zr
C17455	Rem. [16]	0.20	0.25	0.50-1.0	-	-	0.20	0.15-0.50	0.20 Al 0.50 Zr 0.20-0.60 Pb
C17460	Rem. [16]	0.20	0.25	1.0-1.4	-	-	0.20	0.15-0.50	0.20 Al 0.50 Zr
C17465	Rem. [16]	0.20	0.25	1.0-1.4	-	-	0.20	0.15-0.50	0.20 Al 0.50 Zr 0.20-0.60 Pb
C17500	Rem. [16]	0.10	-	-	2.4-2.7	-	0.20	0.40-0.7	0.20 Al
C17510	Rem. [16]	0.10	-	1.4-2.2	0.30	-	0.20	0.20-0.6	0.20 Al
C17530	Rem. [16]	0.20	-	1.8-2.5 [18]	-	-	0.20	0.20-0.40	0.6 Al
C18000	Rem. [16]	0.15	-	1.8-3.0 [18]	-	0.10-0.8	0.40-0.8	-	-
C18030	Rem. [19]	-	0.08-0.12	-	-	0.10-0.20	-	-	0.005-0.015 P
C18040	Rem. [14]	-	0.20-0.30	-	-	0.25-0.35	-	-	0.005-0.015 P 0.05-0.15 Zn
C18045	99.1 min. [19]	-	0.20-0.30	-	-	0.20-0.35	0.05	-	0.15-0.30 Zn

(Continued)

Table 4. *(Continued)* **Chemical Composition of Wrought High Copper Alloys.** *(Source, Copper Development Association.)*

Copper Alloy Number	Cu + Ag	Fe	Sn	Ni	Co	Cr	Si	Be	Other (named)
C18050	Rem. [20]	-	-	-	-	0.05-0.15	-	-	0.005-0.015 Te
C18070	Rem. [20]	-	-	-	-	0.15-0.40	0.02-0.07	-	0.01-0.40 Ti
C18080	Rem. [20]	0.02-0.20	-	-	-	0.20-0.7	0.01-0.10	-	0.01-0.30 Ag 0.01-0.15 Ti
C18100	98.7 min. [16]	-	-	-	-	0.40-1.2	-	-	0.03-0.06 Mg 0.08-0.20 Zr
C18135	Rem. [16]	-	-	-	-	0.20-0.6	-	-	0.20-0.6 Cd
C18140	Rem. [16]	-	-	-	-	0.15-0.45	0.005-0.05	-	0.05-0.25 Zr
C18145	Rem. [16]	-	-	-	-	0.10-0.30	-	-	0.10-0.30 Zn 0.05-0.15 Zr
C18150	Rem. [21]	-	-	-	-	0.50-1.5	-	-	0.05-0.25 Zr
C18200	Rem. [16]	0.10	-	-	-	0.6-1.2	0.10	-	0.05 Pb
C18400	Rem. [16]	0.15	-	-	-	0.40-1.2	0.10	-	0.005 As 0.005 Ca 0.05 Li 0.05 P 0.7 Zn
C18600	Rem. [16]	0.25-0.8	-	0.25	0.10	0.10-1.0	-	-	0.05-0.50 Ti 0.05-0.40 Zr
C18610	Rem. [16]	0.10	-	0.25	0.25-0.8	0.10-1.0	-	-	0.05-0.50 Ti 0.05-0.45 Zr
C18665	99.0 min.	-	-	-	-	-	-	-	0.40-0.9 Mg 0.002-0.04 P
C18700	99.5 min. [63]	-	-	-	-	-	-	-	0.80-1.50 Pb
C18835	99.0 min. [16]	0.10	0.15-0.55	-	-	-	-	-	0.01 P, 0.05 Pb 0.02 0.30 Zn

(Continued)

Table 4. *(Continued)* **Chemical Composition of Wrought High Copper Alloys.** *(Source, Copper Development Association.)*

Copper Alloy Number	Cu + Ag	Fe	Sn	Ni	Co	Cr	Si	Be	Other (named)
C18900	Rem. [16]	-	0.6-0.9	-	-	-	0.15-0.40	-	0.05 P, 0.02 Pb 0.10-0.30 Mn 0.10 Zn 0.10 Al
C18980	Rem. [16]	-	1.0	-	-	-	0.50	-	0.50 Mn 0.15 P, 0.02 Pb
C18990	Rem. [19]	-	1.8-2.2	-	-	0.10-0.20	-	-	0.005-0.015 P
C19000	Rem. [16]	0.10	-	0.9-1.3	-	-	-	-	0.8 Zn, 0.05 Pb 0.15-0.35 P
C19010	Rem. [16]	-	-	0.8-1.8	-	-	0.15-0.35	-	0.01-0.05 P
C19015	Rem. [20]	-	-	0.50-2.4	-	-	0.10-0.40	-	0.02-0.20 P 0.02-0.15 Mg
C19020	Rem. [20]	-	0.30-0.9	0.50-3.0	-	-	-	-	0.01-0.20 P
C19025	Rem. [21]	0.10	0.7-1.1	0.8-1.2	-	-	-	-	0.03-0.07 P 0.20 Zn
C19030	Rem. [21]	0.10	1.0-1.5	1.5-2.0	-	-	-	-	0.01-0.03 P 0.02 Pb
C19100	Rem. [16]	0.20	-	0.9-1.3	-	-	-	-	0.50 Zn, 0.10 Pb 0.35-0.6 Te 0.15-0.35 P
C19140	Rem. [16]	0.05	0.05	0.8-1.2	-	-	-	-	0.50 Zn 0.40-0.80 Pb 0.15-0.35 P
C19150	Rem. [16]	0.05	0.05	0.8-1.2	-	-	-	-	0.50 Zn 0.50-1.0 Pb 0.15-0.35 P
C19160	Rem. [16]	0.05	0.05	0.8-1.2	-	-	-	-	0.50 Zn 0.80 to 1.2 Pb 0.15-0.35 P

(Continued)

Table 4. *(Continued)* **Chemical Composition of Wrought High Copper Alloys.** *(Source, Copper Development Association.)*

-	Cu	Fe	Sn	Zn	Al	Pb	P	Other (named)
C19200	98.5 min. [20]	0.8-1.2	-	0.20	-	-	0.01-0.04	-
C19210	Rem. [20]	0.05-0.15	-	-	-	-	0.025-0.040	-
C19215	Rem. [20]	0.05-0.20	-	1.1-3.5	-	0.02	0.025-0.050	-
C19220	Rem. [20]	0.10-0.30	0.05-0.10	-	-	-	0.03-0.07	0.005-0.015 B 0.10-0.25 Ni
C19260	98.5 min. [19]	0.40-0.8	-	-	-	-	-	0.20-0.40 Ti 0.02-0.15 Mg
C19280	Rem. [20]	0.50-1.5	0.30-0.7	0.30-0.7	-	-	0.005-0.015	-
C19400	97.0 min.	2.1-2.6	-	0.05-0.20	-	0.03	0.015-0.15	-
C19410	Rem. [20]	1.8-2.3	0.6-0.9	0.10-0.20	-	-	0.015-0.050	-
C19450	Rem. [20]	1.5-3.0	0.8-2.5	-	-	-	0.005-0.05	-
C19500	96.0 min. [20]	1.0-2.0	0.10-1.0	0.20	0.02	0.02	0.01-0.35	0.30-1.3 Co
C19520	96.6 min. [20]	0.50-1.5	-	-	-	0.01-3.5	-	-
C19700	Rem. [20]	0.30-1.2	0.20	0.20	-	0.05	0.10-0.40	0.01-0.20 Mg 0.05 Ni 0.05 Co 0.05 Mn
C19710	Rem. [16]	0.05-0.40	0.20	0.20	-	0.05	0.07-0.15	0.10 Ni + Co 0.05 Mn 0.03-0.06 Mg
C19720	Rem. [16]	0.05-0.50	0.20	0.20	-	0.05	0.05-0.15	0.06-0.20 Mg 0.05 Mn 0.10 Ni + Co
C19750	Rem. [20]	0.35-1.2	0.05-0.40	0.20	-	0.05	0.10-0.40	0.01-0.20 Mg 0.05 Ni 0.05 Co 0.05 Mn

(Continued)

Table 4. *(Continued)* **Chemical Composition of Wrought High Copper Alloys.** *(Source, Copper Development Association.)*

-	Cu	Fe	Sn	Zn	Al	Pb	P	Other (named)
C19800	Rem. [20]	0.02-0.50	0.10-1.0	0.30-1.5	-	-	0.01-0.10	0.10-1.0 Mg
C19810	Rem. [20]	1.5-3.0	-	1.0-5.0	-	-	0.10	0.10 Cr 0.10 Mg 0.10 Ti 0.10 Zr
C19900	Rem. [16]	-	-	-	-	-	-	2.9-3.4 Ti

Notes: [14] Includes oxygen-free or deoxidized grades with deoxidizers (such as phosphorus, boron, lithium or others) in an amount agreed upon.
[16] Cu + Sum of Named Elements, 99.5% min.
[17] Ni + Co, 0.20% min.; Ni + Fe + Co, 0.6% max.
[18] Ni values include Co.
[19] Cu + Sum of Named Elements, 99.9% min.
[20] Cu + Sum of Named Elements, 99.8% min.
[21] Cu + Sum of Named Elements, 99.7% min.
[63] Includes Pb.

Table 5. Chemical Composition of Wrought Brasses and Tin Brasses.
(Source, Copper Development Association.)

Copper Alloy No.	Cu	Pb	Fe	Zn	Other (named)
Copper-Zinc Alloys *(Brasses)*					
C21000	94.0-96.0 [20]	0.03 [40]	0.05	Rem.	-
C22000	89.0-91.0 [20]	0.05	0.05	Rem.	-
C22600	86.0-89.0 [20]	0.05	0.05	Rem.	-
C23000	84.0-86.0 [20]	0.05	0.05	Rem.	-
C23030	83.5-85.5 [20]	0.05	0.05	Rem.	0.20-0.40 Si
C23400	81.0-84.0 [20]	0.05	0.05	Rem.	-
C24000	78.5-81.5 [20]	0.05	0.05	Rem.	-
C24080	78.0-82.0 [20]	0.20	-	Rem.	0.10 Al
C25600	71.0-73.0 [21]	0.05	0.05	Rem.	-
C26000	68.5-71.5 [21]	0.07	0.05	Rem.	-
C26130	68.5-71.5 [21]	0.05	0.05	Rem.	0.02-0.08 As
C26200	67.0-70.0 [21]	0.07	0.05	Rem.	-
C26800	64.0-68.5 [21]	0.15	0.05	Rem.	-
C27000	63.0-68.5 [21]	0.10	0.07	Rem.	-
C27200	62.0-65.0 [21]	0.07	0.07	Rem.	-
C27400	61.0-64.0 [21]	0.10	0.05	Rem.	-
C28000	59.0-63.0 [21]	0.30	0.07	Rem.	-
Copper-Zinc-Lead Alloys *(Brasses)*					
C31200	87.5-90.5 [22]	0.7-1.2	0.10	Rem.	0.25 Ni
C31400	87.5-90.5 [22]	1.3-2.5	0.10	Rem.	0.7 Ni
C31600	87.5-90.5 [22]	1.3-2.5	0.10	Rem.	0.7-1.2 Ni 0.04-0.10 P
C32000	83.5-86.5 [22]	1.5-2.2	0.10	Rem.	0.25 Ni
C33000	65.0-68.0 [22]	.25-0.7	0.07	Rem.	-
C33200	65.0-68.0 [22]	1.5-2.5	0.07	Rem.	-
C33500	62.0-65.0 [22]	0.25-0.7	0.15 [24]	Rem.	-
C34000	62.0-65.0 [22]	0.8-1.5	0.15 [24]	Rem.	-
C34200	62.0-65.0 [22]	1.5-2.5	0.15 [24]	Rem.	-
C34500	62.0-65.0 [22]	1.5-2.5	0.15	Rem.	-
C35000	60.0-63.0 [22, 25]	0.8-2.0	0.15 [24]	Rem.	-
C35300	60.0-63.0 [16, 25]	1.5-2.5	0.15 [24]	Rem.	-
C35330	59.5-64.0 [16]	1.5-3.5 [26]	-	Rem.	0.02-0.25 As
C35600	60.0-63.0 [16]	2.0-3.0	0.15 [24]	Rem.	-
C36000	60.0-63.0 [16]	2.5-3.7	0.35	Rem.	-
C36500	58.0-61.0 [22]	0.25-0.7	0.15	Rem.	0.25 Sn
C37000	59.0-62.0 [22]	0.8-1.5	0.15	Rem.	-
C37100	58.0-62.0 [22]	0.6-1.2	0.15	Rem.	-
C37700	58.0-61.0 [16]	1.5-2.5	0.30	Rem.	-

(Continued)

Table 5. *(Continued)* **Chemical Composition of Wrought Brasses and Tin Brasses.**
(Source, Copper Development Association.)

Copper Alloy No.	Cu	Pb	Fe	Zn	Other (named)
Copper-Zinc-Lead Alloys *(Brasses)*					
C37710	56.5-60.0 [16]	1.0-3.0	0.30	Rem.	-
C38000	55.0-60.0 [16]	1.5-2.5	0.35	Rem.	0.50 Al 0.30 Sn
C38500	55.0-59.0 [16]	2.5-3.5	0.35	Rem.	-

Alloy No.	Cu	Pb	Fe	Sn	Zn	P	Other (named)
Copper-Zinc-Tin Alloys *(Tin Brasses)*							
C40400	Rem. 21	-	-	0.35-0.7	2.0-3.0	-	-
C40500	94.0-96.0 [21]	0.05	0.05	0.7-1.3	Rem.	-	-
C40810	94.0-96.0 [21]	0.05	0.08-0.12	1.8-2.2	Rem.	0.028-.04	0.11-0.20 Ni
C40820	94.0 min. [16]	0.02	-	1.0-2.5	.20-2.5	0.05	0.10-0.50 Ni
C40850	94.5-96.5 [21]	0.05	0.05-0.20	2.6-4.0	Rem.	0.02-0.04	0.05-0.20 Ni
C40860	94.0-96.0 [21]	0.05	0.01-0.05	1.7-2.3	Rem.	0.02-0.04	0.05-0.20 Ni
C41000	91.0-93.0 [21]	0.05	0.05	2.0-2.8	Rem.	-	-
C41100	89.0-92.0 [21]	0.10	0.05	0.30-0.7	Rem.	-	-
C41120	89.0-92.0 [21]	0.05	0.05-0.20	0.30-0.7	Rem.	0.02-0.05	0.05-0.20 Ni
C41300	89.0-93.0 [21]	0.10	0.05	0.7-1.3	Rem.	-	-
C41500	89.0-93.0 [21]	0.10	0.05	1.5-2.2	Rem.	-	-
C42000	88.0-91.0 [21]	-	-	1.5-2.0	Rem.	0.25	-
C42200	86.0-89.0 [21]	0.05	0.05	0.8-1.4	Rem.	0.35	-
C42220	88.0-91.0 [21]	0.05	0.05-0.20	0.7-1.4	Rem.	0.02-0.05	0.20 Ni
C42500	87.0-90.0 [21]	0.05	0.05	1.5-3.0	Rem.	0.35	-
C42520	88.0-91.0 [21]	0.05	0.05-0.20	1.5-3.0	Rem.	0.02-0.04	0.05-0.20 Ni
C42600	87.0-90.0 [21]	0.05	0.05-0.20	2.5-4.0	Rem.	0.02-0.05	0.05-0.20 Ni (incl. Co)
C43000	84.0-87.0 [21]	0.10	0.05	1.7-2.7	Rem.	-	-
C43400	84.0-87.0 [21]	0.05	0.05	0.40-1.0	Rem.	-	-
C43500	79.0-83.0 [21]	0.10	0.05	0.6-1.2	Rem.	-	-
C43600	80.0-83.0 [21]	0.05	0.05	0.20-0.50	Rem.	-	-
C44300	70.0-73.0 [22]	0.07	0.06	0.8-1.2 [27]	Rem.	-	0.02-0.06 As
C44400	70.0-73.0 [22]	0.07	0.06	0.8-1.2 [27]	Rem.	-	0.02-0.10 Sb
C44500	70.0-73.0 [22]	0.07	0.06	0.8-1.2 [27]	Rem.	0.02-0.10	-
C46200	62.0-65.0 [22]	0.20	0.10	0.50-1.0	Rem.	-	-
C46400	59.0-62.0 [22]	0.20	0.10	0.50-1.0	Rem.	-	-
C46500	59.0-62.0 [22]	0.20	0.10	0.50-1.0	Rem.	-	0.02-0.06 As
C47000	57.0-61.0 [22]	0.05	-	0.25-1.0	Rem.	-	0.01 Al
C47940	63.0-66.0 [22]	1.0-2.0	0.10-1.0	1.2-2.0	Rem.	-	0.10-0.50 Ni (incl. Co)
C48200	59.0-62.0 [22]	0.40-1.0	0.10	0.50-1.0	Rem.	-	-

(Continued)

Table 5. *(Continued)* **Chemical Composition of Wrought Brasses and Tin Brasses.**
(Source, Copper Development Association.)

Alloy No.	Cu	Pb	Fe	Sn	Zn	P	Other (named)
Copper-Zinc-Tin Alloys *(Tin Brasses)*							
C48500	59.0-62.0 [22]	1.3-2.2	0.10	0.50-1.0	Rem.	-	-
C48600	59.0-62.0 [22]	1.0-2.5	-	0.30-1.5	Rem.	-	0.2-0.25 As

Notes: [16] Cu + Sum of Named Elements, 99.5% min.
[20] Cu + Sum of Named Elements, 99.8% min.
[21] Cu + Sum of Named Elements, 99.7% min.
[22] Cu + Sum of Named Elements, 99.6% min.
[24] For flat products, the iron shall be .10% max.
[25] Cu, 61.0% min. for rod.
[26] Pb may be reduced to 1.0% by agreement.
[27] For tubular products, the minimum Sn content may be .9%.
[40] 0.05% Pb, max., for rod wire and tube.

Table 6. Chemical Composition of Wrought Bonzes and Copper Zinc Alloys.
(Source, Copper Development Association.)

Copper Alloy No.	Cu[16]	Pb	Fe	Sn	Zn	P	Other (named)
Copper-Tin-Phosphorus Alloys *(Phosphor Bronzes)*							
C50100	Rem.	0.05	0.05	0.50-0.8	-	0.01-0.05	-
C50200	Rem.	0.05	10	1.0-1.5	-	0.04	-
C50500	Rem.	0.05	0.10	1.0-1.7	0.30	0.03-0.35	-
C50510	Rem. [21]	-	-	1.0-1.5	0.10-0.25	0.02-0.07	0.15-0.40 Ni
C50580	Rem.	0.05	0.05-0.20	1.0-1.7	0.30	0.02-0.10	0.05-0.20 Ni
C50590	97.0 min.	0.02	0.05-0.40	0.50-1.5	0.50	0.02-0.15	-
C50700	Rem.	0.05	0.10	1.5-2.0	-	0.30	-
C50705	96.5 min.	0.02	0.10-0.40	1.5-2.0	0.50	0.04-0.15	-
C50710	Rem.	-	-	1.7-2.3	-	0.15	0.10-0.40 Ni
C50715	Rem. [28]	0.02	0.05-0.15	1.7-2.3	-	0.025-0.04	-
C50725	94.0 min.	0.02	0.05-0.20	1.5-2.5	1.5-3.0	0.02-0.06	-
C50780	Rem.	0.05	0.05-0.20	1.7-2.3	0.30	0.02-0.10	0.05-0.20 Ni
C50900	Rem.	0.05	0.10	2.5-3.8	0.30	0.03-0.30	-
C51000	Rem.	0.05	0.10	4.2-5.8	0.30	0.03-0.35	-
C51080	Rem.	0.05	0.05-0.20	4.8-5.8	0.30	0.02-0.10	0.05-0.20 Ni
C51100	Rem.	0.05	0.10	3.5-4.9	0.30	0.03-0.35	-
C51180	Rem.	0.05	0.05-0.20	3.5-4.9	0.30	0.02-0.10	0.11-0.20 Ni
C51190	Rem.	0.02	0.05-0.15	3.0-6.5	-	0.025-0.045	0.15 Co
C51800	Rem.	0.02	-	4.0-6.0	-	0.10-0.35	0.01 Al
C51900	Rem.	0.05	0.10	5.0-7.0	0.30	0.03-0.35	-
C51980	Rem.	0.05	0.05-0.20	5.5-7.0	0.30	0.02-0.10	0.05-0.20 Ni
C52100	Rem.	0.05	0.10	7.0-9.0	0.20	0.03-0.35	-
C52180	Rem.	0.05	0.05-0.20	7.0-9.0	0.30	0.02-0.10	0.05-0.20 Ni

(Continued)

Table 6. *(Continued)* **Chemical Composition of Wrought Bonzes and Copper Zinc Alloys.**
(Source, Copper Development Association.)

Copper Alloy No.	Cu[16]	Pb	Fe	Sn	Zn	P	Other (named)
Copper-Tin-Phosphorus Alloys *(Phosphor Bronzes)*							
C52400	Rem.	0.05	0.10	9.0-11.0	0.20	0.03-0.35	-
C52480	Rem.	0.05	0.05-0.20	9.0-11.0	0.30	0.02-0.10	0.05-0.20 Ni
Alloy No.		**Cu[16]**	**Pb**	**Fe**	**Sn**	**Zn**	**P**
Copper-Tin-Lead-Phosphorus Alloys *(Leaded Phosphor Bronzes)*							
C53400		Rem.	0.8-1.2	0.10	3.5-5.8	0.30	0.03-0.35
C54400		Rem.	3.5-4.5	0.10	3.5-4.5	1.5-4.5	0.01-0.50
Alloy No.		**Cu[29]**		**Ag**		**P**	
Copper-Phosphorus and Copper-Silver-Phophorus Alloys *(Brazing Alloys)*							
C55180		Rem.		-		4.8-5.2	
C55181		Rem.		-		7.0-7.5	
C55280		Rem.		1.8-2.2		6.8-7.2	
C55281		Rem.		4.8-5.2		5.8-6.2	
C55282		Rem.		4.8-5.2		6.5-7.0	
C55283		Rem.		5.8-6.2		7.0-7.5	
C55284		Rem.		14.5-15.5		4.8-5.2	
Alloy No.		**Cu**		**Ag**		**Zn**	
Copper-Silver-Zinc Alloys							
C56000		Rem. [16]		29.0-31.0		30.0-34.0	

(Continued)

Table 6. *(Continued)* **Chemical Composition of Wrought Bonzes and Copper Zinc Alloys.** *(Source, Copper Development Association.)*

Alloy No.	Cu[16] incl. Ag	Pb	Fe	Sn	Zn	Al	Mn	Si	Ni	Other (named)
Copper-Aluminum Alloys *(Aluminum Bronzes)*										
C60800	Rem.	0.10	0.10	-	-	5.0-6.5	-	-	-	0.02-0.35 As
C61000	Rem.	0.02	0.50	-	0.20	6.0-8.5	-	0.10	-	-
C61300	Rem. [20]	0.01	2.0-3.0	0.20-0.50	0.10 [30]	6.0-7.5	0.20	0.10	0.15	0.015 P [30]
C61400	Rem.	0.01	1.5-3.5	-	0.20	6.0-8.0	1.0	-	-	0.015 P
C61500	Rem.	0.015	-	-	-	7.7-8.3	-	-	1.8-2.2	-
C61550	Rem.	0.05	0.20	0.05	0.8	5.5-6.5	1.0	-	1.5-2.5	-
C61800	Rem.	0.02	0.50-1.5	-	0.02	8.5-11.0	-	0.10	-	-
C61900	Rem.	0.02	3.0-4.5	0.6	0.8	8.5-10.0	-	-	-	-
C62200	Rem.	0.02	3.0-4.2	-	0.02	11.0-12.0	-	0.10	-	-
C62300	Rem.	-	2.0-4.0	0.6	-	8.5-10.0	0.50	0.25	1.0	-
C62400	Rem.	-	2.0-4.5	0.20	-	10.0-11.5	0.30	0.25	-	-
C62500	Rem.	-	3.5-5.5	-	-	12.5-13.5	2.0	-	-	-
C62580	Rem.	0.02	3.0-5.0	-	0.02	12.0-13.0	-	0.04	-	-
C62581	Rem.	0.02	3.0-5.0	-	0.02	13.0-14.0	-	0.04	-	-
C62582	Rem.	0.02	3.0-5.0	-	0.02	14.0-15.0	-	0.04	-	-
C63000	Rem.	-	2.0-4.0	0.20	0.30	9.0-11.0	1.5	0.25	4.0-5.5	-
C63010	78.0 min. [20]	-	2.0-3.5	0.20	0.30	9.7-10.9	1.5	-	4.5-5.5	-
C63020	74.5 min.	0.03	4.0-5.5	0.25	0.30	10.0-11.0	1.5	-	4.2-6.0	0.20 Co 0.05 Cr
C63200	Rem.	0.02	3.5-4.3 [31]	-	-	8.7-9.5	1.2-2.0	0.10	4.0-4.8 [31]	-
C63280	Rem.	0.02	3.0-5.0	-	-	8.5-9.5	0.6-3.5	-	4.0-5.5	-
C63380	Rem.	0.02	2.0-4.0	-	0.15	7.0-8.5	11.0-14.0	0.10	1.5-3.0	-

(Continued)

Table 6. *(Continued)* **Chemical Composition of Wrought Bonzes and Copper Zinc Alloys.** *(Source, Copper Development Association.)*

Alloy No.	Cu[16] incl. Ag	Pb	Fe	Sn	Zn	Al	Mn	Si	Ni	Other (named)
Copper-Aluminum Alloys *(Aluminum Bronzes)*										
C63400	Rem.	0.05	0.15	0.20	0.50	2.6-3.2	-	0.25-0.45	0.15	0.15 As
C63600	Rem.	0.05	0.15	0.20	0.50	3.0-4.0	-	0.7-1.3	0.15	0.15 As
C63800	Rem.	0.05	0.20	-	0.8	2.5-3.1	0.10	1.5-2.1	0.20 [32]	0.25-0.55 Co
C64200	Rem.	0.05	0.30	0.20	0.50	6.3-7.6	0.10	1.5-2.2	0.25	0.15 As
C64210	Rem.	0.05	0.30	0.20	0.50	6.3-7.0	0.10	1.5-2.0	0.25	0.15 As
Alloy No.	**Cu[16] incl. Ag**	**Pb**	**Fe**	**Sn**	**Zn**	**Mn**	**Si**	**Ni**	**Other (named)**	
Copper-Silicon Alloys *(Silicon Bronzes)*										
C64700	Rem.	0.10	0.10	-	0.50	-	0.40-8	1.6-2.2	-	
C64710	95.0 min.	-	-	-	0.20-0.50	0.10	0.50-9	2.9-3.5	-	
C64725	95.0 min.	0.01	0.25	0.20-0.8	0.50-1.5	-	0.20-8	1.3-2.7	0.01 Ca 0.20 Mg 0.20 Cr	
C64730	93.5 min.	-	-	1.0-1.5	0.20-0.50	0.10	0.50-9	2.9-3.5	-	
C64740	95.0 min.	0.10 max.	-	1.5-2.5	0.20-1.0	-	0.05-0.50	1.0-2.0	0.01 Ca 0.05 Mg	
C64750	Rem.	-	1.0	0.05-0.8	1.0	-	0.10-0.7	0.50-3.0	0.10 Mg 0.10 P 0.10 Zr	
C64760	93.5 min.	0.02	-	0.30	0.20-2.5	-	0.05-0.6	0.40-2.5	0.05 Mg	
C64780	90.0	0.02	-	0.10-2.0	0.20-2.5	0.01-1.0	0.20-0.9	1.0-3.5	0.01 Cr 0.01 Mg 0.01 Ti 0.01 Zr	

(Continued)

Table 6. *(Continued)* **Chemical Composition of Wrought Bonzes and Copper Zinc Alloys.** *(Source, Copper Development Association.)*

Alloy No.	Cu[16] incl. Ag	Pb	Fe	Sn	Zn	Mn	Si	Ni	Other (named)
Copper-Silicon Alloys *(Silicon Bronzes)*									
C64900	Rem.	0.05	0.10	1.2-1.6	1.20	-	0.8-1.2	0.10	0.10 Al
C65100	Rem.	0.05	0.8	-	1.5	0.7	0.8-2.0	-	-
C65400	Rem.	0.05	-	1.2-1.9	0.50	-	2.7-3.4	-	0.01-0.12 Cr
C65500	Rem.	0.05	0.8	-	1.5	0.50-1.3	2.8-3.8	0.6	-
C65600	Rem.	0.02	0.50	1.5	1.5	1.5	2.8-4.0	-	0.01 Al
C66100	Rem.	0.20-8	0.25	-	1.5	1.5	2.8-3.5	-	-
Alloy No.	**Cu[16] incl. Ag**	**Pb**	**Fe**	**Sn**	**Zn**	**Ni incl. Co**	**Al**	**Si**	**Other (named)**
Other Copper-Zinc Alloys									
C66200	86.6-91.0	0.05	0.05	0.20-0.7	Rem.	0.30-1.0	-	-	0.05-0.20 P
C66300	84.5-87.5	0.05	1.4-2.4 [61]	1.5-3.0	Rem.	-	-	-	0.35-0.20 Co
C66400	Rem.	0.015	1.3-1.7 [33]	0.05	11.0-12.0	-	-	-	0.30-0.70 Co[33]
C66410	Rem.	0.015	1.8-2.3	0.05	11.0-12.0	-	-	-	-
C66420	Rem.	-	0.50-1.5	-	12.0-17.0	-	-	-	-
C66430	Rem.	0.05	0.6-0.9	0.6-0.9	13.0-15.0	-	-	-	0.10 P
C66700	68.5-71.5	0.07	0.10	-	Rem.	-	-	-	0.8-1.5 Mn
C66800	60.0-63.0	0.50	0.35	0.30	Rem.	0.25	0.25	0.50-1.5	2.0-3.5 Mn
C66900	62.5-64.5 [20]	0.05	0.25	-	Rem.	-	-	-	11.5-12.5 Mn
C66950	Rem.	0.01	0.50	-	14.0-15.0	-	1.0-1.5	-	14.0-15.0 Mn
C67000	63.0-68.0	0.20	2.0-4.0	0.50	Rem.	-	3.0-6.0	-	2.5-5.0 Mn
C67300	58.0-63.0	0.40-3.0	0.50	0.30	Rem.	0.25	0.25	0.50-1.5	2.0-3.5 Mn
C67400	57.0-60.0	0.50	0.35	0.30	Rem.	0.25	0.50-2.0	0.50-1.5	2.0-3.5 Mn

(Continued)

Table 6. *(Continued)* **Chemical Composition of Wrought Bonzes and Copper Zinc Alloys.** *(Source, Copper Development Association.)*

Alloy No.	Cu[16] incl. Ag	Pb	Fe	Sn	Zn	Ni incl. Co	Al	Si	Other (named)
Other Copper-Zinc Alloys									
C67420	57.0-58.5	0.25-8	0.15-55	0.35	Rem.	0.25	1.0-2.0	0.25-7	1.5-2.5 Mn
C67500	57.0-60.0	0.20	0.8-2.0	0.50-1.5	Rem.	-	0.25	0.50-1.5	0.05-0.50 Mn
C67600	57.0-60.0	0.50-1.0	0.40-1.3	0.50-1.5	Rem.	-	-	0.50-1.5	0.05-0.50 Mn
C68000	56.0-60.0	0.05	0.25-1.25	0.75-1.10	Rem.	0.20-8	0.01	0.04-0.15	0.01-0.50 Mn
C68100	56.0-60.0	0.05	0.25-1.25	0.75-1.10	Rem.	-	0.01	0.04-0.15	0.01-0.50 Mn
C68700	76.0-79.0	0.07	0.06	-	Rem.	-	1.8-2.5	-	0.02-0.06 As
C68800	Rem.	0.05	0.20	-	21.3-24.1 [34]	-	3.0-3.8 [34]	-	0.25-0.55 Co
C69050	70.0-75.0	-	-	-	Rem.	0.50-1.5	3.0-4.0	0.10-0.6	0.01-0.20 Zr
C69100	81.0-84.0	0.05	0.25	0.10	Rem.	0.8-1.4	0.7-1.2	0.8-1.3	0.10 min. Mn
C69400	80.0-83.0	0.30	0.20	-	Rem.	-	-	3.5-4.5	-
C69430	80.0-83.0	0.30	0.20	-	Rem.	-	-	3.5-4.5	0.03-0.06 As
C69700	75.0-80.0	0.50-1.5	0.20	-	Rem.	-	-	2.5-3.5	0.40 Mn
C69710	75.0-80.0	0.50-1.5	0.20	-	Rem.	-	-	2.5-3.5	0.03-0.06 As 0.040 Mn

Notes: [16] Cu + Sum of Named Elements, 99.5% min.
[20] Cu + Sum of Named Elements, 99.8% min.
[21] Cu + Sum of Named Elements, 99.7% min.
[28] Cu + Sn + Fe + P, 99.5% min.
[29] Cu + Sum of Named Elements, 99.85% min.
[30] When the products is for subsequent welding applications and is so specified by the purchaser, Cr, Cd, Zr and Zn shall each be 0.05% max.
[31] Fe content shall not exceed Ni content.
[32] Not including Co.
[33] Fe + Co, 1.8-2.3%.
[34] Al + Zn, 25.1-27.1%.
[61] Fe + Co, 1.4-2.4%.

Table 7. Chemical Composition of Wrought Copper-Nickel Alloys. *(Source, Copper Development Association.)*

Copper Alloy No.	Cu incl. Ag	Pb	Fe	Zn	Ni	Sn	Mn	Other (named)
C70100	Rem. [16]	-	.05	.25	3.0-4.0	-	.50	-
C70200	Rem. [16]	.05	.10	-	2.0-3.0	-	.40	-
C70230	Rem. [16]	-	-	.50-2.0	2.2-3.2	.10-.50	-	.40-.80 Si .10 Ag+B
C70250	Rem. [16]	.05	.20	1.0	2.2-4.2	-	.10	.05-.30 Mg .25-1.2 Si
C70260	Rem. [16]	-	-	-	1.0-3.0	-	-	.20-.70 Si .005 P
C70270	Rem. [16]	.05	.28-1.0	1.0	1.0-3.0	.10-1.0	.15	.20-1.0 Si
C70280	Rem. [16]	.02	.015	.30	1.3-1.7	1.0-1.5	-	.02-.04 P .22-.30 Si
C70290	Rem. [16]	.02	.015	.30	1.3-1.7	2.1-2.7	-	.02-.04 P .22-.30 Si
C70400	Rem. [16]	.05	1.3-1.7	1.0	4.8-6.2	-	.30-.80	-
C70500	Rem. [16]	.05	.10	.20	5.8-7.8	-	.15	-
C70600	Rem. [16]	.05	1.0-1.8	1.0	9.0-11.0	-	1.0	-
C70610	Rem. [16]	.01	1.0-2.0	-	10.0-11.0	-	.50-1.0	.05 S .05 C
C70620	86.5 min. [16]	.02	1.0-1.8	.50	9.0-11.0	-	1.0	.05 C .02 P .02 S
C70690	Rem. [16]	.001	.005	.001	9.0-11.0	-	.001	[36]
C70700	Rem. [16]	-	.05	-	9.5-10.5	-	.50	-
C70800	Rem. [16]	.05	.10	.20	10.5-12.5	-	.15	-
C71000	Rem. [16]	.05	1.0	1.0	19.0-23.0	-	1.0	-
C71100	Rem. [16]	.05	.10	.20	22.0-24.0	-	.15	-
C71300	Rem. [16]	.05	.20	1.0	23.5-26.5	-	1.0	-

(Continued)

Table 7. *(Continued)* **Chemical Composition of Wrought Copper-Nickel Alloys.** *(Source, Copper Development Association.)*

Copper Alloy No.	Cu incl. Ag	Pb	Fe	Zn	Ni	Sn	Mn	Other (named)
C71500	Rem. [16]	.05	.40-1.0	1.0	29.0-33.0	-	1.0	-
C71520	65.0 min. [16]	.02	.40-1.0	.50	29.0-33.0	-	1.0	.05 C .02 P .02 S
C71580	Rem. [16]	.05	.50	.50	29.0-33.0	-	.30	[37]
C71581	Rem. [16]	.02	.40-.7	-	29.0-32.0	-	1.0	[38]
C71590	Rem.	.001	.15	.001	29.0-31.0	.001	.50	[36]
C71640	Rem. [16]	.01	1.7-2.3	-	29.0-32.0	-	1.5-2.5	.03 S .06 C
C71700	Rem. [16]	-	.40-1.0	-	29.0-33.0	-	-	.30-.7 Be
C71900	Rem. [16]	.015	.50	.05	28.0-33.0	-	.20-1.0	2.2-3.0 Cr .02-.35 Zr .01-.20 Ti .04C .25 Si .015 S .02 P
C72150	Rem. [16]	.05	.10	.20	43.0-46.0	-	.05	.10 C .50 Si
C72200	Rem. [20]	.05 [35]	.50-1.0	1.0 [35]	15.0-18.0	-	1.0	.30-.70 Cr .03 Si .03Ti [35]
C72420	Rem. [21]	.02	.7-1.2	.20	13.5-16.5	.10	3.5-5.5	1.0-2.0 A1 .50 Cr .15 Si .05 Mg .15 S .01 P .05 C

(Continued)

Table 7. *(Continued)* **Chemical Composition of Wrought Copper-Nickel Alloys.** *(Source, Copper Development Association.)*

Copper Alloy No.	Cu incl. Ag	Pb	Fe	Zn	Ni	Sn	Mn	Other (named)
C72500	Rem. [20]	.05	.6	.50	8.5-10.5	1.8-2.8	.20	-
C72650	Rem. [21]	.01	.10	.10	7.0-8.0	4.5-5.5	.10	-
C72700	Rem. [21]	.02 [39]	.50	.50	8.5-9.5	5.5-6.5	.05-.30	.10 Nb .15 Mg
C72800	Rem. [21]	.005	.50	1.0	9.5-10.5	7.5-8.5	.05-.30	.10 A1 .001 B .001 Bi .10-.30 Nb .005-.15 Mg .005 P .0025 S .02 Sb .05 Si .01 Ti
C72900	Rem. [21]	.02 [39]	.50	.50	14.5-15.5	7.5-8.5	.30	.10 Nb .15 Mg
C72950	Rem. [21]	.05	.6	-	20.0-22.0	4.5-5.7	.6	-

Notes: [16] Cu + Sum of Named Elements, 99.5% min.
[20] Cu + Sum of Named Elements, 99.8% min.
[21] Cu + Sum of Named Elements, 99.7% min.
[35] The following additional maximum limits shall apply: When the product is for subsequent welding applications and is so specified by the purchaser, .50% Zn, .02% P, .02% Pb, .02% S (.008% S for C71110) and .05% C.
[36] The following additional maximum limits shall apply: .02% C, .015% Si, .003% S, .002% Al, .001% P, .0005% Hg, .001% Ti, .001% Sb, .001% As, .001% Bi, .05% Co, .10% Mg, and .005% Oxygen. For C70690, Co shall be .02% max.
[37] The following additional maximum limits shall apply: .07% C, .15% Si, .024% S, .05% Al, and .03% P.
[38] .02% P, max.; .25% Si, max.; .01% S, max.; .20-.50% Ti.
[39] .005% Pb, max., for hot rolling.

Table 8. Chemical Composition of Wrought Nickel-Silvers. *(Source, Copper Development Association.)*

Copper Alloy No.	Cu incl. Ag	Pb	Fe	Zn	Ni	Mn	Other (named)
C73500	70.5-73.5 [16]	.10	.25	Rem.	16.5-19.5	.50	-
C74000	69.0-73.5 [16]	.10	.25	Rem.	9.0-11.0	.50	-
C74300	63.0-66.0 [16]	.10	.25	Rem.	7.0-9.0	.50	-
C74400	62.0-66.0 [21]	.05	.05	Rem.	2.0-4.0	-	-
C74500	63.5-66.5 [16]	.10 [40]	.25	Rem.	9.0-11.0	.50	-
C75200	63.5-66.5 [16]	.05	.25	Rem.	16.5-19.5	.50	-
C75400	63.5-66.5 [16]	.10	.25	Rem.	14.0-16.0	.50	-
C75700	63.5-66.5 [16]	.05	.25	Rem.	11.0-13.0	.50	-
C76000	60.0-63.0 [16]	.10	.25	Rem.	7.0-9.0	.50	-
C76200	57.0-61.0 [16]	.10	.25	Rem.	11.0-13.5	.50	-
C76400	58.5-61.5 [16]	.05	.25	Rem.	16.5-19.5	.50	-
C76700	55.0-58.0 [16]	-	-	Rem.	14.0-16.0	.50	-
C77000	53.5-56.5 [16]	.05	.25	Rem.	16.5-19.5	.50	-
C77300	46.0-50.0 [16]	.05	-	Rem.	9.0-11.0	-	.01A1 .25P .04- .25Si
C77400	43.0-47.0 [16]	.20	-	Rem.	9.0-11.0	-	-
C78200	63.0-67.0 [16]	1.5-2.5	.35	Rem.	7.0-9.0	.50	-
C79000	63.0-67.0 [16]	1.5-2.2	.35	Rem.	11.0-13.0	.50	-
C79200	59.0-66.5 [16]	.8-1.4	.25	Rem.	11.0-13.0	.50	-
C79800	45.5-48.5 [16]	1.5-2.5	.25	Rem.	9.0-11.0	1.5-2.5	-
C79830	45.5-47.0 [16]	1.0-2.5	.45	Rem.	9.0-10.5	.15- .55	-

Notes: [16] Cu + Sum of Named Elements, 99.5% min.
[21] Cu + Sum of Named Elements, 99.7% min.
[40] .05% Pb, max., for rod wire and tube.

Table 9. Selected Mechanical Properties of Copper Alloy Rod [1] for Machined Products. *(Source, Copper Development Association)*

Copper Alloy No.	Temper	Tensile Strength		Yield Strength[2]		% Elongation[3]	Hardness	Shear Strength		Fatigue Strength[4]	
		KSI	MPa	KSI	MPa			KSI	MPa	KSI	MPa
C10100 C10200	H04 [5]	48	331	44	303	16	HRF87 HRB47	27	186	17	117
C10400 C10500 C10700 C11000 C11300 C11400 C11500 C11600 C12000 C12100	M20	32	221	10	69	55	HRF40	22	152	-	-
C10300 C10800	H04 [5]	48	331	44	303	16	HRF87 HRB47	27	186	-	-
C12200	H04	45	310	40	276	20	HRF85 HRB45	26	179	-	-
C12900	H04 [5]	48	331	44	303	16	HRF87 HRB47	27	186	-	-
	M20	32	221	10	69	55	HRF40	22	152	-	-
C14500 C14520	H04 [5]	48	331	44	303	20	HRB48	27	186	-	-
C14700	H04 [6]	46	317	43	296	11	HRB46	27	186	-	-
C15000	Note [7]	62	427	60	414	15	-	-	-	-	-
C15715	M30	57	393	-	-	27	HRB57	-	-	-	-
C15725	M30	60	414	-	-	19	HRB68	-	-	-	-
C15760	M30 [9]	80	552	-	-	22	HRB80	-	-	-	-
C16200	H04 [10]	73	503	69	474	9	HRB73	56	386	30	207

(Continued)

Table 9. *(Continued)* **Selected Mechanical Properties of Copper Alloy Rod [1] for Machined Products.** *(Source, Copper Development Association)*

Copper Alloy No.	Temper	Tensile Strength		Yield Strength[2]		% Elongation[3]	Hardness	Shear Strength		Fatigue Strength[4]	
		KSI	MPa	KSI	MPa			KSI	MPa	KSI	MPa
C16500	H04 [5, 10]	65	448	55	379	15	HRB75	-	-	-	-
C17200 C17300	TH04 [11]	205	1,415	-	-	3	HRC42	120	825	-	-
C17410	TH04 [12]	120	825	-	-	8	HRB102	70	483	-	-
C17500	TB00	45	310	-	-	28	HRB40	25	172	-	-
C17510	TF00 [13]	120	827	-	-	18	HRB95	70	483	40	276
C18000	TH04	100	690	-	-	13	-	-	-	-	-
C18135	TH01 [10, 14]	62	430	53	368	30	HRB71	-	-	25	172
C18150	TH04	78	538	-	-	13	-	-	-	-	-
C18200 C18400	TF00	72	496	65	448	18	HRB80	-	-	-	-
C18700	H04 [5]	48	331	42	290	15	HRB48	27	186	-	-
C19100	TH04 [15]	78	538	68	469	27	HRB84	41	283	33	228
C19150	TH04 [16]	90	621	-	-	-	HRB80	50	345	-	-
C21000	H04 [17]	55	379	35	241	10	HRB60	37	255	-	-
C22000	H00 [10]	45	310	35	241	25	HRB42	33	228	-	-
C22600	H04 [17]	67	462	50	345	10	HRB65	40	276	-	-
C23000	H04 [17]	70	483	52	359	10	HRB65	43	296	-	-
C24000	H04 [17]	80	552	63	434	12	HRB74	46	317	-	-
C26000 C26130 C26200	H02 [18]	70	483	52	359	30	HRB80	42	290	22 [19]	152 [19]
C26800 C27000	H00	55	379	40	276	48	HRB55	36	248	-	-

(Continued)

Table 9. *(Continued)* **Selected Mechanical Properties of Copper Alloy Rod [1] for Machined Products.** *(Source, Copper Development Association)*

Copper Alloy No.	Temper	Tensile Strength		Yield Strength[2]		% Elongation[3]	Hardness	Shear Strength		Fatigue Strength[4]	
		KSI	MPa	KSI	MPa			KSI	MPa	KSI	MPa
C28000	H01	72	496	50	345	25	HRB78	45	310	-	-
	M30	52	359	20	138	52	HRF78	39	269	-	-
	060	54	372	21	145	50	HRF80	40	276	-	-
C31400	H02 [18]	52	359	45	310	18	HRB58	30	207	-	-
C31600	H04 [5]	65	448	57	393	15	HRB70	39	269	-	-
C32000	H04	70	483	60	414	10	-	-	-	-	-
C34000	H01	55	379	42	290	40	HRB60	36	248	-	-
C34200 C35300 C35330	H02 [18]	58	400	45	310	25	HRB75	34	234	-	-
C34500	H02	70	483	58	400	22	HRB80	42	290	-	-
C35000	H02 [10, 18]	70	483	52	359	22	HRB80	42	290	-	-
C35600	H02 [18]	58	400	45	310	25	HRB78	34	234	-	-
C36000	060	49	338	18	124	53	HRF68	30	207	-	-
C37700	M30	52	359	20	138	45	HRF78	30	207	-	-
C46200	H02	70	483	50	345	2	HRB75	44	303	-	-
C46400	H02 [18]	75	517	53	365	20	HRB82	44	303	-	-
	H01 [20]	69	476	46	317	27	HRB78	43	296	-	-
	050	63	434	30	207	40	HRB60	42	290	-	-
	060	57	393	25	172	47	HRB55	40	276	-	-
C48200	060	57	393	25.0	172	40	HRB55	38	262	-	-
	050	63	434	30.0	207	35	HRB60	39	269	-	-
	H01 [20]	69	476	46	317	20	HRB78	40	276	-	-
	H02 [18]	75	517	53	365	15	HRB82	41	283	-	-

(Continued)

Table 9. *(Continued)* **Selected Mechanical Properties of Copper Alloy Rod [1] for Machined Products.** *(Source, Copper Development Association)*

Copper Alloy No.	Temper	Tensile Strength		Yield Strength[2]		% Elongation[3]	Hardness	Shear Strength		Fatigue Strength[4]	
		KSI	MPa	KSI	MPa			KSI	MPa	KSI	MPa
C48500	H02 [18]	75	517	53	365	15	HRB82	40	276	-	-
	H01 [20]	69	476	46	317	20	HRB78	39	269	-	-
	060	57	393	25	172	40	HRB55	36	248	-	-
C51000	H02 [18]	70	483	58	400	25	HRB78	50	345	-	-
C52100	H02 [10, 18]	80	552	65	448	33	HRB85	54	372	-	-
C54400	H04 [5, 10]	75	517	63	434	15	HRB83	52	359	-	-
	H04 [22]	68	469	57	393	20	HRB80	54	372	-	-
C61000	H04 [21]	70	483	-	-	-	HRB80	-	-	-	-
C61300	H04 [22]	82	565	55	379	35	HRB90	45	310	-	-
C61400	H04	82	565	40	276	35	HRB90	45	310	-	-
C61800	H02 [23]	85	586	42	293	23	HRB89	47	324	28	193
C62300	H02 [23]	95	655	50	345	25	HRB88	50	345	30	207
	M30 [24]	75	517	35	241	35	HRB80	35	241	25	172
C62400	H02 [25]	105	724	52	359	14	HRB92	65	448	36	248
	M30 [24]	90	621	40	276	18	HRB87	58	407	32	221
C62500	M30	100	690	55	379	1	HRC29	60	414	67	462
C63000	H02 [25]	118	814	75	517	15	HRB98	70	483	38	262
	M30 [24]	100	690	60	414	15	HRB96	62	427	36	248
C63020	TQ30	145	1,000	-	-	8	HRC29	78	538	51	352
C63200	050	105	724	53	365	22	HRB96	-	-	-	-
C64200	H04 [21, 23]	102	703	68	469	22	HRB94	59	407	50	345
	M30 [21]	75	517	35	241	32	HRB77	-	-	-	-
	050 [21]	90	621	55	379	28	HRB89	-	-	-	-

(Continued)

Table 9. *(Continued)* **Selected Mechanical Properties of Copper Alloy Rod [1] for Machined Products.** *(Source, Copper Development Association)*

Copper Alloy No.	Temper	Tensile Strength		Yield Strength[2]		% Elongation[3]	Hardness	Shear Strength		Fatigue Strength[4]	
		KSI	MPa	KSI	MPa			KSI	MPa	KSI	MPa
C65100	H04 [5]	70	483	55	379	15	HRB80	45	310	-	-
C65500	H04 [5]	92	634	55	379	22	HRB90	58	400	-	-
	H02 [18]	78	538	45	310	35	HRB85	52	359	-	-
C65600	H04	92	634	55	379	20	HRB90	58	400	-	-
C66100	H04	90	621	55	379	15	HRB90	58	400	-	-
C67400	H02 [21]	92	634	55	379	20	HRB87	48	331	29	200
	060 [21]	70	483	34	234	28	HRB78	-	-	-	-
C67500	H02 [18]	84	579	60	414	19	HRB90	48	331	-	-
	H01 [25]	77	531	45	310	23	HRB83	47	324	-	-
	060	65	448	30	207	33	HRB65	42	290	-	-
C67600	H02	75	517	45	310	18	HRB85	-	-	-	-
C69400	060	85	586	43	296	25	HRB85	-	-	-	-
C69430	H04	85	586	50	345	20	HRB90	-	-	-	-
C69710	H04	70	483	40	276	22	HRB90	-	-	-	-
C70600	061 [27]	47	324	34	234	-	-	-	-	-	-
C71500	H02 [18]	75	517	70	483	15	HRB80	42	290	-	-
C75200	H02 [10, 18]	70	483	60	414	20	HRB78	-	-	-	-
C75700	H04	80	550	53	365	23	-	-	-	-	-
C79200	H04 [21]	72	496	62	427	15	HRB78	-	-	-	-

(Continued)

Table 9. *(Continued)* **Selected Mechanical Properties of Copper Alloy Rod [1] for Machined Products.** *(Source, Copper Development Association)*

Notes: All properties as measured at room temperature (68° F / 20° C).

1. Section size diameter one-inch (25.4 mm) unless otherwise specified.
2. Yield strength at 0.5% extension under load unless otherwise specified.
3. Elongation measured in two-inches (50.8 mm).
4. Fatigue strength for 100×10^6 cycles unless otherwise specified.
5. Specimen 35% cold worked.
6. Specimen 29% cold worked.
7. Specimen condition solution heat treated, cold worked 48%, aged and cold worked 47%.
8. Yield strength for this alloy measured at 0.2% offset.
9. Section size diameter 0.54 inch (13.7 mm).
10. Section size diameter 0.5 inch (12.7 mm).
11. Specimen condition hard and precipitation heat treated.
12. Section size diameter <0.375 inch (<9.5 mm).
13. Specimen precipitation hardened.
14. Specimen heat treated, cold worked 40%, and aged.
15. Specimen 35% cold worked, and heat treated.
16. Section size diameter 0.375-0.500 inch (9.5-12.7 mm).
17. Section size diameter 0.312 inch (7.9 mm).
18. Specimen 20% cold worked.
19. Fatigue strength recorded at 50×10^6 cycles.
20. Specimen 8% cold worked.
21. Section size diameter 0.75 inch (19.0 mm).
22. Specimen 25% cold worked.
23. Specimen 15% cold worked.
24. Section size diameter 4 inches (101.6 mm).
25. Specimen 10% cold worked.
26. Fatigue strength recorded at 300×10^6 cycles.
27. Section size diameter <2.5 inches (<63.5 mm).

Table 10. Selected Physical Properties of Wrought Copper Alloys for Machined Products. *(Source, Copper Development Association.)*

Copper Alloy No.	Coef. Thermal Exp.[1]		Melting Point-Liquidus		Melting Point-Solidus		Elastic Modulus		Shear Modulus	
	10^{-6}/°F	10^{-6}/°C	°F	°C	°F	°C	KSI	MPa	KSI	MPa
C10100	9.6	17.4	1,981	1,083	1,981	1,083	17,000	117,000	6,400	44,100
C10200	9.6	17.4	1,981	1,083	1,981	1,083	17,000	117,000	6,400	44,100
C10300	9.6	17.4	1,981	1,083	1,981	1,083	17,000	117,000	6,400	44,100
C10400	9.6	17.4	1,981	1,083	1,981	1,083	17,000	117,000	6,400	44,100
C10500	9.6	17.4	1,981	1,083	1,981	1,083	17,000	117,000	6,400	44,100
C10700	9.6	17.4	1,981	1,083	1,981	1,083	17,000	117,000	6,400	44,100
C10800	9.6	17.4	1,981	1,083	1,981	1,083	17,000	117,000	6,400	44,100
C11000	9.6	17.4	1,981	1,083	1,949	1,065	17,000	117,000	6,400	44,100
C11300	-	-	1,980	1,082	-	-	17,000	117,000	6,400	44,100
C11400	-	-	1,980	1,082	-	-	17,000	117,000	6,400	44,100
C11500	-	-	1,980	1,082	-	-	17,000	117,000	6,400	44,100
C11600	-	-	1,980	1,082	-	-	17,000	117,000	6,400	44,100
C12000	9.6	17.4	1,981	1,083	1,981	1,083	17,000	117,000	6,400	44,100
C12100	9.6	17.4	1,981	1,083	1,981	1,083	17,000	117,000	6,400	44,100
C12200	9.5	17.2	1,981	1,083	-	-	17,000	117,000	6,400	44,100
C12900	9.6	17.4	1,981	1,083	-	-	17,000	117,000	6,400	44,100
C14500	9.7	17.5	1,976	1,080	1,924	1,051	17,000	117,000	6,400	44,100
C14520	9.7	17.5	1,967	1,075	1,924	1,051	17,000	117,000	6,400	44,100
C14700	9.6	17.4	1,969	1,076	1,953	1,067	17,000	117,000	6,400	44,100
C15000	-	-	1,976	1,080	1,796	980	18,700	129,000	-	-
C15715	9.7	17.5	1,981	1,083	1,981	1,083	19,000	130,000	-	-
C15725	9.9	17.9	1,981	1,083	-	-	19,000	130,000	-	-
C15760	10.0	18.0	1,981	1,083	1,981	1,083	19,000	130,000	-	-

(Continued)

Table 10. *(Continued)* **Selected Physical Properties of Wrought Copper Alloys for Machined Products.** *(Source, Copper Development Association.)*

Copper Alloy No.	Coef. Thermal Exp.[1]		Melting Point-Liquidus		Melting Point-Solidus		Elastic Modulus		Shear Modulus	
	10^{-6}/°F	10^{-6}/°C	°F	°C	°F	°C	KSI	MPa	KSI	MPa
C16200	9.6	17.4	1,969	1,076	1,886	1,030	17,000	117,000	6,400	44,100
C16500	9.8	17.7	1,958	1,070	-	-	17,000	117,000	-	-
C17000	9.9	17.9	1,800	982	1,590	866	18,500	128,000	7,300	50,300
C17200	9.9	17.9	1,800	982	1,590	866	18,500	128,000	7,300	50,300
C17300	9.9	17.9	1,800	982	1,590	866	18,500	128,000	7,300	50,300
C17410	-	-	1,950	1,066	1,875	1,024	19,000	131,000	-	-
C17500	9.8	17.7	1,955	1,068	1,885	1,029	19,000	131,000	7,500	51,700
C17510	9.8	17.7	1,955	1,068	1,885	1,029	19,000	131,000	7,500	51,700
C18000	-	-	-	-	-	-	-	-	-	-
C18100	10.2	18.4	1,967	1,075	-	-	18,200	125,000	6,800	46,900
C18135	9.8	17.7	1,976	1,080	-	-	20,000	141,180	7,500	51,700
C18150	-	-	-	-	-	-	-	-	-	-
C18200	-	-	1,967	1,075	1,958	1,070	17,000	117,000	7,200	50,630
C18400	-	-	1,967	1,075	1,958	1,070	17,000	117,000	7,200	50,630
C18700	9.8	17.7	1,976	1,080	1,747	953	17,000	117,000	6,400	44,100
C19100	9.8	17.7	1,980	1,082	1,900	1,038	16,000	110,000	6,000	41,400
C19150	-	-	1,980	1,082	-	-	-	-	-	-
C21000	-	-	1,949	1,065	1,920	1,049	17,000	117,000	6,400	44,100
C22000	10.2	18.4	1,910	1,043	1,870	1,021	17,000	117,000	6,400	44,100
C22600	10.2	18.4	1,895	1,035	1,840	1,004	17,000	117,000	6,400	44,100
C23000	-	-	1,877	1,027	1,810	988	17,000	117,000	6,400	44,100
C24000	-	-	1,832	1,000	1,770	966	16,000	110,000	6,000	41,400
C26000	-	-	1,750	954	1,680	916	16,000	110,000	6,000	41,400

(Continued)

Table 10. *(Continued)* **Selected Physical Properties of Wrought Copper Alloys for Machined Products.** *(Source, Copper Development Association.)*

Copper Alloy No.	Coef. Thermal Exp.[1]		Melting Point-Liquidus		Melting Point-Solidus		Elastic Modulus		Shear Modulus	
	10^{-6}/°F	10^{-6}/°C	°F	°C	°F	°C	KSI	MPa	KSI	MPa
C26130	-	-	1,750	954	1,680	916	16,000	110,000	6,000	41,400
C26200	-	-	1,750	954	1,680	916	16,000	110,000	6,000	41,400
C26800	-	-	1,710	932	1,660	904	15,000	103,000	5,600	38,600
C27000	-	-	1,710	932	1,660	904	15,000	103,000	5,600	38,600
C28000	-	-	1,660	904	1,650	899	15,000	103,000	5,600	38,600
C31400	-	-	1,900	1,038	1,850	1,010	17,000	117,000	6,400	44,100
C31600	-	-	1,900	1,038	1,850	1,010	17,000	117,000	-	-
C32000	-	-	1,875	1,024	1,820	993	17,000	117,000	6,400	44,100
C33500	-	-	1,700	927	1,650	899	15,000	103,000	5,600	38,600
C34000	-	-	1,700	927	1,630	888	15,000	103,000	5,600	38,600
C34200	-	-	1,670	910	1,630	888	15,000	103,000	5,600	38,600
C34500	-	-	1,688	920	-	-	-	-	-	-
C35000	-	-	1,680	916	1,640	893	15,000	103,000	5,600	38,600
C35300	-	-	1,670	910	1,630	888	15,000	103,000	5,600	38,600
C35330	-	-	1,650	899	1,630	888	14,000	96,500	5,300	36,500
C35600	-	-	1,650	899	1,630	888	14,000	96,500	5,300	36,500
C36000	-	-	1,650	899	1,630	888	14,000	96,500	5,300	36,500
C37000	-	-	1,650	899	1,630	888	15,000	103,000	5,600	38,600
C37700	-	-	1,640	893	1,620	882	15,000	103,000	5,600	38,600
C38500	-	-	1,630	888	1,610	877	14,000	96,500	5,300	36,500
C46200	-	-	1,670	910	-	-	-	-	-	-
C46400	-	-	1,650	899	1,630	888	15,000	103,000	5,600	38,600
C48200	-	-	1,650	899	1,630	888	15,000	103,000	5,600	38,600

(Continued)

Table 10. *(Continued)* **Selected Physical Properties of Wrought Copper Alloys for Machined Products.** *(Source, Copper Development Association.)*

Copper Alloy No.	Coef. Thermal Exp.[1]		Melting Point-Liquidus		Melting Point-Solidus		Elastic Modulus		Shear Modulus	
	10^{-6}/°F	10^{-6}/°C	°F	°C	°F	°C	KSI	MPa	KSI	MPa
C48500	-	-	1,650	899	1,630	888	15,000	103,000	5,600	38,600
C50700	-	-	1,958	1,070	-	-	-	-	-	-
C51000	-	-	1,920	1,049	1,750	954	16,000	110,000	6,000	41,400
C52100	-	-	1,880	1,027	1,620	882	16,000	110,000	6,000	41,400
C54400	-	-	1,830	999	1,700	927	15,000	103,000	5,600	38,600
C61000	-	-	1,904	1,040	-	-	17,000	117,000	6,400	44,100
C61300	-	-	1,915	1,046	1,905	1,041	17,000	117,000	-	-
C61400	-	-	1,915	1,046	1,905	1,041	17,000	117,000	6,400	44,100
C61800	-	-	1,913	1,045	1,904	1,040	17,000	117,000	6,400	44,100
C62300	-	-	1,915	1,046	1,905	1,041	17,000	117,000	6,400	44,100
C62400	-	-	1,900	1,038	1,880	1,027	17,000	117,000	6,400	44,100
C62500	-	-	1,925	1,052	1,917	1,047	16,000	110,000	-	-
C63000	-	-	1,930	1,054	1,895	1,035	17,500	121,000	6,400	44,100
C63020	-	-	-	-	-	-	18,000	124,000	-	-
C63200	-	-	1,940	1,060	1,905	1,041	17,000	117,000	6,400	44,100
C64200	-	-	1,840	1,004	1,800	982	16,000	110,000	6,000	41,400
C64700	-	-	1,994	1,090	1,940	1,060	-	-	-	-
C65100	-	-	1,940	1,060	1,890	1,032	17,000	117,000	6,400	44,100
C65500	-	-	1,880	1,027	1,780	971	15,000	103,000	5,600	38,600
C65600	-	-	1,866	1,019	-	-	-	-	-	-
C66100	-	-	1,866	1,019	-	-	-	-	-	-
C67000	-	-	1,652	900	-	-	-	-	-	-
C67300	-	-	-	-	-	-	-	-	-	-

(Continued)

Table 10. *(Continued)* **Selected Physical Properties of Wrought Copper Alloys for Machined Products.** *(Source, Copper Development Association.)*

Copper Alloy No.	Coef. Thermal Exp.[1]		Melting Point-Liquidus		Melting Point-Solidus		Elastic Modulus		Shear Modulus	
	10^{-6}/°F	10^{-6}/°C	°F	°C	°F	°C	KSI	MPa	KSI	MPa
C67400	-	-	1,625	885	1,590	866	16,000	110,000	6,000	41,400
C67500	-	-	1,630	888	1,590	866	15,000	103,000	5,600	38,600
C67600	-	-	1,652	900	-	-	-	-	-	-
C69400	-	-	1,685	918	1,510	821	16,000	110,000	-	-
C69430	-	-	1,685	918	1,510	821	16,000	110,000	-	-
C69710	-	-	1,706	930	-	-	-	-	-	-
C70600	9.3	16.7	2,093	1,145	-	-	18,000	124,000	6,800	46,900
C71500	9.0	16.3	2,260	1,238	2,140	1,171	22,000	15,200	8,300	57,200
C74500	9.1	16.5	1,870	1,021	-	-	17,500	121,000	6,600	45,500
C75200	9.0	16.3	2,030	1,110	1,960	1,071	18,000	124,000	6,800	46,900
C75400	9.0	16.3	1,970	1,077	1,900	1,038	18,000	124,000	6,800	46,900
C75700	9.0	16.3	1,900	1,038	-	-	18,000	124,000	6,800	46,900
C79200	9.0	16.2	-	-	-	-	-	-	-	-

Note: [1] Coefficient of thermal expansion for temperature range 68 to 392° F (20 to 200° C).

exposure (1650° F / 900° C). Physical properties similar to pure copper. Excellent elevated temperature mechanical properties. C15760 rod used primarily for RWMA Class II and Class III resistance welding applications. C15715 used for electrical conductors and springs.

High Copper Alloys. C16200 – C19900 are high copper alloys with designated copper content of less than 99.3%, but more than 96%, that do not fall into any other copper alloy group.

C16200, C16500: Moderate strength, good wear resistance. Mostly used for trolley wire, occasionally specified for machined products. C16500 is slightly stronger than C16200. C16200 used for RWMA Class I applications.

C17000, C17200, C17300, C17410, C17500, C17510: Both C17200 and C17300 can develop more than 200 ksi (1380 MPa) tensile strength. C17410, C17500, and C17510 have conductivities of about 50% IACS and moderate strength. These beryllium copper alloys are available in ductile heat treatable (hardenable) tempers as well as in mill hardened tempers. Alloy C17300, available in either rod or wire, is leaded for increased machinability. These alloys are used in a wide range of applications requiring very high strength and stiffness with good conductivity. Typical uses include electrical/electronic connectors, current carrying springs, precision screw machined parts, welding electrodes, bearings, plastic molds, and corrosion resistant components.

C18000, C18100, C18135, C18150: Good wear resistance and high strength at elevated temperatures. Used for high stress, high temperature applications such as welding electrodes, electrical components, contacts, and studs.

C18200, C18400: Moderately high elevated temperature strength. Conductivity 80% IACS. Used for resistance welding equipment components such as tips, clamps, and wheels, and for other electrical and mechanical power transmission devices, circuit breaker parts, and high strength fasteners for elevated temperature service.

C18700: Free machining copper with 96% IACS conductivity and good corrosion resistance. Used for electrical/electronic connectors, welding torch tips, and products requiring a significant amount of machining.

C19100, C19150: Moderately high strength alloys with good conductivity (55% IACS) and corrosion resistance. Highly machinable: C19100 uses tellurium (Te) to achieve free machining properties, and C19150 uses lead (Pb). Used for cylindrical connector contacts, bolts, bushings, electrical and mechanical components, gears, and marine hardware.

Copper-Zinc Alloys (Brasses). C21000 – C28000 are brasses with Cu and Zn as the principal alloying agents.

C21000: Good formability and corrosion resistance. More often cold formed than machined. Used for ammunition components, coinage, and medallions.

C22000: Similar to C21000, but stronger. Used for corrosion resistant products, decorative items, and fasteners. Better suited to cold forming than machining.

C22600, C23000: Similar to, but stronger than, C22000. Used for cold formed electrical and hardware components, jewelry, decorative chain. Sometimes specified for dezincification-resistant fasteners and fittings.

C24000: Similar to, but stronger than, C23000.

C26000, C26130, C26200, C26800, C27700, C28000: C26000 has the highest ductility in the yellow brass series, and is also known as Jewelry brass. It is easily machined, but more often cold formed. C27000 hot forms more readily and is stronger than C26000. Used for both hot and cold formed products.

Copper-Zinc-Lead Alloys (Leaded Brasses). C31200 – C38500 are leaded brasses with Cu, Zn, and Pb as the principal alloying agents.

C31400, C31600: Free machining versions of the 20000 series yellow brasses. Moderate corrosion resistance. Used for architectural hardware, pole line hardware, and fasteners. C31600 is somewhat stronger than C31400.

C32000, C33500, C34000, C34200, C34500, C35000, C35300, C35330, C35600: Free machining brasses with good formability. C34500 has the best combination of machining and formability. C35300 is dezincification resistant, and better for acid waters. These brasses are used for screw machine products requiring some cold formability, as for knurling, peening, crimping, thread rolling. Typical uses include hose fittings; watch, clock, and lock parts; bicycle spoke nipples; and plumbing valve components.

C36000: Highest machinability of all copper alloys. Used for screw machine products in a wide range of applications. Commonly called free cutting brass.

C37000: Normally plate and tube alloy, but also produced as rods, bars, and shapes.

C37700: Known as forging brass, it has excellent forgability, and is free cutting. Used for forgings requiring extensive machining, architectural hardware, fittings, mechanical components, and specialty fasteners.

C38500: Free machining, used as-extruded or as-forged, but best applied where products require machining as the secondary operation. Used mostly for architectural hardware and trim. Most easily hot worked. Light pink in color. Often called Architectural bronze.

Copper-Zinc-Tin Alloys (Tin Brasses). C40400 – C48600 are tin brasses, with Cu, Zn, Sn, and Pb as the principal alloying agents.

C46200, C46400, C48200, C48500: These are commonly known as Naval brass. Excellent hot forgability, good corrosion resistance. Used for fasteners and hardware for corrosion resistant service, including marine applications. C46200 has higher dezincification resistance than C46400. These brasses are used for high strength cold headed products, including fasteners.

Copper-Tin-Phosphorus Alloys (Phosphor Bronzes). C50100 – C52480 are phosphor bronzes with Cu, Sn, and P as the principal alloying agents.

C50700, C51000, C52100: Good combination of strength and ductility. Excellent spring qualities.

Copper-Tin-Phosphorus-Lead Alloys (Leaded Phosphor Bronze). C53400 – C54400 are leaded phosphor bronzes with Cu, Sn, Pb, and P as the principal alloying agents.

C55180 – C55299 are Copper Phosphorus and Copper Silver Phosphorus *Brazing Alloys* with Cu, P, and Ag as the principal alloying agents.

C55300 – C60799 are Copper Silver Zinc Bronzes with Cu, Ag, and Zn as the principal alloying agents.

C54400: Free machining, with good cold working properties. Very good wear resistance. Often used for bearings, cam followers, and similar products.

Copper-Aluminum Alloys (Aluminum Bronzes). C60800 – C64210 are aluminum bronzes, and Cu, Al, Ni, Fe, Si, and Sn are the alloying agents.

C61000, C61300, C61400, C61800, C62300, C62400, C62500, C63000, C63020, C63200, C64200: Aluminum bronzes and their alloyed modifications are best known for moderate to high strength, and excellent corrosion resistance. C61300 has good formability; C61800 has good corrosion resistance; C62300 has good acid resistance; C62400 is heat treatable to high strength; C63000 and C63020 have typical tensile strengths of 118 and 145 ksi (814 and 1000 MPa), respectively; C63400 has moderate strength and corrosion resistance; and C63400 is free machining and has good antigalling and antiseizing properties. These alloys are used for valve and pump components for industrial process streams, and for marine equipment, high strength fasteners, and pole line hardware.

Copper-Silicon Alloys (Silicon Bronzes). C64700 – C67000 are silicon bronzes with Cu, Si, and Sn as the principal alloying agents.

C64700, C65100, C65500, C65600, C66100, C67000: Moderate strength, good corrosion resistance, similar to C64200 in properties and applications. Used for valve guides, valve stems, fasteners, pole line hardwire, marine fittings. Moderate to high strength, and good corrosion resistance. Choice of alloy depends on corrosive environment and manufacturing process. C66100 is free cutting and adaptable to high speed screw machines. Applications similar to aluminum bronzes: high strength fasteners, marine and pole hardware.

Miscellaneous Copper-Zinc Alloys. C67200 – C69710 is reserved for "other copper alloys," and no alloys are specified.

C67300, C67400, C67500, C67600, C69400, C69430, C69710: High strength, hot forgable alloys. Some with free machining qualities that allow them to run on high speed machining centers. Used in heavy duty mechanical components including bearings. Arsenical alloys C69430 and C69710 resist dezincification and are better for use in acid waters. Used principally for valve stems and pump components.

Copper-Nickel Alloys. C70100 – C72950 are copper nickel alloys with Cu, Ni, and Fe as the principal alloying agents.

C70600, C71500: Excellent corrosion resistance, especially in marine environments. Moderately high strength, good creep resistance at elevated temperatures. Properties generally increase with nickel content. Copper-nickels are used mainly for seawater service as forged and machined valve and pump components, fittings, and hardware. Relatively high in cost compared to copper-aluminum and other alloys with similar mechanical properties. Often specified for use where high corrosion resistance is required and where concern over chloride stress-corrosion cracking prevents use of stainless steels.

Copper-Nickel-Zinc Alloys (Nickel Silvers). C73500 – C79830 are nickel silvers with Cu, Ni, and Zn as the principal alloying agents.

C74500, C75200, C75400, C75700, C79200: Good formability, good corrosion- and tarnish resistance. Alloys have pleasing silver-like color. C75400 and C76390 are similar to, but stronger than, C75200. Uses include decorative items, jewelry, musical instrument valves and components, optical instrument components, fittings for food and dairy equipment, screws, rivets, and slide fasteners.

Description of cast copper alloys

Characteristics and popular uses of cast copper alloys are given below. Composition of cast copper alloys are given in **Tables 11** through **17**; **Table 18** provides mechanical properties; and **Table 19** provides physical properties.

Coppers. C80100 – C81200 are cast coppers with a minimum copper content of 99.3%.

These coppers are used almost exclusively for their unsurpassed electrical and thermal conductivities in products such as terminals, connectors, and (water cooled) hot metal handling equipment. They also have high corrosion resistance, but this is usually a secondary consideration.

High Copper Alloys. C81400 – C82800 are high copper alloys with designated copper content in excess of 94%, to which Ag may be added for special properties.

These alloys are used primarily for their combination of high strength and good conductivity. Chromium coppers (C81400 and C81500) with tensile strength of 45 ksi (310 MPa) and conductivity of 82% IACS (as heat treated) are used in electrical contacts, clamps, welding gear, and similar electromechanical hardware. At more than 160 ksi (1,100 MPa), the beryllium coppers (C82800) have the highest tensile strength of all the copper

(Text continued on p. 249)

Table 11. Chemical Composition of Cast Coppers and Cast High Copper Alloys. *(Source, Copper Development Association.)*

Copper Number			Cu (incl. Ag) % (min.)			P			Other (named)		
Coppers											
C80100			99.95			-			-		
C80410			99.9			-			.10		
C81100			99.70			-			-		
C81200			99.9			.045-.065			-		
Copper Alloy No.	**Cu[16]**	**Be**	**Co**	**Si**	**Ni**	**Fe**	**Al**	**Sn**	**Pb**	**Zn**	**Cr**
High Copper Alloys											
C81400	Rem.	.02-.10	-	-	-	-	-	-	-	-	.6-1.0
C81500	Rem.	-	-	.15	-	.10	.10	.10	.02	.10	.40-1.5
C81540	95.1 min. [41]	-	-	.40-.8	2.0-3.0 [42]	.15	.10	.10	.02	.10	.10-.60
C82000	Rem.	.45-.80	2.40-2.70 [42]	.15	.20	.10	.10	.10	.02	.10	.10
C82200	Rem.	.35-.80	.30	-	1.0-2.0	-	-	-	-	-	-
C82400	Rem.	1.60-1.85	.20-.65	-	.20	.20	.15	.10	.02	.10	.10
C82500	Rem.	1.90-2.25	.35-.70 [42]	.20-.35	.20	.25	.15	.10	.02	.10	.10
C82510	Rem.	1.90-2.15	1.0-1.2	.20-.35	.20	.25	.15	.10	.02	.10	.10
C82600	Rem.	2.25-2.55	.35-.65	.20-.35	.20	.25	.15	.10	.02	.10	.10
C82700	Rem.	2.35-2.55	-	.15	1.0-1.5	.25	.15	.10	.02	.10	.10
C82800	Rem.	2.50-2.85	.35-.70 [42]	.20-.35	.20	.25	.15	.10	.02	.10	.10

Notes: [16] Cu + Sum of Named Elements, 99.5% min.
[41] Includes Ag.
[42] Ni + Co.

Table 12. Chemical Composition of Cast Brasses. *(Source, Copper Development Association.)*

Copper Alloy No.	Cu [43, 44]	Sn	Pb	Zn	Ni (incl. Co)	Other (named)
Copper-Tin-Zinc and Copper-Tin-Zinc-Lead Alloys *(Red and Leaded Red Brasses)*						
C83300	92.0-94.0	1.0-2.0	1.0-2.0	2.0-6.0	-	-
C83400	88.0-92.0	0.20	0.50	8.0-12.0	1.0	0.25 Fe, 0.25 Sb, 0.08 S, 0.03 P [45], 0.005 Al, 0.005 Si
C83450	87.0-89.0	2.0-3.5	1.5-3.0	5.5-7.5	0.8-2.0	0.30 Fe, 0.25 Sb, 0.08 S, 0.03 P [45], 0.005 Al, 0.005 Si
C83500	86.0-88.0	5.5-6.5	3.5-5.5	1.0-2.5	0.50-1.0	0.25 Fe, 0.25 Sb, 0.08 S, 0.03 P [45], 0.005 Al, 0.005 Si
C83600	84.0-86.0	4.0-6.0	4.0-6.0	4.0-6.0	1.0	0.30 Fe, 0.25 Sb, 0.08 S, 0.05 P [45], 0.005 Al, 0.005 Si
C83800	82.0-83.8	3.3-4.2	5.0-7.0	5.0-8.0	1.0	0.30 Fe, 0.25 Sb, 0.08 S, 0.03 P [45], 0.005 Al, 0.005 Si
C83810	Rem.	2.0-3.5	4.0-6.0	7.5-9.5	2.0	0.50 Fe [46], Sb [46], As [46], 0.005 Al, .10 Si
Copper-Tin-Zinc and Copper-Tin-Zinc-Lead Alloys *(Semi-Red and Leaded Semi-Red Brasses)*						
C84200	78.0-82.0	4.0-6.0	2.0-3.0	10.0-16.0	0.8	0.40 Fe, 0.25 Sb, 0.08 S, 0.05 P [45], 0.005 Al, 0.005 Si
C84400	78.0-82.0	2.3-3.5	6.0-8.0	7.0-10.0	1.0	0.40 Fe, 0.25 Sb, 0.08 S, 0.02 P [45], 0.005 Al, 0.005 Si
C84410	Rem.	3.0-4.5	7.0-9.0	7.0-11.0	1.0	Fe [48], Sb [48], .01 Al, .20 Si, .05 Bi
C84500	77.0-79.0	2.0-4.0	6.0-7.5	10.0-14.0	1.0	0.40 Fe, 0.25 Sb, 0.08 S, 0.02 P [45], 0.005 Al, 0.005 Si
C84800	75.0-77.0	2.0-3.0	5.5-7.0	13.0-17.0	1.0	0.40 Fe, 0.25 Sb, 0.08 S, 0.02 P [45], 0.005 Al, 0.005 Si
Alloy No.	**Cu [43]**	**Sn**	**Pb**	**Zn**	**Ni + Co**	**Other (named)**
Copper-Tin-Zinc and Copper-Tin-Zinc-Lead Alloys *(Yellow and Leaded Yellow Brasses)*						
C85200	70.0-74.0 [49]	0.7-2.0	1.5-3.8	20.0-27.0	1.0	0.6 Fe, 0.20 Sb, 0.05 S, 0.02 P, 0.005 Al, 0.05 Si
C85400	65.0-70.0 [52]	0.50-1.5	1.5-3.8	24.0-32.0	1.0	0.7 Fe, 0.35 Al, 0.05 Si
C85500	59.0-63.0 [49]	0.20	0.20	Rem.	0.20	0.20 Fe, 0.20 Mn
C85700	58.0-64.0 [53]	0.50-1.5	0.8-1.5	32.0-40.0	1.0	0.7 Fe, 0.8 Al, 0.05 Si
C85800	57.0 min. [53]	1.5	1.5	31.0-41.0	0.50	0.50 Fe, 0.05 Sb, 0.25 Mn, 0.05 As, 0.05 S, 0.01 P, 0.55 Al, 0.25 Si

(Continued)

Table 12. *(Continued)* **Chemical Composition of Cast Brasses.** *(Source, Copper Development Association.)*

Alloy No.	Cu [43, 50]	Sn	Pb	Zn	Fe	Ni + Co	Al	Mn	Si [45]
Manganese Bronze and Leaded Manganese Bronze Alloys *(High Strength and Leaded High Strength Yellow Brasses)*									
C86100	66.0-68.0	0.20	0.20	Rem.	2.0-4.0	-	4.5-5.5	2.5-5.0	-
C86200	60.0-66.0	0.20	0.20	22.0-28.0	2.0-4.0	1.0	3.0-4.9	2.5-5.0	-
C86300	60.0-66.0	0.20	0.20	22.0-28.0	2.0-4.0	1.0	5.0-7.5	2.5-5.0	-
C86400	56.0-62.0	0.50-1.5	0.50-1.5	34.0-42.0	0.40-2.0	1.0	0.50-1.5	0.10-1.5	-
C86500	55.0-60.0	1.0	0.40	36.0-42.0	0.40-2.0	1.0	0.50-1.5	0.10-1.5	-
C86550	57.0 min.	1.0	0.50	Rem.	0.7-2.0	1.0	0.50-2.5	0.10-3.0	0.10
C86700	55.0-60.0	1.5	0.50-1.5	30.0-38.0	1.0-3.0	1.0	1.0-3.0	0.10-3.5	-
C86800	53.5-57.0	1.0	0.20	Rem.	1.0-2.5	2.5-4.0	2.0	2.5-4.0	-
Alloy No.	**Cu [16]**	**Pb**	**Zn**	**Fe**	**Al + Co**	**Si**	**Other (named)**		
Copper Silicon Alloys *(Silicon Bronzes and Silicon Brasses)*									
C87300	94.0 min.	0.20	0.25	0.20	-	3.5-4.5	0.8-1.5 Mn		
C87400	79.0 min. [47]	1.0	12.0-16.0	-	0.8	2.5-4.0	-		
C87500	79.0 min.	0.50	12.0-16.0	-	0.50	3.0-5.0	-		
C87600	88.0 min.	0.50	4.0-7.0	0.20	-	3.5-5.5	0.25 Mn		
C87610	90.0 min.	0.20	3.0-5.0	0.20	-	3.0-5.0	0.25 Mn		
C87800	80.0 min.	0.15	12.0-16.0	0.15	0.15	3.8-4.2	0.25 Sn, 0.15 Mn, 0.01 Mg, 0.20 Ni + Ag, 0.05 S, 0.01 P, 0.05 As, 0.05 Sb		
Alloy No.	**Cu**	**Sn**	**Zn**	**Ni + Co**	**Bi**	**Other (named)**			
Copper-Bismuth and Copper-Bismuth-Selenium Alloys *(High Strength and Leaded High Strength Yellow Brasses)*									
C89320	87.0-91.0 [16]	5.0-7.0	1.0	1.0	4.0-6.0	0.09 Pb, 0.20 Fe, 0.35 Sb, 0.08 S, 0.30 P, 0.005 Al, 0.005 Si			
C89325	84.0-88.0 [50]	9.0-11.0	1.0	1.0	2.7-3.7	0.10 Pb, 0.15 Fe, 0.50 Sb, 0.08 S, 0.10 P, 0.005 Al, 0.005 Si, [54]			
C89510	86.0-88.0 [16]	4.0-6.0	4.0-6.0	1.0	0.50-1.5 [66]	0.25 Pb, 0.20 Fe, 0.25 Sb, 0.08 S, 0.05 P, 0.005 Al, 0.005 Si, 0.35-0.7 Se [66]			

(Continued)

Table 12. *(Continued)* **Chemical Composition of Cast Brasses.** *(Source, Copper Development Association.)*

Alloy No.	Cu	Sn	Zn	Ni + Co	Bi	Other (named)
Copper-Bismuth and Copper-Bismuth-Selenium Alloys *(High Strength and Leaded High Strength Yellow Brasses)*						
C89520	85.0-87.0 [16]	5.0-6.0	4.0-6.0	1.0	1.6-2.2	0.25 Pb, 0.20 Fe, 0.25 Sb, 0.08 S, 0.05 P, 0.005 Al, 0.005 Si, 0.8-1.1 Se, [64]
C89550	58.0-64.0 [16]	0.00-1.2	32.0-38.0	1.0	0.6-1.2	0.10 Pb, 0.50 Fe, 0.05 Sb, 0.05 S, 0.01 P, 0.10-0.6 Al, 0.25 Si, 0.01-0.10 Se
C89831	87.0-91.0 [50]	2.7-3.7	2.0-4.0	1.0	2.7-3.7	0.10 Pb, 0.30 Fe, 0.25 Sb, 0.08 S, 0.050 P, 0.005 Al, 0.005 Si, [54]
C89833	87.0-91.0 [50]	4.0-6.0	2.0-4.0	1.0	1.7-2.7	0.10 Pb, 0.30 Fe, 0.25 Sb, 0.08 S, 0.050 P, 0.005 Al, 0.005 Si, [54]
C89835	85.0-89.0 [50]	6.0-7.5	2.0-4.0	1.0	1.7-2.7	0.10 Pb, 0.20 Fe, 0.35 Sb, 0.08 S, 0.10 P, 0.005 Al, 0.005 Si, [54]
C89837	84.0-88.0 [50]	3.0-4.0	6.0-10.0	1.0	0.7-1.2	0.10 Pb, 0.30 Fe, 0.25 Sb, 0.08 S, 0.050 P, 0.005 Al, 0.005 Si, [54]
C89844	83.0-86.0 [44]	3.0-5.0	7.0-10.0	1.0	2.0-4.0	0.20 Pb, 0.30 Fe, 0.25 Sb, 0.08 S, 0.05 P, 0.005 Al, 0.005 Si
C89940	64.0-68.0 [16]	3.0-5.0	3.0-5.0	20.0-23.0	4.0-5.5	0.01 Pb, 0.7-2.0 Fe, 0.10 Sb, 0.05 S, 0.10-0.15 P, 0.005 Al, 0.15 Si, 0.20 Mn

Notes:
[16] Cu +Sum of Named Elements, 99.5% min.
[43] In determining copper min., copper may be calculated as Cu + Ni.
[44] Cu +Sum of Named Elements, 99.3% min.
[45] For continuous castings, P shall be 1.5%, max.
[46] Fe + Sb + as shall be .50% max.
[47] Cu +Sum of Named Elements, 99.2% min.
[48] Fe + Sb + as shall be .8% max.
[49] Cu +Sum of Named Elements, 99.1% min.
[50] Cu +Sum of Named Elements, 99.0% min.
[52] Cu +Sum of Named Elements, 98.9% min.
[53] Cu +Sum of Named Elements, 98.7% min.
[54] .01-2.0% as any single or combination of Ce, La or other rare earth* elements, as agreed upon. *ASM International definition: one of the group of chemically similar metals with atomic numbers 57 through 71, commonly referred to as lanthanides.
[64] Bi : Se ^{3} 2:1.
[66] Experience favors Bi : Se ^{3} 2:1.

Table 13. Chemical Composition of Cast Bronzes. *(Source, Copper Development Association.)*

Copper Alloy No.	Cu[43, 51]	Sn	Zn	Ni (incl. Co)	P[45]	Other (named)
Copper-Tin Alloys *(Tin Bronzes)*						
C90200	91.0-94.0	6.0-8.0	0.50	0.50	0.05	0.30 Pb, 0.20 Fe, 0.20 Sb, 0.05 S, 0.005 Al, 0.005 Si
C90300	86.0-89.0	7.5-9.0	3.0-5.0	1.0	0.05	0.30 Pb, 0.20 Fe, 0.20 Sb, 0.05 S, 0.005 Al, 0.005 Si
C90500	86.0-89.0 [21]	9.0-11.0	1.0-3.0	1.0	0.05	0.30 Pb, 0.20 Fe, 0.20 Sb, 0.05 S, 0.005 Al, 0.005 Si
C90700	88.0-90.0	10.0-12.0	0.50	0.50	0.30	0.50 Pb, 0.15 Fe, 0.20 Sb, 0.05 S, 0.005 Al, 0.005 Si
C90710	Rem.	10.0-12.0	0.05	0.10	0.05-1.2	0.25 Pb, 0.10 Fe, 0.20 Sb, 0.05 S, 0.005 Al, 0.005 Si
C90800	85.0-89.0	11.0-13.0	0.25	0.50	0.30	0.25 Pb, 0.15 Fe, 0.20 Sb, 0.05 S, 0.005 Al, 0.005 Si
C90810	Rem.	11.0-13.0	0.30	0.50	0.15-0.8	0.25 Pb, 0.15 Fe, 0.20 Sb, 0.05 S, 0.005 Al, 0.005 Si
C90900	86.0-89.0	12.0-14.0	0.25	0.50	0.05	0.25 Pb, 0.15 Fe, 0.20 Sb, 0.05 S, 0.005 Al, 0.005 Si
C91000	84.0-86.0	14.0-16.0	1.5	0.8	0.05	0.20 Pb, 0.10 Fe, 0.20 Sb, 0.05 S, 0.005 Al, 0.005 Si
C91100	82.0-85.0	15.0-17.0	0.25	0.50	1.0	0.25 Pb, 0.25 Fe, 0.20 Sb, 0.05 S, 0.005 Al, 0.005 Si
C91300	79.0-82.0	18.0-20.0	0.25	0.50	1.0	0.25 Pb, 0.25 Fe, 0.20 Sb, 0.05 S, 0.005 Al, 0.005 Si
C91600	86.0-89.0	9.7-10.8	0.25	1.2-2.0	0.30	0.25 Pb, 0.20 Fe, 0.20 Sb, 0.05 S, 0.005 Al, 0.005 Si
C91700	84.0-87.0	11.3-12.5	0.25	1.2-2.0	0.30	0.25 Pb, 0.20 Fe, 0.20 Sb, 0.05 S, 0.005 Al, 0.005 Si
Alloy No.	**Cu[43, 44]**	**Sn**	**Pb**	**Zn**	**Ni + Co**	**Other (named)**
Copper-Tin Lead Alloys *(Leaded Tin Bronzes)*						
C92200	86.0-90.0	5.5-6.5	1.0-2.0	3.0-5.0	1.0	0.25 Fe, 0.25 Sb, 0.05 S, 0.05 P [45], 0.005 Al, 0.005 Si
C92210	86.0-89.0	4.5-5.5	1.7-2.5	3.0-4.5	0.7-1.0	0.25 Fe, 0.20 Sb, 0.05 S, 0.03 P [45], 0.005 Al, 0.005 Si
C92220	86.0-88.0 [43]	5.0-6.0	1.5-2.5	3.0-5.5	0.50-1.0	0.25 Fe, 0.05 P [45]
C92300	85.0-89.0	7.5-9.0	0.30-1.0	2.5-5.0	1.0	0.25 Fe, 0.25 Sb, 0.05 S, 0.05 P [45], 0.005 Al, 0.005 Si
C92310	Rem.	7.5-8.5	0.30-1.5	3.5-4.5	1.0	0.005 Al, 0.005 Si, 0.03 Mn
C92400	86.0-89.0	9.0-11.0	1.0-2.5	1.0-3.0	1.0	0.25 Fe, 0.25 Sb, 0.05 S, 0.05 P [45], 0.005 Al, 0.005 Si
C92410	Rem.	6.0-8.0	2.5-3.5	1.5-3.0	0.20	0.20 Fe, 0.25 Sb, 0.005 Al, 0.005 Si, 0.05 Mn
C92500	85.0-88.0	10.0-12.0	1.0-1.5	0.50	0.8-1.5	0.30 Fe, 0.25 Sb, 0.05 S, 0.30 P [45], 0.005 Al, 0.005 Si
C92600	86.0-88.5	9.3-10.5	0.8-1.5	1.3-2.5	0.7	0.20 Fe, 0.25 Sb, 0.05 S, 0.03 P [45], 0.005 Al, 0.005 Si

(Continued)

Table 13. *(Continued)* **Chemical Composition of Cast Bronzes.** *(Source, Copper Development Association.)*

Alloy No.	Cu[43, 44]	Sn	Pb	Zn	Ni + Co	Other (named)
Copper-Tin Lead Alloys *(Leaded Tin Bronzes)*						
C92610	Rem.	9.5-10.5	0.30-1.5	1.7-2.8	1.0	0.15 Fe, 0.005 Al, 0.005 Si, 0.03 Mn
C92700	86.0-89.0	9.0-11.0	1.0-2.5	0.7	1.0	0.20 Fe, 0.25 Sb, 0.05 S, 0.25 P [45], 0.005 Al, 0.005 Si
C92710	Rem.	9.0-11.0	4.0-6.0	1.0	2.0	0.20 Fe, 0.25 Sb, 0.05 S, 0.10 P [45], 0.005 Al, 0.005 Si
C92800	78.0-82.0	15.0-17.0	4.0-6.0	0.8	0.8	0.20 Fe, 0.25 Sb, 0.05 S, 0.05 P [45], 0.005 Al, 0.005 Si
C92810	78.0-82.0	12.0-14.0	4.0-6.0	0.50	0.8-1.2	0.50 Fe, 0.25 Sb, 0.05 S, 0.05 P [45], 0.005 Al, 0.005 Si
C92900	82.0-86.0	9.0-11.0	2.0-3.2	0.25	2.8-4.0	0.20 Fe, 0.25 Sb, 0.05 S, 0.05 P [45], 0.005 Al, 0.005 Si
Copper-Tin-Lead Alloys *(Hi-Leaded Tin Bronzes)*						
C93100	Rem. [43, 50]	6.5-8.5	2.0-5.0	2.0	1.0	0.25 Fe, 0.25 Sb, 0.05 S, 0.30 P [45], 0.005 Al, 0.005 Si
C93200	81.0-85.0 [43, 50]	6.3-7.5	6.0-8.0	1.0-4.0	1.0	0.20 Fe, 0.35 Sb, 0.08 S, 0.15 P [45], 0.005 Al, 0.005 Si
C93400	82.0-85.0 [43, 50]	7.0-9.0	7.0-9.0	0.8	1.0	0.20 Fe, 0.50 Sb, 0.08 S, 0.50 P [45], 0.005 Al, 0.005 Si
C93500	83.0-86.0 [43, 50]	4.3-6.0	8.0-10.0	2.0	1.0	0.20 Fe, 0.30 Sb, 0.08 S, 0.05 P [45], 0.005 Al, 0.005 Si
C93600	79.0-83.0 [44]	6.0-8.0	11.0-13.0	1.0	1.0	0.20 Fe, 0.55 Sb, 0.08 S, 0.15 P [45], 0.005 Al, 0.005 Si
C93700	78.0-82.0 [50]	9.0-11.0	8.0-11.0	0.8	0.50	0.7 [55] Fe, 0.50 Sb, 0.08 S, 0.10 P [45], 0.005 Al, 0.005 Si
C93720	83.0 min. [50]	3.5-4.5	7.0-9.0	4.0	0.50	0.7 Fe, 0.50 Sb, 0.10 P [45]
C93800	75.0-79.0 [50]	6.3-7.5	13.0-16.0	0.8	1.0	0.15 Fe, 0.8 Sb, 0.08 S, 0.05 P [45], 0.005 Al, 0.005 Si
C93900	76.5-79.5 [52]	5.0-7.0	14.0-18.0	1.5	0.8	0.40 Fe, 0.50 Sb, 0.08 S, 1.5 P [45], 0.005 Al, 0.005 Si
C94000	69.0-72.0 [53]	12.0-14.0	14.0-16.0	0.50	0.50-1.0	0.25 Fe, 0.50 Sb, 0.08 [56] S, 0.05 P [45], 0.005 Al, 0.005 Si
C94100	72.0-79.0 [53]	4.5-6.5	18.0-22.0	1.0	1.0	0.25 Fe, 0.8 Sb, 0.08 [56] S, 0.05 P [45], 0.005 Al, 0.005 Si
C94300	67.0-72.0 [50]	4.5-6.0	23.0-27.0	0.8	1.0	0.15 Fe, 0.8 Sb, 0.08 [56] S, 0.08 P [45], 0.005 Al, 0.005 Si
C94310	Rem. [50]	1.5-3.0	27.0-34.0	0.50	0.25-1.0	0.50 Fe, 0.50 Sb, 0.05 P [45]
C94320	Rem. [50]	4.0-7.0	24.0-32.0	-	-	0.35 Fe
C94330	68.5-75.5 [50]	3.0-4.0	21.0-25.0	3.0	0.50	0.7 Fe, 0.50 Sb, 0.10 P [45]
C94400	Rem. [50]	7.0-9.0	9.0-12.0	0.8	1.0	0.15 Fe, 0.8 Sb, 0.08 S, 0.50 P [45], 0.005 Al, 0.005 Si
C94500	Rem. [50]	6.0-8.0	16.0-22.0	1.2	1.0	0.15 Fe, 0.8 Sb, 0.08 S, 0.05 P [45], 0.005 Al, 0.005 Si

(Continued)

Table 13. *(Continued)* **Chemical Composition of Cast Bronzes.** *(Source, Copper Development Association.)*

Alloy No.	Cu (incl. Ag)	Su	Pb	Zn	Ni + Co	Other (named)
Copper-Tin-Nickel Alloys *(Tin-Nickel Bronzes)*						
C94700	85.0-90.0 [53]	4.5-6.0	0.10 [57]	1.0-2.5	4.5-6.0	0.25 Fe, 0.15 Sb, 0.20 Mn, 0.05 S, 0.05 P, 0.005 Al, 0.005 Si
C94800	84.0-89.0 [53]	4.5-6.0	0.30-1.0	1.0-2.5	4.5-6.0	0.25 Fe, 0.15 Sb, 0.20 Mn, 0.05 S, 0.05 P, 0.005 Al, 0.005 Si
C94900	79.0-81.0 [51]	4.0-6.0	4.0-6.0	4.0-6.0	4.0-6.0	0.30 Fe, 0.25 Sb, 0.10 Mn, 0.08 S, 0.05 P, 0.005 Al, 0.005 Si
Alloy No.	**Cu**	**Fe**	**Ni + Co**	**Al**	**Mn**	**Other (named)**
Copper-Aluminum-Iron and Copper-Aluminum-Iron-Nickel Alloys *(Aluminum Bronzes)*						
C95200	86.0 min. [50]	2.5-4.0	-	8.5-9.5	-	-
C95210	86.0 min. [50]	2.5-4.0	1.0	8.5-9.5	1.0	0.05 Pb, 0.05 Mg, 0.25 Si, 0.50 Zn, 0.10 Sn
C95220	Rem. [16]	2.5-4.0	2.5	9.5-10.5	0.50	-
C95300	86.0 min. [50]	0.8-1.5	-	9.0-11.0	-	-
C95400	83.0 min. [16]	3.0-5.0	1.5	10.0-11.5	0.50	-
C95410	83.0 min. [16]	3.0-5.0	1.5-2.5	10.0-11.5	0.50	-
C95420	83.5 min. [16]	3.0-4.3	0.50	10.5-12.0	0.50	-
C95500	78.0 min. [16]	3.0-5.0	3.0-5.5	10.0-11.5	3.5	-
C95510	78.0 min. [20]	2.0-3.5	4.5-5.5	9.7-10.9	1.5	0.30 Zn, 0.20 Sn
C95520	74.5 min. [16]	4.0-5.5	4.2-6.0	10.5-11.5	1.5	0.03 Pb, 0.15 Si, 0.30 Zn, 0.25 Sn, 0.20 Co, 0.05 Cr
C95600	88.0 min. [50]	-	0.25	6.0-8.0	-	1.8-3.2 Si
C95700	71.0 min. [16]	2.0-4.0	1.5-3.0	7.0-8.5	11.0-14.0	0.10 Si
C95710	71.0 min. [16]	2.0-4.0	1.5-3.0	7.0-8.5	11.0-14.0	0.05 Pb, 0.15 Si, 0.50 Zn, 1.0 Sn, 0.05 P
C95720	73.0 min. [16]	1.5-3.5	3.0-6.0	6.0-8.0	12.0-15.0	0.03 Pb, 0.10 Si, 0.10 Zn, 0.10 Sn, 0.20 Cr
C95800	79.0 min. [16]	3.5-4.5 [31]	4.0-5.0 [31]	8.5-9.5	0.8-1.5	0.03 Pb, 0.10 Si
C95810	79.0 min. [16]	3.5-4.5 [31]	4.0-5.0 [31]	8.5-9.5	0.8-1.5	0.10 Pb, 0.05 Mg, 0.10 Si, 0.50 Zn
C95820	77.5 min. [47]	4.0-5.0	4.5-5.8	9.0-10.0	1.5	0.02 Pb, 0.10 Si, 0.20 Zn, 0.20 Sn
C95900	Rem. [16]	3.0-5.0	0.50	12.0-13.5	1.5	-

Notes: [16] Cu + Sum of Named Elements, 99.5% min.
[20] Cu + Sum of Named Elements, 99.8% min.
[21] Cu + Sum of Named Elements, 99.7% min.
[31] Fe content shall not exceed Ni content.
[43] In determining copper min., copper may be calculated as Cu + Ni.
[44] Cu + Sum of Named Elements, 99.3% min.
[45] For continuous castings, P shall be 1.5%, max.
[47] Cu + Sum of Named Elements, 99.2% min.
[50] Cu + Sum of Named Elements, 99.0% min.
[51] Cu + Sum of Named Elements, 99.4% min.

(Continued)

Table 13. *(Continued)* **Chemical Composition of Cast Bronzes.** *(Source, Copper Development Association.)*

Notes: [52] Cu + Sum of Named Elements, 98.9% min.
[53] Cu + Sum of Named Elements, 98.7% min.
[55] Fe shall be 0.35% max., when used for steel-backed bearings.
[56] For continuous castings, S shall be 0.25%, max.
[57] The mechanical properties of C94700 (heat treated) may not be attainable if the lead content exceeds 0.01%.

Table 14. Chemical Composition of Cast Copper-Nickels. *(Source, Copper Development Association.)*

Copper Alloy No.	Cu[16]	Fe	Ni (incl. Co)	Mn	Nb	Other (named)
Copper-Nickel-Zinc Alloys *(Copper Nickels)*						
C96200	Rem.	1.0-1.8	9.0-11.0	1.5	1.0 [58]	.01 Pb, .50 Si, .10 C, .02 S, .02 P
C96300	Rem.	.50-1.5	18.0-22.0	.25-1.5	.50-1.5	.01 Pb, .50 Si, .15 C, .02 S, .02 P
C96400	Rem.	.25-1.5	28.0-32.0	1.5	.50-1.5	.01 Pb, .50 Si, .15 C, .02 S, .02 P
C96600	Rem.	.8-1.1	29.0-33.0	1.0	-	.01 Pb, .15 Si, .40-.7 Be
C96700	Rem.	.40-1.0	29.0-33.0	.40-1.0	-	.01 Pb, .15 Si, 1.1-1.2 Be, .15-.35 Zr, .15-.35 Ti
C96800	Rem.	.50	9.5-10.5	.05-.30	.10-.30	.005 Pb, .05 Si, [59]
C96900	Rem.	.50	14.5-15.5	.05-.30	.10	.02 Pb, .15 Mg, 7.5-8.5 Sn, .50 Zn
C96950	Rem.	.50	11.0-15.5	.05-.40	.10	.02 Pb, .30 Si, 5.8-8.5 Sn, .15 Mg

Notes: [16] Cu + Sum of Named Elements, 99.5% min.
[58] When the product or casting is intended for subsequent welding applications, and so specified by the purchaser, the Nb content shall be 0.40% max.
[59] The following additional maximum impurity limits shall apply: 0 .10% Al, 0.001% B, 0 .001% Bi, 0.005-.15% Mg, 0.005% P, 0.0025% S, 0.02% Sb, 7.5-8.5% Sn, 0.01% Ti, 1.0% Zn.

Table 15. Chemical Composition of Cast Nickel Silvers. *(Source, Copper Development Association.)*

Copper Alloy No.	Cu	Sn	Pb	Zn (incl. Co)	No (incl. Co)	Other (named)
Copper-Nickel-Zinc Alloys *(Nickel Silvers)*						
C97300	53.0-58.0 [50]	1.5-3.0	8.0-11.0	17.0-25.0	11.0-14.0	1.5 Fe, .35 Sb, .08 S, .05 P, .005 Al, .50 Mn, .15 Si
C97400	58.0-61.0 [50]	2.5-3.5	4.5-5.5	Rem.	15.5-17.0	1.5 Fe, .50 Mn
C97600	63.0-67.0 [21]	3.5-4.5	3.0-5.0	3.0-9.0	19.0-21.5	1.5 Fe, .25 Sb, .08 S, .05 P, .005 Al, 1.0 Mn, .15 Si
C97800	64.0-67.0 [22]	4.0-5.5	1.0-2.5	1.0-4.0	24.0-27.0	1.5 Fe, .20 Sb, .08 S, .05 P, .005 Al, 1.0 Mn, .15 Si

Notes: [21] Cu + Sum of Named Elements, 99.7% min.
[22] Cu + Sum of Named Elements, 99.6% min.
[50] Cu + Sum of Named Elements, 99.0% min.

Table 16. Chemical Composition of Cast Leaded Coppers. *(Source, Copper Development Association.)*

Copper Alloy No.	Cu	Sn	Pb	P (incl. Co)	Fe	Other (named)
Copper-Lead Alloys						
C98200	Rem. [16]	.60-2.0	21.0-27.0	.10	.70	.50 Zn, .50 Ni, .50 Sb
C98400	Rem. [16]	.50	26.0-33.0	.10	.70	1.5 Ag, .50 Zn, .50 Ni, .50 Sb
C98600	60.0-70.0	.50	30.0-40.0	-	.35	1.5 Ag
C98800	56.5-62.5 [41]	.25	37.5-42.5 [60]	.02	.35	5.5 [60] Ag, .10 Zn
C98820	Rem.	1.0-5.0	40.0-44.0	-	.35	-
C98840	Rem.	1.0-5.0	44.0-58.0	-	.35	-

Notes: [16] Cu + Sum of Named Elements, 99.5% min.
[41] Includes Ag.
[60] Pb and Ag may be adjusted to modify the alloy hardness.

Table 17. Chemical Composition of Cast Special Copper Alloys. *(Source, Copper Development Association.)*

Copper Alloy No.	Cu[21]	Ni	Fe	Al	Mn	Other (named)
C99300	Rem.	13.5-16.5	4.0-1.0	10.7-11.5	-	.05 Sn, .02 Pb, 1.0-2.0 Co, .02 Si
C99350	Rem.	14.5-16.0 [18]	1.0	9.5-10.5	.25	.15 Pb, 7.5-9.5 Zn
C99400	Rem.	1.0-3.5	1.0-3.0	.50-2.0	.50	.25 Pb, .50-2.0 Si, .50-5.0 Zn
C99500	Rem.	3.5-5.5	3.0-5.0	.50-2.0	.50	.25 Pb, .50-2.0 Si, .50-5.0 Zn
C99600	Rem.	.20	.20	1.0-2.8	39.0-45.0	.10 Sn, .02 Pb, .20 Co, .10 Si, .20 Zn, .05 C
C99700	54.0 min.	4.0-6.0	1.0	.50-3.0	11.0-15.0	1.0 Sn, 2.0 Pb, 19.0-25.0 Zn
C99750	55.0-61.0	5.0	1.0	.25-3.0	17.0-23.0	.50-2.5 Sn, 17.0-23.0 Zn

Notes: [18] Includes Co.
[21] Cu + Sum of named elements, 99.7% min.

Table 18. Mechanical Properties of Selected Copper Casting Alloys. *(Source, Copper Development Association.)*

UNS Number	App. [1]	Temper [2]	Tensile Strength [3]		Yield Strength [4]		% Elong. [5]	Hardness
			ksi	MPa	ksi	MPa		
C80100	S	M01	25	172	9	62	40	-
C81100	S	M01	25	172	9	62	40	-
C81400	S	M01	30	207	12 [6]	83 [6]	35	R_B 62
C81400	S	TF00	53	365	36 [6]	248 [6]	11	R_B 69
C81500	S	TF00	51	352	40	276	17	-
C82000	S	M01	50	345	20 [6]	138 [6]	20	R_B 55
C82000	S	O11	65	448	37 [6]	255 [6]	12	-
C82000	S	TB00	47	324	15 [6]	103 [6]	25	R_B 40
C82000	S	TF00	96	662	75 [6]	517 [6]	6	R_B 96
C82200	S	M01	50	345	25 [6]	172 [6]	20	R_B 55

(Continued)

Table 18. *(Continued)* **Mechanical Properties of Selected Copper Casting Alloys.**
(Source, Copper Development Association.)

UNS Number	App. [1]	Temper [2]	Tensile Strength [3]		Yield Strength [4]		% Elong. [5]	Hardness
			ksi	MPa	ksi	MPa		
C82200	S	011	65	448	40 [6]	276 [6]	15	R_B 75
C82200	S	TB00	45	310	12 [6]	83 [6]	30	R_B 30
C82200	S	TF00	95	655	75 [6]	517 [6]	7	R_B 96
C82400	S	O11	100	690	80 [6]	551 [6]	3	R_C 21
C82400	S	TB00	60	414	20 [6]	138 [6]	40	R_B 59
C82400	S	TF00	155	1,068	145 [6]	1,000 [6]	1	R_C 38
C82500	S	M01	75	517	40 [6]	276 [6]	15	R_B 81
C82500	S	O11	120	827	105 [6]	724 [6]	2	R_C 30
C82500	S	TB00	60	414	25 [6]	172 [6]	35	R_B 63
C82500	S	TF00	160	1,103	150 [6]	1,034 [6]	1	R_C 43
C82600	S	M01	80	552	50 [6]	345 [6]	10	R_B 68
C82600	S	O11	120	827	105 [6]	724 [6]	2	R_C 31
C82600	S	TB00	70	483	30 [6]	207 [6]	12	R_B 75
C82600	S	TF00	165	1,138	155 [6]	1,069 [6]	1	R_C 45
C82700	S	TF00	155 [7]	1,069 [7]	130 [8]	896 [8]	2	R_C 39
C82800	S	M01	80	552	50 [6]	345 [6]	10	R_B 88
C82800	S	O11	125	862	110 [6]	758 [6]	2	R_C 31
C82800	S	TB00	80	552	35 [6]	241 [6]	10	R_B 58
C82800	S	TF00	165	1,138	155 [6]	1,069 [6]	1	R_C 46
C83300	S	M01	32	221	10	69	35	R_B 35
C83400	S	M01	35	241	10	69	30	R_F 50
C83600	S, CL	M01, M02	37	255	17	117	30	-
C83600	C	M07	36 [7]	248 [7]	19 [9]	131 [9]	15 [10]	-
C83600	C	M07	50 [7]	345 [7]	25 [9]	170 [9]	12 [10]	-
C83800	S, CL	M01, M02	35	241	16	110	25	-
C83800	C	M07	30 [7]	207 [7]	15 [9]	103 [9]	16 [10]	-
C84200	S	M01	35	241	14	97	27	-
C84400	S	M01	34	234	15	103	26	-
C84500	S	M01	35	241	14	97	28	-
C84800	S	M01	37	255	14	97	35	-
C85200	S, CL	M01, M02	38	262	13	90	35	-
C85400	S, CL	M01, M02	34	234	12	83	35	-
C85500	S	M01	60	414	23	159	40	R_B 55
C85700	S, CL	M01, M02	50	345	18	124	40	-
C85800	D	M04	55	379	30 [6]	207 [6]	15	R_B 55
C86100	S	M01	95	655	50 [6]	345 [6]	20	-
C86200	S, CL, C	M01, M02, M07	95	655	48 [6]	331 [6]	20	-

(Continued)

Table 18. *(Continued)* **Mechanical Properties of Selected Copper Casting Alloys.**
(Source, Copper Development Association.)

UNS Number	App. [1]	Temper [2]	Tensile Strength [3]		Yield Strength [4]		% Elong. [5]	Hardness
			ksi	MPa	ksi	MPa		
C86300	S	M01	119	821	67 [6]	462 [6]	18	-
C86300	S, CL	M01, M02	110 [7]	758 [7]	60 [8]	414 [8]	12 [10]	-
C86300	C	M07	110 [7]	758 [7]	62 [8]	427 [8]	14 [10]	-
C86400	S	M01	65	448	25 [6]	172 [6]	20	-
C86500	S, CL	M01, M02	71	490	29	200	30	-
C86500	C	M07	70 [7]	483 [7]	25 [8]	172 [8]	25 [10]	-
C86700	S	M01	85	586	42	290	20	R_B 80
C86800	S	M01	82	565	38	262	22	-
C87300	S, CL	M01, M02	55	379	25	172	30	-
C87400	S, CL	M01, M02	55	379	24	165	30	-
C87500	S, CL	M01, M02	67	462	30	207	21	-
C87600	S	M01	66	455	32	221	20	R_B 76
C87800	D	M04	85	586	50 [6]	345 [6]	25	R_B 85
C90200	S	M01	38	124	16	110	30	-
C90300	S, CL	M01, M02	45	310	21	145	30	-
C90300	C	M07	44 [7]	303 [7]	22 [9]	152 [9]	18 [10]	-
C90500	S, CL	M01, M02	45	310	22	152	25	-
C90500	C	M07	44 [7]	303 [7]	25 [9]	172 [9]	10 [10]	-
C90700	S	M01	44	303	22	152	20	-
C90700	CL, PM	M02, M05	55	379	30	207	16	-
C90700	C	M07	40 [7]	276 [7]	25	172	10	-
C90900	S	M01	40	276	20	138	15	-
C91000	S	M01	32	221	25	172	2	-
C91100	S	M01	35	241	25	172	2	-
C91300	S	M01	35	241	30	207	0.5	-
C91600	S	M01	44	303	22	152	16	-
C91600	CL, PM	M02, M05	60	414	32	221	16	-
C91700	S	M01	44	303	22	152	16	-
C91700	CL, PM	M02, M05	60	414	32	221	16	-
C92200	S, CL	M01, M02	40	276	20	138	30	-
C92200	C	M07	38 [7]	262 [7]	19 [9]	131 [9]	18 [10]	-
C92300	S, CL	M01, M02	40	276	20	138	25	-
C92300	C	M07	40 [7]	276 [7]	19 [9]	131 [9]	16 [10]	-
C92500	S	M01	44	303	20	138	20	-
C92500	C	M07	40 [7]	276 [7]	24 [9]	166 [9]	10 [10]	-
C92600	S	M01	44	303	20	138	30	R_F 78
C92700	S	M01	42	290	21	145	20	-
C92700	C	M07	38 [7]	262 [7]	20 [9]	138 [9]	8	-

(Continued)

Table 18. *(Continued)* **Mechanical Properties of Selected Copper Casting Alloys.**
(Source, Copper Development Association.)

UNS Number	App. [1]	Temper [2]	Tensile Strength [3]		Yield Strength [4]		% Elong. [5]	Hardness
			ksi	MPa	ksi	MPa		
C92800	S	M01	40	276	30	207	1	R_B 80
C92900	S, PM, C	M01, M05, M07	47	324	26	179	20	-
C93200	S, CL	M01, M02	35	241	18	124	20	-
C93200	C	M07	35 [7]	241 [7]	20 [9]	138 [9]	10 [10]	-
C93400	S	M01	32	221	16	110	20	-
C93500	S, CL	M01, M02	32	221	16	110	20	-
C93500	C	M07	30 [7]	207 [7]	16 [9]	110 [9]	12 [10]	-
C93700	S, CL	M01, M02	35	241	18	124	20	-
C93700	C	M07	35 [7]	241 [7]	20 [9]	138 [9]	6 [10]	-
C93700	C	M07	40 [7]	276 [7]	25 [9]	172 [9]	6 [10]	-
C93800	S, CL	M01, M02	30	207	16	110	18	-
C93800	CL	M02	33	228	20	138	12	-
C93800	C	M07	25 [7]	172 [7]	-	-	5 [10]	-
C93900	C	M07	32	221	22	152	7	-
C94300	S	M01	27	186	13	90	15	-
C94300	S, CL	M01, M02	21 [7]	145 [7]	-	-	10 [10]	-
C94300	C	M07	21 [7]	145 [7]	15 [9]	103 [9]	7 [10]	-
C94400	S	M01	32	221	16	110	18	-
C94500	S	M01	25	172	12	83	12	-
C94700	S, C	M01, M07	50	345	23	159	35	-
C94700	S, C	TX00	85	586	60	414	10	-
C94800	S, C	M01, M07	45	310	23	159	35	-
C94800	S	TX00	60	414	30	207	8	-
C95200	S, CL	M01, M02	80	552	27	186	35	R_B 64
C95200	C	M07	68 [7]	469 [7]	26 [9]	179 [9]	20 [10]	-
C95300	S, CL	M01, M02	75	517	27	186	25	R_B 67
C95300	C	M07	70 [7]	483 [7]	26 [9]	179 [9]	25 [10]	-
C95300	S, CL, C	TQ50	85	586	42	290	15	R_B 81
C95400	S, CL	M01, M02	85	586	35	241	18	-
C95400	C	M07	85 [7]	586 [7]	32 [9]	221 [9]	12 [10]	-
C95400	S, CL	TQ50	105	724	54	372	8	-
C95400	C	TQ50	95 [7]	655 [7]	45 [9]	310 [9]	10 [10]	-
C95410	S	M01	85	586	35	241	18	-
C95410	S	TQ50	105	724	54	372	8	-
C95500	S, CL	M01, M02	100	690	44	303	12	R_B 87
C95500	C	M07	95 [7]	665 [7]	42 [9]	290 [9]	10 [10]	-

(Continued)

Table 18. *(Continued)* **Mechanical Properties of Selected Copper Casting Alloys.**
(Source, Copper Development Association.)

UNS Number	App. [1]	Temper [2]	Tensile Strength [3]		Yield Strength [4]		% Elong. [5]	Hardness
			ksi	MPa	ksi	MPa		
C95500	S, CL	TQ50	120	827	68	469	10	R_B 96
C95600	S	M01	75	517	34	234	18	-
C95700	S	M01	95	655	45	310	26	-
C95800	S, CL	M01, M02	95	655	38	262	25	-
C95800	C	M07	90 [7]	621 [7]	38 [9]	262 [9]	18 [10]	-
C96200	S	M01	45 [7]	310 [7]	25 [9]	172 [9]	20 [10]	-
C96300	S	M01	75 [7]	517 [7]	55 [9]	379 [9]	10 [10]	-
C96400	S	M01	68	469	37	255	28	-
C96600	S	TB00	75	517	38	262	12	R_B 74
C96600	S	TF00	120	827	75	517	12	R_C 24
C97300	S	M01	35	241	17	117	20	-
C97400	S	M01	38	262	17	117	20	-
C97600	S	M01	45	310	24	165	20	-
C97800	S	M01	55	379	30	207	15	-
C99300	S	M01	95	655	55	379	2	-
C99400	S	M01	66	455	34	234	25	-
C99400	S	TF00	79	545	54	372	-	-
C99500	S	M01	70 [7]	483 [7]	40 [9]	276 [9]	12 [10]	-
C99500	S	TF00	86	593	62	427	8	-
C99700	S	M01	55	379	25	172	25	-
C99700	D	M04	65	448	27	186	15	-
C99750	S	M01	65	448	32	221	30	R_B 77
C99750	S	TQ50	75	517	40	276	20	R_B 82

Notes: [1] Application abbreviations are: S = Sand. C = Continuous. CL = Centrifugal. D = Die. I = Investment. P = Plaster. PM = Permanent Mold.
[2] Temper designations given in text.
[3] Typical tensile strength given unless otherwise indicated.
[4] Typical yield strength at 0.5% extension unless otherwise indicated.
[5] Typical percent elongation in 2 inches (50.8 mm) given unless otherwise indicated.
[6] Typical yield strength at 0.2% offset.
[7] Minimum tensile strength.
[8] Minimum yield strength at 2% offset.
[9] Minimum yield strength at 0.5% extension.
[10] Minimum percent elongation in 2 inches (50.8 mm).

Table 19. Physical Properties of Selected Copper Casting Alloys. *(Source, Copper Development Association)*

UNS Number	Liquidus Point		Density lb/cu in.	Coef. Thermal Expan.[1]		Thermal Conduct.[2]	Electrical Conduct.[3]	Elastic Modulus	
	°F	°C		µin. °F	µm °C			ksi	MPa
C80100	1,981	1,083	0.323	9.4	16.9	226	100	17,000	117,000
C81100	1,981	1,083	0.323	9.4	16.9	200	92	17,000	117,000
C81400	2,000	1,093	0.318	10.0	18.0	150	60	16,000	110,000
C81500	1,985	1,085	0.319	9.5	17.1	182	82	16,000	114,000
C82000	1,990	1,088	0.311	9.9	17.8	150	45	17,000	117,000
C82200	2,040	1,116	0.316	9.0 [4]	16.2 [4]	106	45	16,000	114,000
C82400	1,825	996	0.304	9.4 [4]	16.9 [4]	76.9	25	18,000	128,000
C82500	1,800	982	0.302	9.4 [4]	16.9 [4]	74.9	20	18,000	128,000
C82600	1,750	954	0.302	9.4 [4]	16.9 [4]	73.0	19	19,000	131,000
C82700	1,750	954	0.292	9.4 [4]	16.9 [4]	74.9	20	19,000	132,000
C82800	1,710	932	0.294	9.4 [4]	16.9 [4]	70.8	18	19,000	133,000
C83300	1,940	1,060	0.318	-	-	-	32	15,000	103,000
C83400	1,910	1,043	0.318	10.0	18.0	109	44	15,000	103,000
C83600	1,850	1,010	0.318	10.0 [4]	18.0 [4]	41.6	15	13,000	93,000
C83800	1,840	1,004	0.312	10.0 [4]	18.0 [4]	41.8	15	13,000	91,000
C84200	1,820	993	0.311	10.0 [4]	18.0 [4]	41.8	16	14,000	96,000
C84400	1,840	1,004	0.314	10.0	18.0	41.8	16	13,000	89,000
C84500	1,790	977	0.312	10.0	18.0	41.6	16	14,000	96,000
C84800	1,750	954	0.310	10.0 [4]	18.0 [4]	41.6	16	15,000	103,000
C85200	1,725	941	0.307	11.5 [5]	20.8 [5]	48.5	18	11,000	75,000
C85400	1,725	941	0.305	11.1 [5]	20.0 [5]	50.8	20	12,000	82,000
C85500	1,652	900	0.304	11.8 [5]	21.3 [5]	67.0	26	15,000	103,000
C85700	1,725	941	0.304	12.0	21.6	48.5	22	14,000	96,000
C85800	1,650	899	0.305	-	-	48.5	20	15,000	103,000
C86100	1,725	941	0.288	12.0	21.6	20.5	8	15,000	103,000
C86200	1,725	941	0.288	12.0	21.6	20.5	8	15,000	103,000
C86300	1,693	923	0.283	12.0	21.6	20.5	8	14,000	97,000
C86400	1,616	880	0.301	11.0 [4]	19.8 [4]	51.0	19	14,000	96,000
C86500	1,616	880	0.301	11.3 [5]	20.4 [5]	49.6	22	15,000	103,000
C86700	1,616	880	0.301	11.0 [4]	19.8 [4]	-	17	15,000	103,000
C86800	1,652	900	0.290	-	-	-	9	15,000	103,000
C87300	1,780	971	0.302	10.9	19.6	16.4	6	15,000	103,000
C87400	1,680	916	0.300	10.9	19.6	16.0	7	15,000	106,000
C87500	1,680	916	0.299	10.9	19.6	16.0	7	15,000	106,000
C87600	1,780	971	0.300	-	-	16.4	6	17,000	117,000
C87800	1,680	916	0.300	10.9	19.6	16.0	7	20,000	138,000
C90200	1,915	1,046	0.318	10.1	18.2	36.0	13	16,000	110,000
C90300	1,832	1,000	0.318	10.0 [4]	18.0 [4]	43.2	12	14,000	96,000

(Continued)

Table 19. *(Continued)* **Physical Properties of Selected Copper Casting Alloys.**
(Source, Copper Development Association.)

UNS Number	Liquidus Point		Density lb/cu in.	Coef. Thermal Expan.[1]		Thermal Conduct.[2]	Electrical Conduct.[3]	Elastic Modulus	
	°F	°C		μin. °F	μm °C			ksi	MPa
C90500	1,830	999	0.315	11.0	19.8	43.2	11	15,000	103,000
C90700	1,830	999	0.317	10.2 [4]	18.4 [4]	40.8	10	15,000	103,000
C90900	1,792	978	-	-	-	-	-	16,000	110,000
C91000	1,760	960	-	-	-	-	9	16,000	110,000
C91100	1,742	950	-	-	-	-	8	15,000	103,000
C91300	1,632	889	-	-	-	-	7	16,000	110,000
C91600	1,887	1,031	0.320	9.0 [4]	16.2 [4]	40.8	10	16,000	110,000
C91700	1,859	1,015	0.316	9.0 [4]	16.2 [4]	40.8	10	15,000	103,000
C92200	1,810	988	0.312	10.0	18.0	40.2	14	14,000	96,000
C92300	1,830	999	0.317	10.0 [4]	18.0 [4]	43.2	12	14,000	96,000
C92500	-	-	0.317	-	-	-	-	16,000	110,000
C92600	1,800	982	0.315	10.0 [4]	18.0 [4]	-	9	15,000	103,000
C92700	1,800	982	0.317	10.0 [4]	18.0 [4]	27.2	11	16,000	110,000
C92800	1,751	955	-	-	-	-	-	16,000	110,000
C92900	1,887	1,031	0.320	9.5 [4]	17.1 [4]	33.6	9	14,000	96,000
C93200	1,790	977	0.322	10.0 [5]	18.0 [5]	33.6	12	14,000	100,000
C93400	-	-	0.320	10.0 [4]	18.0 [4]	33.6	12	11,000	75,000
C93500	1,830	999	0.320	9.9 [4]	17.8 [4]	40.7	15	14,000	100,000
C93700	1,705	929	0.320	10.3 [4]	18.5 [4]	27.1	10	11,000	75,000
C93800	1,730	943	0.334	10.3 [4]	18.5 [4]	30.2	11	10,000	72,000
C93900	1,730	943	0.334	10.3 [4]	18.5 [4]	30.2	11	11,000	75,000
C94300	-	-	0.336	-	-	36.2	9	10,000	72,000
C94400	1,725	941	0.320	10.3 [4]	18.5 [4]	30.2	10	11,000	75,000
C94500	1,475	802	0.340	10.3 [4]	18.5 [4]	30.2	10	10,000	72,000
C94700	1,660	904	0.320	10.9 [4]	19.6 [4]	31.2	-	15,000	103,000
C94800	1,660	904	0.320	10.9	19.6	22.3	12	15,000	103,000
C95200	1,913	1,045	0.276	9.0	16.2	29.1	11	15,000	103,000
C95300	1,913	1,045	0.272	9.0	16.2	36.3	13	16,000	110,000
C95400	1,900	1,038	0.269	9.0	16.2	33.9	13	15,000	107,000
C95410	1,900	1,038	0.269	9.0	16.2	33.9	13	15,000	107,000
C95500	1,930	1,054	0.272	9.0	16.2	24.2	8	16,000	110,000
C95600	1,840	1,004	0.278	9.2	16.6	22.3	8	15,000	103,000
C95700	1,814	990	0.272	9.8	17.6	7.0	3	18,000	124,000
C95800	1,940	1,060	0.276	9.0	16.2	20.8	7	16,000	114,000
C96200	2,100	1,149	0.323	9.5	17.1	26.1	11	18,000	124,000
C96300	2,190	1,199	0.323	9.1	16.4	21.3	6	20,000	138,000
C96400	2,260	1,238	0.323	9.0	16.2	16.4	5	21,000	145,000
C96600	2,160	1,182	0.318	9.0	16.2	17.4	4	22,000	152,000

(Continued)

Table 19. *(Continued)* **Physical Properties of Selected Copper Casting Alloys.**
(Source, Copper Development Association.)

UNS Number	Liquidus Point		Density lb/cu in.	Coef. Thermal Expan.[1]		Thermal Conduct.[2]	Electrical Conduct.[3]	Elastic Modulus	
	°F	°C		μin. °F	μm °C			ksi	MPa
C97300	1,904	1,040	0.321	9.0	16.2	16.5	6	16,000	110,000
C97400	2,012	1,100	0.320	9.2	16.6	15.8	6	16,000	110,000
C97600	2,089	1,143	0.321	9.3	16.7	13.0	5	19,000	131,000
C97800	2,156	1,180	0.320	9.7	17.5	14.7	4	19,000	131,000
C99300	1,970	1,077	0.275	9.2	16.6	25.4	9	18,000	124,000
C99400	-	-	0.300	-	-	-	12	19,000	133,000
C99500	-	-	0.300	8.3	14.9	-	10	19,000	131,000
C99700	1,655	902	0.296	-	-	-	3	16,000	114,000
C99750	1,550	843	0.290	13.5 [4]	24.3 [4]	-	2	17,000	117,000

Notes. [1] At 68 to 572° F (20 to 300° C) unless otherwise indicated.
[2] Btu/ft²/ft/h/°F.
[3] % IACS At 68° F (20° C).
[4] Coefficient of thermal expansion measured at 68 to 392° F (20 to 200° C).
[5] Coefficient of thermal expansion measured at 68 to 212° F (20 to 100° C).

alloys—they are used in heavy duty mechanical and electromechanical equipment requiring ultrahigh strength and good electrical and/or thermal conductivity, but require reliable lubrication and well-aligned shafts when used as bearing material. Corrosion resistance for high copper alloys is as good or better than that of pure copper.

Brasses. C83300 – C83810 are red and leaded red brasses with Cu, Zn, Sn, and Pb as the principal alloying agents.

C84200 – C84800 are semi-red and leaded semi-red brasses with Cu, Zn, Sn, and Pb as the principal alloying agents.

C85200 – C85800 are yellow and leaded yellow brasses with Cu, Zn, Sn, and Pb as the principal alloying agents.

C86100 – C86800 are high strength and leaded high strength yellow brasses with Cu, Zn, Mn, Fe, and Pb as the principal alloying agents.

C87300 – C87900 are silicon bronzes and silicon brasses with Cu, Zn, and Si as the principal alloying agents.

C89320 – C89940 are high strength and leaded high strength yellow brasses with Cu, Sn, and Bi as the principal alloying agents.

The most important brasses in terms of tonnage poured are the leaded red brass C83600, and the leaded semi-red brasses C84400, C84500, and C84800. All of these alloys are widely used in water valves, pumps, pipe fittings, and plumbing hardware. Leaded yellow brasses such as C85400, C85700, and C85800 are relatively low in cost and have excellent castability, high machinability, and favorable machining characteristics. These materials are commonly used for mechanical products such as gears and machine components. The high strength and leaded high strength yellow brasses are the strongest, as cast, of all the copper alloys. They are used primarily for heavy duty mechanical products requiring moderately good corrosion resistance at a reasonable cost. The silicon bronzes/brasses have moderate strength and good corrosion resistance, and are well suited for die, permanent mold, and investment casting methods. Applications range from bearings and gears to plumbing goods and intrically shaped pump and valve components.

Tin Bronzes. C90200 – C91700 are tin bronzes with Cu and Sn as the principal alloying agents.

C92200 – C92900 are leaded tin bronzes with Cu, Sn, Zn, and Pb as the principal alloying agents.

C93100 – C94500 are high leaded tin bronzes with Cu, Sn, and Pb as the principal alloying agents.

Tin bronzes offer excellent corrosion resistance, reasonably high strength, and good wear resistance. Unleaded tin bronze C90300 is used for bearings, pump impellers, piston rings, valve fittings, and other mechanical products. C92300, which is leaded, has similar uses and is specified when better machinability and/or pressure tightness are needed. C90500 is hard and strong, and is especially resistant to seawater corrosion. C93200 is the best known bronze bearing alloy and has unsurpassed wear performance against steel journals. C93500 combines favorable antifriction properties with good load carrying capabilities, and it conforms well to slight shaft misalignments. Lead weakens all these bearing alloys, but imparts the ability to tolerate interrupted lubrication. It also allows dirt particles to become harmlessly imbedded in the bearing's surface, thereby protecting the journal. The "premier" bearing alloys, C93800 and C94300, also wear well with steel and are best known for their ability to conform to slightly misaligned shafts.

Nickel Tin Bronzes. C94700 – C94900 are nickel tin bronzes with Cu, Sn, Zn, and Ni as the principal alloying agents.

These alloys have moderate strength and very good corrosion resistance, especially in aqueous media. One member of the group, C94700, can be age hardened to tensile strengths as high as 75 ksi (517 MPa). Nickel tin bronzes are used for bearings, but are more frequently used as valve and pump components, gears, shifter forks, and circuit breaker parts.

Aluminum Bronzes. C95200 – C95500 are aluminum bronzes with Cu, Al, Fe, and Ni as the principal alloying agents.

Aluminum strengthens copper and imparts oxidation resistance. This group is best known for high corrosion and oxidation resistance combined with exceptionally good mechanical properties. The alloys are readily fabricated and welded and are used in cast structures.

Copper Nickels. C96200 – C96950 are copper nickels with Cu, Ni, and Fe as the principal alloying agents.

These alloys offer excellent resistance to seawater corrosion, high strength, and good fabricability. They are widely used in marine equipment as pumps, impellers, valves, tailshaft sleeves, and centrifugally cast pipe.

Nickel Silvers. C97300 – C97800 are nickel silvers with Cu, Ni, Zn, Pb, and Sn as the principal alloying agents.

These alloys have excellent corrosion resistance and very good machinability. Valves, fittings, and hardware cast in nickel silvers are used in food and beverage handling equipment, and as seals and labyrinth rings in steam turbines.

Leaded Coppers. C98200 – C98840 are leaded coppers with Cu and Pb as the primary alloying agents.

These alloys provide the high corrosion resistance of copper along with the favorable lubricity and low friction characteristics of high leaded bronzes. They operate well under intermittent, unreliable, or dirty lubrication, and can operate underwater with water lubrication. Used for light to moderate loads and high speeds, as in rod bushings and main bearings for refrigeration compressors, or hydraulic pump bushings.

Special Alloys. C99300 – C99750 are special alloys whose compositions do not fall into any of the above categories.

Copper product forms

Specific terminology has been assigned to the physical forms of copper materials, and should be specified when ordering. *Rod* products may have round, hexagonal, or octagonal shapes, and they are supplied in straight lengths (not coiled). *Bar* products are square or rectangular, and supplied in straight lengths. *Shapes* are also supplied in straight lengths, but may have oval, half-round, geometric, or custom ordered cross sections. *Wire* products may have any cross section, but they are always supplied in coils or on spools.

Wrought copper alloy selection based on machinability

All copper alloys are machinable in the sense that they can be cut with standard machine tooling. In fact, high speed steel tools work well with all but the hardest alloys. Since chip appearance is a good indicator of a material's machinability, copper alloys are divided into three groups based on chip characteristics. Small, fragmented chips are produced by the leaded free cutting alloys, which are designated "Type I Free Cutting" materials. Curled but brittle chips are produced by those alloys designated "Type II Short-Chip" materials. Long, stringy chips are associated with metals with homogeneous, or single phase, structures, and these alloys are designated "Type III Long-Chip" materials. Comparative machinability of wrought copper alloys is shown in **Table 20**.

Type I Free Cutting Alloys

The lead bearing copper alloys are more like composite structures than true alloys because lead is insoluble in copper and appears as a dispersion of microscopic globules. The stress raising effect of these lead particles causes chips to break up into tiny flakes as the metal passes over the tool face. Chips from leaded alloys remain in contact with the tool face for a very short time before the energy of fracture propels them away from the cutting tool. The short contact time reduces friction, which in turn minimizes tool wear and energy consumption. It is also thought that the lead globules may actually act as an internal lubricant as they pass over the tool face. The beneficial effect of lead on free cutting behavior increases with lead content, but the rate of improvement decreases as lead content rises. Significantly improved machinability can be detected and measured in alloys containing less than 0.5% Pb, but maximum free cutting behavior occurs at concentrations between 0.5% and 3.5%.

Although lead does not have a significant effect on strength, leaded alloys can be difficult to cold work extensively. The effect becomes more pronounced as lead content rises. Therefore, alloys such as C34500 and C35300, which have lower lead content than C36000 (Pb = 3.1%), may be better choices for products that require both high speed machining and extensive cold deformation. There are no hard and fast rules, unfortunately, because free cutting brass's nominal lead content is usually low enough to permit even deep knurling and coarse rolled threads. Other detrimental effects of lead include an impairment in welding and brazing properties (for further information about welding copper alloys, see the Welding section of this book), as well as related problems in welding electrodes and cutting torch tips that operate at high temperatures.

Lead does not significantly alter an alloy's corrosion behavior, but trace quantities can be dissolved form machined surfaces by appropriately aggressive media, including some potable waters. If lead exposure is a problem, wetted surfaces can be protected with electroplated or organic coatings. Surface lead may also be removed by relatively simple etching processes that leave only "pure" brass exposed to the environment.

Additions of tellurium, sulfur, and bismuth also promote free cutting behavior in copper alloys. These elements can actually be more efficient than lead in terms of the amounts needed to produce optimum improvements, but they are not without shortcomings.

Table 20. Wrought Copper Alloys Ranked by Machinability. *(Source, Copper Development Association.)*

Copper Alloy No.	Traditional Name *Descriptive Name*	Machinability Index Rating	Copper Alloy No.	Traditional Name *Descriptive Name*	Machinability Index Rating
Type I: Free Cutting Copper Alloys for Screw Machine Production (Suitable for automatic machining at the highest available cutting speeds.)					
C36000	Free-Cutting Brass	100	C37700	Forging Brass	80
C35600	Extra High Leaded Brass	100	C54400	Phosphor Bronze, B-2	80
C32000	Leaded Red Brass	90	C19100	*Nickel Copper with Tellurium*	75
C34200	High Leaded Brass, 64 1/2%	90	C19150	*Leaded Nickel Copper*	75
C35300	High Leaded Brass, 62%	90	C34000	Medium Leaded Brass, 64 1/2%	70
C35330	*Arsenical Free- Cutting Brass*	90	C35000	Medium Leaded Brass, 62%	70
C38500	Architectural Bronze	90	C37000	Free-Cutting Muntz Metal	70
C14500	Tellurium-Bearing Copper	85	C48500	Naval Brass, High Leaded	70
C14520	*Phosphorus Deoxidized, Tellurium-Bearing Copper*	85	C67300	*Leaded Silicon-Manganese-Bronze*	70
			C69710	*Leaded Arsenical Silicon Red Brass*	70
C14700	Sulfur Bearing Copper	85	C17300	*Leaded Beryllium Copper*	60
C18700	*Leaded Copper*	85	C33500	Low Leaded Brass	60
C31400	Leaded Commercial Bronze	80	C67600	*Leaded Manganese Bronze*	60
C31600	Leaded Commercial Bronze (Nickel-Bearing)	80	C48200	Naval Brass, Medium Leaded	50
			C66100	*Leaded Silicon Bronze*	50
C34500	*Leaded Brass*	80	C79200	*Leaded Nickel Silver, 12%*	50
Type II: Short Chip Copper Alloys (Usually multiphase alloys. Short, curled, or serrated chips. Screw machine production depends on type of cutting operation.)					
C64200	*Arsenical Silicon-Aluminum Bronze*	60	C67400	*Silicon-Manganese-Aluminum-Brass*	30
C62300	*Aluminum Bronze, 9%*	50	C67500	Manganese Bronze A	30
C62400	*Aluminum Bronze, 10 1/2%*	50	C69400	Silicon Red Brass	30
C17410	Beryllium Copper	40	C69430	*Arsenical Silicon Red Brass*	30

(Continued)

Table 20. *(Continued)* **Wrought Copper Alloys Ranked by Machinability.** *(Source, Copper Development Association.)*

Copper Alloy No.	Traditional Name *Descriptive Name*	Machinability Index Rating	Copper Alloy No.	Traditional Name *Descriptive Name*	Machinability Index Rating
Type II: Short Chip Copper Alloys (Usually multiphase alloys. Short, curled, or serrated chips. Screw machine production depends on type of cutting operation.)					
C17500	Beryllium Copper	40	C15715	*Aluminum Oxide Dispersion-Strengthened Copper*	20
C17510	Beryllium Copper	40			
C28000	Muntz Metal, 60%	40	C15725	*Aluminum Oxide Dispersion-Strengthened Copper*	20
C61800	*Aluminum Bronze, 10%*	40			
C63000	*Nickel-Aluminum Bronze, 10%*	40	C15760	*Aluminum Oxide Dispersion-Strengthened Copper*	20
C63020	*Nickel-Aluminum Bronze, 11%*	40			
C63200	*Nickel-Aluminum Bronze*	40	C17000	Beryllium Copper	20
C46200	Naval Brass, 63 1/2%	30	C17200	Beryllium Copper	20
C46400	Naval Brass, Uninhibited	30	C62500	*Aluminum Bronze, 13%*	20
Type III: Long Chip Copper Alloys (Usually single phase alloys. Stringy or tangled chips; somewhat gummy behavior. Not suitable for screw machine work.)					
C63200	*Nickel-Aluminum Bronze, 9%*	40	C12100	*Phosphorus-Deoxided, Low Residual Phosphorus Copper*	20
C22600	Jewelry Bronze, 87 1/2%	30			
C23000	Red Brass, 85%	30	C12200	Phosphorus-Deoxidized, High Residual Phosphorus Copper	20
C24000	Low Brass, 80%	30			
C26000	Cartridge Brass, 70%	30	C12900	Fire-Refined Tough Pitch Copper with Silver	20
C26130	*Arsenical Cartridge Brass, 70%*	30			
C26800	Yellow Brass, 66%	30	C15000	Zirconium Copper	20
C27000	Yellow Brass, 65%	30	C16200	Cadmium Copper	20
C61000	*Aluminum Bronze, 7%*	30	C16500	*Cadmium-Tin Copper*	20
C61300	*Aluminum Bronze, 7%*	30	C18000	*Nickel-Chromium-Silicon Copper*	20

(Continued)

Table 20. *(Continued)* **Wrought Copper Alloys Ranked by Machinability.** *(Source, Copper Development Association.)*

Copper Alloy No.	Traditional Name *Descriptive Name*	Machinability Index Rating	Copper Alloy No.	Traditional Name *Descriptive Name*	Machinability Index Rating
Type III: Long Chip Copper Alloys (Usually single phase alloys. Stringy or tangled chips; somewhat gummy behavior. Not suitable for screw machine work.)					
C61400	*Aluminum Bronze, 7%*	30	C18100	*Chromium-Zirconium-Magnesium Copper*	20
C65100	Low Silicon Bronze B	30			
C65500	High Silicon Bronze A	30	C18135	*Chromium-Cadmium-Copper*	20
C65600	*Silicon Bronze*	30	C18150	*Chromium-Zirconium-Copper*	20
C67000	Manganese Bronze B	30	C18200	Chromium Coppers	20
C10100	Oxygen-Free Electronic Copper	20	C18400	Chromium Coppers	20
C10200	Oxygen-Free Copper	20	C21000	Gilding, 95%	20
C10300	Oxygen-Free Extra Low Phosphorus Copper	20	C22000	Commercial Bronze, 90%	20
			C50700	*Tin (Signal) Bronze*	20
C10400	Oxygen-Free Coppers with Silver	20	C51000	Phosphor Bronze, 5% A	20
C10500	Oxygen-Free Coppers with Silver	20	C52100	Phosphor Bronze, 8% B-2	20
C10700	Oxygen-Free Coppers with Silver	20	C61000	*Aluminum Bronze, 7%*	20
C10800	Oxygen-Free Low Phosphorus Copper	20	C64700	*Silicon-Bronze*	20
C11000	Electrolytic Tough Pitch Copper	20	C70600	Copper-Nickel, 10%	20
C11300	Tough Pitch Coppers with Silver	20	C71500	Copper-Nickel, 30%	20
C11400	Tough Pitch Coppers with Silver	20	C74500	Nickel-Silver, 65-10	20
C11500	Tough Pitch Coppers with Silver	20	C75200	Nickel-Silver, 65-18	20
C11600	Tough Pitch Coppers with Silver	20	C75400	Nickel-Silver, 65-15	20
C12000	Phosphorus-Deoxidized, Low Residual Phosphorus Copper	20	C75700	Nickel-Silver, 65-12	20

Bismuth and tellurium can cause serious embrittlement and/or directionally sensitive ductility when present in uncontrolled concentrations, and sulfur cannot be used in many alloy systems due to its reactivity. Generally, unleaded free cutting alloys are considerably more expensive than leaded versions, and therefore should be considered only when lead must be avoided entirely.

Type II Short-Chip Alloys

Two or more phases may appear in the microstructure when alloy concentration is sufficiently high. The beta phase in heterogeneous (two or more phases) Type II alloys makes cold work more difficult, but considerably improves the capacity for hot deformation. Type II alloys tend to be stronger than single-phase materials, but ductility is correspondingly reduced.

As Type II alloys pass over the cutting tool, the metal tends to shear laterally in a series of closely spaced steps that produce ridges on the short, helical chips. This intermittent shear process raises the potential for tool chatter and poor surface quality, but these problems can be avoided by adjusting machining parameters appropriately. Chip breakers can be used to reduce the length of the coiled chips.

The machinability of Type II alloys is dependent on the complex relationships between alloy microstructure and mechanical properties. Power consumption varies with mechanical properties and work hardening rate, while the shape of the chips depends on the metal's ductility. In part because of their relatively manageable chips, multiphase alloys are considered to have better machinability than ductile Type III metals. In fact, the higher power consumption for the harder Type II alloys is generally taken as being less important than tool wear or chip management—both of which are less favorable in Type III alloys. Type II alloys are often processed on automatic screw machines, although production rates are considerably lower than those attainable with free cutting alloys.

Type III Long-Chip Alloys

The copper alloys with essentially uniform microstructures are the simplest. In pure copper, the structure contains only one phase (or crystal form), commonly designated "alpha." This single-phase structure is retained within fairly broad limits when alloying elements are added, but the alloy content at which additional phases begin to appear differs with the individual alloying element and with processing conditions. For example, up to 39% zinc can be added to copper to form a single-phase alpha brass. The alpha structure can also accommodate up to 9% aluminum and remain homogeneous. The equilibrium room temperature solubility limit for tin in copper is quite low (near 1%, by weight), but copper-tin alloys (tin bronzes) that are heated during processing remain structurally homogeneous up to almost 15% tin because the transformation that produces the second phase is comparatively sluggish. Copper-nickel alloys have homogeneous alpha structures no matter how much nickel is present.

Type III alloys are soft and ductile in the annealed state. Their mechanical properties are governed by alloying and the degree of cold work. Pure copper, for example, can be strengthened only by cold working, while the strength of single-phase brasses, bronzes, aluminum bronzes, and copper-nickels is derived from the combined effects of cold work and alloying. Even highly alloyed single-phase alloys retain a considerable degree of ductility, and while their chips are long and stringy, they have a smooth and uniform surface that reflects the uninterrupted passage of the cutting tool. The chip is thicker than the feed rate because the copper is upset as it passes over the face of the tool, while the accompanying cold work makes the chip hard and springy while consuming energy that, when converted to heat in the chip and cutting tool, increases tool wear.

Initial hardness, work hardening rate, and chip appearance can be related to the machinability of Type III alloys.

Initial hardness, either from prior cold work or from alloy content. Soft materials generally consume less energy than harder alloys, and usually produce less tool wear. However, softer materials tend to deflect more under the pressure of the cutting tool, which can reduce dimensional accuracy. When improperly ground tools are employed on softer materials, chatter and poor surface can result from the tool's tendency to dig into these metals.

Work hardening rate, as a function of deformation. High work hardening rates result in high energy consumption, hard chips, and high tool wear. Severely work hardened chips will tear away from the underlying soft matrix, causing smearing and poor surface finish.

Chip appearance. Although Type III metals are machinable, their tendency to form long, stringy chips makes them an unlikely choice for high speed production on automatic screw machines where clearing chip tangles can cause difficulties.

Other factors influence the machinability of all three types of copper alloys.

Finer grain sizes are generally beneficial to material strength and ductility. The effect is most pronounced in leaded multiphase alloys, although some improvements in machinability will be experienced in fine-grained Type III materials as well.

Texture. The grain structure in wrought, rolled, or extruded metals will reflect the direction of deformation—the metal's grains, grain boundaries, and second phases, if any, tend to become elongated in the direction of hot or cold work, leading to the fibrous texture of heavily wrought metals. Since any nonuniformity in a metal's structure can influence chip behavior, machinability is somewhat enhanced by cutting in a direction that crosses the direction of deformation. For example, rods and bars machine better circumferentially than longitudinally, and plates machine better in the transverse or through-thickness direction than parallel to the rolling direction.

Temper. Generally, the harder the metal, the higher the power needed, but the economic effect of temper on attainable cutting speeds and surface finish is usually more important than power consumption. Temper is discussed in more detail later in this section.

Heat treatment. Beryllium coppers are rough machined best in the solution annealed state, then heat treated to full hardness before finish machining, or, if necessary, grinding. This will also avoid the possibility of distortion caused by the slight (0.5%) volume change that occurs during heat treatment.

Tempers for wrought copper alloys

Copper alloys are said to have a harder temper if they have been cold worked and/or heat treated, and a softer temper when they are in an as-hot-formed state or when the effects of cold work and/or heat treatment have been removed by annealing. As can be expected, higher strength and hardness (harder tempers) are gained at the cost of reduced ductility. As applied to heat treated copper alloys, temper carries exactly the opposite meaning than for heat treated steels where tempering generally implies softening.

Temper designations for wrought copper alloys are defined in ASTM B 601, and the basic designations are given in **Table 21**. The designations imply specific mechanical properties only when used in association with a particular alloy, product form and size. In order to specify a specific copper alloy, it is necessary to provide the UNS Number, the product form and size (for example, 1/4 inch [6.3 mm] round rod), and the desired temper. Familiarity with the temper system is important because properties vary considerably for different forms and tempers of the same alloy. For example, a one-inch rod of the free cutting brass alloy C3600 in the Half Hard (H02) temper has a yield strength of 45 ksi (310

Table 21. Temper Designations for Wrought Copper Alloys.

Annealed - O	Temper Names	Cold Worked with Added Treatments - HR	Temper Name
O025	Hot Rolled & Annealed		
O030	Hot Extruded & Annealed	HR50	Drawn & Stress Relieved
O050	Light Anneal	**As Manufactured - M**	**Temper Names**
O60	Soft Anneal	M10	As Hot Forged- Air Cooled
O61	Annealed (Also Mill Annealed)	M11	As Hot Forged- Quenched
O70	Dead Soft Anneal	M20	As Hot Rolled
Annealed with Grain Size Prescribed - OS	**Nominal Average Grain Size, mm**	M30	As Extruded
		Solution Heat Treated - TB	**Temper Name**
OS005	0.005	TB00	Solution Heat Treated (A)
OS010	0.010	**Solution Treated and Cold Worked- TD**	**Temper Names**
OS015	0.015		
OS020	0.020	TD00	Solution Treated & Cold Worked: 1/8 Hard
OS025	0.025		
OS035	0.035	TD01	Solution Treated & Cold Worked: 1/4 Hard
OS050	0.050		
OS060	0.060	TD02	Solution Treated & Cold Worked: 1/2 Hard
OS070	0.070		
OS100	0.100	TD03	Solution Treated & Cold Worked: 3/4 Hard
OS120	0.120		
OS150	0.150	TD04	Solution Treated & Cold Worked: Hard (H)
OS200	0.200		
Cold Worked. Based on Cold Rolling or Drawing - H	**Temper Names**	**Solution Heat Treated and Precipitation Heat Treated - TF**	**Temper Name**
H00	1/8 Hard	TF00	Solution Heat Treated & Aged (AT)
H01	1/4 Hard		
H02	1/2 Hard	**Quench Hardened - TQ**	**Temper Names**
H03	3/4 Hard	TQ00	Quenched Hardened
H04	Hard	TQ30	Quenched Hardened and Tempered
H06	Extra Hard		
H08	Spring	**Solution Heat Treated, Cold Worked, & Precipitation Heat Treated - TH**	**Temper Names**
H10	Extra Spring		
H12	Special Spring		
H13	Ultra Spring	TH01	1/4 Hard and Precipitation Heat Treated (1/4 HT)
H14	Super Spring		

(Continued)

Table 21. *(Continued)* **Temper Designations for Wrought Copper Alloys.**

Cold Worked. Based on Particular Products (Wire) - H	Temper Names	Solution Heat Treated, Cold Worked, & Precipitation Heat Treated - TH	Temper Names
H60	Cold Heading, Forming	TH02	1/2 Hard and Precipitation Heat Treated (1/2 HT)
H63	Rivet		
H64	Screw	TH03	3/4 Hard and Precipitation Heat Treated (3/4 HT)
H66	Bolt		
H70	Bending	TH04	Hard and Precipitation Heat Treated (HT)
H80	Hard Drawn		
Cold Worked and Stress Relieved - HR	**Temper Names**	**Mill Hardened Tempers - TM**	**Manufacturing Designation**
HR01	1/4 Hard & Stress Relieved	TM00	AM
HR02	1/2 Hard & Stress Relieved	TM01	1/4 HM
HR04	Hard & Stress Relieved	TM02	1/2 HM
HR08	Spring & Stress Relieved	TM04	HM
HR010	Extra Spring & Stress Relieved	TM06	XHM
		TM08	XHMS
		Temper Designations Based on ASTM B 601:Temper Designations for Copper and Wrought Alloys - Wrought and Cast.	

MPa), while the same alloy produced as a 1 in. × 6-in. (25.4 mm × 152.4 mm) bar in the Soft Anneal (O60) temper has a yield strength of only 20 ksi (138 MPa).

The choice of temper depends on the properties required and the type of processing to be done. The Half Hard (H02) cold worked temper is the most frequently specified for screw machine products because it combines the best levels of strength and ductility to suit both machinability and functional requirements. Annealed tempers such as O50 and O60 refer to soft, formable structures that are usually specified for cold formed rather than machined products. Annealed alloys may have inferior machinability, with poor surface finishes, because of a tendency for chips to tear away from the work during cutting. Lightly cold worked tempers such as Eighth Hard (H00) and Quarter Hard (H01) improve machinability yet retain sufficient ductility for forming operations. Hard (H04), Extra Hard (H06), and Spring (H08) tempers produce maximum strength at the expense of ductility.

Electrical and thermal conductivity vary with the degree of temper, but the nature and extent of the effect depend strongly on the type of alloy and its metallurgical condition. Chemical properties such as corrosion resistance and plateability are not strongly affected by temper, but residual cold work-tensile stresses render some copper alloys more susceptible to stress corrosion cracking than would be the case in the annealed state.

Tempers for copper casting alloys are given in **Table 22**.

Cold forming and heat treatment of wrought copper alloys

Cold forming. As previously discussed, the highest machinability ratings for copper belong to the leaded copper alloys, particularly to the free cutting brasses. It was also pointed out that lead lowers ductility and can make alloys difficult to cold form. For this

Table 22. Standard Temper Designations for Copper Casting Alloys.

Annealed - O	Temper Names
O10	Cast & Annealed (Homogenized)
O11	As Cast and Precipitation Heat Treated
As Manufactured - M	**Temper Names**
M01	As Sand Cast
M02	As Centrifugal Cast
M03	As Plaster Cast
M04	As Pressure Die Cast
M05	As Permanent Mold Cast
M06	As Investment Cast
M07	As Continuous Cast
Temper Designations Based on ASTM B 601	

Heat Treated - TQ	Temper Names
TQ00	Quench Hardened
TQ30	Quench Hardened & Tempered
TQ50	Quench Hardened & Temper Annealed
Solution Heat Treated & Spinodal Heat Treated - TX	**Temper Name**
TX00	Spinodal Hardened
Solution Heat Treated - TB	**Temper Name**
TB00	Solution Heat Treatment
Solution Heat Treated & Precipitation Heat Treated - TF	**Temper Name**
TF00	Precipitation Hardened

reason, parts requiring deep knurls or high pitch rolled threads should be specified in alloys with reduced lead content. Alloys C34000 (0.80–1.5% Pb), C34500, and C35300 (both 1.5–2.5% Pb) are good choices. Similar conditions apply to parts such as hose fittings and automobile sensor housings, since both frequently incorporate spun flanges to crimp the parts in place. Alloy C36000 is ductile enough to permit such bends in most instances, but severe deformation may require a lower lead free cutting alloy. The Half Hard temper normally used for screw machine products offers adequate formability for routine work, but a softer temper may be required when extensive cold work must be performed.

Electrical crimp connectors made from free cutting coppers frequently demand a high degree of cold formability. In this case, it is important to take the directionality of the metal's ductility into account. Tellurium bearing coppers and alloys can be significantly less ductile in the radial or transverse direction (with regard to extrusion and drawing direction) than along the long axis of the tool.

Heat treatment. The mechanical properties of several of the machining rod copper alloys can be improved through heat treatment. Heat treatments are sometimes combined with mechanical working to enhance the strengthening effect. Zirconium copper, C15000, can be strengthened by a solution heat treatment at 1,650–1,740° F (900–950° C), followed by quenching, and aging at 930–1,020– F (500–550° C). Cold working following the solution anneal increases the final strength. Cold work also benefits the heat treatable beryllium coppers. Low-beryllium copper alloys (<1% Be, known as the "high conductivity" alloys), such as C17500 and its derivatives, are heat treated by solution annealing at 1,700° F (925° C), rapid quenching, and aging at 900° F (480° C) for two to three hours. Intermediate cold working, up to about 40%, results in an additional strength increase of about 5% above that attainable by heat treatment alone. So-called high strength beryllium coppers are solution annealed at 1,450° F (788° C), quenched, and precipitation hardened at 600° F (315° C).

Chromium-zirconium-magnesium copper (C18100) is solution annealed at 1,650–1,790° F (900–977° C), quenched and hardened at 750–930° F (400–500° C). Chromium-cadmium copper (C18135) is solution heat treated at 1,700–1,760° F (925–960° C), and aged at 850–1,050° F (455–565° C). The same temperatures apply to solution annealed and cold worked material. Chromium coppers (C18200 and C18400) are solution treated at 1,800–1,850° F (980–1010° C), and aged at 800–930° F (425–500° C). A minimum of 50% cold work can be applied after solution annealing to increase the alloy's final strength. Nickel-tellurium copper (C19100) can be age hardened at 800–900° F (425–480° C) after solution annealing at 1,300–1,450° F (705–790° C), or hot working at slightly higher temperatures.

Copper casting alloy selection based on fabricability

Castings usually require further processing after shake-out and cleaning. Machining is the most common secondary operation, and welding is often required to repair minor defects or to join several castings into a larger assembly. Factors to be considered before selecting cast copper alloys for a process or application are shown in **Table 23**.

Machinability

As a class, cast copper alloys can be described as being relatively easy to machine, when compared to steel, and far easier to machine when compared to stainless steels, nickel base alloys, and titanium. The easiest casting alloys to machine are those that contain more than 2% lead. These alloys are free cutting, tool wear is minimal, and surface finishes are generally excellent. High speed steel is the accepted tooling material for these alloys, but carbides are often used for the stronger leaded compositions. Cutting fluids help in the reduction of airborne lead-bearing particles, but they are not otherwise needed when cutting the highly leaded brasses and bronzes. Leaded cast copper alloys have machinability ratings greater than 70 (with wrought free cutting brass C36000 being 100).

Next in order of machinability are moderate to high strength alloys that contain sufficient alloying elements to form second phases in their microstructure (often called duplex or multiphase alloys). Examples include unleaded yellow brasses, manganese bronzes, and silicon brasses and bronzes. These alloys have short, brittle, tightly curled chips that tend to break into manageable segments, and using chipbreaker geometries facilitates this action. Surface finishes can be quite good for the duplex alloys, but cutting speeds will be lower, and tool wear higher, than with the free cutting grades.

The unleaded, single phase alpha alloys (which include high conductivity coppers, high copper alloys such as chromium coppers, beryllium coppers, tin bronzes, red brasses, aluminum bronzes, and copper nickels), as a group, have a general tendency to form long, stringy chips that interfere with machining operations. In addition, pure copper and high nickel alloys tend to weld to the tool face, which leads to impaired surface finishes. Cutting tools used with these alloys should be highly polished and ground with generous rake angles to help ease the flow of chips away from the workpiece. Adequate relief angles will help avoid trapping particles between the tool and workpiece, where they might scratch the freshly machined surface, and cutting fluids should be used to provide adequate lubrication.

Table 23. Technical Factors in the Choice of Casting Method for Copper Alloys. *(Source, Copper Development Association.)*

Casting Method	Copper Alloys	Size Range	General Tolerances	Surface Finish	Min. Section Thickness	Ordering Quantities	Relative Cost[1]
Sand	All	All sizes, depends on foundry capability.	±1/32 in up to 3 in.; ±3/64 in 3-6 in.; add ±0.003 in./in above 6 in.; add ±0.020 to ±0.060 in across parting line.	150-500 μin. rms	1/8-1/4 in.	All	1-3
No-Bake	All	All sizes, but usually > 10 lbs.	Same as sand casting.	Same as sand casting	Same as sand casting	All	1-3
Shell	All	Typical maximum mold area = 550 in. ^{2} typical maximum thickness = 6 in.	±0.005-0.010 in up to 3 in.; add ±0.002 in/in. above 3 in.; add ±0.005 to 0.010 in. across parting line.	125-200 μin. rms	3/32 in.	≥100	2-3
Permanent Mold	Coppers, high copper alloys, yellow brasses, high strength brasses, silicon bronze, high zinc silicon brass, most tin bronzes, aluminum bronzes, some nickel silvers.	Depends on foundry capability; best ≈ 50 lbs. Best max. thickness ≈ 2 in.	Usually ±0.010 in.; optimum ±0.005 in., ±0.002 in part-to-part.	150-200 μin. rms, best ≈ 70 μin. rms	1/8-1/4 in.	100-1,000, depending on size	2-3
Die	Limited to C85800, C86200, C86500, C87800, C87900, C99700, C99750 & some proprietary alloys.	Best for small, thin parts; max. area ≤ 3 ft. ^{2}.	±0.002 in/in; not less than 0.002 in. on any one dimension; add ±0.010 in. on dimensions affected by parting line.	32-90 μin. rms	0.05-0.125 in.	≥1000	1
Plaster	Coppers, high copper alloys, silicon bronze, manganese bronze, aluminum bronze, yellow brass.	Up to 800 in. ^{2}, but can be larger.	One side of parting line, ±0.015 in up to 3 in.; add ±0.002 in./in above 3 in.; add 0.010 in across parting line, and allow for parting line shift of 0.015 in.	63-125 μin. rms, best ≈ 32 μin. rms	0.060 in.	All	4

(Continued)

Table 23. *(Continued)* **Technical Factors in the Choice of Casting Method for Copper Alloys.** *(Source, Copper Development Association.)*

Casting Method	Copper Alloys	Size Range	General Tolerances	Surface Finish	Min. Section Thickness	Ordering Quantities	Relative Cost[1]
Investment	Almost all	Fraction of an ounce to 150 lbs., up to 48 in.	±0.003 in less than 1/4; ±0.004 in between 1/4 to 1/2 in.; ±0.005 in./in between 1/2-3 in.; add ±0.003 in./in above 3 in.	63-125 μin. rms	0.030 in.	>100	5
Centrifugal	Almost all	Ounce to 25,000 lbs. Depends on foundry capacity.	Castings are usually rough machined by foundry.	Not applicable	1/4 in.	All	1-3

[1] 1 = low cost, 5 = high cost.

Magnesium Alloys

Magnesium is usually one of the first materials considered when light weight is a designer's priority—it is the lightest mass produced metal that remains stable under normal operating conditions. It is also the eighth most abundant element in the earth's crust, and the third most abundant element in seawater. As with other metals, magnesium alloys increase in tensile strength, yield strength, and hardness in low temperature environments, but ductility suffers. Tensile properties reverse at elevated temperatures, and most suffer a reduction of 40 to 75% when operating temperatures increase from room temperature (68° F, 20° C) to 600° F (315° C). The density of the magnesium alloys ranges from a low of 1.74 gr/cm^3 (alloy K1A, which is often used for its dampening characteristics) to highs of 1.86 gr/cm^3 (alloy ZH62A), but most are in the range of 1.80 to 1.83. Electrical conductivity of magnesium is 38.6 IACS. Both wrought and cast alloys are available. The wrought is available in extrusions, forgings, sheet, plate, and bar, and the cast for sand, permanent mold, investment, and die castings.

Pure magnesium ignites easily. The alloys are less susceptible to ignition, but care must be taken during machining, and fine chips should be regarded as fire hazards. The larger chips produced by roughing cuts and medium finishing cuts present very little danger of igniting, and even fine chip fires are unlikely at cutting speeds below 700 sfm (213 smm). Feed rates should never be in excess of 0.001 ipr (0.02 mm/rev). Low viscosity cutting fluids minimize risk and reduce the possibility of workpiece distortion due to overheating. Cutting solutions containing water should never be used as water increases the risk of scrap igniting during shipping and storage, and also reduces the value of the scrap.

Composition and properties of magnesium alloys

Although UNS designations have been assigned to magnesium alloys, they are more commonly identified by a four-part ASTM number that reflects the composition of the alloy in the first three parts, and the temper designation in the fourth. For example, the alloy AZ31B-F has aluminum and zinc (AZ) as its principal alloying agents, in amounts of 3% aluminum and 1% zinc (stated as 31 in the alloy number). The B indicates that this is the second alloy standardized with 6% Al and 3% Zn, and the F indicates that the alloy is as fabricated. For a description of the ASTM designation system, see **Table 1**. A more detailed explanation of the temper system can be found below.

Tables 2 and **3** provide chemical compositions for casting and wrought magnesium alloys, and **Tables 4** and **5** show their mechanical and thermal properties. Characteristics of some popular alloys follow. References below to AMS stand for Aerospace Materials Specifications, and those to ASTM stand for American Society for Testing and Materials.

Wrought Alloys

AZ31B. A wrought magnesium base alloy containing aluminum and zinc. It is available as sheet and plate (AMS 4375 and AMS 4377), plate (AMS 4376), extrusion (ASTM B 107), and forging (ASTM B 91). It has good room temperature strength and ductility and is used primarily for applications where the temperature does not exceed 300° F (149° C). Increased strength is obtained in the sheet and plate form by strain hardening with a subsequent partial anneal (H24 and H26 tempers). No treatments are available for increasing the strength of this alloy after fabrication, but AZ31B has moderate strength and is widely used. Forming this alloy must be done at elevated temperatures if small radii or deep draws are required. If the temperatures used are too high or the time is too great, H24 and H26 temper will be softened. It is readily welded but must be stress relieved after welding to prevent stress corrosion cracking.

AZ61A. A wrought magnesium base alloy containing aluminum and zinc. It is

Table 1. Standard ASTM Designation System for Magnesium Alloys and Tempers.

Part	Purpose	Form	Symbols Used
First Part	Indicates the two principal alloying elements.	Two code letters represent the two main alloying elements (if only one alloying element is present, only one letter is used), with the element with the highest percentage listed first. If percentage content of the two primary alloying agents are equal, they are listed alphabetically.	A, Aluminum; B, Bismuth; C, Copper; D, Cadmium; E, Rare Earth; F, Iron; G, Magnesium; H, Thorium; K, Zirconium; L, Lithium; M, Manganese; N, Nickel; P, Lead; Q, Silver; R, Chromium; S, Silicon; T, Tin; W, Yttrium; Y, Antimony; Z, Zinc
Second Part	Indicates the percentage content of the two principal alloying elements.	Two numbers correspond to rounded off percentages of the two main alloying elements. They are arranged in the same order as the alloy designations in the first part.	Whole numbers are used.
Third Part	Distinguishes between different alloys with the same percentages of the two principal alloying elements.	One letter is used.	A, First composition, registered with ASTM; B, Second composition, registered; C, Third composition, registered; D, High purity, registered; E, High corrosion resistance, registered; X, Not registered with ASTM
Fourth Part	Indicates Temper. See text for full and expanded description of tempers.	One letter and, as required, one number, separated from the third part by a hyphen.	F, As fabricated; O, Annealed; H10, H11, Slightly strain hardened; H23, H24, H26, Strain hardened and partially annealed; T4, Solution heat treated; T5, Artificially aged only; T6, Solution heat treated and artificially aged; T8, Solution heat treated, cold worked, and artificially aged

Table 2. Chemical Composition of Magnesium Casting Alloys.

ASTM No.	UNS No.	Nominal Composition — Expressed as Percent by Weight
Sand and Permanent Mold Castings		
AM100A	M10100	Al 9.3-10.7; Cu 0.1 max.; Mg 90; Mn 0.1 min.; Ni 0.01 max.; Si 0.3 max.; Zn 0.3 max.; Other 0.3 max.
AZ63A	M11630	Al 5.3-6.7; Cu 0.25 max.; Mg 91; Mn 0.15 min.; Ni 0.01 max.; Si 0.3 max.; Zn 2.5-3.5.; Other 0.3 max.
AZ81A	M11810	Al 7-8.1; Cu 0.1 max.; Mg 92; Mn 0.13 min.; Ni 0.01 max.; Si 0.3 max.; Zn 0.4-1.; Other 0.3 max.
AZ91C	M11914	Al 8.1-9.3; Cu 0.1 max.; Mg 90; Mn 0.13 min.; Ni 0.01 max.; Si 0.3 max.; Zn 0.4-1.; Other 0.3 max.
AZ91E	M11921	Al 8.1-9.3; Cu 0.015 max.; Fe 0.005 max.; Mg 90; Mn 0.17-0.35; Ni 0.001 max.; Si 0.2 max.; Zn 0.35-1.; Other 0.3 max.

(Continued)

Table 2. *(Continued)* **Chemical Composition of Magnesium Casting Alloys.**

ASTM No.	UNS No.	Nominal Composition — Expressed as Percent by Weight
Sand and Permanent Mold Castings		
AZ92A	M11920	Al 8.3-9.7; Cu 0.25 max.; Mg 89; Mn 0.1 min.; Ni 0.01 max.; Si 0.3 max.; Zn 1.6-2.4.; Other 0.3 max.
EZ33A	M12330	Cu 0.1 max.; Mg 93; Nd 2.5-4; Ni 0.01 max.; Zn 2-3.1; Zr 0.5-1; Other 0.3 max.
HK31A	M13310	Cu 0.1 max.; Mg 96; Ni 0.01 max.; Th 2.5-4; Zn 0.3 max.; Zr 0.4-1; Other 0.3 max.
HZ32A	M13320	Cu 0.1 max.; Mg 94; Ni 0.01 max.; Rare Earths 0.1 max.; Th 2.5-4; Zn 1.7-2.5; Zr 0.5-1; Other 0.3 max.
K1A	M18010	Mg 99; Zr 0.4-1; Other 0.3 max.
QE22A	M18220	Ag 2-3; Cu 0.1 max.; Mg 95; Nd 1.7-2.5; Ni 0.01 max.; Zr 0.4-1; Other 0.03 max.
QH21A	-	Ag 2-3; Cu 0.1 max.; Mg 95; Nd 0.6-1.5; Ni 0.01 max.; Th 0.6-1.6; Zn 0.2 max.; Zr 0.4-1; Other 0.03 max.
ZE41A	M16410	Cu 0.1 max.; Mg 94; Mn 0.15 max.; Nd 0.75-1.75; Ni 0.01 max.; Zn 3.5-5; Zr 0.4-1; Other 0.3 max.
ZE63A	M16630	Cu 0.1 max.; Mg 91; Nd 2.1-3; Ni 0.01 max.; Zn 5.5-6; Zr 0.4-1; Other 0.3 max.
ZH62A	M16620	Cu 0.1 max.; Mg 92; Ni 0.01 max.; Th 1.4-2.2; Zn 5.2-6.2; Zr 0.5-1; Other 0.3 max.
ZK61A	M16610	Cu 0.1 max.; Mg 93; Ni 0.01 max.; Zn 5.5-6.5; Zr 0.6-1; Other 0.3 max.
Die Casting		
AM60A	M10600	Al 5.5-6.5; Cu 0.35 max.; Mg 94; Mn 0.13 min.; Ni 0.03 max.; Si 0.5 max.; Zn 0.22 max.
AM60B	M10603	Al 5.5-6.5; Cu 0.01 max.; Fe 0.005 max.; Mg 94; Mn 0.25 min.; Ni 0.002 max.; Si 0.1 max.; Zn 0.22 max.; Other 0.003 max.
AS21X1	-	Al 1.7; Mg 97; Mn 0.4 min.; Si 1.1
AS41XA	M10410	Al 3.5-5; Cu 0.06 max.; Mg 94; Mn 0.2-0.5; Ni 0.03 max.; Si 0.5-1.5; Zn 0.12 max.; Other 0.3 max.
AZ91A	M11910	Al 8.3-9.7; Cu 0.1 max.; Mg 90; Mn 0.13 min.; Ni 0.03 max.; Si 0.5 max.; Zn 0.35-1 max.; Other 0.3 max.
AZ91B	M11912	Al 8.3-9.7; Cu 0.35 max.; Mg 90; Mn 0.13 min.; Ni 0.03 max.; Si 0.5 max.; Zn 0.35-1; Other 0.3 max.

Table 3. Chemical Composition of Magnesium Extruded, Forging, and Sheet and Plate Alloys.

ASTM No.	UNS No.	Nominal Composition — Expressed as Percent by Weight
Extruded		
AZ10A	M11100	Al 1-1.5; Ca 0.04 max.; Cu 0.1 max.; Fe 0.005 max.; Mg 98; Mn 0.2 min.; Ni 0.005 max.; Si 0.1 max.; Zn 0.2-0.6
AZ31B	M11311	Al 2.5-3.5; Ca 0.04 max.; Cu 0.05 max.; Fe 0.005 max.; Mg 97; Mn 0.2 min.; Ni 0.005 max.; Si 0.1 max.; Zn 0.6-1.4; Other 0.3 max.
AZ61A	M11610	Al 5.8-7.2; Cu 0.05 max.; Fe 0.005 max.; Mg 92; Mn 0.15 min.; Ni 0.005 max.; Si 0.1 max.; Zn 0.4-1.5; Other 0.3 max.
AZ80A	M11800	Al 7.8-9.2; Cu 0.05 max.; Fe 0.005 max.; Mg 91; Mn 0.12 min.; Ni 0.005 max.; Si 0.1 max.; Zn 0.2-0.8; Other 0.3 max.
EA55RS	-	Al 5; Mg 85; Nd 4.9; Zn 5
HM31A	M13312	Mg 96; Mn 1.2 min.; Th 2.5-3.5; Other 0.3 max.

(Continued)

Table 3. *(Continued)* **Chemical Composition of Magnesium Extruded, Forging, and Sheet and Plate Alloys.**

ASTM No.	UNS No.	Nominal Composition — Expressed as Percent by Weight
Extruded		
M1A	M15100	Ca 0.3 max.; Cu 0.05 max.; Mg 99; Mn 1.2 min.; Ni 0.01 max.; Si 0.1 max.; Other 0.3 max.
ZK40A	M16400	Mg 95; Zn 3.5-4.5; Zr 0.45 min.; Other 0.3 max.
ZK60A	M16600	Mg 94; Zn 4.8-6.2; Zr 0.45 min.; Other 0.3 max.
Forgings		
AZ31B	M11311	Al 2.5-3.5; Ca 0.04 max.; Cu 0.05 max.; Fe 0.005 max.; 97 min 0.2.; Max. 0.005; Si 0.1 max.; Zn 0.6-1.4; Other 0.3 max.
AZ61A	M11610	Al 5.8-7.2; Cu 0.05 max.; Fe 0.005 max.; 92 min.0.15; 0.005 max,; Si 0.1 max.; Zn 0.4-1.5; Other 0.3 max.
AZ80A	M11800	Al 7.8-9.2; Cu 0.05 max.; Fe 0.005 max.; 91 min.0.12; 0.005 max,; Si 0.1 max.; Zn 0.2-0.8; Other 0.3 max.
HM21A	M13210	Mg 97; Mn 0.45-1.1; Th.1.5-2.5; Other 0.3 max.
M1A	M15100	Ca 0.3 max.; Cu 0.05 max.; Mg 99; Mn 1.2 min.; Ni 0.01 max.; Si 0.1 max.; Other 0.3 max.
ZK60A	M16600	Mg 94; Zn 4.8-6.2; Zr 0.45 min.; Other 0.3 max.
Sheet and Sheet and Plate		
AZ31B	M11311	Al 2.5-3.5; Ca 0.04 max.; Cu 0.05 max.; Fe 0.005 max.; 97 min.0.2; 0.005 max.; Si 0.1 max.; Zn 0.6-1.4; Other 0.3 max.
HK31A	M13310	Cu 0.1 max.; Mg 96; Ni 0.01 max.; Th 2.5-4; Zn 0.3 max.; Zr 0.4-1; Other 0.3 max.
HM21A	M13210	Mg 97; Mn 0.45-1.1; Th 1.5-2.5; Other 0.3 max.
M1A	M15100	Ca 0.3 max.; Cu 0.05 max.; Mg 99; Mn 1.2 min.; Ni 0.01 max.; Si 0.1 max.; Other 0.3 max.

available as extrusion (ASM 4350) and forging (ASTM B91). Much like AZ31B in general characteristics, but the increased aluminum content slightly increases the strength while decreasing the ductility. Artificial aging will also increase strength but reduce ductility. Severe forming must be done at elevated temperatures. It is readily welded but must be stress relieved after welding to prevent stress corrosion cracking.

ZK60A. A wrought magnesium base alloy containing zinc and zirconium. It is available as extrusion (ASTM B107 and AMS 4352), and die and hand forgings (AMS 4362). Increased strength is obtained by artificial aging (T5) from the as fabricated (F) temper. It has the best combination of high room temperature strength and ductility of the wrought magnesium base alloys, and is used primarily at temperatures below 300° F (149° C). ZK60A has good ductility as compared with other high strength magnesium alloys, and can be formed or bent cold into shapes not possible with those alloys having less ductility. It is not considered weldable.

Casting Alloys

AM100A. A magnesium base casting alloy containing aluminum and a small amount of manganese. It is available as investment casting (AMS 4455), permanent mold casting (AMS 4483), and as a casting alloy as specified by MIL-M-46062. Primarily used for permanent mold castings, AM100A has about the same characteristics as AZ92A. It has less tendency to microshrinkage and hot shortness than the Mg-Al-Zn alloys, plus good weldability and fair pressure tightness.

(Text continued on p. 273)

Table 4. Mechanical Properties of Magnesium Alloys.

ASTM No.	Ultimate Tensile Strength		Yield Strength		Ultimate Bearing Strength		Bearing Yield Strength		Modulus of Elasticity		Elong. % [1]	HB[2]
	PSI	MPa	PSI	MPa	PSI	MPa	PSI	MPa	ksi	GPa		
Sand and Permanent Mold Castings												
AM100A-T6	39,885	275	15,954	110	-	-	-	-	6,527	45	4	69
AZ63A-T6	39,885	275	18,855	130	-	-	52,214	360	6,527	45	5	73
AZ81A-T4	39,885	275	12,038	85	58,015	400	34,809	240	6,527	45	15	55
AZ91C-T6	39,885	275	21,031	145	74,695	515	52,214	360	6,498	44.8	6	70
AZ91E-T6	39,885	275	21,031	145	74,695	515	52,214	360	6,498	44.8	6	70
AZ92A-T6	39,885	275	21,756	150	79,771	550	65,267	450	6,527	45	3	84
EZ33A-T5	23,206	160	15,954	110	57,290	395	39,885	275	6,527	45	2	50
HK31A-T6	31,908	220	15,229	105	60,916	420	39,885	275	6,527	45	8	55
HZ32A-T5	26,832	185	13,053	90	59,466	410	36,985	255	6,527	45	4	55
K1A-F	26,107	180	7,977	55	45,977	317	18,130	125	5,511	38	19	-
QE22A-T6	37,710	260	28,282	195	-	-	-	-	6,527	45	3	80
QH21A-T6	39,885	275	29,733	205	-	-	-	-	6,527	45	4	-
ZE41A-T5	29,733	205	20,305	140	70,343	485	50,763	350	6,527	45	3.5	62
ZE63A-T6	43,511	300	27,557	190	-	-	-	-	6,527	45	10	70
ZH62A-T5	34,809	240	24,656	170	71,794	495	49,313	340	6,527	45	4	70
ZK61A-T5	44,962	310	26,832	185	-	-	-	-	6,527	45	10	68
ZK61A-T5	44,962	310	28,282	195	-	-	-	-	6,527	45	10	70
Die Castings												
AM60A-F	29,733	205	16,679	115	-	-	-	-	6,527	45	6	-
AM60B-F	29,733	205	16,679	115	-	-	-	-	6,527	45	6	-
AS21X1	-	-	18,855	130	-	-	-	-	6,527	45	9	-

Notes: [1] Elongation in 50 mm (2 inches). [2] Hardness Brinell with 500 kg load and 10 mm ball.

(Continued)

Table 4. *(Continued)* **Mechanical Properties of Magnesium Alloys.**

ASTM No.	Ultimate Tensile Strength		Yield Strength		Ultimate Bearing Strength		Bearing Yield Strength		Modulus of Elasticity		Elong. % [1]	HB [2]
	PSI	MPa	PSI	MPa	PSI	MPa	PSI	MPa	ksi	GPa		
Die Castings												
AS41XA-F	31,908	220	21,756	150	-	-	-	-	6,527	45	4	-
AZ91A-F	33,359	230	21,756	150	-	-	-	-	6,498	44.8	3	63
AZ91B-F	33,359	230	21,756	150	-	-	-	-	6,498	44.8	3	63
K1A-F	23,931	165	12,038	83	-	-	-	-	5,511	38	8	-
Extruded Bars, Rods and Shapes												
AZ10A-F	34,809	240	22,481	155	-	-	-	-	6,527	45	10	-
AZ31B-F	37,710	260	29,008	200	55,840	385	33,359	230	6,527	45	15	49
AZ61A-F	44,962	310	33,359	230	68,168	470	41,336	285	6,527	45	16	60
AZ80A-F	49,313	340	36,260	250	79,771	550	50,763	350	6,527	45	7	67
AZ80A-T5	55,114	380	39,885	275	-	-	-	-	6,527	45	7	82
EA55RS-T4	61,641	425	53,664	370	-	-	-	-	6,527	45	14	-
HM31A-T5	43,511	300	39,160	270	68,168	470	47,863	330	6,527	45	10	-
M1A-F	36,985	255	26,107	180	50,763	350	28,282	195	6,527	45	12	44
ZK40A-T5	40,030	276	36,985	255	-	-	-	-	6,527	45	4	-
ZK60A-F	49,313	340	37,710	260	79,771	550	55,114	380	6,527	45	11	75
ZK60A-T5	52,939	365	44,237	305	84,847	585	58,740	405	6,527	45	11	88
Forgings												
AZ31B	37,710	260	24,656	170	-	-	-	-	6,527	45	15	-
AZ61A-F	42,786	295	26,107	180	72,519	500	41,336	285	6,527	45	12	55
AZ80A-F	47,863	330	33,359	230	-	-	-	-	6,527	45	11	69
AZ80A-T5	50,038	345	36,260	250	-	-	-	-	6,527	45	11	72

Notes: [1] Elongation in 50 mm (2 inches). [2] Hardness Brinell with 500 kg load and 10 mm ball.

(Continued)

Table 4. *(Continued)* **Mechanical Properties of Magnesium Alloys.**

ASTM No.	Ultimate Tensile Strength		Yield Strength		Ultimate Bearing Strength		Bearing Yield Strength		Modulus of Elasticity		Elong. %[1]	HB[2]
	PSI	MPa	PSI	MPa	PSI	MPa	PSI	MPa	ksi	GPa		
Forgings												
AZ80A-T6	49,313	340	36,260	250	-	-	-	-	6,527	45	5	72
HM21A-F	33,359	230	20,305	140	-	-	-	-	6,527	45	15	-
HM21A-T5	33,359	230	21,756	150	36,260	250	23,206	160	6,527	45	9	50
M1A	36,260	250	23,206	160	-	-	-	-	6,527	45	7	47
ZK60A-T5	44,237	305	31,183	215	60,916	420	41,336	285	6,527	45	16	65
ZK60A-T6	47,137	325	39,160	270	65,267	450	46,412	320	6,527	45	11	75
Sheet and Sheet and Plate												
AZ31B-O	36,985	255	21,756	150	70,343	485	42,061	290	6,527	45	21	56
AZ31B-H24	42,061	290	31,908	220	71,794	495	47,137	325	6,527	45	15	73
AZ31B-H26	39,885	275	27,557	190	71,794	495	39,885	275	6,527	45	10	-
HK31A-O	29,008	200	18,130	125	-	-	-	-	6,527	45	30	-
HK31A-H24	36,260	250	27,557	190	60,916	420	41,336	285	6,527	45	15	58
HM21A-T8	34,084	235	24,656	170	60,191	415	39,160	270	6,527	45	11	-
HM21A-T81	36,985	255	30,458	210	65,992	455	46,412	320	6,527	45	6	-
M1A-F	33,359	230	18,130	125	50,763	350	29,008	200	6,527	45	17	48

Notes: [1] Elongation in 50 mm (2 inches). [2] Hardness Brinell with 500 kg load and 10 mm ball.

Table 5. Thermal Properties of Magnesium Alloys.

ASTM No.	Process Temperature[1]	CTE[2]		CTE[3]		Thermal Conductivity[4]		Solidus		Liquidus	
	°C	μin/in.	μm/m	μin/in.	μm/m	Engl.	metric	°F	°C	°F	°C
Sand and Permanent Mold Castings											
AM100A-T6	735-845	14	25	16	28	507	73	865	463	1,103	595
AZ63A-T6	705-845	15	26.1	16	28	534	77	851	455	1,130	610
AZ81A-T4	705-845	14	25	16	28	355	51.1	914	490	1,130	610
AZ91C-T6	625-700	14	26	16	28	505	72.7	878	470	1,103	595
AZ91E-T6	625-700	14	26	16	28	505	72.7	878	470	1,103	595
AZ92A-T6	700-845	14	26	16	28	500	72	833	445	1,103	595
EZ33A-5	750-820	15	26.4	16	28	694	100	1,013	545	1,193	645
HK31A-6	-	15	26.8	16	28	638	92	1,094	590	1,202	650
HZ32A-5	750-820	15	26.7	16	28	763	110	1,022	550	1,202	650
K1A-F	750-820	15	27	16	28	847	122	1,202	650	1,202	650
QE22A-T6	750-820	15	26.7	16	28	784	113	1,022	550	1,193	645
QH21A-T6	750-820	15	26.7	16	28	784	113	995	535	1,184	640
ZE41A-T5	750-820	14	26	16	28	784	113	977	525	1,193	645
ZE63A-T6	750-820	15	26.5	16	28	756	109	950	510	1,175	635
ZH62A-T5	750-820	15	27.1	16	28	763	110	968	520	1,166	630
ZK61A-T5	705-815	15	27	16	28	833	120	986	530	1,175	635
ZK61A-T6	705-815	15	27	16	28	833	120	986	530	1,175	635

Notes: [1] Process temperature is the casting temperature for sand, permanent mold, and die castings, and the hot-working temperature for extruded materials, forgings, sheet, and sheet and plate. [2] Coefficient of thermal expansion, from 32 to 212° F (0 to 100° C) measured in μin./in.-°F, and μm/m-°C. [3] Coefficient of thermal expansion, from 32 to 480° F (0 to 250° C) measured in μin./in.-°F, and μm/m-°C. [4] Metric measurement in W/m-K. English measurement in Btu-in./hr-ft²-°F.

(Continued)

Table 5. *(Continued)* **Thermal Properties of Magnesium Alloys.**

ASTM No.	Process Temperature[1]	CTE[2]		CTE[3]		Thermal Conductivity[4]		Solidus		Liquidus	
	°C	µin/in.	µm/m	µin/in.	µm/m	Engl.	metric	°F	°C	°F	°C
Die Castings											
AM60A-F	650-695	14	25.6	16	28	430	62	1,004	540	1,139	615
AM60B-F	650-695	14	25.6	16	28	430	62	1,004	540	1,139	615
AS21X1	-	14	26	16	28	-	-	-	-	-	-
AS41XA	650-695	15	26.1	16	28	472	68	1,049	565	1,148	620
AZ91A-F	625-700	14	26	16	28	505	72.7	878	470	1,103	595
AZ91B-F	625-700	14	26	16	28	505	72.7	878	470	1,103	595
K1A-F	750-820	15	27	16	28	847	122	1,202	650	1,202	650
Extruded Bars, Rods and Shapes											
AZ10A-F	-	15	26.6	16	28	763	110	1,166	630	1,193	645
AZ31B-F	230-425	14	26	16	28	666	96	1,121	605	1,166	630
AZ61A-F	230-400	14	26	16	28	486	70	977	525	1,148	620
AZ80A-F	320-400	14	26	16	28	527	76	914	490	1,130	610
AZ80A-T5	320-400	14	26	16	28	527	76	914	490	1,130	610
HM31A-T5	370-540	14	26	16	28	722	104	1,121	605	1,202	650
M1A-F	-	14	26	16	28	-	-	1,198	648	1,200	649
ZK40A-T5	-	14	26	16	28	763	110	-	-	-	-
ZK60A-F	-	14	26	16	28	833	120	968	520	1,175	635
ZK60A-T5	-	14	26	16	28	833	120	968	520	1,175	635

Notes: [1] Process temperature is the casting temperature for sand, permanent mold, and die castings, and the hot-working temperature for extruded materials, forgings, sheet, and sheet and plate. [2] Coefficient of thermal expansion, from 32 to 212° F (0 to 100° C) measured in µin./in.-°F, and µm/m-°C. [3] Coefficient of thermal expansion, from 32 to 480° F (0 to 250° C) measured in µin./in.-°F, and µm/m-°C. [4] Metric measurement in W/m-K. English measurement in Btu-in./hr-ft^2-°F.

(Continued)

Table 5. *(Continued)* **Thermal Properties of Magnesium Alloys.**

ASTM No.	Process Temperature[1]	CTE[2]		CTE[3]		Thermal Conductivity[4]		Solidus		Liquidus	
	°C	μin/in.	μm/m	μin/in.	μm/m	Engl.	metric	°F	°C	°F	°C
Forgings											
AZ31B	230-425	14	26	16	28	666	96	1,121	605	1,166	630
AZ61A-F	230-400	14	26	16	28	486	70	977	525	1,148	620
AZ80A-F	320-400	14	26	16	28	527	76	914	490	1,130	610
AZ80A-T5	320-400	14	26	16	28	527	76	914	490	1,130	610
AZ80A-T6	320-400	14	26	16	28	527	76	914	490	1,130	610
HM21A-F	455-595	15	26.8	16	28	937	135	1,121	605	1,202	650
HM21A-T5	455-595	15	26.8	16	28	937	135	1,121	605	1,202	650
M1A	-	14	26	16	28	-	-	1,198	648	1,200	649
ZK60A-T5	-	14	26	16	28	833	120	968	520	1,175	635
ZK60A-T6	-	15	27	16	28	833	120	968	520	1,175	635
Sheet and Sheet and Plate											
AZ31B-O	230-425	14	26	16	28	666	96	1,121	605	1,166	630
AZ31B-H24	230-425	14	26	16	28	666	96	1,121	605	1,166	630
AZ31B-H26	230-425	14	26	16	28	666	96	1,121	605	1,166	630
HK31A-O	-	15	26.8	16	28	638	92	1,094	590	1,202	650
HK31A-H24	-	15	26.8	16	28	638	92	1,094	590	1,202	650
HK21A-T8	455-595	15	26.8	16	28	937	135	1,121	605	1,202	650
HM21A-T81	455-595	15	26.8	16	28	937	135	1,121	605	1,202	650
M1A-F	-	14	26	16	28	-	-	1,198	648	1,200	649

Notes: [1] Process temperature is the casting temperature for sand, permanent mold, and die castings, and the hot-working temperature for extruded materials, forgings, sheet, and sheet and plate. [2] Coefficient of thermal expansion, from 32 to 212° F (0 to 100° C) measured in μin./in.-°F, and μm/m-°C. [3] Coefficient of thermal expansion, from 32 to 480° F (0 to 250° C) measured in μin./in.-°F, and μm/m-°C. [4] Metric measurement in W/m-K. English measurement in Btu-in./hr-ft^2-°F.

(Continued)

AZ91C/AZ91D/AZ91E. These alloys are available as sand casting (AMS 4437 and AMS 4446), investment casting (AMS 4452), and as a casting alloy as specified by MIL-M-46062. AZ91C is a magnesium base casting alloy containing aluminum and zinc. A slightly purer version, AZ91D, has excellent castability and saltwater resistance and has overtaken the once dominant AZ91B as the workhorse for die castings. AZ91E is a version that contains a significantly lower level of impurities resulting in improved corrosion resistance. These alloys have good castability with a good combination of ductility and strength. AZ91C and AZ91E are the most commonly used sand castings for temperatures under 300° F (149° C). Both have fair weldability and pressure tightness.

AZ92A. A magnesium base casting alloy containing aluminum and zinc. It is available as sand casting (AMS 4434), permanent mold casting (AMS 4484), and as a casting alloy as specified by MIL-M-46062. It is slightly stronger and less ductile than AZ91C, but they are much alike in other characteristics. It has fair weldability and pressure tightness.

EZ33A. A magnesium base casting alloy containing rare earths, zinc, and zirconium. It is available as sand casting (AMS 4442) in the artificially aged (T5) temper. EZ33A has lower strength than the Mg-Al-Zn alloys at room temperature but is less affected by increasing temperatures. It is generally used for applications in the temperature range of 300 to 500° F (149 to 260° C). EZ33A castings are very sound and are sometimes specified for pressure tightness. It has good stability in the T5 temper and excellent weldability. It is sometimes used in applications requiring good dampening ability.

QE22A. A magnesium base casting alloy containing silver, rare earths in the form of didymium, and zirconium. It is available as sand casting (AMS 4418) and as a casting alloy as specified by MIL-M-46062 for sand and permanent type mold castings. It is used in the solution heat treated and artificially aged (T6) condition where a high yield strength is needed at temperatures up to 600° F (316° C). QE22A has good weldability and fair pressure tightness.

ZE41A. A magnesium base casting alloy containing zinc, zirconium, and rare earth elements. It is available as sand (AMS 4439) or permanent mold castings in the artificially aged temper (T5). ZE41A has a higher yield strength than the Mg-Al-Zn alloys at room temperature, and is more stable at elevated temperatures. It is useful for applications at temperatures up to 320° F (160° C). ZE41A castings possess good weldability and are pressure tight.

Temper designation system for magnesium

The following temper designation system is used on all forms of wrought and cast magnesium and magnesium alloy products except ingots. Only operations recognized as significantly influencing the characteristics of the product are indicated.

–F *As Fabricated.* Applies to the products of shaping in which no special control over thermal conditions or strain hardening is employed.

–O *Annealed Recrystallized (Wrought Products Only).* Applies to wrought products that are annealed to obtain the lowest strength temper.

–H *Strain Hardened (Wrought Products Only).* Applies to products that have their strength increased by strain hardening, with or without supplementary thermal treatments to produce some reduction in strength. The H is always followed by two or more digits.

–H1 *Strain Hardened Only.* Applies to products that are strain hardened to obtain the desired strength without supplementary thermal treatment. The number following this designation indicates the degree of strain hardening.

–H2 *Strain Hardened and Partially Annealed.* Applies to products that are strain hardened more than the desired final amount and then reduced in strength to the desired level by partial annealing. The number following this designation indicated the degree of strain hardening remaining after the product has been partially annealed.

–H3 *Strain Hardened and Stabilized.* Applies to products that are strain hardened and whose mechanical properties are stabilized by a low temperature thermal treatment to slightly lower strength and increase ductility. The number following this designation indicates the degree of strain hardening remaining after the stabilization treatment.

The digit following the designations –H1, –H2, and –H3 indicates the final degree of strain hardening. Tempers between 0 (annealed) and 8 (full hard) are designated by numerals 1 through 7. Material having an ultimate tensile strength about midway between that of the 0 temper and that of the 8 temper is designated by the numeral 4; about midway between the 0 and 4 tempers by the numeral 2; and about midway between the 4 and 8 tempers by the numeral 6, etc. Number 9 designates tempers whose minimum ultimate tensile strength exceeds that of the 8 temper.

The third digit, when used, indicates a variation of a two-digit temper. It is used when the degree of control of temper or the mechanical properties of both differ from, but are close to, that (or those) for the two digit H temper designation to which it is added. Numerals 1 through 9 may be arbitrarily assigned as the third digit for an alloy and product to indicate a specific degree of control of temper or special mechanical property limits.

–W *Solution Heat Treated.* An unstable temper applicable only to alloys that spontaneously age at room temperature after solution heat treatment. This designation is specific only when the period of natural aging is indicated. For example, W ½ hour.

–T *Thermally Treated to Product Stable Temperatures Other Than –F, –O, or –H.* Applies to products that are thermally treated, with or without supplementary strain hardening, to product stable tempers. The T is always followed by one or more digits.

Numerals 1 through 10 following the –T indicate specific sequences of basic treatments, as follows.

–T1 *Cooled from an Elevated Temperature Shaping Process and Naturally Aged to a Substantially Stable Condition.* Applies to products for which the rate of cooling from an elevated temperature shaping process, such as casting or extrusion, is such that their strength is increased by room temperature aging.

–T2 *Annealed (Castings Only).* Applies to a type of annealing treatment used to improve ductility and increase stability.

–T3 *Solution Heat Treated and Cold Worked.* Applies to products that are cold worked to improve strength after solution heat treatment, or in which the effect of cold work in flattening or straightening is recognized in mechanical property limits.

–T4 *Solution Heat Treated and Naturally Aged to Substantially Stable Condition.* Applies to products that are not cold worked after solution heat treatment, or in which the effect of cold work in flattening or straightening may not be recognized in mechanical property limits.

–T5 *Cooled from an Elevated Temperature Shaping Process and Artificially Aged.* Applies to products that are cooled from an elevated temperature shaping process, such as casting or extrusion, and artificially aged to improve mechanical properties or dimensional stability, or both.

–T6 *Solution Heat Treated and Artificially Aged.* Applies to products that are not cold worked after solution heat treatment, or in which the effect of cold work in flattening or straightening may not be recognized in mechanical property limits.

–T7 *Solution Heat Treated and Stabilized.* Applies to products that are stabilized after solution heat treatment to carry them beyond a point of maximum strength to provide control of some special characteristic.

–T8 *Solution Heat Treated, Cold Worked, and Artificially Aged.* Applies to products that are cold worked to improve strength, or in which the effect of cold work in flattening or straightening is recognized in mechanical property limits.

–T9 *Solution Heat Treated, Artificially Aged, and Cold Worked.* Applies to products that are cold worked to improve strength.

–T10 *Cooled from an Elevated Temperature Shaping Process, Artificially Aged, and Cold Worked.* Applies to products that are artificially aged after cooling from an elevated temperature shaping process, such as extrusion, and cold worked to further improve strength.

Additional digits, the first of which shall not be zero, may be added to designations –T1 through –T10 to indicate a variation in treatment that significantly alters the product characteristics (other than mechanical properties) that are or would be obtained using the basic treatment.

Heat treatment procedures for magnesium alloys

A potential fire hazard exists in the heat treatment of magnesium alloys. If, through oversight or failure of the temperature control equipment, the temperature of the furnace appreciably exceeds the maximum solution heat treating temperature of the alloy, the alloy may ignite and burn. A suitable sulfur dioxide (0.5 – 1.0%) or carbon dioxide (3.0 – 3.5%) atmosphere prevents ignition until the upper allowable limits have been exceeded by a considerable amount. If a fire does ignite, the sulfur dioxide or carbon dioxide supply to the furnace should be shut off, because the burning magnesium unites with the oxygen in these materials. An effective method of extinguishing a magnesium fire in a gas tight furnace is to introduce boron trifluoride gas (BF_3) through a small opening into the closed furnace.

The temperatures for solution treatment shown in **Table 6** are the maximum temperatures to which the alloys may be heated without danger of high temperature deterioration or fusion of the eutectic. The alloys may be heat treated at lower temperatures, but in such cases a longer time at temperature than shown in this table will be necessary in order to develop satisfactory properties.

AZ63A, AZ81A, AZ91C, AM100A, and ZK61A castings will be irreversibly harmed if not brought slowly to the solution heat treating temperature. Certain eutectic constituents in these alloys melt at a temperature lower than that required for the solution heat treatment. Consequently, time should be allowed for the constituents to dissolve before their melting point is reached.

The aging treatments recommended in **Table 7** for "as cast" materials are used to improve mechanical properties, to provide stress relief, and to stabilize the alloys in order to prevent dimensional change later (especially during machining). Both yield strength and hardness are increased somewhat by this treatment at the expense of a slight amount of ductility. This treatment is often recommended for those applications where "as cast" mechanical properties suffice, and dimensional stability is essential.

Table 6. Temperatures for Solution Heat Treating Magnesium Alloys. *(Source, MIL-M-6857D.)*

Alloy	Temp. Range °F (°C)	Time at Temp. (Hours)[1]	Max. Permissible Temp. °F (°C)
AM100A	790 - 810 (421 - 432)	16 - 24	810 (432)
AZ63A	720 - 735 (382 - 390)	10 - 14	735 (390)
AZ81A	770 - 785 (410 - 418)	16 - 24	785 (418)
AZ91C and E	770 - 785 (410 - 418)	16 - 24	785 (418)
AZ92A	760 - 775 (404 - 413)	16 - 24	775 (413)
ZE63A	900 - 910 (482 - 488)	10 - 72	915 (488)
ZK61A	920 - 935 (493 - 502)	2	935 (502)
HK31A	1,040 - 1,060 (560 - 571)	2	1,060 (571)
QH21A	980 - 1,000 (527 - 538)	4 - 8	1,000 (538)
QE22A [2]	975 - 995 (524 - 535)	4 - 8	1,000 (538)
EQ21A [2]	960 - 980 (516 - 527)	4 - 8	990 (532)
WE54A [2]	970 - 990 (521 - 532)	6 - 8	1,000 (538)

Notes: [1] Heavy section castings, one inch (25.4 mm) thick or over, may require longer times than indicated. [2] Must be quenched in water held at 140 - 196°F (60 - 91° C) or other suitable media.

Table 7. Aging Treatments for Magnesium Alloys. *(Source, MIL-M-6857D.)*

Alloy and Temper	Time and Temperature Required to Produce Temper °F (°C)
AM100A-T5 [1]	5 hr. at 450° F (232° C)
AM100A-T6 [2]	5 hr. at 450° F (232° C) or 24 hr. at 400° F (204° C)
AZ63A-T5 [1]	4 hr. at 500° F (260° C) or 5 hr. at 450° F (232° C)
AZ63A-T6 [2]	5 hr. at 435° F (224° C) or 5 hr. at 450° F (232° C)
AZ91C-T5 [1]	4 hr. at 425° F (219° C) or 16 hr. at 335° F (169° C)
AZ91C-T6 [2]	4 hr. at 425° F (219° C) or 16 hr. at 335° F (169° C)
AZ91E-T6 [2]	4 hr. at 425° F (219° C) or 16 hr. at 335° F (169° C)
AZ92A-T5 [1]	5 hr. at 450° F (232° C)
AZ92A-T6 [2]	5 hr. at 425° F (219° C)
ZK51A-T5 [1]	8 hr. at 425° F (219° C) or 12 hr. at 350° F (177° C)
ZK61A-T5 [1]	48 hr. at 300° F (149° C)
ZK61A-T6 [2]	48 hr. at 265° F (130° C)
ZE41A-T5 [1]	24 hr. at 480° F (249° C) or 1-6 hr. at 620-680° F (327-360° C) + air cool
ZE63A-T6 [2]	48 hr. at 285° F (141° C)
EZ33A-T5 [1]	5 hr. at 425° F (219° C) or 2 hr. at 650° F (343° C) + 5 hr. at 425° F (219° C)
HK31A-T6 [2, 3]	16 hr. at 400° F (204° C)
HZ32A-T5 [1]	16 hr. at 600° F (316° C)
QH21A-T6 [2]	8 hr. at 400° F (204° C)
QE22A-T6 [2]	8 hr. at 400° F (204° C)
EQ21-T6	8 to 16 hr. at 400° F (204° C)
WE54A-T6 [2]	10 to 20 hr. at 480° F (249° C)

Notes: [1] The T5 temper is obtained by artificially aging from the as-cast (F) temper. [2] The T6 temper is obtained by artificially aging from the solution heat treated (T4) temper. [3] HK318-T4 should be brought to aging temperature as rapidly as possible to minimize grain growth.

Titanium

Material properties

The material properties of titanium and its alloys are mainly determined by their alloy content and heat treatment, both of which are influential in determining the form in which the material is bound. Titanium is one of the few allotropic metals, which means that it can exist in two different crystallographic forms. Under equilibrium conditions, pure titanium has an "alpha" (the "α" symbol is sometimes used) structure up to 1,620° F (900° C), above which it transforms to a "beta" (or "β") structure. The inherent properties of these two structures are quite different. The alpha structure is closely packed and hexagonally shaped, and the beta form is a body centered cubic structure. Through alloying and heat treatment, one or the other or a combination of these two structures can be made to exist at service temperatures, and the properties of the material vary accordingly.

Titanium and titanium alloys of the alpha and alpha-beta type exhibit crystallographic textures in sheet form in which certain crystallographic planes or directions are closely aligned with the direction of prior working. The presence of textures in these materials leads to anisotrophy (exhibiting different values of a property in different directions with respect to a fixed reference system in the material) with respect to mechanical and physical properties—especially Poisson's ratio and Young's modulus. Wide variations experienced in these properties both within and between sheets of titanium alloys have been qualitatively related to variations of texture. In general, the degree of texturing, and hence the variations of Poisson's ratio and Young's modulus, that is developed for alpha-beta alloys tends to be less than that developed in all alpha titanium alloys. Rolling temperature has a pronounced effect on the texturing of titanium alloys which may not in general be affected by subsequent thermal treatments. At present, these variations in mechanical properties have been documented on sheet only, and may not be present in other products. **Tables 1**, **2**, and **3** give composition, mechanical properties, and thermal properties of selected titanium alloys.

Beta transus temperature. The descriptions of individual unalloyed and alloyed titanium materials in this section often contain references to their beta transus temperature. This temperature is determined by metallographically examining quenched specimens after solution heat treating. Magnification of 500X is used to determine the amount of primary phase still present. The temperature at which this phase is no longer present is deemed the beta transus temperature. Beta transus temperatures are given in **Table 3**, but it should be remembered that the composition of the alloy will impact the temperature, and that beta transus temperatures may, and often do, vary from lot to lot.

An important engineering characteristics of titanium is that it is about 40% lighter than steel and 60% heavier than aluminum, but its high strength and moderate weight give it the highest strength-to-weight ratio of any structural metal. Another plus is that titanium is naturally corrosion resistant, due to a tough surface film of oxide.

Titanium alloy product designations reflect the alloy's constituent elements, and the percentage of each element contained in the alloy. Since these designations can become rather complex (as in Ti-6Al-2Sn-4Zr-6Mo) they are often identified by abbreviated designations (such as Ti-6246 for the example just given) or by their UNS number (R56260 for this example). The source of much of the information in this section was MIL-HDBK-5H.

Environmental considerations

Below 300° F (149° C), and above 700° F (371° C), creep deformation of titanium alloys can be expected at stresses below the yield strength. Room temperature creep

Table 1. Chemical Composition of Titanium and Titanium Alloys.

Designation	UNS No.	Nominal Composition — Expressed as Percent by Weight
Unalloyed Grades		
Grade 1	R50250	C 0.1 max.; Fe 0.2 max.; H 0.015 max.; N 0.03 max.; O 0.18 max.; Ti 99.5
Grade 2	R50400	C 0.1 max.; Fe 0.3 max.; H 0.015 max.; N 0.03 max.; O 0.25 max.; Ti 99.2
Grade 3	R50550	C 0.1 max.; Fe 0.3 max.; H 0.015 max.; N 0.05 max.; O 0.35 max.; Ti 99.1
Grade 4	R50700	C 0.1 max.; Fe 0.5 max.; H 0.015 max.; N 0.05 max.; O 0.4 max.; Ti 99
Grade 7	R52400	C 0.1 max.; Fe 0.3 max.; H 0.015 max.; N 0.03 max.; O 0.25 max.; Pd 0.2; Ti 99
Grade 11	R52250	Pd 0.2; Ti 99
Grade 12	R53400	Mo 0.3; Ni 0.8; Ti 99
Alpha and Near-Alpha Alloys		
Ti-3Al-2.5V	R56320	Al 3; Ti 95; V 2.5
Ti-5Al-2.5Sn	R54520	Al 5; Fe 0.5 max.; O 0.2 max.; Sn 2.5; Ti 92.5
Ti-5Al-2.5Sn, ELI	R54521	Al 5; Fe 0.25 max.; O 0.12 max.; Sn 2.5; Ti 92.5
Ti-6Al-2Nb-1Ta-0.8Mo	R56210	Al 6; Mo 0.8; Nb 2; Sn 2.5; Ta 1; Ti 90
Ti-6Al-2Sn-4Zr-2Mo	R54620	Al 6; Mo 2; Ti 88; Zr 4
Ti-8Al-1Mo-1V	R54810	Al 8; Mo 1; Ti 90; V 1
Alpha-Beta Alloys		
Ti-4Al-4Mo-2Sn-0.5Si	-	Al 4; Mo 4; Si 0.5; Sn 2; Ti 89
Ti-6Al-4V	R65400	Al 6; Fe 0.25 max.; O 0.2 max.; Ti 90; V 4
Ti-6Al-4V, ELI	R65400	Al 6; Fe 0.14 max.; O 0.13 max.; Ti 90; V 4
Ti-6Al-6V-2Sn	R56620	Al 6; Sn 2; Ti 86; V 6
Ti-6Al-2Sn-2Zr-2Mo-2Cr-.25Si	-	Al 6; Cr 2; Mo 2; Si 0.25; Sn 2; Ti 86; Zr 2
Beta Alloys		
Ti-8Mo-8V-2Fe-3Al	-	Al 3; Fe 2; Mo 8; Ti 79; V 8
Ti-15Mo-5Zr-3Al	-	Al 3; Mo 15; Ti 77; Zr 5
Ti-10V-2Fe-3Al	-	Al 3; Fe 2; Ti 85; V 10
Ti-13V-2.7Al-7Sn-2Zr	-	Al 2.7; Sn 7; Ti 75; V 13; Zr 2
Beta III	R58030	Mo 11.5; Sn 4.5; Ti 78; Zr 6

of unalloyed titanium may exceed 0.2 percent creep-strain in 1,000 hours at stresses that exceed 50% of the design tensile yield stress. While the characteristics of the alloys are generally superior, creep should be a consideration when specifying titanium or titanium alloys.

The use of titanium and its alloys in contact with either liquid oxygen or gaseous oxygen at cryogenic temperatures should be avoided, since either the presentation of a fresh surface (such as produced by tensile rupture) or impact may initiate a violent reaction. Impact of the surface in contact with liquid oxygen will result in a reaction even at very low energy levels. In gaseous oxygen, a partial pressure of 50 PSI is sufficient to ignite a fresh titanium surface over the temperature range of –250° F (–157° C) to room temperature

(Text continued on p. 284)

Table 2. Mechanical Properties of Titanium Alloys.

Designation	Ultimate Tensile Strength		Yield Strength		Compressive Yield Strength		Modulus of Elasticity		Shear Modulus		Elong. % [1]	Hardness
	PSI	MPa	PSI	MPa	PSI	MPa	ksi	MPa	ksi	GPa		
Unalloyed Grades												
Grade 1	34,809	240	24,656	170	-	-	15,229	105	6,525	45	24	RA 70
Grade 1, Annealed	47,863	330	34,809	240	49,313	340	14,504	100	5,510	38	30	HB 120
Grade 2	49,893	344	39,885	275	-	-	15,229	105	6,525	45	20	RB 80
Grade 3	63,817	440	55,114	380	65,267	450	15,229	105	6,525	45	18	RC 16
Grade 4	79,771	550	69,618	480	-	-	15,229	105	5,800	40	15	RC 23
Grade 7	49,893	344	39,885	275	-	-	15,229	105	6,525	45	20	RB 75
Grade 11	34,809	240	24,656	170	-	-	15,229	105	6,525	45	24	-
Grade 12	65,267	450	55,114	380	-	-	15,229	105	6,525	45	12	HB 200
Alpha and Near-Alpha Alloys												
Ti-3Al-2.5V [2]	89,924	620	72,519	500	100,076	690	14,504	100	6,380	44	15	RC 24
Ti-5Al-2.5Sn	124,878	861	119,946	827	120,382	830	16,679	115	6,960	48	15	RC 36
Ti-5Al-2.5Sn, ELI [3]	112,404	775	104,427	720	-	-	15,954	110	6,960	48	12	RC 33
Ti-6Al-2Nb-1 Ta-0.8Mo	120,382	830	110,229	760	118,931	820	16,969	117	7,395	51	10	RC 30
Ti-6Al-2Sn-4Zr-2Mo [4]	146,488	1,010	143,588	990	156,641	1,080	17,405	120	7,540	52	3	RC 34
Ti-8Al-1Mo-1 V	135,901	937	131,985	910	-	-	17,405	120	6,670	46	18	RC 36
Alpha-Beta Alloys												
Ti-4Al-4Mo-2 Sn-0.5Si [5]	159,542	1,100	136,336	940	-	-	16,679	115	7,105	49	7	360

Notes: [1] Elongation percentage at break. [2] Alpha annealed. [3] Annealed. [4] Sheet. [5] Solution treated and aged (STA). [6] Alpha-beta processed. [7] Solution treated. [8] Annealed and aged.

(Continued)

Table 2. *(Continued)* **Mechanical Properties of Titanium Alloys.**

Designation	Ultimate Tensile Strength		Yield Strength		Compressive Yield Strength		Modulus of Elasticity		Shear Modulus		Elong. % [1]	Hardness
	PSI	MPa	PSI	MPa	PSI	MPa	ksi	MPa	ksi	GPa		
Alpha-Beta Alloys												
Ti-5Al-2Sn-2Zr-4Mo-4Cr [6]	171,870	1,185	165,343	1,140	-	-	16,679	115	7,105	49	8	RC 40
Ti-6Al-4V [7]	137,786	950	127,633	880	140,687	970	16,505	113.8	6,380	44	14	RC 36
Ti-6Al-4V, ELI [3]	124,733	860	114,580	790	124,733	860	16,505	113.8	6,380	44	15	RC 35
Ti-6Al-6V-2Sn [3]	152,290	1,050	142,137	980	-	-	15,998	110.3	6,525	45	14	RC 38
Ti-6Al-2Sn-4Zr-6Mo [5]	229,160	1,580	204,504	1,410	-	-	16,534	114	7,105	49	4	RC 39
Ti-6Al-2Sn-2Zr-2Mo-2Cr-.25Si [3]	149,389	1,030	140,687	970	-	-	17,695	122	6,670	46	-	-
Beta Alloys												
Ti-8Mo-8V-2Fe-3Al [5]	188,549	1,300	-	-	-	-	15,954	110	6,815	47	3	410
Ti-15Mo-5Zr-3Al [5]	213,931	1,475	-	-	-	-	14,504	100	6,235	43	14	412
Ti 10V-2Fe-3Al [7]	145,038	1,000	134,885	930	-	-	15,954	110	5,945	41	17	-
Ti-13V-2.7Al-7Sn [5]	166,794	1,150	159,542	1,100	-	-	15,229	105	6,525	45	6	RC 39
Ti-13V-11Cr-3Al [7]	143,588	990	120,382	830	-	-	14,359	99	6,235	43	23	RC 30
Ti-13V-11Cr-3Al [8]	175,496	1,210	165,343	1,140	155,191	1,070	15,954	110	6,235	43	7	RC 43

Notes: [1] Elongation percentage at break. [2] Alpha annealed. [3] Annealed. [4] Sheet. [5] Solution treated and aged (STA). [6] Alpha-beta processed. [7] Solution treated. [8] Annealed and aged.

(Continued)

Table 2. *(Continued)* **Mechanical Properties of Titanium Alloys.**

Designation	Ultimate Tensile Strength		Yield Strength		Compressive Yield Strength		Modulus of Elasticity		Shear Modulus		Elong. % [1]	Hardness
	PSI	MPa	PSI	MPa	PSI	MPa	ksi	MPa	ksi	GPa		
Beta Alloys												
Ti-15V-3Cr-3Al-3Sn [7]	114,580	790	111,679	770	-	-	11,893	82	5,075	35	22	RB 95
Beta III [5]	182,023	1,255	171,145	1,180	162,443	1,120	15,664	108	5,945	41	5	325

Notes: [1] Elongation percentage at break. [2] Alpha annealed. [3] Annealed. [4] Sheet. [5] Solution treated and aged (STA). [6] Alpha-beta processed. [7] Solution treated. [8] Annealed and aged.

Table 3. Thermal Properties of Titanium Alloys.

Designation	Beta Transus		CTE		CTE		Thermal Conductivity[1]		Solidus		Liquidus	
	°F	°C	µin/in	µm/m	µin/in	µm/m	Engl.	metric	°F	°C	°F	°C
Unalloyed Grades												
Grade 1	1630	888	4.8 [2]	8.6 [2]	5.4 [3]	9.7 [3]	111	16	-	-	3,038	1,670
Grade 1, Annealed	1630	888	4.8 [4]	8.6 [4]	5.1 [5]	9.2 [5]	111	16	-	-	3,038	1,670
Grade 2	1675	913	4.8 [2]	8.6 [2]	5.4 [5]	9.7 [5]	114	16.4	-	-	3,029	1,665
Grade 3	1688	920	4.8 [2]	8.6 [2]	5.1 [5]	9.2 [5]	138	19.9	-	-	3,020	1,660
Grade 4	1742	950	4.8 [2]	8.6 [2]	5.1 [5]	9.2 [5]	119	17.2	-	-	3,020	1,660
Grade 7	1675	913	4.8 [2]	8.6 [2]	5.1 [5]	9.2 [5]	114	16.4	-	-	3,029	1,665
Grade 11	1630	888	4.8 [2]	8.6 [2]	5.1 [5]	9.2 [5]	118	17	-	-	3,038	1,670
Grade 12	1634	890	4.8 [2]	8.6 [2]	5.3 [5]	9.5 [5]	132	19	-	-	3,020	1,660

Notes: **CTE** is Coefficient of Thermal Expansion measured in µin./in.-°F, and µm/m-°C. [1] Metric measurement for thermal conductivity in W/m-K; English measurement in Btu-in./hr-ft2-°F. [2] CTE from 32 to 212° F (0 to 100° C). [3] CTE from 32 to 1,000° F (0 to 540° C). [4] CTE from -7 to 200° F (20 to 93° C). [5] CTE from 32 to 600° F (0 to 315° C). [6] Alpha annealed. [7] CTE from -7 to 600° F (20 to 315° C). [8] Annealed. [9] CTE from -7 to 1,200° F (20 to 650° C) [10] Sheet. [11] CTE from 32 to 480° F (0 to 250° C). [12] CTE from 32 to 930° F (0 to 500° C). [13] Solution treated and aged (STA). [14] Alpha-beta processed. [15] CTE from -7 to 750° F (20 to 400° C). [16)] Solution treated. [17] Annealed and aged.

(Continued)

Table 3. *(Continued)* **Thermal Properties of Titanium Alloys.**

Designation	Beta Transus		CTE		CTE		Thermal Conductivity[1]		Solidus		Liquidus	
	°F	°C	µin/in	µm/m	µin/in	µm/m	Engl.	metric	°F	°C	°F	°C
Alpha and Near-Alpha Alloys												
Ti-3Al-2.5V [6]	1715	935	5.3 [4]	9.61 [4]	5.5 [7]	9.86 [7]	58	8.3	-	-	1,700	3,092
Ti-5Al-2.5Sn	1904-1994	1040-1090	5.2 [2]	9.4 [2]	5.3 [7]	9.5 [7]	54	7.8	-	-	2,894	1,590
Ti-5Al-2.5Sn, ELI [8]	1904-1994	1040-1090	5.2 [2]	9.4 [2]	5.3 [7]	9.5 [7]	54	7.8	-	-	2,894	1,590
Ti-6Al-2Nb-1Ta-0.8Mo	2030	1110	5.1 [4]	9.2 [4]	5 [9]	9 [9]	44	6.4	-	-	3,002	1,650
Ti-6Al-2Sn-4Zr-2Mo [10]	1814	990	4.3 [4]	7.7 [4]	4.5 [7]	8.1 [7]	49	7.1	-	-	3,092	1,700
Ti-8Al-1Mo-1V	1904	1040	5.1 [11]	9.2 [11]	5.6 [12]	10.2 [12]	42	6	-	-	2,804	1,540
Alpha-Beta Alloys												
Ti-4Al-4Mo-2Sn-0.5Si [13]	1787	975	4.9 [4]	8.8 [4]	5.1 [7]	9.2 [7]	52	7.5	-	-	3,002	1,650
Ti-5Al-2Sn-2Zr-4Mo-4Cr [14]	1634	890	4.7 [4]	8.5 [4]	5.4 [15]	9.7 [15]	54	7.8	-	-	-	-
Ti-6Al-4V [16]	1796	980	4.8 [4]	8.6 [4]	5.1 [7]	9.2 [7]	46	6.7	2,919	1,604	3,020	1,660
Ti-6Al-4V, ELI [8]	1796	980	5.1 [7]	9.2 [7]	5.4 [9]	9.7 [9]	46	6.7	2,919	1,604	3,020	1,660
Ti-6Al-6V-2Sn [8]	1733	945	5 [4]	9 [4]	5.2 [7]	9.4 [7]	46	6.6	2,961	1,627	3,000	1,649
Ti-6Al-2Sn-4Zr-6Mo [13]	1715	935	5.5 [11]	9.9 [11]	5.8 [3]	10.4 [3]	53	7.7	2,903	1,595	3,047	1,675
Ti-6Al-2Sn-2Zr-2Mo-2Cr-.25Si [8]	1760	960	5.2 [4]	9.4 [4]	5.1 [7]	9.2 [7]	49	7	-	-	3,002	1,650

Notes: **CTE** is Coefficient of Thermal Expansion measured in µin./in.-°F, and µm/m-°C. [1] Metric measurement for thermal conductivity in W/m-K; English measurement in Btu-in./hr-ft2-°F. [2] CTE from 32 to 212° F (0 to 100° C). [3] CTE from 32 to 1,000° F (0 to 540° C). [4] CTE from -7 to 200° F (20 to 93° C). [5] CTE from 32 to 600° F (0 to 315° C). [6] Alpha annealed. [7] CTE from -7 to 600° F (20 to 315° C). [8] Annealed. [9] CTE from -7 to 1,200° F (20 to 650° C) [10] Sheet. [11] CTE from 32 to 480° F (0 to 250° C). [12] CTE from 32 to 930° F (0 to 500° C). [13] Solution treated and aged (STA). [14)] Alpha-beta processed. [15)] CTE from -7 to 750° F (20 to 400° C). [16] Solution treated. [17] Annealed and aged.

(Continued)

Table 3. *(Continued)* **Thermal Properties of Titanium Alloys.**

Designation	Beta Transus		CTE		CTE		Thermal Conductivity[1]		Solidus		Liquidus	
	°F	°C	µin/in	µm/m	µin/in	µm/m	Engl.	metric	°F	°C	°F	°C
Beta Alloys												
Ti-8Mo-8V-2Fe-3Al [13]	1427	775	4.7 [4]	8.5 [4]	-	-	52	7.5	-	-	-	-
Ti-15Mo-5Zr-3Al [13]	1445	785	4.7 [4]	8.5 [4]	-	-	52	7.5	-	-	-	-
Ti 10V-2Fe-3Al [16]	1400	760	5.4 [4]	9.7 [4]	-	-	54	7.8	-	-	-	-
Ti-13V-2.7Al-7Sn [13]	1382	750	4.7 [4]	8.5 [4]	-	-	54	7.8	-	-	-	-
Ti-13V-11Cr-3Al [16]	1292	700	5.4 [4]	9.67 [4]	5.6 [7]	9.99 [7]	48	6.9	-	-	-	-
Ti-13V-11Cr-3Al [17]	1292	700	5.4 [4]	9.67 [4]	-	-	48	6.9	-	-	-	-
Ti-15V-3Cr-3Al-3Sn [16]	1400	760	4.7 [4]	8.5 [4]	5.1 [11]	9.1 [11]	56	8.08	-	-	-	-
Beta III [13]	1400	760	4.2 [4]	7.6 [4]	4.5 [7]	8.1 [7]	44	6.28	2,863	1,572	3,074	1,690

Notes: **CTE** is Coefficient of Thermal Expansion measured in µin./in.-°F, and µm/m-°C. [1] Metric measurement for thermal conductivity in W/m-K; English measurement in Btu-in./hr-ft2-°F. [2] CTE from 32 to 212° F (0 to 100° C). [3] CTE from 32 to 1,000° F (0 to 540° C). [4] CTE from -7 to 200° F (20 to 93° C). [5] CTE from 32 to 600° F (0 to 315° C). [6] Alpha annealed. [7] CTE from -7 to 600° F (20 to 315° C). [8] Annealed. [9] CTE from -7 to 1,200° F (20 to 650° C) [10)] Sheet. [11] CTE from 32 to 480° F (0 to 250° C). [12] CTE from 32 to 930° F (0 to 500° C). [13] Solution treated and aged (STA). [14] Alpha-beta processed. [15] CTE from -7 to 750° F (20 to 400° C). [16] Solution treated. [17] Annealed and aged.

or higher. Also, under certain conditions, titanium, when in contact with cadmium, silver, mercury, or certain of their compounds, may become embrittled.

Titanium is susceptible to stress corrosion cracking in certain anhydrous chemicals including methyl alcohol and nitrogen tetroxide. Traces of water tend to inhibit the reaction in either environment. Red fuming nitric acid with less than 1.5% water and 10 to 20% NO_2 can crack the metal and result in a pyrophoric reaction. Titanium alloys are also susceptible to stress corrosion by dry sodium chloride at elevated temperatures. Cleaning with a nonchlorinated solvent to remove salt deposits (including fingerprints) is recommended on parts used above 450° F (232° C).

Unalloyed titanium

Unalloyed (commercially pure) titanium is available in all familiar product forms and is noted for its excellent formability, but it is used primarily where strength is not the primary requirement. It has excellent corrosion resistance properties. These materials are supplied in the annealed condition, which permits extensive forming at room temperature. Severe forming operations can also be accomplished at elevated temperatures (300 to 900° F [149 to 482° C]). Property degradation can be experienced after severe forming if as-received material properties are not restored by reannealing. They can be readily welded by several methods. Atmospheric shielding is preferable although spot or seam welding may be accomplished without shielding. Brazing requires protection from the atmosphere, which may be obtained by fluxing as well as by inert gas or vacuum shielding.

Titanium has an unusually high affinity for oxygen, nitrogen, and hydrogen at temperatures above 1,050° F (566° C). As this results in embrittlement of the material, usage should be limited to temperatures below that indicated. Additional chemical reactivity between titanium and selected environments such as methyl alcohol, chloride salt solutions, hydrogen, and liquid metal can take place at lower temperatures.

Commercially pure titanium is fully annealed by heating to 1,000 to 1,300° F (538 to 704° C) for 10 to 30 minutes. These materials cannot be hardened by heat treatment.

Alpha and near-alpha titanium alloys

The alpha titanium alloys contain essentially a single phase at room temperature, similar to that of unalloyed titanium. Alloys identified as near-alpha titanium have principally an all-alpha structure but contain small quantities of a beta phase because the composition contains some beta stabilizing elements. In both alloy types, alpha phase is stabilized by aluminum, tin, and zirconium. These elements, especially aluminum, contribute greatly to strength. The beta stabilizing additions (molybdenum and vanadium) improve fabricability and metallurgical stability of highly alpha alloyed materials.

These alloys have toughness at extremely low temperatures as well as long-term strength at elevated temperatures—they are well suited to cryogenic applications and uses requiring good elevated temperature creep strength. The characteristics of near-alpha alloys are predictably between those of all-alpha and alpha-beta alloys in regard to fabricability, weldability, and elevated temperature strength. The hot workability of both alpha and near-alpha alloys is inferior to that of the alpha-beta or beta alloys, and the cold workability is very limited at the high strength level of these grades. However, considerable forming is possible if correct forming temperatures and procedures are used.

Selected Alloys

Ti-5Al-2.5Sn. This all-alpha alloy is available in many product forms and at two purity levels. The high purity grade is used principally for cryogenic applications and may be characterized as having lower strength but higher ductility and toughness than the standard grade. The normal purity grade may also be used at low temperatures but is primarily suited for room-to-elevated-temperature applications (up to 900° F [482° C] or to 1,100° F [593° C] for short periods), where weldability is an important consideration. *Manufacturing Considerations*: Ti-5Al-2.5Sn is not so readily formed into complex shapes as other alloys with similar room temperature properties, but it far surpasses them in weldability—inert gas, vacuum shielded arc, or spot and seam accomplished without atmospheric shielding. Brazing requires protection from the atmospheres. Except for some forging operations, fabrication is conducted at temperatures where the structure remains all-alpha. Severe forming operations may be accomplished at temperatures up to 1,200° F (649° C). Moderately severe forming can be done at 300 to 600° F (149 to 316° C0, and simple forming can be done at room temperature. Most forming and welding operations are followed by an annealing treatment to relieve residual stress. Standard grade has been used at moderately low cryogenic temperatures, but the ELI (extra low interstitial) grade has higher toughness and has been used in temperatures down to –423° F (–253° C). *Heat Treatment*: Anneal by heating to 1,400° F (760° C) for 60 minutes and 1,600° F (871° C) for 10 minutes and cooling in air. Stress relieving requires 1 to 2 hours at 1,000 to 1,200° F (538 to 649° C). This alloy cannot be hardened by heat treatment.

Ti-8Al-1Mo-1V. This near-alpha composition was developed for improved creep resistance and thermal stability up to about 850° F (454° C). It is available a billet, bar, plate, sheet, strip, extrusion, and forgings. *Manufacturing Considerations*: Room temperature forming of sheet is somewhat more difficult than with Ti-6Al-4V, and hot forming is required for severe operations. Ti-8Al-1Mo-1V can be readily fusion welded with inert gas protection or spot welding without atmosphere protection. Weld strengths are comparable to those of the parent metal although ductility is somewhat lower in the weldment. Oxidation resistance and thermal stability up to 850° F (454° C) are good, but extended exposure to temperatures in excess of 600° F (316° C) adversely affects room temperature spot weld tension strength. Not recommended for structural applications at liquid hydrogen temperatures (–423° F [–253° C]), and the alloy is susceptible to chloride stress corrosion attack in either elevated temperature or ambient temperature chloride environments. *Heat Treatment*: Three treatments are used with this alloy. 1) Single anneal: 1,450° F (788° C) for 8 hours, surface cool. 2) Duplex anneal: 1,450° F for 8 hours, furnace cool, then 1,450° F for 15 to 20 minutes, air cool. 3) Solution treated and stabilized. 1,825° F (996° C) for 1 hour, air cool, then 1,075°F (579° C) for 8 hours, air cool. The single anneal is used to obtain the highest room temperature mechanical properties. The duplex anneal is used to obtain the highest fracture toughness. Both the single anneal and the duplex anneal are compatible with hot forming operations. The solution treated and stabilized condition is used for forgings.

Ti-6Al-2Sn-4Zr-2Mo(+Si). This near-alpha titanium composition was developed for improved elevated temperature performance. Initially, it was used in advanced performance gas turbine engine applications, and in the Si version alloy contains approximately 0.08% silicon to improve creep strength without affecting thermal stability. With silicon, it is creep resistant and relatively stable to about 1,050°F (565° C). It is available in bar, billet, plate, sheet, strip, and extrusion. This is the most commonly used titanium alloy for elevated temperature operations. *Manufacturing Considerations*. Elevated temperatures may be used for severe sheet forming operations, while room temperature forming may be used for mild contouring. The material can be welded using TIG or MIG fusion processes to achieve 100% joint efficiencies but with limited weld zone ductility. As in welding any titanium alloy, shielding from atmospheric contamination is required except for spot and seam welding. Forging at temperatures below the beta transus temperature is recommended—for optimum creep properties, beta forging or a modification of the process is recommended with some loss of ductility likely. *Heat Treatment*: Several treatments are available, as follows. For sheet and strip: 1) Duplex anneal: 1,650° F (899° C) for $^{1}/_{2}$ hour, air cool, then 1,450° F (788° C) for $^{1}/_{4}$ hour, air cool. 2) Triplex anneal: 1,650° F for $^{1}/_{2}$ hour, air cool, then 1,450° F for $^{1}/_{4}$ hour, air cool, then 1,100° F (593° C) for 2 hours, air cool. For plate: 1) Duplex anneal: 1,650° F for 1 hour, air cool, then 1,100° F for 8 hours, air cool. 2) Triplex anneal: 1,650° F for $^{1}/_{2}$ hour, air cool, then 1,450° F for 2 hours, air cool. For bars and forgings: 1) Duplex anneal: Solution anneal 25 to 50° (14 to 28° C) below beta transus temperature for 1 hour, air cool (or faster), then 1,100° F for 8 hours, air cool.

Alpha-beta titanium alloys

These alloys contain both alpha and beta phases at room temperature. The alpha phase is similar to that of unalloyed titanium but is strengthened by alpha stabilizing additions (aluminum). The beta phase is the high temperature phase of titanium, but is stabilized to room temperature by sufficient quantities of beta stabilizing elements such as vanadium, molybdenum, iron, or chromium. In addition to strengthening of titanium by alloying additions, alpha-beta alloys may be further strengthened by heat treatment. Alpha-beta alloys have good strength at room temperature and for short times at elevated temperatures, but they are not noted for long-time creep strength. With the exception of annealed Ti-6Al-4V, these alloys are not recommended for cryogenic applications. Because of their two-phase microstructure, the weldability of most alloys in this group is relatively poor.

Selected Alloys

Ti-6Al-4V. This alloy is available in all mill product forms as well as castings and powder metallurgy forms. It can be used in either the annealed or solution treated plus aged (STA) conditions. For maximum toughness, it should be used in the annealed or duplex annealed conditions. For maximum strength, the STA condition should be used. The useful temperature range for this alloy is –320 to +750° F (–196 to 399° C). This alloy is responsible for about 45% of all industrial applications of titanium alloys. *Manufacturing Considerations*: Forging should be done above the beta transus temperature using procedures to promote a high toughness material. It is routinely finished below the beta transus temperature for good combinations of fabricability, strength, ductility, and toughness. Elevated temperatures are used to form flat rolled products, but extensive forming may be accomplished at room temperature. This alloy can be spot welded and, in certain applications, fusion welded. Stress relief annealing after welding is recommended. It is resistant to hot-salt stress corrosion to about its maximum use temperature, but it is marginally susceptible to aqueous chloride solution stress corrosion. *Heat Treatment*: This alloy is commonly specified in either the annealed condition or in the fully heat treated condition. Annealing requires 1 hour at 1,300° F (704° C) followed by furnace cooling if maximum ductility is required. The specified fully heat treated, or solution treated and aged (STA), condition for 1) sheet is as follows. Solution treat at 1,700° F (927° C) for 5 to 25 minutes, quench in water. Age at 975° F (524° C) for 4 to 6 hours, air cool. For 2) bars and forgings, the procedure is: Solution treat at 1,700° F for one hour, quench in water. Age at 1,000° F (538° C) for 3 hours, air cool.

Ti-6Al-6V-2Sn. Although similar to Ti-6Al-4V in many respects, this alloy has higher strength and deeper hardenability. Available forms include billet, bar, plate, sheet, strip, and extrusions, and these may be used in either the annealed or the STA condition. Maximum strength is developed in SAT condition in sections up to about two inches in thickness. *Manufacturing Considerations*. To ensure optimum properties for forgings, at least 50% reduction should be done at temperatures below the beta transus temperature (<1,730°

F [<945° C]). This alloy is readily formable in the annealed condition. In sheet or plate forms, the alloy is generally used in the annealed condition, although it is capable of heat treatment to higher strength levels with some loss of toughness. When sheet or plate is hot formed at any temperature over 1,000° F (538° C) and air cooled, it should be restabilized by reheating to 1,000° F followed by air cooling. Welding is not recommended although limited weld joining operations are possible if the assembly can be thermally treated after welding to restore ductility to the weld and heat affected zones. Creep strength above 650° F (343° C) and long-term stability at temperatures above 800° F (427° C) are not good. Oxidation resistance is satisfactory in short-term exposures to 1,000° F, and this alloy is nearly equivalent to Ti-6Al-4V in terms of hot salt and aqueous chloride solution stress corrosion resistance. *Heat Treatment*: This alloy is commonly specified in either the annealed condition or the STA condition. The STA procedure is as follows: Solution treat at 1,625° F (885° C) for $^1/_2$ to 1 hour, quench in water. Age at 975 to 1,025° F (524 to 552° C) for 4 to 8 hours, air cool.

Beta, near-beta, and metastable-beta titanium alloys

Although there is no clear-cut definition for beta titanium alloys, conventional terminology usually refers to near-beta alloys and metastable-beta alloys as classes of beta titanium alloys. A near-beta alloy is generally one that has appreciably higher beta stabilizer content than a conventional alpha-beta alloy, but is not quite sufficiently stabilized to readily retain an all beta structure with an air cool of thin sections. Instead, a water quench is required, even for thin sections. Metastable alloys are even more heavily alloyed with beta stabilizers than near-beta alloys and readily retain an all beta structure upon air cooling of thin sections. Due to the added stability of these alloys, it is not necessary to heat treat below the beta transus to enrich the beta phase. Therefore, these alloys do not normally contain primary alpha since they are usually solution treated above the beta transus. They are termed "metastable" because the resultant beta phase is not really stable—it can be aged to precipitate alpha for strengthening purposes.

Stable beta alloys are so heavily alloyed with beta stabilizers that the beta phase will not decompose to alpha plus beta upon subsequent aging. An example would be Ti-30Mo, but there are presently no such alloys being produced commercially.

Selected Alloys

Ti-13V-11Cr-3Al. This is a heat treatable alloy possessing good workability and toughness in the annealed condition and high strength in the heat treated condition. It is noted for its exceptional ability to harden in heavy sections (up to 6 inch [15.24 cm] or greater) to tensile strengths of 170 ksi (1172 MPa). *Manufacturing Considerations*: It has good formability at room temperature, and stretch forming is conducted at 500° F (260° C). It is readily spot welded, and arc welding will provide very ductile joints that have low strength. Stability is good up to 550° F (288° C) in the annealed condition, and up to 600° F (316° C) in the SAT condition. Prolonged exposures above these temperatures may result in loss of ductility. Hot salt stress corrosion is possible at temperatures as low as 500° F (260° C) in highly stressed applications such as rivet heads. *Heat Treatment*: This alloy is commonly specified in either the annealed condition or in the fully heat treated condition. The full heat treatment procedure is: Solution heat at 1,450° F (788° C) for 15 to 60 minutes, air cool (water quench if material is over 2 inches (50.8 mm) thick). Age at 900° F (482° C) for 2 to 60 hours depending on strength level.

Ti-15V-3Cr-3Sn-3Al (Ti-15-3). This is a metastable beta-titanium alloy. It was developed primarily to lower the cost of titanium sheet metal parts by reducing materials and processing costs. Unlike conventional alpha-beta alloys, this alloy is strip producible and has excellent room temperature formability characteristics. It can be aged to a wide range of strength levels to meet a variety of application needs. Although originally developed as a sheet alloy, it has expanded into other areas such as fasteners, foil, tubing, castings, and forgings. *Manufacturing Considerations*: Usually supplied in the solution annealed form, this alloy is easily cold formed due to its single phase (beta) structure. After cold forming, it can be resolution treated in the 1,450 to 1,500° F (788 to 816° C) range and subsequently aged in the 900 to 1,100° F (482 to 593° C) range, depending upon desired strength. Care should be taken to ensure that no surface contamination results from the solution treatment. The alloy can also be hot formed, but heating time prior to forming should be minimized to prevent aging prior to forming. It is readily welded by standard titanium welding techniques. This alloy should not be used in the solution treated condition. Long time exposure of solution treated and cold worked material to service temperatures of approximately 300° F (149° C), or solution treated material to service temperatures above approximately 400° F (204° C), can result in low ductility. *Heat Treatment*: This alloy should be solution treated for 10 to 30 minutes in the 1,450 to 1,550° F (788 to 816° C) range, cooled at a rate approximating an air cool of 0.125 inch (3.175 mm) sheet and subsequently aged. Aging is generally conducted in the 900 to 1,100° F (482 to 593° C) range, followed by an air cool.

Ti-10V-2Fe-3Al (Ti-10-2-3). This is a solute lean beta (near-beta) titanium alloy that was developed primarily as a high strength forging alloy. Its forging characteristics are excellent, possessing flow properties at 1,500° F (816° C) similar to Ti-6Al-4V at 1,700° F (927° C), allowing for the possibility of lower die costs and better die fill capacity. This alloy provides the best combination of strength and toughness of any commercially available titanium alloy, and it is considered to be deep hardenable, capable of generating high strengths in section thicknesses up to approximately 5 inches (12.5 cm). *Manufacturing Considerations*: Ti-10-2-3 is usually supplied as bar or billet that has been finish forged or rolled in the alpha-beta field. In order to optimize the microstructure for the high strength condition, the forging is usually given a pre-form forge above the beta transus, followed by a 15 to 25% reduction below the beta transus. Ideally, the beta forging operation is finished through the beta transus, followed by a quench. The intent of the two-step forging process is to develop a structure without grain boundary alpha, but with elongated primary alpha needles in an aged beta matrix. The alloy is readily weldable by conventional titanium welding techniques. In the solution treated plus aged condition, the material exhibits excellent resistance to stress corrosion cracking. In the solution treated condition, it should not be subjected to long-term exposure in the 500 to 800° F (260 to 427° C) range, since such exposure could result in high strength, low ductility conditions. *Beta Flecks*: This is a segregation-prone alloy that can exhibit a microstructural phenomenon known as beta flecking. Certain areas may possess a lower beta transus than the matrix (due primarily to beta stabilizer enrichment) and, as such, can fully transform during heat treatment just below the matrix transus. In severe cases, this can lead to lower ductility and a reduction in fatigue strength due to grain boundary alpha formation in the flecked region. *Heat Treatment*: For the high strength condition, the alloy is generally solution treated approximately 65° F (36° C) below the beta transus (which is typically 1,400 to 1,480° F [760 to 804° C]), followed by a water quench and an 8 hour age at 900 to 950° F (482 to 510° C). Overaging in the 950 to 1,150° F (510 to 621° C) range may also be used to obtain lower strength levels.

Heat treating titanium alloys

The Tables in this section give temperature ranges and schedules for solution heat treating, aging, stress relieving, and annealing various alloys. These ranges encompass the more exact temperatures given above in descriptions of specific materials. For complete accuracy, the more specific temperatures and times should be followed, and the values in the Tables should be used as reliable ranges. When heat treating titanium, maximum quench delays can be critical and should not be exceeded, even though the beta-stabilized alloys are more tolerant to delay times than the alpha materials. Quench delay time begins when the furnace door starts to open, and ends when the last corner of the load is immersed in the quenchant. Schedules are as follows.

Nominal Thickness	Maximum Delay Time
Up to 0.25 inch (6.4 mm)	6 seconds
0.26 to 0.99 inch (6.5 to 25.3 mm)	8 seconds
1.00 inch and over (25.4 mm and over)	10 seconds

Table 4 gives solution heat treating schedules, **Table 5** provides aging schedules, **Table 6** gives the stress relief schedule, and **Table 7** shows annealing temperatures and times.

Table 4. Solution Heat Treating Schedules for Titanium Alloy Raw Materials and Semi-finished Parts. *(Source, MIL-H-81200B.)*

Designation	Solution Heat Treating Temperature				Soaking Time-in Minutes[1]		Cooling Method[8]
	Sheet, Strip, Plate		Bars, Forgings, Castings		Sheet, Strip, Plate	Bars, Forgings, Castings[2]	
	°F	°C	°F	°C			
Alpha Alloys							
Ti-8Al-1Mo-1V [6]	-	-	1650-1850	900-1010	-	20-90	Note 3
Alpha-Beta Alloys							
Ti-6Al-4V	1650-1775	900-970	1650-1775	900-970	2-90	20-120	Water quench
Ti-6Al-6V-2Sn [5]	1550-1700	870-925	1550-1700	870-925	2-60	20-90	Water quench
Ti-6Al-2Sn-4Zr-2Mo	1500-1675	815-915	1650-1800	900-980	2-90	20-120	Air cooled
Ti-6Al-2Sn-4Zr-6Mo	1500-1675	815-915	1500-1675	815-915	2-90	20-120	Note 5
Ti-11Sn-5Zr-2A1-1Mo [6]	-	-	1625-1675	885-915	-	20-120	Air cooled
Ti-6Al-2Sn-2Zr-2Mo-2Cr-0.25Si	1600-1700	870-925	1600-1700	870-925	2-60	20-120	Water quench
Ti-5A1-2Sn-2Zr-4Mo-4Cr [6]	-	-	1450-1500	790-815	-	20-120	Water quench
Beta Alloys							
Ti-13V-11Cr-3A1 [4]	1400-1500	760-815	1400-1500	760-815	2-60	2-60	Note 9
Ti-15V-3A1-3Cr-3Sn [7]	1400-1500	760-815	-	-	2-30	20-90	Note 3
Ti-10V-2Fe-3A1 [6]	-	-	1300-1425	705-775	-	60-120	Water quench

Notes: [1] Soaking time shall be considered to begin as soon as the lowest reading control thermocouple is at the lower limit of the specified solution treating temperature range. [2)] Longer soaking times may be necessary for specific forgings. Shorter soaking times are satisfactory when soak time is accurately determined by thermocouples attached to the load. Soaking time should be measured from the time the furnace charge reaches the soaking temperature.

(Continued)

Table 4. *(Continued)* **Solution Heat Treating Schedules for Titanium Alloy Raw Materials and Semi-finished Parts.** *(Source, MIL-H-81200B.)*

Notes: [3] Air cool or faster. [4] For material less than 0.100 inch (2.54 mm). For thickness greater than 0.100 inch, duplex solution treatment is applicable as follows: 1300°F to 1375°F (705 to 745°C) for 50 to 80 minutes, air-cooled, then 1400°F to 1450°F (760 to 790°C), 20 to 60 minutes. [5] Air cooling may be applied in relatively thin sections. Water quench is required for thick sections. [6] No solution heat treating cycle is specified for sheet, strip and plate. [7] No solution heat treating cycle is specified for bars, forgings and castings. [8] When vacuum furnace equipment is used, inert gas cooling may be applied in lieu of air cooling. [9] An air cool may be applied to sections up to 0.50 inch thick (12.7 mm). A water quench shall be applied to sections greater than 0.50 inch thick.

Table 5. Aging Schedule for Titanium Alloys. *(Source, MIL-H-81200B.)*

Designation	Aging Temperature		Soaking Times Hours[1]
	°F	°C	
Alpha Alloys			
Ti-8Al-1Mo-1V	1000-1150	540-620	8-24 [2]
Ti-11Sn-5Zr-2A1-1Mo	900-1000	480-540	20-30 [2, 3]
Alpha-Beta Alloys			
Ti-6A1-4V	900-1275	480-690	2-8 [2, 3]
Ti-6A1-6V-2Sn	875-1150	470-620	2-10
Ti-6A1-2Sn-4Zr-2Mo	1050-1150	565-620	2-8 [2]
Ti-6A1-2Sn-4Zr-6Mo	1050-1250	480-675	4-8
Ti-6A1-2Sn-2Zr-2Mo-2Cr-0.25Si	900-1250	480-675	2-10
Ti-5A1-2Sn-2Zr-4Mo-4Cr	1100-1250	480-675	4-8 [3]
Beta Alloys			
Ti-13V-11Cr-3A1 [4]	825-975	440-525	2-60 [3]
Ti-15V-3A1-3Cr-3Sn	900-1250	480-675	2-24 [3]
Ti-10V-2Fe-3A1	900-1150	480-620	8-10 [3]

Notes: [1] Soaking time shall be considered to begin as soon as the lowest reading control thermocouple is at the lower limit of the specified solution treating temperature range. [2] See Table V for duplex annealing. An 8-hour stabilizing treatment at 1050-1100°F (565-595°C) can be considered an aging treatment. [3] Aging time and temperature depends on strength level desired. [4] Springs may be aged at 800°F (425°C).

Table 6. Stress Relief Schedule for Titanium and Titanium Alloys. *(Source, MIL-H-81200B.)*

Designation	Stress Relief Temperature		Soaking Times Hours
	°F	°C	
Unalloyed Titanium			
Commercially Pure, All Grades	900-1100	480-595	15-240
Alpha Alloys			
Ti-5Al-2.5Sn	1000-1200	540-650	15-360
Ti-5Al-2.5Sn ELI	1000-1200	540-650	15-360
Ti-6A1-2Cb-1Ta-0.8Mo	1000-1200	540-650	15-60
Ti-8A1-1Mo-1V	1100-1400	595-760	10-75
Ti-11Sn-5Zr-2A1-1Mo	900-1000	480-540	120-480

(Continued)

Table 6. *(Continued)* **Stress Relief Schedule for Titanium and Titanium Alloys.** *(Source, MIL-H-81200B.)*

Designation	Stress Relief Temperature		Soaking Times Hours
	°F	°C	
Alpha-Beta Alloys			
Ti-3A1-2.5V	700-1100	370-595	15-240
Ti-6A1-4V [1]	900-1200	480-650	60-240
Ti-6A1-4V ELI [1]	900-1200	480-650	60-240
Ti-6A1-6V-2Sn	900-1200	480-650	60-240
Ti-6A1-2Sn-4Zr-2Mo	900-1200	480-650	60-240
Ti-5A1-2Sn-2Zr-4Mo-4Cr	900-1200	480-650	60-240
Ti-6A1-2Sn-2Zr-2Mo-2Cr-0.25Si	900-1200	480-650	60-240
Alpha-Beta Alloys			
Ti-13V-11Cr-3A1	1300-1350	480-730	30-60
Ti-15V-3A1-3Cr-3Sn	1450-1500	790-815	30-60
Ti-10V-2Fe-3A1	1250-1300	675-705	30-60

Note: [1] Stress relief may be accomplished simultaneously with hot forming at temperatures of 1400°F (760°C) to 1450°F (790°C). Caution should be exercised in stress relieving titanium alloys that have been strengthened by solution treating and aging. The stress relieving temperature should not exceed the aging temperature used in heat treatment.

Table 7. Annealing Schedule for Titanium and Titanium Alloys. *(Source, MIL-H-81200B.)*

Designation	Annealing Temperature				Soak Time[11, 12]	
	Sheet, Strip, Plate		Bars, Forgings		Sheet, Strip, Plate	Bars, Forgings
	°F	°C	°F	°C		
Unalloyed Titanium						
Commercially Pure, All Grades	1200-1500 [5]	650-815 [5]	1200-1450 [8]	650-815 [8]	15-120 min.	1-2 hrs.
Alpha Alloys						
Ti-5Al-2.5Sn	1300-1550 [4]	705-845 [4]	1300-1550 [4]	705-845 [4]	10-120 min.	1-4 hrs.
Ti-5Al-2.5Sn ELI	1300-1650 [4]	700-900 [4]	1300-1550 [4]	705-845 [4]	10-120 min.	1-4 hrs.
Ti-6Al-2Cb-1Ta-0.8Mo	1450-1650 [5]	790-900 [5]	1450-1650 [5]	790-900 [5]	30-120 min.	1-2 hrs.
Ti-8Al-1Mo-1V	1400-1500 [1]	760-815 [1]	1650-1850 [2, 3]	900-1000 [2, 3]	1-8 hrs.	1-2 hrs.
Alpha-Beta Alloys						
Ti-3Al-2.5V	1200-1450 [5]	650-790 [5]	1200-1450 [5]	650-790 [5]	30-120 min.	1-3 hrs.
Ti-6Al-4V [10]	1300-1600 [5, 6]	705-870 [4]	1300-1450 [4]	705-790 [4]	15-60 min.	1-2 hrs.
Ti-6Al-4V ELI [10]	1300-1600 [5, 6]	705-870 [4]	1300-1450 [4]	705-790 [4]	15-60 min.	1-2 hrs.
Ti-6Al-6V-2Sn	1300-1500 [5]	705-815 [5]	1300-1450 [5]	705-790 [5]	10-120 min.	1-2 hrs.
Ti-6Al-2Sn-4Zr-2Mo	1600-1700 [7, 13] 1600-1700 [8, 14]	870-925 [7] 870-925 [8]	1764-1781 [4, 8] -	962-976 [4, 8] -	10-60 min. [13] 30-120 min. [14]	1-2 hrs. -
Ti-6Al-2Sn-4Zr-6Mo	-	-	1500-1675 [4, 9]	815-915 [4, 9]	-	1-2 hrs.

(Continued)

Table 7. *(Continued)* **Annealing Schedule for Titanium and Titanium Alloys.** *(Source, MIL-H-81200B.)*

Designation	Annealing Temperature				Soak Time[11, 12]	
	Sheet, Strip, Plate		Bars, Forgings		Sheet, Strip, Plate	Bars, Forgings
	°F	°C	°F	°C		
Alpha-Beta Alloys						
Ti-11Sn-5Zr-2A-l-1Mo	-	-	1625-1675 [2]	885-915 [2]	-	1-2 hrs.
Ti-6Al-2Sn-2Zr-2Cr-2Mo	1275-1600 [4]	690-870 [4]	1275-1600 [4]	690-870 [4]	15-360 min.	15-360 min.
Beta Alloys						
Ti-13V-11Cr-3A1	1400-1500 [2, 9]	760-815 [2, 9]	1400-1500 [2, 9]	760-815 2, 9	10-60 min.	30-120 min.
Ti-15V-3A1-3Cr-3Sn	1400-1500 [2, 9]	760-815 [2, 9]	-	-	3-30 min.	-
Ti-10V-2Fe-3A1	-	-	1300-1340 [2, 9]	705-727 [2, 9]	-	30-120 min.

Notes: BT denotes that the annealing temperature is the respective beta transus temperature minus the temperature range given at the right. [1] Furnace cool to below 900°F (480°C) - duplex anneal requires second anneal at 1450°F (790°C) for 15 min. followed by air cool. [2] Air cool or faster. [3] Followed by stabilization at 1100°F (595°C) for 8 hrs. and air cool. [4] Air cool. [5] Air cool or slower. [6] When duplex anneal (solution treat and anneal) is specified for 6A1-4V, the annealing treatment is as follows: Heat to beta transus temperature minus 50 to 75°F (14 to 32°C), hold for 1 to 2 hours for bars/forgings/plate, air cool or faster, reheat 1300-1400°F (705-760°C) for 1 or 2 hrs. and air cool. [7] Air cool followed by 1450°F (790°C) for 15 min. and air cool. [8] Air cool followed by 1100°F (595°C) for 8 hrs. and air cool. [9] Followed by aging at a temperature to develop required properties. [10] When recrystallization anneal is specified to optimize fracture toughness, it is generally as follows: Heat to beta transus temperature minus 50 to 75°F (14 to 32°C), hold for 1 to 4 hrs., air cool or slower, reheat between 1300 to 1400°F (705-760°C) for 1 or 2 hrs. and air cool. [11] Soaking time shall be considered to begin as soon as the furnace recovers to the specified soaking temperature. The table below offers guidance in selecting soaking times:

Thickness, inches (mm)	Soaking time, minutes ± 5
0.125 (3.19) and under	20
0.126-0.250 (3.20-6.35)	30
0.251-0.500 (6.36-12.70)	40
0.501-0.750 (12.71-19.07)	50
0.751-1.000 (19.08-25.40)	60
Over 1.000 (25.40)	1 hour minimum plus add fifteen minutes for each quarter inch of thickness over one inch.

[12] Annealing times for flat-rolled product processed as a continuous coil can be shortened to 2 minutes minimum. [13] For sheet only. [14] For plate only.

Heat-resistant Superalloys and Other Exotic Alloys

Heat-resistant, or "super," alloys

Heat-resistant, or "Super," nickel type alloys (superalloys) are arbitrarily defined as iron alloys richer in alloy content than the 18% chromium, 8% nickel types, or as alloys with a base element other than iron, and which are intended for elevated temperature service. Superalloys may be nickel, iron-chromium-nickel, or cobalt based, and they are usually specified for applications where operating temperatures are 1,000° F (540° C) or higher. Oxidation resistance at elevated temperatures is a requirement, and they must be able to withstand elevated temperatures without special surface protection.

Mechanical properties

Most of these alloys are available from the mill in a wide variety of heat treatments, and their mechanical properties are normally affected by relatively minor variations in chemistry (composition is given in **Table 1**), processing, and heat treatment. Consequently, the mechanical properties, given in **Table 2** or in the descriptions of individual alloys, apply only to the form (shape), size (thickness), and heat treatment indicated.

Strength properties of the heat-resistant alloys generally decrease with increasing temperatures or increasing time at temperature. There are exceptions, particularly in the case of age hardening alloys that may actually show an increase in strength with temperature or time, within a limited range, as a result of further aging. In most cases, however, this increase in strength is temporary and cannot usually be taken advantage of in service. At cryogenic temperatures, the strength properties of the heat-resistant alloys are generally higher than at room temperature, provided that some ductility is retained at the low temperatures.

Ductility varies with temperature and is somewhat erratic. Generally, ductility decreases with increasing temperature from room temperature up to about 1,200 to 1,400° F (649 to 760° C), where it reaches a minimum value, then increases with higher temperatures. Prior creep exposure may also adversely affect ductility.

Physical properties are shown in **Table 3**.

Iron-chromium-nickel base heat-resistant alloys

The alloys in this group, in terms of cost and in maximum service temperature, generally fall between the austenitic stainless steels and the nickel-cobalt based alloys. They are specified for operating conditions in the temperature range of 1,000 to 1,200° F (538 to 649° C), in applications where the stainless steels are inadequate but service conditions do not justify the more expensive nickel or cobalt alloys.

The complex-base alloys comprising this group range from those in which iron is considered the base element, to those which border on the nickel-base alloys. All contain sufficient alloying elements to place them in the superalloy category, yet contain enough iron to reduce their cost considerably. Chromium, in amounts ranging from 10 to 20% or higher, primarily increases oxidation resistance and contributes to strengthening these alloys. Nickel and cobalt strengthen these materials. Molybdenum, tungsten, and columbium contribute to hardness and strength, particularly at higher temperatures. Titanium and aluminum are added to provide age hardening. Heat treatment is performed with conventional equipment and fixtures such as would be used for austenitic stainless steels. Since these alloys are susceptible to carburization during heat treatment, it is good practice to remove all grease, oil, or cutting lubricant from the surface before treating. A low-sulfur and natural or slightly oxidizing furnace atmosphere is recommended for heating.

Table 1. Nominal Composition of Selected Wrought "Superalloys."

Material	UNS No.	Composition %
A-286	S66286	Ni 26, Cr 15, Ti 2, Mo 1.3, Al 0.2, B 0.015, Fe bal.
Hastelloy G	N06007	Cr 22, Fe 19.5, Mo 6.5, Co 2.5 max., Cb+Ta 2, Cu 2, Mn 1.5, Si 1 max., W 1 max., C 0.05 max., Ni bal.
Hastelloy C-276	N10276	Mo 16, Cr 15.5, Fe 5, W 4, Co 2.5 max., Mn 1 max., V 0.35 max., Si 0.08 max., C 0.02 max., Ni bal.
Hastelloy X	N06002	Cr 22, Fe 18.5, Mo 9, Co 1.5, Mn 1 max., Si 1 max., W 0.6, C 0.10, Ni bal.
214	N07214	Ni 75, Cr 16, Al 4.5, Fe 3, Mn 0.2, Si 0.1, C 0.05, Other Y.01
230	N06230	Ni 57, Cr 22, W 14, Co 5 max., Mo 2, Mn 0.5, Si 0.4, Al 0.3, Fe 0.3 max., C 0.10, La 0.02
Haynes 188	R30188	Cr 22, Ni 22, W 14, Fe 3 max., Mn 1.25 max., Si 0.35, La 0.04, C 0.10, Co bal.
IN-102	N06102	Cr 15, Fe 7, Mo 3, W 3, Cb 3, Ti 0.6, Al 0.4, Zr 0.03, B 0.005, Ni bal.
Incoloy 800	N08800	Fe 45.5, Ni 32.5, Cr 21, Al 0.4, Ti 0.4, C 0.05
800HT	N08811	Fe 46, Ni 33, Cr 21, Mn 0.8, Si 0.5, Al 0.4, Ti 0.4, C 0.08
Incoloy 825	N08825	Ni 38-46, Cr 19.5-23.5, Fe 22 min, Mo 2.5-3.5, Cu 1.5-3, Ti 0.6-1.2, Mn 1 max, Si 0.5 max
907	N19907	Fe 42, Ni 38, Co 13, Nb 4.7, Ti 1.5, Si 0.15, Al 0.03
909	N19909	Fe 42, Ni 38, Co 13, Nb 4.7, Ti 1.5, Si 0.4, C 0.01, B 0.001
Inconel 600	N06600	Ni 76, Cr 15, Fe 8, Mn 0.5, Cu 0.2, Si 0.2, C 0.08
Inconel 625	N06625	Ni 61, Cr 21.5, Mo 9, Fe 4, Cb 3.6, Mn 0.2, Si 0.2, C 0.05
Inconel 690	N06690	Ni 61, Cr 29, Fe 9, Cu 0.2, Mn 0.2, Si 0.2, C 0.02
Inconel 718	N07718	Ni 52.5, Cr 19, Fe 18.5, Cb 5.1, Mo 3.0, Ti 0.9, Al 0.5, Mn 0.2, Si 0.2, C 0.04
Inconel X750	N07750	Ni 73, Cr 15.5, Fe 7, Ti 2.5, Cb 1, Ti 0.9, Al 0.7, Mn 0.5, Si 0.2, C 0.04
L-605	R30605	Cr 20, W 15, Ni 10, Fe 3 max., Mn 1.5, Si 1 max., C 0.10, Co bal.
M-252	N07252	Cr 20, Co 10, Mo 10, Ti 2.6, Al 1, B 0.005, Ni bal.
N-155	R30155	Cr 21, Co 20, Ni 20, Mo 3, W 2.5, Mn 1.5, Cb+Ta 1, Si 1 max., N 0.15, C 0.10, Fe bal.
Nimonic 80A	N07080	Ni 74.5, Cr 20.5, Ti 2.4, Al 1.25, Fe 0.5, C 0.05
901	N09901	Ni 43, Fe 35, Cr 12, Mo 6, Ti 3, Co 1 max., Mn 0.5 max., Si 0.4 max., Al 0.2, C 0.05
Rene 41	N07041	Cr 19, Co 11, Mo 10, Ti 3.1, Fe 2, Al 1.5, B 0.005, Ni bal.
Udimet 500	N07500	Co 18.5, Cr 18, Mo 4, Al 2.9, Ti 2.9, Zr 0.05, B 0.006, Ni bal.
Ultimet	R31233	Co 54, Cr 26, Ni 9, Mo 5, Fe 3, W 2, Mn 0.8, Si 0.3, C 0.05, B 0.015
Waspaloy	N07001	Cr 19.5, Co 13.5, Mo 4.3, Ti 3, Fe 2, Al 1.3, Zr 0.04, B 0.006, Ni bal.

Note: Y denotes Y_2O_3

Table 2. Mechanical Properties of Selected Wrought "Superalloys."

Material	Form and Condition	Yield Strength[1]		Tensile Strength[2]		% Elong.[3]	Hardness
		ksi	MPa	ksi	MPa		
A-286	Bar 1800° F OQ + 1325° F AC	105	724	146	1007	25	R_C 28
Hastelloy G	Sheet Solution Heat Treated	46	317	102	703	62	R_B 84
Hastelloy C-276	Sheet Solution Heat Treated	52	359	115	793	61	R_B 90

(Continued)

Table 2. *(Continued)* **Mechanical Properties of Selected Wrought "Superalloys."**

Material	Form and Condition	Yield Strength[1]		Tensile Strength[2]		% Elong.[3]	Hardness
		ksi	MPa	ksi	MPa		
Hastelloy X	Sheet Solution Heat Treated	52	359	114	786	43	R_B 90
214	Typical as Supplied Condition	83	573	135	930	42	-
230	Typical as Supplied Condition	57	393	125	861	49	-
Haynes 188	Sheet 2150° F WQ	70	483	139	958	56	R_B 96
IN-102	Bar 1800° F AC	73	503	139	958	47	HB 215
Incoloy 800	Rod Hot Rolled Annealed	42	290	87	600	44	HB 150
800HT	Typical as Supplied Condition	35	241	83	572	49	-
Incoloy 825	Typical as Supplied Condition	45	310	100	690	45	
907	Typical as Supplied Condition	162	1116	195	1344	15	-
909	Typical as Supplied Condition	148	1020	190	1310	16	-
Inconel 600	Rod Hot Rolled Annealed	41	283	96	662	45	HB 150
Inconel 625	Rod Hot Rolled Annealed	70	483	140	965	50	HB 180
Inconel 690	Rod Hot Rolled Annealed	51	352	104	717	47	R_B 90
Inconel 718	Rod Hot Rolled Annealed Aged	180	1,241	208	1,434	20	R_C 45
Inconel X750	Rod Hot Rolled,1625° F + Aged, Sheet Annealed + Aged	126 122	869 841	184 177	1,269 1,220	25 27	R_C 36 R_C 36
L-605	Sheet Solution Heat Treated	67	462	146	1,007	59	R_C 24
M-252	Bar 1900° F AC + 1400° F AC	122	841	180	1,241	16	R_C 39
N-155	Sheet Solution Heat Treated	58	400	118	814	49	R_B 92
Nimonic 80A	1975° F, Air Cool + 1300° F	90	621	154	1,062	39	R_C 34
901	Typical as Supplied Condition	130	896	175	1207	15	-
Rene 41	Bar 1950° F AC + 1400° F AC	154	1,062	206	1,420	14	-
Udimet 500	Bar 1975° F AC + 1550° F AC + 1400° F AC	110	758	176	1,213	16	-
Ultimet	Typical as Supplied Condition	80	551	145	1000	35	-
Waspaloy	Bar 1975° F AC + 1550° F AC + 1400° F AC	115	793	185	1,276	25	-

Notes: Codes used for condition are WQ = Water Quench; AC = Air Cool; OQ = Oil Quench. [1] Yield strength at 0.2% offset, measured at room temperature. [2] Ultimate tensile strength at room temperature. [3] Elongation in 2 inches (50.4 mm).

Table 3. Physical Properties of Selected Wrought "Superalloys."

Material[1]	Density		CTE[2]		Modulus of Elasticity		Melting Point	
	lbs/cu in.	gr/cu cm	Eng.	Metric	ksi	GPa	°F	°C
A-286	0.286	7.92	9.17 [3]	16.49 [3]	29.1	201	2550	1399
Hastelloy G	0.300	8.31	7.5	13.49	27.8	191.5	2450	1343
Hastelloy C-276	0.321	8.89	6.2	11.15	29.8	205.3	2500	1371
Hastelloy X	0.297	8.22	7.70	13.84	28.5	196.4	2470	1354

(Continued)

Table 3. *(Continued)* **Physical Properties of Selected Wrought "Superalloys."**

Material[1]	Density		CTE[2]		Modulus of Elasticity		Melting Point	
	lbs/cu in.	gr/cu cm	Eng.	Metric	ksi	GPa	°F	°C
214	0.291	8.06	-	-	31.6	217.7	2471	1355
230	0.324	8.97	7.0	12.59	30.6	210.8	2462	1350
Haynes 188	0.324	8.97	6.6	11.87	33.6	231.5	2425	1329
IN-102	0.309	8.56	7.32	13.16	29.7	204.6	2530	1388
Incoloy 800	0.287	7.95	7.9	14.20	28	192.9	2525	1385
800 HT	0.287	7.95	7.9	14.20	28.5	196.4	2498	1370
Incoloy 825	0.294	8.14	7.7	14.00	-	-	2525	1385
907	0.299	8.28	4.4	7.91	23.9	164.7	2462	1350
909	0.296	8.20	4.4	7.91	23.0	158.5	2570	1410
Inconel 600	0.304	8.42	7.4	13.31	30	207	2575	1413
Inconel 625	0.305	8.45	7.1	12.77	30.1	207.4	2460	1349
Inconel 690	0.296	8.20	7.8	14.02	30.6	210.8	2510	1377
Inconel 718	0.296	8.20	7.2	12.95	29	199.8	2437	1336
Inconel X750	0.298	8.25	7.0	12.59	31	213.6	2600	1427
L-605	0.330	9.14	6.8	12.23	32.6	224.6	2570	1410
M-252	0.298	8.25	5.9	10.61	29.8	205.3	2500	1371
N-155	0.296	8.20	7.9	14.20	29.3	201.9	2470	1354
Nimonic 80A	0.297	8.22	6.2	11.15	32.1	221.2	2500	1371
901	0.294	8.14	7.5	13.49	30	207	2409	1321
Rene 41	0.298	8.25	6.6	11.87	31.9	219.8	2500	1371
Udimet 500	0.290	8.03	6.8	12.14	32.1	221.2	2540	1393
Ultimet	0.305	8.45	7.1	12.77	-	-	2534	1390
Waspaloy	0.296	8.20	6.8	12.23	30.9	212.9	2475	1357

Notes: [1] For material condition, see "Form and Condition" column of the mechanical properties Table. [2] Thermal Coefficient of Expansion (CTE) measured from 32 to 212° F (0 to 100° C) and expressed in μin./°F for English units, and μm/°C for Metric units. [3] CTE measured between 70 and 200° F (21 and 93° C).

These alloys closely resemble the austenitic stainless steels insofar as forging, cold forming, machining, welding, and brazing are concerned. Their higher strength may require the use of heavier forging or forming equipment, and machining is somewhat more difficult than for the stainless steels.

Selected Alloys

Alloy A-286. This precipitation hardening iron-base alloy was developed for parts requiring high strength up to 1,300° F (704° C) and oxidation resistance to 1,500°F (816° C). It is often specified for jet engines and gas turbines for parts such as turbine buckets, bolts, and discs, and for sheet metal assemblies. It is available in the usual mill forms. *Manufacturing Considerations:* A-286 is somewhat harder to hot or cold work than the austenitic stainless steels. Its forging range is 1,800 to 2,150° F (982 to 1,176° C). When finishing below 1,800° F, light reductions (under 15%) must be avoided to prevent grain coarsening during subsequent heat treatment. A-286 is readily machined in the partially or fully aged condition, but is soft and gummy in the solution treated condition. It should be welded in the solution treated condition, but fusion welding is difficult with large sections sizes and moderately difficult for small cross sections and sheet. Cracking may be encountered in the welding of heavy sections or parts under high restraint. A dimensional contraction of 0.0008 inch (0.02 mm) per inch (per 25.4 mm) is experienced during aging. Oxidation resistance is equivalent to Type 310 stainless steel up to 1,800° F (982° C). *Heat Treatment:* Sheet, strip and plate are normally solution treated (1,800° F/982° C). Bar, forging, tubing, and ring are either solution heat treated (1,800° F), or solution heat treated (1,800° F) and aged. Bar, forging, and tubing is either solution treated (1,650° F/899° C), or solution treated (1,650° F) and aged.

Alloy N-155. Also known as Multimet, N-155 is a solid solution alloy developed to withstand high stresses up to 1,500° F (816° C). It has good oxidation properties and good ductility, and can be readily fabricated by conventional means. It has been used in many aircraft applications including afterburner parts, combustion chambers, exhaust assemblies, turbine parts, and bolting. *Manufacturing Considerations:* N-155 is forged readily between 1,650 and 2,200° F (899 and 1,204° C). It is easily formed by conventional methods, but intermediate anneals may be required to restore ductility. This alloy is machinable in all conditions, using low cutting speeds and ample flow of coolant. Weldability is comparable to austenitic stainless steels, and oxidation resistance of N-155 sheet is good up to 1,500° F (816° C). *Heat Treatment:* Sheet, tubing, bar, and forging materials are normally solution treated, but bar and forging can also be solution treated and aged.

Nickel base heat-resistant alloys

Nickel is the base element for most of the higher temperature heat-resistant alloys. While it is more expensive than iron, nickel provides an austenitic structure that has greater toughness and workability than ferritic structures of the same strength level.

The common alloying agents for nickel are cobalt, iron, chromium, molybdenum, titanium, and aluminum. Cobalt, when substituted for a portion of the nickel in the matrix, improves high temperature strength. Small additions of iron tend to strengthen the nickel matrix and reduce the cost. Chromium is added to increase strength and oxidation resistance at very high temperatures. Molybdenum contributes to solid solution strengthening. Titanium and aluminum are added to most nickel base heat-resistant alloys to permit age hardening by the formation of Ni_3 (Ti, Al) precipitates, and aluminum also contributes to oxidation resistance. The nature of the alloying elements in the age hardening nickel base alloys makes vacuum melting of those alloys advisable. However, the additional cost of vacuum melting is more than compensated for in elevated temperature properties.

Manufacturing Considerations

Forging: The wrought alloys can, to some degree, be forged. The matrix-strengthened alloys can be forged with proper consideration for cooling rates, atmosphere, etc. Most of the precipitation hardening grades can be forged, but heavier equipment is required and a smaller range of reductions can be safely attained. *Cold Forming:* Almost all wrought nickel base alloys in sheet form are cold formable. The lower strength alloys offer few problems, but the higher strength alloys require forming pressures and more frequent anneals. *Machining:* These alloys are readily machinable, provided the optimum conditions of heat treatment, type of tool, speed, feed, depth of cut, etc., are achieved. *Welding:* The matrix strengthening type alloys offer no serious problems in welding. All of the common resistance and fusion welding processes (except submerged arc) have been successfully employed. For the age hardening alloys, it is necessary to observe the following precautions. 1) Welding should be confined to annealed material where design permits. In full aged hardened materials, the hazard of cracking in the weld and/or parent material is great. 2) If design permits, join some portions only after age hardening. The parts to be joined should be "safe ended" with a matrix strengthened type alloy and then age hardened. Welding should then be carried out on the safe ends. 3) Parts that are severely worked or deformed should be annealed before welding. 4) After welding, the weldment will often require stress relieving before aging. 5) Material must be heated rapidly to the stress relieving temperature. 6) In many age hardening alloys, fusion welds fusion welds may exhibit only 70 to 80% of the rupture strength of the parent material. *Brazing:* Solid solution type chromium containing alloys respond well to brazing using techniques and brazing alloys applicable to the austenitic stainless steels. The aluminum-titanium age hardened nickel base alloys are difficult to braze unless some method of fluxing, solid or gaseous, is used, Most of the high-temperature alloys of the nickel base type are brazed with Ni-Cr-Si-B and Ni-Cr-Si types of brazing alloy.

Heat Treatment

The nickel base alloys are heat treated with conventional equipment and fixtures such as would be used with austenitic stainless steels. Since nickel base alloys are more susceptible to sulfur embrittlement than are the iron base alloys, it is essential that sulfur bearing materials such as grease, oil, cutting lubricants, marking paints, etc., be removed before heat treatment. Mechanical cleaning, such as wire brushing, is not adequate and if used should be followed by washing with a suitable solvent or by vapor degreasing. A low-sulfur content furnace atmosphere should be used. Good furnace control with respect to time and temperature is desirable since overheating some of the alloys by as little as 35° F (19.5° C) impairs strength and corrosion resistance. When it is necessary to anneal the age hardening alloys, a protective atmosphere (such as argon) lessens the possibility of surface contamination or depletion of the precipitation hardening elements. This precaution is less critical in heavier sections because the oxidized surface layer is a smaller percentage of the cross section. After solution annealing, the alloys are generally quenched in water. Heavy sections may require air cooling to avoid surface cracking from thermal stress. In stress relief annealing of a structure or assembly composed of an aluminum-titanium hardened alloy, it is vitally important to heat the structure rapidly through the age hardening temperature range of 1,200 to 1,400° F (649 to 760° C), which is also the low ductility range, so that stress relief can be achieved before any aging takes place. Parts that are likely to be used in the fully heat treated condition should be solution treated, air cooled, and subsequently aged. Little difficulty can be expected with distortion under rapid heating conditions, and distortion of weldments of substantial size is less than can be expected with slow heating methods.

Selected Alloys

Hastelloy X. (Hastelloy is a registered trademark of Haynes International, Inc.) This nickel base alloy is used for combustor-liner parts, turbine exhaust weldments, afterburner parts, and other parts requiring oxidation resistance and moderately high strength above 1,450° F (788° C). It is not hardenable except by cold working, and is used in the solution treated (annealed) condition. It is available in all common mill forms. *Machining Considerations:* Hastelloy X is somewhat difficult to forge. Forging should be started at 2,150 to 2,200° F (1,117 to 1,204° C) and continued as long as the material flows freely. It should be in the annealed condition for optimum cold forming, and severely formed detail parts should be solution treated at 2,150° F for seven to ten minutes and cooled rapidly after forming. Machinability is similar to that of austenitic stainless steel. The alloy is tough and requires low cutting speeds and ample cutting fluids. It can be resistance or fusion welded or brazed, but large or complex fusion weldments require stress relief at 1,600° F (871° C) for one hour. Hastelloy X has good oxidation resistance up to 2,100° F (1149° C). It age hardens somewhat during long exposure between 1,200 and 1,800° F (649 and 892° C).

Inconel 625. (Inconel is a registered trademark of INCO International, Inc.) This is a solid solution, matrix strengthened nickel base alloy primarily used for applications requiring good corrosion and oxidation resistance at temperatures up to approximately 1,800° F (892° C) and also where such parts may require welding. The strength of the alloy is derived from the strengthening effects of molybdenum and columbium. Therefore, precipitation hardening is not required and the alloy is used in the annealed condition. Its strength is greatly affected by the amount of cold work prior to annealing, and by the annealing temperature. The material is normally annealed at 1,700 to 1,900° F (927 to 1,038° C) for a period of time commensurate with thickness. *Machining Considerations:* Because the alloy was developed to retain high strength at elevated temperatures, it resists deformation at hot working temperatures but can be readily fabricated with adequate equipment. The combination of strength, corrosion resistance, and ability to be fabricated, including welding by common industrial practices, are the alloy's outstanding features.

Inconel 706. (Inconel is a registered trademark of INCO International, Inc.) This nickel base vacuum melted precipitation hardened alloy has characteristics similar to Inconel 718 (description follows) except that Inconel 706 has greatly improved machinability. It has good formability and weldability, and excellent resistance to postweld strain age cracking. Depending on choice of heat treatment, this alloy can be used for applications requiring either 1) high resistance to creep and stress rupture up to 1,300° F (704° C), or 2) high tensile strength at cryogenic temperatures or elevated temperatures for short periods. The creep resistant heat treatment is characterized by an intermediate stabilizing treatment before precipitation hardening. Inconel 706 also has good resistance to oxidation and corrosion over a broad range of temperatures and environments. Creep resistant solution treatment temperature is 1,750° F (954° C). Tensile strength solution treatment temperature is 1,800° F (982° C).

Inconel 718. (Inconel is a registered trademark of INCO International, Inc.) Inconel 718 is a vacuum melted, precipitation hardened nickel base alloy that can be easily welded and excels in its resistance to strain age cracking. It is also readily formable. Depending on choice of heat treatments, this alloy finds applications requiring either 1) high resistance to creep and stress rupture to 1,300° F (704° C), or 2) high strength at cryogenic temperatures. It also has good oxidation resistance up to 1,800° F (982° C). It is available in all wrought forms and investment castings.

Inconel X-750. (Inconel is a registered trademark of INCO International, Inc.) This high strength oxidation resistant nickel base alloy is used for parts requiring high strength up to 1,000° F (538° C), or high creep strength up to 1,500°F (816° C), and for low stressed parts operating at temperatures up to 1,900° F (1,038° C). It is hardenable by various combinations of solution treatment and aging, depending on its form and application. Inconel X-750 is available in all the usual wrought mill forms. *Machining Considerations:* This alloy can be readily forged between 1,900 and 2,225° F (1,038 and 1,218° C). "Hot-cold" working between 1,200 and 1,600° F (649 and 871° C) is harmful and should be avoided. Inconel X-750 is readily formed but should be solution treated at 1,925° F (1,052° C) for seven to ten minutes following severe forming operations. It is somewhat more difficult to machine than austenitic stainless steels. Rough machining is easier in the solution treated condition. Finish machining is easier in the partially or fully aged condition. Fusion welding is difficult for large section sizes and moderately difficult for small cross sections and sheet. It must be welded in the annealed or solution treated condition, and weldments should be stress relieved at 1,650° F (899° C) for two hours before aging. Nickel brazing, followed by precipitation heat treatment of the brazed assembly, results in strength nearly equal to the fully heat treated material. Oxidation resistance of Inconel X-750 is good to 1,900° F (1,038° C), but the beneficial effects of aging are lost above 1,500° F (816° C). This alloy is subject to attack in sulfur containing atmospheres. *Heat Treatment:* Annealed and aged for strip and plate; mill annealed plus 1,300° F (704° C) for 20 hours, then air cool. Equalized and aged for bar and forging; 1,625° F (885° C) for four hours, air cool, then 1,300° F (704° C) for 24 hours.

Rene 41. (Rene is a registered trademark of the General Electric Corp.) This vacuum precipitation hardening nickel base alloy was developed for highly stressed parts operating between 1,200 and 1,800°F (649 and 982° C). Its applications include afterburner parts, turbine castings, and high temperature bolts and fasteners. It is available in the form of sheet, bars, and forgings. *Machining Considerations:* Rene 41 is forged between 1,900 and 2,150° F (1,038 and 1,177° C). Small reductions must be made when breaking up an as-cast structure as cracking may be encountered in finishing below 1,850° F (1,010° C). This alloy work hardens rapidly and frequent anneals are required. To anneal, heat rapidly to 1,950° F (1,066° C) for 30 minutes and quench. Machining is difficult. In the soft solution annealed condition it is gummy. Therefore, it should be in the fully aged condition for optimum machinability, and tungsten carbide tools should be used. It can be welded satisfactorily in the solution treated condition, and after welding the parts should be solution treated for stress relief. Rene 41 should not be exposed to temperatures above 2,050° F (1,121° C) during latter stages of hot working or during subsequent operations or severe intergranular cracking may be encountered. Oxidation resistance is good to 1,800° F (982° C). Lengthy exposure above the aging temperature of 1,400 to 1,650° F (760 to 899° C) results in loss of strength and room temperature ductility.

Waspaloy. (Waspaloy is a registered trademark of Pratt and Witney Aircraft Div., United Technologies.) The precipitation of titanium and aluminum compounds, plus the solid solution effects of chromium, molybdenum, and cobalt, strengthen this vacuum melted precipitation hardened nickel base alloy. It was designed for highly stressed parts operating at temperatures up to 1,550° F (843° C) such as aircraft gas turbine blades and discs, and rocket engine parts. Waspaloy is available in all the usual mill forms. *Machining Considerations:* The optimum range for forging is 1,900 to 2,050° F (1,038 to 1,121° C). The alloy should not be worked below 1,900° F due to danger of cracking and also decreasing the stress rupture life. Sufficient soaking time between heating is necessary to ensure complete recrystallization, but excessive long-term soaking at the high forging temperature should be avoided. Furnace atmospheres should be either neutral or slightly oxidizing to prevent carburization and to minimize scaling. This alloy is rather difficult to machine. Drilling, turning, etc., can be best accomplished in the solution treated and partially aged condition. Generally, carbide tools are preferred, and slow feeds are required to avoid work hardening. For finish machining, grinding is preferable. Waspaloy is susceptible to hot cracking or "hot shortness" above 2,150° F (1,177° C), so extreme care should be exercised in the design of weldments so that restraint can be minimized. Welding should be done in the annealed condition, with minimum heat input, and cooling should be accelerated by chill bars and gas backup. The alloy has good resistance to oxidation at temperatures up to 1,750° F (954° C), and to combustion products encountered in gas turbines. *Heat Treatment:* Two heat treatments are used for this material. 1) For optimum tensile strength, solution treat at 1,825 to 1,900° F (996 to 1,038° C), stabilize at 1,550° F (843° C), air cool for 24 hours, age for 16 hours at 1,400° F (760° C), air cool. 2) For optimum stress rupture properties, solution treat at

1,975° F (1,079° C), stabilize at 1,550° F (843° C), air cool for 24 hours, age at 1,400° F (760° C) for 16 hours, air cool.

Cobalt base heat-resistant alloys

Very few of the heat-resistant alloys can be considered to have a cobalt base, as cobalt is seldom the predominant element. Common alloying elements for cobalt are chromium, nickel, carbon, molybdenum, and tungsten. Chromium is added to increase strength and oxidation resistance at very high temperatures; nickel to increase toughness; carbon to increase both hardness and strength (especially when combined with chromium and the other carbide formers, molybdenum and tungsten); molybdenum and tungsten also contribute to solid solution strengthening. Vacuum melting is not required for these alloys, allowing the cobalt based alloys to be competitively priced with vacuum melted nickel base alloys, but higher in price than the nickel alloys.

Manufacturing Considerations

Forging: Because these alloys are designed to have very high strength at temperatures near the forging range, they require the use of heavy forging equipment. However, the forgability of these alloys is good over a fairly wide range of temperatures. Hot cold working is neither required nor recommended for these alloys. *Cold Forming:* These alloys, when the solution treated condition, have excellent ductility and are readily cold formed. Because of their capacity for work hardening, they require higher forming pressures and frequent anneals. *Machining:* These alloys are tough and work harden rapidly. Consequently, heavy-duty vibration-free machine tools, sharp cutting tools (high speed steel or carbide tipped), and low cutting speeds are required. *Welding:* The weldability of the cobalt base is comparable to that of the austenitic stainless steels. Welding may be accomplished with all commonly used welding processes, but large or complex weldments require stress relief.

Heat Treatment

The cobalt base alloys are heat treated with conventional equipment and fixtures such as those used with austenitic stainless steels. The use of good heat treating practices is recommended, although not as critical as in the case of the nickel base alloys.

Special Precaution

If the cobalt base alloys have not been exposed to neutron radiation, no special safety precautions in handling are required. However, neutron radiation creates a very dangerous radioactive isotope, cobalt 60, which has a half-life of 5.2 years. Special precautions must be employed to protect personnel from the radioactive material.

Selected Alloys

L-605. (This alloy is also known as Haynes Alloy 25.) L-605 is a corrosion and heat-resistant cobalt base alloy used for moderately stressed parts operating between 1,000 and 1,900° F (538 and 1,038° C). Its applications include gas turbine blades and rotors, combustion chambers, and afterburner parts. It is hardenable only by cold working, and is usually used in the annealed condition. All usual mill forms are available. *Machining Considerations:* L-606 forges moderately well between 1,900 and 2,250° F (1,038 and 1,232° C). In the annealed condition it has excellent formability at room temperature, but severely formed parts should be annealed at 2,225° F (1,218° C) for seven to ten minutes. This alloy is difficult to machine. Its toughness and capacity for work hardening necessitates the use of sharp tools and low cutting speeds. High speed steel or carbide cutting tools are recommended. L-605 can be fusion or resistance welded or brazed, but large or complex fusion weldments should be stress relieved at 1,300° F (704° C) for two hours. Oxidation resistance is excellent to 1,900° F (1,038° C).

HS 188. (This alloy is also known as Haynes 188.) This corrosion and heat-resistant cobalt base alloy is used for moderately stressed parts up to 2,100° F (1,149° C). It has outstanding oxidation resistance up to 2,100° F resulting from the addition of minute amounts of lanthanum (chemical symbol La) to the alloy system. HS 188 exhibits excellent post-aged ductility after prolonged heating of 1,000 hours at temperatures up to 1,600° F (871° C) inclusive. Gas turbine applications include transition ducts, combustion cans, spray bars, flame holders, and liners. It is not hardenable except by cold working, and is used in the solution treated condition. *Machining Considerations:* The alloy can be forged and welded. Welding can be accomplished by both manual and automatic welding methods including electron beam, gas tungsten air, and resistance welding. Like other cobalt base alloys, machining is difficult, necessitating the use of sharp tools and low cutting speeds. High speed steel or carbide cutting tools are recommended.

Other exotic alloys and materials

Additional nickel alloys, plus beryllium, will be discussed in this section.

Nickel Alloys

Monel contains nickel as the base metal, comprising at least 63% of the material content. The crystal structure is face centered cubic, and its high strength and excellent corrosion resistance in freshwater, seawater, chlorinated solvents, many acids (including sulfuric and hydrofluoric), and alkalis makes it a popular choice for pump and marine applications. Monel is a registered trademark of INCO International, Inc.

Monel 400. UNS N04400. Contents: 63 – 66% nickel; 28 – 31.5% copper; 2.5% iron (max.); 2% manganese (max); 0.5% silicon (max); 0.3% carbon (max); and 0.024 sulfur (max). Nickel content includes cobalt. Primarily used for marine engineering, chemical and hydrocarbon processing equipment, valves, pump shafts, fittings, fasteners, and heat exchangers. Standard product forms are round, hexagon, flats, forging stock, pipe, tube, plate, sheet, strip, and wire. This alloy is hardenable only by cold working. In the annealed state, tensile strength at room temperature is 79.8 ksi (550 MPa), and at 800° F (425° C) it is 65.3 ksi (450 MPa). Yield strength is 34.8 ksi (240 MPa) at room temperature, and 24.6 ksi (170 MPa). Density is 0.319 lb/cu in. (8.83 Mb/cu m).

Monel R-405. UNS N04405. Contents: 63 – 66% nickel; 28 – 31.5% copper; 2.5% iron (max.); 2% manganese (max); 0.5% silicon (max); 0.3% carbon (max); and 0.025 – 0.06% sulfur. Nickel content contains cobalt. This is a free machining version of Monel 400. A controlled amount of sulfur is added to the alloy to provide sulfide inclusions that act as chip breakers. Used for meter and valve parts, fasteners, and screw machine products. Standard product forms are round, hexagon, flats, and wire. This alloy is hardenable only by cold working. Mechanical properties are the same as Monel 404.

Monel K-500. UNS N05500. Contents: 63 – 65.5% nickel; 27 – 33% copper; 2.3 – 3.15% aluminum; 2% iron (max.); 1.5% manganese (max); 0.5% silicon (max); 0.35 – 0.85 titanium; 0.25% carbon (max); and 0.01% sulfur. Nickel content contains cobalt. This precipitation hardenable alloy combines the corrosion resistance of Monel 400 with greater strength and hardness. It has low permeability and is nonmagnetic to under –150° F (–101° C). In the age hardened condition, Monel K-500 is more prone to stress corrosion than Monel 400. Used for pump shafts, oil well tools and instruments, springs, valve trim, fasteners, and marine propeller shafts. Product forms are round, hexagon, flats, forging stock, pipe, tube, plate, sheet, strip, and wire. In the precipitation hardened condition, tensile strength at room temperature is 159.5 ksi (1,100 MPa), and at 800° F (425° C) it is 124.7 ksi (860 MPa). Yield strength is 114.5 ksi (790 MPa) at room temperature, and 91.3 ksi (630 MPa). Density is 0.305 lb/cu in. (8.47 Mb/cu m).

MP35N. UNS R30035. (Also known as Carpenter MP35N. MP35N is a registered trademark of SPS Technologies, Inc.) Contents: 33 – 37% nickel; 33% cobalt; 19 – 21% chromium; 9 – 10.5% molybdenum; 1% iron (max.); 1% titanium (max.); 0.15% manganese (max.); 0.15% silicon (max.); 0.025% carbon (max.); 0.015% phosphorus (max.); 0.01% boron; 0.01% sulfur (max.). This alloy is more properly designated a nickel-cobalt-chromium-molybdenum alloy due to the high percentage of each of these elements in its composition. It has ultrahigh tensile strength, good ductility and toughness, and excellent corrosion resistance. It also has excellent resistance to sulfidation, high temperature oxidation, and hydrogen embrittlement. Its properties are developed through work hardening, phase transformation, and aging. In the fully work hardened condition, service temperatures up to 750° F (399° C) are suggested. It is used for such diverse applications as fasteners, springs, nonmagnetic electrical components, and instrument components in medical, seawater, oil well, and chemical and food processing environments. This material is heat treated by aging at 1,000 to 1,200° F (538 to 649° C) for four hours, and air cooling. In its normal state, with 0% cold reduction, tensile strength is 135 ksi (931 MPa), and yield strength is 60 ksi (414 MPa). Cold drawn 35%, and aged at 1,050° F (565° C) for four hours and air cooled, tensile strength is 227 psi (1,565 MPa), and yield strength is 217 ksi (1,496 MPa).

Beryllium

Particles of beryllium and its components, such as beryllium oxide, are toxic. Special precautions must be taken to prevent inhalation. It is the second lightest metal, second only to magnesium, with a density of 0.0067 lb/cu in. (1.855 gr/cu cm). Pure beryllium has a tensile strength of 53.7 ksi (370 MPa), yield strength of 34.8 ksi (240 MPa), and a modulus of elasticity of 43,950 ksi (303 GPa). Its hardness is in the range of R_B 75 – 85. Standard grade beryllium bars, rods, tubing, and machined shapes are produced from vacuum pressed powder with a 1.50% maximum beryllium oxide content. Sheet and plate are fabricated from vacuum hot pressed powder with 2% maximum beryllium oxide content. Beryllium hot pressed block can be forged and rolled, but requires temperatures of 700° F (371° C) and higher because of brittleness. A temperature range of 1,000 to 1,400° F (538 to 760° C) is recommended. Beryllium sheet should be formed at 1,300 to 1,350° F (704 to 732° C), holding at temperature no more than 1.5 hours for maximum springback. Forming above 1,400 C° F (760° C) will result in a reduction in strength.

Carbide tools are most often used when machining beryllium. Mechanical metal removal techniques cause microcracks and metallographic twins. Finishing cuts are usually 0.002 – 0.005 inch (0.05 – 0.127 mm) in depth to minimize surface damage. Finish machining should be followed by chemical etching at least 0.002 inch (0.05 mm) from the surface to remove machining damage. A combination of 1.350° F 732° C) stress relief followed by a 0.0005 inch (0.013 mm) etch may be necessary for close tolerance parts. Damage free machining techniques include chemical milling and electrochemical machining.

Fusion welding is not recommended. Parts may be joined by brazing, soldering, braze welding, adhesive bonding, diffusion bonding, squeeze riveting, and bolting, threading, or press fitting techniques specifically designed to avoid damage.

Beryllium is alloyed in beryllium coppers (covered in the Copper section), and beryllium nickel (approximately 2% Be, 0.5 Ti, and the remainder Ni), and aluminum. Beralcast 191 (trademark of the Starmet Corp.) contains 61 – 68% beryllium; 27.5 – 34.5%

aluminum; 1.65 – 2.35 silver; 1.65 – 2.5% silicon; and 0.2% iron (max.). It is lighter than aluminum and titanium, but has greater ductility than pure beryllium. It is several times stiffer than aluminum, magnesium, aluminum, or aluminum base metal matrix composites. Its uses include satellite components, avionics packaging, aerospace systems, and golf chubs. The density of Beralcast 191 is 2.16 gr/cu cm, tensile strength is 28.5 ksi (196.5 MPa), yield strength is 20 ksi (137.9 MPa), and modulus of elasticity is 29,300 ksi (202 GPa).

Nonmetallic Engineering Materials

Plastics

When applied to materials, "plastics" usually refers to a class of synthetic organic materials which, though solid in the finished form, were at some stage in their processing fluid enough to be shaped. These materials may be divided into two general categories: thermoplastics which, like paraffin, can be softened and resoftened repeatedly without undergoing a change in chemical composition; and thermosetting resins which undergo a chemical change with the application of heat and pressure, and cannot be resoftened. Plastics in finished form are made up of long chain molecules (polymers) that are built by combining single molecules (monomers) under heat and pressure. Cross-linking is a permanent connection between two polymer molecules binding them together through a system involving primary chemical bonds.

Plastics have many desirable characteristics including light weight, high strength-to-weight ratio, and ease and economy of fabrication. Their specific gravity roughly ranges from 0.92 to 2.3, and their weight from 0.033 to 0.079 pounds per cubic inch (by comparison, the specific gravity for aluminum ranges from 0.091 to 0.108, and for steel it is 0.283). In addition, most plastics have excellent electrical resistance, high heat insulation properties, and good resistance to corrosion and chemical action. Many plastics have self-lubricating characteristics, and many can be made transparent, translucent, opaque, and colored. Substantial increases in temperature will cause a decline in the physical properties of most plastics—although some plastics soften at 180° F (82° C), most can resist temperatures up to 300° F (149° C)—but some can be used successfully in temperatures reaching 500 to 600° F (260 to 316° C). While many plastics should not be used at temperatures below –40° F (–40° C), a few can be used at cryogenic temperature. Plastics are subject to higher thermal expansion, creep, cold flow, low temperature embrittlement, and deformation under load than metals, some varieties change dimensions through solvent and moisture absorption, while others are degraded by ultraviolet and nuclear radiation.

The cost per pound for plastics exceeds that of steel and some other materials. However, they may be among the cheapest to use. Despite their low operating temperatures, plastics may be superior to metals as high temperature heat shields for short exposures (thermal conductivity values for plastics are in the range of 0.002; for aluminum the value is 0.500). Some reinforced plastics (glass reinforced epoxies, polyesters, and phenolics) are now nearly as strong and rigid (particularly in relation to weight) as most steels, and they may be more dimensionally stable. Some oriented films and sheets (oriented polyesters) may have greater strength-to-weight ratios than cold rolled steels.

Definitions of plastics terms

Acetal resins. Copolymers containing the acetal linkage (–CH_2–O–), e.g., polyoxymethylene.

Acrylic resin. A synthetic resin prepared from acrylic acid or from a derivative of acrylic acid.

Alkyd resins. Polyesters made from dicarboxylic acids and diols, primarily used as coatings, modified with vegetable oil, fatty acids, etc.

Arc resistance. Time required for a given applied electrical voltage to render the surface of a material conductive because of carbonization by the arc discharge.

Autoclave. A closed vessel producing an environment of fluid pressure, with or without heat, to an enclosed object that is undergoing a chemical reaction or other operation.

Autoclave molding. A process similar to the pressure bag technique. The lay-up is covered by a pressure bag and the entire assembly is placed in an autoclave capable of providing heat and pressure for curing the part. The pressure bag is normally vented to the outside.

Bag molding. A method of molding or laminating that involves the application of a fluid pressure to a flexible material which transmits the pressure to the material being molded or bonded. Fluid pressure is usually applied by means of air, steam, water, or vacuum.

Breakdown voltage. The voltage required, under specific conditions, to cause the failure of an insulating material. See also dielectric strength in the next section on standardized plastics tests.

Butadiene. A diene monomer with the structure CH_2=CH–CH=CH_2. May be copolymerized with styrene and with acrylonitrile. Its homopolymer is used as a synthetic rubber.

Cellulose acetate. An acetic acid ester of cellulose obtained by the action, under rigidly controlled conditions, of acetic acid and acetic anhydride on purified cellulose usually obtained from cotton liners. All three available hydroxyl groups in each glucose unit of the cellulose can be acetylated, but in the preparation of cellulose acetate it is usual to acetylate fully and then to lower the acetyl value (expressed as acetic acid) to 52 to 56% by partial hydrolysis. When compounded with suitable plasticizers, the result is a tough thermoplastic material.

Cellulose acetate butyrate. An ester of cellulose made by the action of a mixture of acetic and butyric acids and their anhydrides on a purified cellulose. It is used in the manufacture of plastics that are similar in their general properties to cellulose acetate but are tougher and have better moisture resistance and dimensional stability.

Clarity. The characteristic of a transparent body that allows distinct high-contrast images or high-contrast objects to be observable through the body.

Crazing. Fine cracks that may extend in a network on or under the surface or through a layer of a plastic material. Usually occurs in the presence of an organic liquid or vapor, with or without the application of mechanical stress.

Creep. The dimensional change with time of a material under load that follows the initial instantaneous elastic deformation. Creep at room temperature is called cold flow.

Cross-linking. The formation of primary valence bonds between polymer molecules. When extensive, as in thermosetting resins, cross-linking makes one infusible, insoluble supermolecule of all the chains.

Crystallinity. A state of molecular structure in some resins that denotes stereo-regularity and compactness of the molecular chains forming the polymer. Normally can be attributed to the formation of solid crystals having a definite geometric form.

Denier. A direct numbering system for expressing linear density, equal to the mass in grams per 9,000 meters of yarn, filament, fiber, or other textile strand.

Dimensional stability. The ability of a plastic part to retain its original dimensions during its service life.

Elastomer. A material that at room temperature stretches under low stress to at least twice its length and snaps back to the original length upon release of the stress.

Encapsulating. Encasing an article (usually an electronic component) in a closed envelope of plastic by immersing the object in a casting resin and allowing the resin to polymerize or, if hot, to cool.

Epoxy resins. Based on ethylene oxide, its derivatives, or homologs, epoxy resins form straight-chain thermoplastics and thermosetting resins, e.g., by the condensation of bisphenol and epichlorohydrin to yield a thermoplastic that is converted to a thermoset by active hydrogen-containing compounds, e.g., polyamines and dianhydrides.

Ethylene-vinyl acetate. Copolymer of ethylene and vinyl acetate having many of the properties of polyethylene but of considerably increased flexibility for its density; elongation and impact resistance are also increased.

Fiber. This term usually refers to relatively short lengths of very small cross sections of various materials. Fibers can be made by chopping filaments (converting). Staple fibers may be 0.5 in. (13 mm) to several inches or centimeters in length and usually from one to five denier.

Fiberglass. A widely used reinforcement for plastics that consists of fibers made from borosilicate and other formulations of glass. The reinforcements are in the form of roving (continuous or chopped), yarns, mat, milled, or woven fabric.

Filament winding. Roving or single strands of glass, metal, or other reinforcement are wound in a predetermined pattern onto a suitable mandrel. The pattern is designed to give maximum strength in the directions required. The strands can be run from a creel through a resin bath before winding, or preimpregnated materials can be used. When the right number of layers has been applied, the wound mandrel is cured at room temperature or in an oven.

Filler. An inexpensive, inert substance added to a plastic to make it less costly. Fillers may also improve physical properties, particularly hardness, stiffness, and impact strength. The particles are usually small in contrast to those of reinforcement, but there is some overlap between the functions of the two types of material.

Film. An optional term for sheeting having a nominal thickness not greater than 0.01 in. (0.254 mm).

Flame-retardant resin. A resin compounded with certain chemicals to reduce or eliminate its tendency to burn. For polyethylene and similar resins, chemicals such as antimony trioxide and chlorinated paraffins are useful.

Furan resins. Dark-colored, thermosetting resins that are primarily liquids ranging from low-viscosity polymers to thick, heavy syrups. Based on furfuryl or furfuryl alcohol.

Heat distortion point. The temperature at which a standard test bar deflects 0.01 in. (0.254 mm) under a stated pressure of either 66 or 264 PSI (0.455 or 1.82 MPa).

Homopolymer. A polymer consisting of only one monomeric species.

Melamine formaldehyde resin. A synthetic resin derived from the reaction of melamine (2,4,6-triamino-1,3,5-triazine) with formaldehyde.

Monomer. A relatively simple compound that can react to form a polymer. See also Polymer.

Nylon. The generic name for all synthetic fiber-forming polyamides. They can be formed into monofilaments and yarns characterized by great toughness, strength and elasticity, high melting point, and good resistance to water and chemicals. The material is widely used for bristles in industrial and domestic brushes and for many textile applications; it is also used in injection molding gears, bearings, combs, etc.

Olefins. A group of unsaturated hydrocarbons of the general formula $C_x H_{2y}$ and named after the corresponding paraffins by the addition of "ene" or "ylene" to the stem. Examples are ethylene and pentene-1.

Organic. A material or compound composed of hydrocarbons or their derivatives, or those materials found naturally or derived from plant or animal origin.

Permeability. (1) The passage or diffusion of vapor, liquid, or solid through a barrier without physically or chemically affecting the barrier. (2) The rate of such passage.

Phenolic resin. A synthetic resin produced by the condensation of phenol with formaldehyde. Phenolic resins form the basis of a family of thermosetting molding materials, laminated sheet and oven-drying varnishes. They are also used as impregnating agents and as components of paints, varnishes, lacquers, and adhesives.

Plasticizer. Chemical agents added to plastic compositions to improve flow and processability and to reduce brittleness. These improvements are achieved by lowering the glass transition temperature.

Polyamide. A polymer in which structural units are linked by amide grouping. Many polyamides are fiber formers.

Polycarbonate resins. Polymers derived from the direct reaction between aromatic and aliphatic dihydroxy compounds with phosgene or by the ester exchange reaction with appropriate phosgene-derived precursors. Structural units are linked by carbonate groups.

Polyester. A resin formed by the reaction between a dibasic acid and a dihydroxyalcohol–both organic–or by the polymerization of a hydroxy carboxylic acid. Modification with multifunctional acids and/or alcohols and some unsaturated reactants permit cross-linking to thermosetting resins.

Polyethylene. A thermoplastic material composed solely of ethylene. It is normally a translucent, tough, waxy solid that is unaffected by water and by a large range of chemicals.

Polyimide resins. Aromatic polyimides made by reacting pyrometallic dianhydride with aromatic diamines. Characterized by high resistance to thermal stress. Applications include components for internal combustion engines.

Polymer. A high-molecular-weight compound, natural or synthetic, whose structure can usually be represented by a repeated small unit, the -mer, e.g., polyethylene, rubber, and cellulose. Synthetic polymers are formed by addition or condensation polymerization of monomers. Some polymers are elastomers, some are plastics, and some are fibers.

Polymerization. A chemical reaction in which the high-molecular-weight molecules are formed from the original substances. When two or more monomers are involved, the process is called copolymerization or heterpolymerization.

Polyurethane resins. A family of resins produced by reacting diisocyanates in excess with glycols to form polymers having free isocyanate groups. Under the influence of heat or certain catalysts, these groups will react with each other or with water, glycols, etc., to form a thermoset.

Prepreg. Ready to mold or cure material in sheet form which may be tow, tape, cloth, or mat impregnated with resin. It may be stored before use.

Reinforcement. A strong inert material put into a plastic to improve its strength, stiffness, and impact resistance. Reinforcements are usually long fibers of glass, sisal, cotton, etc., in woven or nonwoven form. To be effective, the reinforcing material must form a strong adhesive bond with the resin.

Resin. Any of a class of solid or semisolid organic products of natural or synthetic origin, generally of high molecular weight, and with no definite melting point. Most resins are polymers. The term “resin” is often used synonymously with “plastic” or “polymer,” but it more accurately refers to any thermoplastic or thermosetting type plastic existing in either the solid or liquid state before processing.

Resistivity. The ability of a material to resist passage of electrical current either through its bulk or on a surface. The unit of volume resistivity is the ohm-cm, and of surface resistivity it is the ohm.

Roving. A number of strands, tows, or ends collected into a parallel bundle with little or no twist. In spun yarn production, an intermediate state between silver and yarn.

Self-extinguishing. A somewhat loosely used term describing the ability of a material to cease burning once the source of flame has been removed.

Sheet (thermoplastic). A flat section of a thermoplastic resin with the length considerably greater than the width and 0.01 in. (0.254 mm) or greater in thickness.

Stabilizer. An ingredient used in the formulation of some polymers to assist in maintaining the physical and chemical properties of the compounded materials at their initial values throughout the processing and service life of the material, e.g., heat and ultraviolet stabilizers.

Tack. Stickiness of the prepreg.

Thermal expansion (coefficient of). The fractional change in length (sometimes volume specified) of a material for a unit change in temperature. Values for plastics range from 10^{-5} to 2×10^{4} mm/mm °C (1.8×10^{-5} to 36×10^{-5} in./in. °F).

Twist. The number of turns about its axis per unit of length in a yarn or other textile strand. It may be expressed as turns per inch (tpi) or turns per centimeter (tpcm).

Vacuum bag molding. A process in which the lay-up is cured under pressure generated by drawing a vacuum in the space between the lay-up and a flexible sheet placed over it and sealed at the edges.

Volume resistivity (specific insulation resistance). The electrical resistance between opposite faces of a 1-centimeter (0.39-inch) cube of insulating material. It is measured under prescribed conditions using a direct current potential after a specified time of electrification. It is commonly expressed in ohm-centimeters. The recommended test is ASTM D 257.

Weathering. The process of degradation and decompostion that results from exposure to the atmosphere, to chemical action, and to the action of other natural environmental factors.

Standard tests for rating plastic materials

Testing the mechanical, thermal, and electrical properties of plastics is carried out under a series of tests defined primarily by ASTM. The following descriptions of these tests and rating criteria were provided by Quadrant Engineered Plastics Products.

Coefficient of Friction (ASTM D 3702). Coefficient of Friction (COF) is the measure of resistance to the sliding of one surface over another. Testing can be conducted in a variety of ways although thrust washers testing is most common. The results do not have a unit of measure associated with them since the COF is the ratio of sliding force to normal force acting on two mating surfaces. COF values are useful to compare the relative "slickness" of various materials, usually run unlubricated over or against polished steel. Since the value reflects sliding resistance, the lower the value, the "slicker" the bearing material. Two values are usually given for COF. "Static" COF refers to the resistance at initial movement from a bearing "at rest." "Dynamic" COF refers to the resistance once the bearing or mating surface is in motion at a given speed.

Coefficient of Linear Thermal Expansion (E 831 TMA). The coefficient of linear thermal expansion (CLTE) is the ratio of the change in a linear dimension to the original dimensions of the material for a unit change of temperature. It is usually measured in units of in./in./°F. CLTE is a very important consideration if dissimilar materials are to be assembled in applications involving large temperature changes. A thermoplastic's CLTE can be decreased (making it more dimensionally stable) by reinforcing it with glass fibers or other additives. The CLTE of plastics varies widely. The most stable plastics approach the CLTE of aluminum but exceed that of steel by up to ten times.

Compressive Strength (ASTD D 695). Compressive strength measures a material's ability to support a compressive force.

Continuous Service Temperature. This value is most commonly defined as the maximum ambient service temperature (in air) that a material can withstand and retain at least 50% of its initial physical properties after long-term service (approximately 10 years). Most thermoplastics can withstand short-term exposure to higher temperatures without significant deterioration. When selecting materials for high temperature service, both heat deflection temperature and continuous service temperature need to be considered.

Dielectric Constant (ASTM D 150$_{(2)}$). The Dielectric Constant, or permittivity, is a measure of the ability of a material to store electrical energy. Polar molecules and induced dipoles in a plastic will align themselves with an applied electric field. It takes energy to make this alignment occur. Some of the energy is converted to heat in the process. This loss of electrical energy in the form of heat is called dielectric loss, and is related to the dissipation factor. The rest of the electrical energy required to align the electric dipoles is stored in the material. It can be released at a later time to do work.

The higher the dielectric constant, the more electrical energy can be stored. A low dielectric constant is desirable in an insulator, whereas someone wanting to build a capacitor will look for materials with high dielectric constants. Dielectric constants are dependent on frequency, temperature, moisture, chemical contamination, and other factors.

Dielectric Strength (ASTM D 149). When an insulator is subjected to increasingly high voltages, it eventually breaks down and allows a current to pass. The voltage reached before breakdown divided by the sample thickness is the dielectric strength of the material, measured in volts/mil. It is generally measured by putting electrodes on either side of a test specimen and increasing the voltage at a controlled rate. Factors that affect dielectric strength in applications include: temperature, sample thickness, conditioning of the sample, rate of increase in voltage, and duration of test. Contamination or internal voids in the sample also affect dielectric strength.

Dissipation Factor (ASTM D 150). The dissipation factor, or dielectric loss tangent, indicates the ease with which molecular ordering occurs under the applied voltage. It is most commonly used in conjunction with dielectric constant to predict power loss in an insulator.

Elongation (ASTM D 638). Elongation, which is always associated with tensile strength, is the increase in length at fracture.

Flammability. In electrical applications (or any applications where plastic constitutes a significant percentage of an enclosed space), the consequences of exposure to an actual flame must be considered (i.e., plastic panels used in the interior of an aircraft cabin). Flammability tests measure combustibility, smoke generation, and ignition temperatures of materials.

UL 94 Flammability Class (HB, V-2, V-1, V-0, 5V). In this test, specimens are subjected to a specified flame exposure. The relative ability to continue burning after the flame is removed is the basis for classification. In general, the more favorable ratings are given to materials that extinguish themselves rapidly and do not drip flaming particles. Each rating is based on a specific material thickness (i.e., UL 94–V1 @ $^1/_8$" thick). The UL rating scale from highest burn rate (HB) to most flame retardant is HB, V-2, V-1, V-0, 5V.

Flexural Strength (ASTM D 790). Flexural properties measure a material's resistance to bending under load. For plastics, the data is usually calculated at 5% deformation/strain, which is the loading necessary to stretch the outer surface 5%.

Glass Transition (Tg) Temperature (ASTM D 3418). The glass transition temperature, Tg, is the temperature above which an amorphous polymer becomes soft and rubbery. Except when thermoforming, it is important to ensure that an amorphous polymer is used below its Tg if reasonable mechanical performance is expected.

Heat Deflection Temperature (ASTM D 648). The heat deflection temperature is the temperature at which a $^1/_2$" thick test bar, loaded to a specified bending stress, deflects by 0.010 in. It is sometimes called the "heat distortion temperature" (HDT). This value is used as a relative measure of the ability of various materials to perform at elevated temperatures short term, while supporting loads.

Melting Point (ASTM D 3418). The temperature at which a crystalline thermoplastic changes from a solid to a liquid.

Specific Gravity (ASTM D 792). Specific gravity is the ratio of the mass of a given volume of material compared to the mass of the same volume of water, both measured at 73°F (23°C). (Density of a material divided by the density of water.) Since it is a dimensionless quantity, it is commonly used to compare materials. Specific gravity is used extensively to determine part cost and weight. Materials with specific gravities less than 1.0 (such as polyethylene and polypropylene) float in water. This can help with identification of an unknown plastic.

Surface Resistivity (ASTM D 257). This test measures the ability of current to flow over the surface of a material. Unlike the volume resistivity test, the test electrodes are both placed on the same side of the test specimen. However, like volume resistivity, surface resistivity is affected by environmental changes such as moisture absorption. Surface resistivity is used to evaluate and select materials for testing when static charge dissipation or other surface characteristics are critical.

Tensile Strength (ASTM D 638). Ultimate tensile strength is the force per unit area required to break a material under tension. It is expressed in PSI or Pascals. The force required to pull apart one square inch of plastic may range from 1,000 to 50,000 PSI (6.8 to 345 MPa).

Volume Resistivity (ASTM D 257). The volume resistivity of a material is its ability to resist the flow of electricity, expressed in ohms-cm. The more readily the current flows, the lower the volume resistivity. Volume resistivity can be used to predict the current flow from an applied voltage as demonstrated by Ohm's Law.

$$V = IR$$

where: V = Applied voltage (volts)
I = Electrical current (amperes)
R = Resistance of the wire (ohms).

As the Resistivity Continuum in *Figure 1* indicates:

- insulators exhibit resistivities of 10^{12} and higher
- antistatic/partially conductive products exhibit resistivities of 10^5 to 10^{12}
- conductive products exhibit resistivities of 10^6 to 10^5.

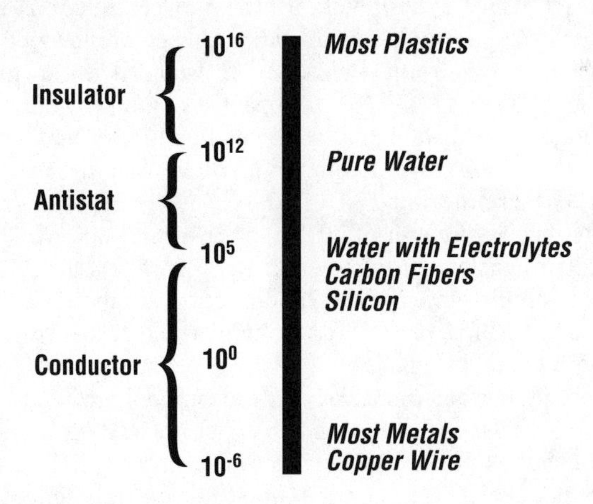

Figure 1. The resistivity continuum. (Source, Quadrant Engineered Plastic Products.)

Water Absorption (ASTM D 570). Water absorption is the percentage increase in weight of a material due to absorption of water. Standard test specimens are first dried then weighed before and after immersion in 73°F (23°C) water. Weight gain is recorded after 24 hours, and again when saturation is reached. Both percentages are important since they reflect absorption rate. Mechanical and electrical properties, and dimensional stability are affected by moisture absorption.

Thermoplastics

Thermoplastics (TP) are a relatively large group of synthetic organic materials that may be divided into three broad categories: 1) rigid load bearing materials, 2) environmental resistant materials, and 3) general purpose materials. Thermoplastics can be repeatedly softened and hardened by heating and cooling, and are susceptible to the effects of temperatures, time, environment, loading rate, and processing. These effects must be charted for each material in order to evaluate engineering usage in relation to performance requirements. **Table 1** gives abbreviations (which are routinely substituted for the full name) for popular thermoplastics and thermoset plastics, and **Table 2** provides selected properties for plastics commonly used in manufacturing.

Rigid Load Bearing Materials

These load bearing plastic materials approach Hooken behavior (Hooke's law relates to elasticity and essentially states that stress is proportional to strain up to the proportional limit of the material). Therefore, reasonable predictability of their behavior may be based on the established equations of state that are used for metals.

ABS (acrylonitrile-butadiene-styrene) resins. The three basic monomers that are used in ABS plastic are combined in varying proportions to form the commercially available resins. The acrylonitrile provides chemical resistance, the butadiene gives greater toughness and impact resistance, and the styrene-acrylonitrile copolymer provides good strength and rigidity. The main variations are as follows. 1) Medium impact, which is a hard, rigid, tough material used for appearance parts requiring high strength, good fatigue resistance, surface hardness, and gloss. 2) High impact, which is used where additional impact strength is required at the expense of hardness and rigidity. 3) Extra high impact, which has the highest impact strength, but further decreases in strength, rigidity, and hardness. 4) Low temperature impact, which is designed for high impact strength at temperatures as low as –40° F (–40° C). Strength, rigidity, and heat resistance are lowered. 5) High strength and heat resistant, which provides maximum heat resistance (heat distortion point at 264 PSI [1820 kPa] is about 229° F [110° C]). Its impact strength is comparable to the high impact type, but it has higher tensile and flexural strength, stiffness, and hardness. ABS resins are available as compounds for injection molding, blow molding, extrusion, and calendaring as well as sheet for thermoforming.

The ultimate tensile strength of ABS ranges from 3,000–9,000 PSI (20.7–62 MPa), and it has excellent resistance to many corrosive materials. It is used for a wide variety of products including instrument panels and interior automotive trim, keyboard housings, portable appliance housings, pipes, fittings, helmets, electrical connectors, luggage shells, and deflectors for hot air systems.

Acetals. There are two basic acetal resins—a homopolymer and a copolymer. In general, the copolymer has better stability and resistance to heat aging, and the homopolymer offers slightly better "as molded" mechanical properties. These are among the strongest and stiffest thermoplastics, and compression and tensile strengths show linearity with temperatures ranging from –45 to 250° F (–43 to 121° C). They have the highest fatigue endurance limits of all thermoplastics, and are only lightly affected by moisture. Good

Table 1. Abbreviations Used for Selected Plastics, Polymers, Elastomers, and Chemicals.

ABA	acrylonitrile-butadiene-acrylate	ABS	acrylonitrile-butadiene-styrene copolymer
ACM	acrylic acid ester rubber	ACS	acrylonitrile-chlorinated PE-styrene
AES	acrylonitrile-ethylene-propylene-styrene	AMMA	acrylonitrile-methyl methacrylate
AN	acrylonitrile	AO	antioxidant
APET	amorphous polyethylene terephthalate	APP	atactic polypropylene
ASA	acrylic-styrene-acrylonitrile	ATH	aluminum trihydrate
AZ(O)	azodicarbonamide	BMI	bismaleimide
BO	biaxially-oriented (film)	BOPP	biaxially-oriented polypropylene
BR	butadiene rubber	BS	butadiene styrene rubber
CA	cellulose acetate	CAB	cellulose acetate butyrate
CAP	cellulose acetate propionate	CAP	controlled atmosphere packaging
CF	cresol formaldehyde	CFC	chlorofluorocarbons
CHDM	cyclohexanedimethanol	CN	cellulose nitrate
COP	copolyester	COPA	copolyamide
COPE	copolyester	CP	cellulose propionate
CPE	chlorinated polyethylene	CPET	crystalline polyethylene terephthalate
CPP	cast polypropylene	CPVC	chlorinated polyvinyl chloride
CR	chloroprene rubber	CS	casein
CTA	cellulose triacetate	DABCO	diazobicyclooctane
DAM	diallyl maleate	DAP	diallyl phthalate
DCPD	dicyclopentadiene	DE	diotamaceous earth
DEA	dielectric analysis	DETDA	diethyltoluenediamine
DMT	dimethyl ester of terephthalate	EAA	ethylene acrylic acid
EB	electron beam	EBA	ethylene butyl acrylate
EC	ethyl cellulose	ECTFE	ethylene-chlorotrifluoroethylene copolymer
EEA	ethylene-ethyl acrylate	EG	ethylene glycol
EMA	ethylene-methyl acrylate	EMAA	ethylene-methacrylic acid
EMAC	ethylene-methyl acrylate copolymer	EMPP	elastomer modified polypropylene
EnBA	ethylene normal butyl acrylate	EP	epoxy resin, also ethylene-propylene
EPDM	ethylene-propylene terpolymer rubber	EPM	ethylene-propylene rubber
EPS	expandable polystyrene	ESI	ethylene-styrene copolymers
ETFE	ethylene-tetrafluoroethylene copolymer	EVA(C)	polyethylene-vinyl acetate
EVOH	polyethylene-vinyl alcohol copolymers	FEP	fluorinated ethylene propylene copolymer
FPVC	flexible polyvinyl chloride	GPC	gel permeation chromotography
GPPS	general purpose polystyrene	GTP	group transfer polymerization
HALS	hindered amine light stabilizer	HAS	hindered amine stabilizers
HCFC	hydrochlorofluorocarbons	HCR	heat-cured rubber
HDI	hexamethylene diisocyanate	HDPE	high-density polyethylene
HDT	heat deflection temperature	HFC	hydrofluorocarbons
HIPS	high-impact polystyrene	HMDI	diisocyanato dicyclohexylmethane
HNP	high nitrile polymer	IPI	isophorone diisocyanate

(Continued)

Table 1. *(Continued)* **Abbreviations Used for Selected Plastics, Polymers, Elastomers, and Chemicals.**

IV	intrinsic viscosity	LCP	liquid crystal polymers
LDPE	low-density polyethylene	LLDPE	linear low-density polyethylene
LP	low-profile resin	MbOCA	3,3'-dichloro-4,4-diamino-diphenylmethane
MBS	methacrylate-butadiene-styrene	MC	methyl cellulose
MDI	methylene diphenylene diisocyanate	MEKP	methyl ethyl ketone peroxide
MF	melamine formaldehyde	MMA	methyl methacrylate
MPE	metallocene polyethylenes	MPF	melamine-phenol-formaldehyde
NBR	nitrile rubber	NDI	naphthalene diisocyanate
NR	natural rubber	ODP	ozone depleting potential
OFS	organofunctional silanes	OPET	oriented polyethylene terephthalate
OPP	oriented polypropylene	O-TPV	olefinic thermoplastic vulcanizate
OEM	original equipment manufacturer	OSA	olefin-modified styrene-acrylonitrile
PA	polyamide	PAEK	polyaryletherketone
PAI	polyamide imide	PAN	polyacrylonitrile
PB	polybutylene	PBA	physical blowing agent
PBAN	polybutadiene-acrylonitrile	PBI	polybenzimidazole
PBN	polybutylene naphthalate	PBS	polybutadiene styrene
PBT	polybutylene terephthalate	PC	polycarbonate
PCC	precipitated calcium carbonate	PCD	polycarboiimide
PCR	post-consumer recyclate	PCT	polycyclohexylenedimethylene terephthalate
PCTA	copolyester of CHDM and PTA	PCTFE	polychlorotrifluoroethylene
PCTG	glycol-modified PCT copolymer	PE	polyethylene
PEBA	polyether block polymide	PEC	chlorinated polyethylene
PEDT	3,4 polyethylene dioxithiophene	PEEK	polyetheretherketone
PEI	polyether imide	PEK	polyetherketone
PEKEKK	polyetherketoneetherketoneketone	PEN	polyethylene naphthalate
PES	polyether sulfone	PET	polyethylene terephthalate
PETG	PET modified with CHDM	PF	phenol formaldehyde
PFA	perfluoroalkoxy resin	PI	polyimide
PID	proportional, integral, derivative	PIBI	butyl rubber
PMDI	polymeric methylene diphenylene	PMMA	polymethyl methacrylate
PMP	polymethylpentene	PO	polyolefins
POM	polyacetal	PP	polypropylene
PPA	polyphthalamide	PPC	chlorinated polypropylene
PPE	polypropylene ether, modified	PPO	polyphenylene oxide
PPS	polyphenylene sulfide	PPSU	polyphenylene sulfone
PS	polystyrene	PSU	polysulfone
PTA	purified terephthalic acid	PTFE	polytetrafluoroethylene
PU	polyurethane	PUR	polyurethane
PVC	polyvinyl chloride	PVCA	polyvinyl chloride acetate
PVDA	polyvinylidene acetate	PVDC	polyvinylidene chloride

(Continued)

Table 1. *(Continued)* **Abbreviations Used for Selected Plastics, Polymers, Elastomers, and Chemicals.**

PVDF	polyvinylidene fluoride	PVF	polyvinyl fluoride
PVOH	polyvinyl alcohol	RHDPE	recycled high density polyethylene
RPET	recycled polyethylene terephthalate	SI	silicone plastic
SAN	styrene acrylonitrile copolymer	SB	styrene butadiene copolymer
SBC	styrene block copolymer	SBR	styrene butadiene rubber
SMA	styrene maleic anhydride	SMC	sheet molding compound
SMC-C	SMC-continuous fibers	SMC-D	SMC-directionally oriented
SMC-R	SMC-randomly oriented	TA	terephthalic acid
TDI	toluene diisocyanate	TEO	thermoplastic elastomeric olefin
TLCP	thermoplastic liquid crystal polymer	T/N	terephthalate/naphthalate
TPA	terephthalic acid	TP	thermoplastic
TPE	thermoplastic elastomer	TPO	thermoplastic olefins
TPU	thermoplastic polyurethane	TPV	thermoplastic vulcanizate
TS	thermoset	UF	urea formaldehyde
ULDPE	ultralow-density polyethylene	UP	unsaturated polyester resin
UR	urethane	VA(C)	vinyl acetate
VC	vinyl chloride	VDC	vinylidene chloride
VLDPE	very low-density polyethylene	ZNC	Ziegler-Natta catalyst

Table 2. Properties of Plastics Used in Manufacturing. *(Source, Plastics Technology Handbook.)*

Application (See Notes)	Melt Flow Rate	Density	Tensile Strength	Elong. at Yield	Flexibility Modulus	Notched Izod Impact[2]	Deflection Temp. °F	UL 94 Rating[3]
	gr/10 min.	gr/cu. cm.	ksi	%	10 E+5	Ft/lb in.	66 psi/264 psi	1/8 in.
ABS (Acrylonitrile-Butadiene-Styrene)								
BM	0.80	1.02-1.05	5.0-6.67	-	2.4-2.95	8.0-8.15	212/192	HB
EX	0.4-2.5	1.02-1.07	4.4-6.7	2-40	2.4-3.5	3.5-8.4	195/187	HB
IM	8.0-5.8	1.03-1.06	5.6-7.7	4-36	3.2-4.0	2.0-6.0	209/185	V-0, HB
ABS/Polycarbonate Alloy								
IM	2.8-10	1.2-1.8	7.7-8.5	3-5	2.7-4.2	3.0-9.0	250/235	V-0, HB
Acetal								
IM	6.5-9.0	1.42-1.54	8.4-9.6	10-30	3.3-6.5	1.2-1.3	325/250	HB
Acrylics								
EX/IM	3.3-7.5	1.17-1.19	7.0-10.0	3-5	3.6-4.7	0.04-0.06	185/170	HB
Cellulose Acetate Butyrate								
IM	-	1.17-1.21	3.5-6.2	-	1.6-2.5	3.0-6.5	195/175	-
PTFE (Polytetrafluoroethtlene)								
CM	-	2.1-2.19	2.5-4.2	200-400	0.7-0.82	3.5	-	V-0
Ionomers								
EBF	1.1-1.7	0.94	-	350-400	-	No Break	-	-

(Continued)

Table 2. *(Continued)* **Properties of Plastics Used in Manufacturing.** *(Source, Plastics Technology Handbook.)*

Application (See Notes)	Melt Flow Rate	Density	Tensile Strength	Elong. at Yield	Flexibility Modulus	Notched Izod Impact[2]	Deflection Temp. °F	UL 94 Rating[3]
	gr/10 min.	gr/cu. cm.	ksi	%	10 E+5	Ft/lb in.	66 psi/264 psi	1/8 in.
Liquid Crystal Polymers								
IM	-	1.3-1.65	24.0-26.0	-	16.0-26.0	1.7-2.3	470/445	V-0
Nylon Type 6								
EX	2.6-4.5	1.12-1.13	3.8-10.5	5-200	3.5-4.6	0.9-2.3	250/150	V-2, HB
IM	-	1.1-1.3	9.0-20.0	3-15	3.4-10.0	1.0-3.5	390/175	V-0, HB
Nylon Type 12								
IM	3.0-9.0	1.02-1.25	5.0-13.5	6-20	1.6-7.0	1.5-4.7	275/135	HB
Nylon Type 66								
IM	-	1.15-1.39	9.5-25	-	4.1-12.0	0.9-4.5	475/450	V-0, HB
Polycarbonate (PC)								
EX	5.0-10.0	1.2-1.17	8.5-9.5	6-7	3.3-3.5	6.0-15.0	290/270	V-0, HB
IM	7.5-16.5	1.2-1.25	8.7-9.5	6-8	3.1-8.5	4.0-15.0	275-265	V-0, HB
Thermoplastic Polyesters (PBT)								
IM	8.0-12.0	1.41-1.65	6.0-17.4	2.5-7	5.1-14.0	1.1-10.0	400/350	V-0, HB
Thermoplastic Polyesters (PET)								
IM	-	1.55-1.70	17.2-22.4	1.5-3.0	10.2-16.8	1.2-1.9	460/440	V-0, HB
Polyethersulfone (PES)								
IM	-	1.4-1.65	10.0-21.5	1.3-2.9	3.8-12.0	1.3-1.6	410/400	V-0
Polyethylene (LDPE-unfilled, 0.910-0.940 gm/cu cm)								
EX-Film	0.25-6.5	0.92-0.924	1.2-1.7	100-600	-	-	-	-
IM	5-65	0.92-0.924	1.25-1.7	190-400	0.18-0.45	-	-	-
Polyethylene (HDPE-unfilled, 0.940 gm/cu cm and higher)								
BM	0.4-9.5	0.95-0.96	3.5-4.5	10-800	1.1-2.2	1.9-8.8	175 @ 66psi	-
IM	9.0-80.0	0.95-0.96	3.5-4.2	10-800	1.2-2.2	0.35-1.2	175 @ 66psi	-
Polyphenylene Ether/Oxide (PPO/PPE) Based Resins								
IM	-	1.07-1.27	6.0-12.5	-	3.3-10.0	2.0-6.5	270/260	V-0, HB
Polyphenylene Sulfide (PPS)								
IM	-	1.5-1.75	4.5-23.0	1.1-3.4	5.5-20.0	0.9-1.5	480/470	V-0
Polypropylene-Unfilled								
EX-Film	3.5-12.0	0.89-0.91	3.1-4.8	6-14	1.5-2.3	-	220 @ 66	-
IM	2.0-65.0	0.9-1.04	3.7-6.2	9-14	2.6-2.4	0.5-2.0	210 @ 66 psi	-
Polypropylene-Filled								
IM	5.0-18.0	1.05-1.2	3.5-10.8	-	2.2-6.5	0.6-1.4	260/170	HB
Polystyrene (PS general purpose)								
IM	4.0-10.0	1.04-1.2	4.2-6.8	-	3.7-4.5	0.3-1.0	200/175	V-0, HB
Polystyrene-Impact								
IM	5.0-9.5	1.03-1.05	3.1-5.7	-	2.7-3.5	1.1-3.2	190/175	V-0, HB

(Continued)

Table 2. *(Continued)* **Properties of Plastics Used in Manufacturing.** *(Source, Plastics Technology Handbook.)*

Application (See Notes)	Melt Flow Rate	Density	Tensile Strength	Elong. at Yield	Flexibility Modulus	Notched Izod Impact[2]	Deflection Temp. °F	UL 94 Rating[3]
	gr/10 min.	gr/cu. cm.	ksi	%	10 E+5	Ft/lb in.	66 psi/264 psi	1/8 in.
Styreneacrylonitrile Copolymers (SAN)								
IM	2.0-8.5	1.07-1.35	9.3-14.0	1.2-1.5	1.3-5.5	0.3-0.7	210/200	HB
Polyvinyl Chloride (PVC-rigid)								
IM	-	1.3-1.4	6.5-9.0	-	3.5-4.3	1.5-15.0	168/156	V-0 / 5V
Polyvinyl Chloride (PVC-flexible)								
EX	-	1.25-1.55	1.7-2.6	250-350	-	-	-	-
IM	-	1.22-1.36	1.4-2.7	275-400	-	-	-	-
Thermoplastic Elastomers-Olefinic Type								
IM	7.0-10.0	0.89-0.91	1.8-2.5	500-700	-	-	200/135	-
Thermoplastic Elastomers-Polyurethane Type								
IM	10.0-30.0	1.13-1.4	5.5-6.9	420-550	-	-	-	-
Thermoplastic Elastomers-Styrene Type								
IM	2.0-11.3	0.9-1.5	-	500-900	-	-	-	-

Notes.
Application abbreviations:
BM = Blow Molding EX = Extrusion IM = Injection Molding CM = Compression Molding
EBF = Extrusion Blown Film

1 Values are medians (not minimums and maximums) expressed in ranges based on properties provided by the manufacturer of the particular plastic and are comparative purposes only. Consult manufacturer for exact specifications for a specific material.

2 Izod impact test conducted with 1/8 in. indenter.

3 Underwriters' Laboratories Rating for fire retardance. Best achievable rating is 5V.

electrical properties, which are relatively unaffected by aging, make them suitable for electrical parts. Acetals are available as compounds for extrusion, injection, and blow molding. Fiberglass reinforced acetals are also available.

Acetels are used as replacements for metals where the higher ultimate strength of metal is not required, and costly finishing and assembly operations can be eliminated. Typical parts are pump impellers, molded and machined rollers, gears, appliance cases, automobile dashes, pipes, fittings, bearings, and zippers.

Polyallomers are highly crystalline polymers prepared from two or more monomers (namely propylene and ethylene) by a polymerization process. Formulations are available for high stiffness, medium impact strength and moderately high stiffness, high impact strength, and extra high impact strength. They may be injection molded, extruded, and thermoformed and are used in wire and cable jacketing, bottles, pipe fittings, and utility boxes and containers. Properties are very similar to HDPE and PP.

Polyamides (nylons). Polyamide (PA), or nylon, was the first of the thermoplastic engineering plastics. Some of the nylons are identified by the number of carbon atoms in the parent diamine and dibasic acid (i.e., Nylon 6, Nylon 12, etc.). Nylons have high strength and elongation, high modulus of flexure, good impact strength, and good resistance to nonpolar solvents. They are softened by polar materials such as alcohol and glycols, and are attacked by strong aids and oxidizing agents.

General purpose nylon molding materials are available for extrusion, injection molding, blow molding, and casting. Sheet and film are also available. Molybdenum disulfide, a solid lubricant, is used as a filler for type 6, 6/6, 6/10, and 6/12 nylons to improve wear and abrasion resistance, frictional characteristics, flexure strength, stiffness, and heat resistance. Glass fiber reinforced nylons show substantial improvement in tensile strength, heat distortion temperatures, and impact strength (in some cases). Nylon 11 and Nylon 12 have exceptional dimensional stability because of their low moisture rates and they offer complete resistance to zinc chloride.

Nylons are used extensively in the automobile industry, and for gears, living hinges, gaskets, wire insulation, high pressure flexible tubing, and chemical containers.

Polycarbonates. Polycarbonate (PC) resins are derived from aromatic and aliphatic dihydroxy compounds. They have properties that include excellent rigidity, toughness, and impact strength coupled with dimensional stability over a wide temperature range. Their heat and flame resistance is excellent: heat distortion temperatures range from 270 to 280° F (132 to 138° C) when stressed at 264 PSI (1820 kPa). Other features are unusually good electrical properties, but they should not be used where high arcing is involved. They are stain resistant and have been approved by the U.S. F.D.A. for safe use with all types of foodstuffs. Fatigue resistance is low, and PCs should not be stressed above 10% of their tensile or compressive strength in long term loading.

Polycarbonate molding compounds are available for extrusions, injection molding, and blow molding. Glass reinforced PCs have twice the tensile strength and tensile modulus, and four times the flexure modulus of nonreinforced resins. Typical applications include safety shields, lenses, electrical relay covers, pump impellers, fuse caps, electrical switch components, and appliance housings.

Polyethylenes (high-density). In volume, polyethylene (PE) is the most widely used plastic. The terms low-, medium-, and high-density refer to ASTM designations, and the low- and medium- grades are discussed under "Environmental Resistant Materials," which follows this section. As density increases, hardness, heat distortion point thickness, ultimate strength, and impermeability to liquids and gases increases. As density decreases, impact strength, mold shrinkage, and stress crack resistance increases. High-density PE has a density range of 0.941 to 0.965 gr/cu cm; stiffness in flexure of 81,000 to 150,000 PSI (558 to 1034 MPa), and a maximum service temperature of 180° F (82° C). Most high-density materials are classified as homopolymers (0.96 densities) and copolymers (0.95 densities). The copolymers are usually butene or hexane copolymers. The homopolymers are stiffer and better suited for thin wall containers where stress is not likely, while the copolymers are more stress crack resistant and more suitable for other applications (molded containers and blown bottles).

Polyethylenes have excellent chemical resistance and electrical properties, but their load bearing characteristics and weatherability are not good. It is flammable, and heat distortion temperatures are fairly low. Although it rates well in processing ability, mold shrinkage is above average. PE compounds can be extruded, injection molded, blow molded, and centrifugally cast. Film and sheet are available. Cellular PE, rigid foam, and fine powder are also available. Applications include containers, electrical insulation, pipes and tubing, grocery sacks, and fuel tanks for automobiles.

Polypropylenes. Polypropylene (PP) offers a good balance of properties. It has a good balance of strength, rigidity, and hardness, making it a good choice for parts requiring stiffness in thick sections, and flexibility in thin areas, and it retains strength up to 184° F (140° C), at which point it quickly looses stiffness. Chemical and biological resistance are outstanding, it has good resistance to environmental stress cracking, does not absorb moisture, and its electrical properties are excellent at high frequencies, heat, and humidity. As the lightest of plastics, it offers a high strength to weight ratio, high surface hardness, and excellent dimensional stability, but its load bearing ability is significantly affected by time. Rigidity, strength, dimensional stability, and heat resistance characteristics can be significantly increased with glass fiber reinforcements.

Polypropylenes are available as molding compounds for extrusion, injection molding, thermoforming, and blow molding, as well as in film and sheet. PPs are used in houseware containers, "living hinges" for a variety of applications, unbreakable medical and hygienic equipment that can be sterilized, appliance parts such as washer tubs, appliance cabinets, pipes and fittings, automotive interior parts, electrical connectors, extrusion coating, and packaging film.

Polystyrenes (modified). Polystyrene (PS) is available in many grades, and modified versions feature altered properties that are influenced by the addition of modifying agents such as rubber (for impact strength), methyl or alpha methyl styrene (for heat resistance), methyl methacrylate (for light stability), and acrylonitrile (for chemical resistance). The high heat resistant materials are produced as copolymers, while the high impact materials are produced by blending.

Polystyrenes are relatively inexpensive, and their clear, colorless properties permit an almost infinite range of color possibilities. They feature low moisture absorption, excellent dimensional stability, high hardness, good fatigue life. Their mechanical properties are affected by temperature, time, and environment, and, though flammable, PSs burn at a slow rate. They are available for extrusion, injection molding, compression, rotational and blow molding, and as polystyrene film, sheet, and foam. Uses include pipes and fittings, automobile parts, large appliance panels, battery housings, packaging materials, and housewares.

Polyvinyl chloride. Polyvinyl chloride (PVC) represents a major portion of the plastic consumed worldwide. PVC and chlorinated polyvinyl chloride (CPVC) retain good strength properties over their temperature range. CPVC has higher tensile strength and modulus, plus a high heat resistance that allows it to resist heat deflection under load to a temperature range of 180 to 220° F (82 to 102° C), versus 158° F (70° C) for PVC. At 212° F (100° C) CPVC retains a strength of 2,100 PSI (14 MPa). These rigid vinyls withstand strong acids and alkalies, and a wide variety of other corrosive liquids. They can be extruded, cast, coated, injection molded, blow molded, rotationally molded, and calendered, and are available as both film and sheet as well as rigid and flexible foams. Primary uses are pipe and fittings, insulation, electrical wire, chemical storage tanks, house siding, and window sash.

Environmental Resistant Materials

These materials do not possess outstanding load bearing characteristics and are non-Hookean (stress is *not* proportional to strain). They are, however, extremely durable in one or more properties such as wear and abrasion resistance, hardness, chemical resistance, heat resistance, and weathering.

Acrylics. These resins feature crystal clarity, outstanding weatherability, and excellent dimensional stability and stain resistance. Impact strength is good to –40° F (–40° C), and resistance to cracking after thermal shock is also good, as are electrical properties. Acrylics are also immune to attack by fungal organisms. Negatives are that they have a tendency to cold flow, a low softening point (160 to 200° F [66 to 93° C]), low scratch resistance, and they are affected by oxidizing acids and other chemicals.

Acrylics are available as compounds for extrusion, injection molding, blow molding, and casting, and in sheet or film forms. They are used for transparent enclosures in aircraft, lenses, containers, optical systems, architectural panels and facia, and outdoor signs and displays.

Cellulosics (cellulose acetate butyrate and ethyl cellulose). Cellulose plastics are not synthesized by the usual polymerization of a monomer—they are prepared by the chemical modification of the natural polymer cellulose. In general, these materials are soluble in ketones and esters, will slightly soften in alcohol, and will decompose in strong acids and bases. They do, however, have good resistance to oil and grease and they have very good molding characteristics. Cellulose acetate butyrate (CAB) can withstand sharp blows without shattering, has exceptional clarity, and can be colored in a wide range of transparent to translucent colors. It is used in warning lights, light filters, instrument panels, protective cases, tool handles, tubing, and developing tanks. Ethyl cellulose (EC) has good low temperature impact resistance, performs well at a wide range of ambient temperatures, has good gloss, colorability, clarity, heat resistance, and excellent dimensional stability. It is used in helmets, cases, gears, refrigerators, slides, tubing, and tool handles. Cellulosics are available for extrusion, injection molding, and blow molding, and sheet and film are available.

Chlorinated polyethers. Chlorinated polyether (CP) is a corrosion resistant thermoplastic that is widely used for production equipment in the chemical and processing industries. They provide excellent chemical and corrosion resistance up to 275° F (135° C), and maintain their strength and mechanical properties to 280° F (138° C). Their thermal conductivity is lower than most thermoplastics, and they have excellent creep resistance and unusually low water absorption properties. They are, however, susceptible to attack by fuming nitric and sulfuric acids.

Chlorinated polyethers are available for injection molding, extrusion, and fluidizing bed coatings. Typical applications are pumps, linings, pipes and fittings, valves, tanks, and meters used in the production of chemicals.

Flexible vinyls are included in this category. These may be plasticized polyvinyl chloride (PVC) or polyvinyl plastic (PVB). The properties of PVC were covered in the previous section.

Fluorocarbons. Also called fluoroplastics or fluoropolymers, these materials are based on polymers made with monomers composed of fluorine and carbon. There are several varieties.

Polytetrafluoroethylene (PTFE or TFE) and *fluorinated ethylene propylene* (FEP) resins are relatively impermeable to many chemicals. Only molten alkali metals, gaseous fluoride at high temperature and pressure, and a few halogenated compounds show any effect on these florocarbons. They are strong and tough down to –450° F (–268° C), and are rated for continuous service to 550° F (288° C) for TFE, and 400° F (204° C) for FEP. Other features are excellent weather resistance and electrical characteristics. However, due to its high viscosity, TFE cannot be processed with conventional melt extrusion and injection molding techniques—ram extrusion and paste extrusion are feasible. TFE is available as granular powder for compression molding or ram extrusion, as powders for lubricated extrusion, and as dispersions for dip coating and impregnating. TFE is used in nonlubricated bearings, chemical resistant pipe and pump parts, high temperature electrical parts, wire, and cable. FEP is supplied in pellets for melt extrusion and molding, and as dispersions. FEP is used for wire insulation and jackets, high frequency connectors, microwave components, electrical terminals, and terminal insulators.

Polychlorotrifluoroethylene (PCTFE) resists most corrosive chemicals and has a temperature range of –400 to + 400° F (–240 to + 204° C) with a low rate of thermal expansion/contraction (it also remains flexible and tough at cryogenic temperatures). However, its hardness falls off rapidly after 300° F (149° C). Its permeability to water and other fluids is among the very lowest for any plastic, and ultraviolet and weather resistance is excellent. PCTFE is available for compression injection, transfer molding, extrusion, and dispersion coatings. It is used for corrosion resistant liners, oxygen and liquid oxygen seals, pump stators, valve linings, beakers, test tubes, syringes, printed circuits, self-locking bolts, O-rings, jacketed cables, and many other applications.

Polyvinylidene fluoride (PVDF) is higher in tensile strength, has a lower specific gravity, and lower service temperature (300° F [149° C]) than TFE. Its useful temperature range is –420 to + 300° F (–251 to + 149° C), and it is flame and weather resistant. It is attacked by sulfuric acid. PVDF resins are available for compression, injection, transfer, and blow molding, and for extrusion and dispersion coatings. It is used in processing tanks, valves, tubes, pump liners, and pump impellers. Films and coatings are used in the packaging industry.

Ethylene-tetrafluoroethylene (ETFE) has a molding rate faster than the other fluoropolymers, high flex life, and exceptional impact strength, even at very low temperatures. Its thermal properties are good, with a long term continuous service temperature of 300° F (149°C), and intermittent service ceiling of 400° F (204° C). Electrical properties are outstanding, and flammability rating is very good. ETFE resins are available for extrusion and injection molding, and are available as fiber and film. It is used in labware, valve liners, and as electrical insulation on wire for most applications.

Ethylene-chlorotrifluoroethylene (ECTFE) has high impact resistance and tensile strength. Its useful properties are available from cryogenic temperatures to 325° F (163° C) and it is highly resistant to corrosives even at elevated temperatures. Electrical properties are good, as are weather and abrasion resistance. ECTFE is available in pellet and powder forms and can be extruded, injection, transfer, compression molded, as well as dispersion coated, rotationally molded, and powder coated. It is used for wire and cable coatings, and for chemically resistant linings in mixing tanks.

Polyethylenes (low- and medium-density). General polyethylene (PE) characteristics were covered earlier under "Polyethylenes (high-density). Low-density PE has a density range of 0.910–0.925 gr/cu cm, and a stiffness modulus of 13,000–30,000 PSI (90–207 MPa). These materials have high impact strength but relatively low heat resistance—the maximum recommended continuous service temperature is 140° F (60° C). Usually available in cube and pellet form, it is now available in powders. Medium-density PE has a density range of 0.926–0.940 gr/cu cm, and a stiffness modulus of 31,000–80,000 PSI (214–552 MPa). It has a maximum service temperature of 160, and a stiffness modulus of 31,000–80,000 PSI (214–552 MPa). It has a maximum service temperature of 160° F (71° C). Film and sheeting is the largest single use for low- and medium-density PEs.

General Purpose Materials

These materials are not outstanding in load bearing or environmental resistance, but they feature ease of processing and good appearance factors.

Cellulosics (cellulose acetate and cellulose propionate). Cellulose acetate (CA) has exceptional clarity and is used in photo and x-ray film, safety glasses and shields, optical frames, knobs, and handles. Cellulose propionate (CP) maintains a hard glossy surface, and is used in telephone headsets, pens, typing keys on keyboards, toothbrush handles, and face shields. See the section above on Cellulosics for more information on cellulose plastics.

Polystyrenes (general). General polystyrenes are clear, and can be colored to an infinite range of hues. They are commonly used for indoor lighting. See the section above on "Polystyrenes (modified)" for more information.

Thermoset plastics

Once thermoset (TS) plastics are cross-linked, they will not soften under heat. With the application of heat, they undergo a series of changes that are irreversible. The chemical change that takes place is known as polymerization. Thermosetting molding and casting materials permit a high degree of freedom for enhancing the existing properties of the basic polymers, and the use of reinforcements and fillers are the rule rather than the exception because the tightly cross-linked, high molecular weight thermosetting polymers are inherently weak and easily fractured. They may not be used alone in structural applications and must be filled or reinforced with strengthening materials. The reinforcements and fillers may be organic or inorganic, and may take the form of powders, discrete fibers, or woven or macerated fabric. The particle geometry and the effects of processing play an important role in the final properties of the compound. **Table 1** gives abbreviations (which are routinely substituted for the full name) for popular thermosets and thermoplastics, and **Table 2** provides selected properties for plastics commonly used in manufacturing.

Alkyds. Structurally, alkyd resins are modified polyesters, but the term "polyester" is used almost exclusively for linear polyesters derived from dihydric alcohols and dibasic acids, whereas alkyds (a word derived from alcohol+acid) are formed by the condensation of a dibasic acid or anhydride with a polyhydric alcohol. The formulations for alkyd resins can be extensively varied by the introduction of different polyhydric alcohols and anhydrides. Due to their excellent weatherability, toughness, adhesion, flexibility, and ease of application, alkyd resins have become the major synthetic resin for use in surface coatings. Alkyd molding compounds use an unsaturated polymerizable alkyd as their base resin, and these compounds cure at high speed and require low molding pressure.

Alkyds have excellent dimensional stability, excellent fungus resistance, and are suitable for use at 400° F (204° C) without degradation. They possess excellent dielectric strength, arc, and dry insulation resistance. Their most notable disadvantages are cost, very low impact strength, and difficulty in molding to close tolerances. Fillers can be used to improve these characteristics. Alkyds are available as flakes, ropes, slugs, and sheets, granular powder, or putty, and they are used in electrical insulation, auto and boat bodies, transformer components, and automobile ignition systems.

Allylics. These are among the most versatile, yet unappreciated, of the thermoset plastics. Low pressure is generally adequate for their molding and curing as no water or other volatiles are formed during their polymerization. Diallyl phthalate (DAP) is the most widely used allylic, and it offers good shelf life, excellent laminating qualities, and good surface hardness. Allylics have the best moisture resistance of the thermosets, exhibit very low mold and post mold shrinkage, and the resin is completely inert in the presence of metals. Heat resistance in continuous service (up to 350° F [177° C]) for DAP is outstanding, and it remains stable in storage at ambient temperatures while cure at temperatures above 200° F (93° C) is normal and rapid. Common fillers include Dacron for improved shock and moisture resistance; Orlon for the same properties but lower heat resistance; and Nylon to increase shock and wear resistance at slight sacrifice to moisture absorption properties.

Compression transfer molding compounds are available in a range of colors. Typical applications include encapsulating shells, insulators, decorative laminates, heat resistant handles, protective insulating coatings, and air ducts.

Amino resins. Amino plastics are obtained by a condensation reaction between formaldehyde and such compounds as urea, melamine, dicyandiamide, ethylene urea, and sulfonamide. The resins with the lowest formaldehyde are used in molding compound production. The urea and melamine amino compounds are the most widely used, and the following is restricted to a review of these two materials.

These materials can be produced in an unlimited range of stable colors, but the melamine products undergo color changes at temperatures above 210° F (99° C) and the ureas will change color at 170° F (77° C). Pronounced color change and blistering occur after one-half hour at 300° F (149° C). These plastics have excellent insulation and arc resistance properties and are classified as self-extinguishing. Physical properties of melamine based products are relatively unaffected between –70 and + 210° F (–57 and + 99° C), while maximum extended temperature range for the ureas is –70 to + 170° F (–57 to + 77° C), and below –70° F urea moldings are subject to embrittlement. Both materials are subject to initial mold shrinkage as well as after mold shrinkage. They are among the hardest plastics available, and their high compression strength provides high resistance to deformation under load. The molding compounds are available in a large range of plasticities, and the rate of cure can be controlled during manufacture. Amino plastics are used in closures for glass and metal containers, buttons, handles, toaster bases, food service trays, ignition parts, and electrical housings and sockets.

Epoxy resins. Epoxy thermoset resins (EPs) are generally derived by reacting bisphenol-A and epichlorohydrin. In fact, approximately 85% of the world's epoxy resins are formed in this way. Other popular forms are epoxidized novolacs, cycloaliphatic resins, and phenoxy resins. The entire family offers an extremely broad capability for blending specific properties through the use of resin systems, fillers, and additives. Epoxies have high dimensional stability over long periods of time and humidity, excellent mechanical, chemical, and thermal shock resistance, excellent tensile, flexural, and compressive strengths, and have a service range of –85 to + 275° F (–65 to + 135° C). Applications are wide and include pressure bottles, oil storage tanks, printed circuit boards, boat bodies, encapsulation material for electrical components, body sealers, and surface coatings.

Phenolics. These are the oldest and least expensive of the thermosetting plastics. They have excellent electrical properties, good impact strength and high flexural modulus, and have high resistance to deformation under load. Most phenolics can operate to temperatures up to 300° F (149° C), and thermal conductivity is low. Colors are restricted to black, dark browns, and tans. Applications include camera cases, electrical sockets, soldering gun handles, small motor housings, distributor caps and rotors, insulation for cables, and pump impellers.

Polyesters. There are a many polyester plastics, and three classifications will be covered here: unsaturated polyester, high temperature aromatic polyester, and thermoplastic aromatic polyester.

Unsaturated polyester. These plastics are produced from an unsaturated dibasic acid and a glycol, and are usually cross-linked with styrene, vinyl toluene, methyl methacrylate, or diallyl phthalate. They can be formed to a wide range of physical properties and may be brittle and hard, tough and resilient, or soft and flexible. They have good weathering and chemical resistance, and fire retardance is achievable. They can be used in open mold casting, vacuum bag molding, die molding, and injection molding. Commonly used in furniture, bowling balls, boats, and vehicle bodies.

High temperature aromatic polyester. These materials have excellent stability to air above 600° F (316° C). It has good thermal conductivity and insulating properties. Machinability of molded shapes is excellent, and it has self-lubricating characteristics. It can be used for self-lubricating bearings and seals, as well as other parts of processing pumps where wear resistance and corrosion resistance are critical.

Thermoplastic aromatic polyester. These polyesters are produced in pellet form for injection molding, general purpose extrusion, film extrusion, and extrusion coating. Powders are available for electrostatic spray coatings. They are tough, even at low temperatures, and have outstanding surface hardness, lubricity, and abrasion resistance. Dimensional stability is excellent. Commonly used in pump components, gears, spray nozzles, sunglass lenses, alternator components, and control knobs.

Polyurethanes. Polyurethane plastic (PUR) products are available as flexible foams, rigid foams, and as an elastomer. Generally, flexible urethane foams are prepared commercially from polyether or polyester resins, diisocyanates, and water in the presence of catalysts. The reaction between the water and the isocyantes liberates carbon dioxide which functions as a blowing agent to create an open cellular structure. By juggling reactants, catalysts, and emulsifiers, either rigid or flexible foams with a wide range of properties can be created. *Flexible foams* are effective sound deadeners and energy absorbers. The polyether type is slightly more flammable than the polyester version, but both can be self-extinguishing with the addition of retardants. *Rigid foams* can also be made self-extinguishing, and have excellent low and high temperature insulating characteristics. Their dimensional stability is good, but they can be attacked by strong acids and bases. Foams are used in cushions, upholstery, padding, clothing, gaskets, insulation in cabin ventilating systems, nonsinkable boats and pontoons, and refrigerator insulation. PUR RIM (reaction injection molding) foams have been widely used in automobile bumpers. *Elastomers* are used in tires, seals and gaskets, bushings, soles and heels, and rollers used in printing. Polyurethane elastomers have the highest abrasion resistance of all elastomers, and higher load bearing capacity than other elastomers of comparable hardness. Their low temperature properties are excellent and continuous service temperature range is –30 to + 180° F (–34 to + 82° C).

Silicones. Silicones come in a very wide variety of forms, including rigid thermosetting, laminating, and molding formulas that may be available in fluids, elastomers, gels, or resins. They have an expansive temperature ranging from –70 to + 500° F (–57 to 260° C), but some varieties can withstand even more extreme temperatures. Although these materials have superior temperature ratings and resistance to chemicals, their tensile strength, and abrasion resistance are poor, and they are expensive.

Process technology

A major factor contributing to the growth of plastics use is the innovations made in processing technology. For example, injection molding of thermosetting materials and the use of structural foams formed from existing injection molding equipment have become commonplace. Each process method is virtually unique in its operation and tooling, and secondary processes may be required. The most common primary processes are outlined in **Table 3** and discussed in more detail in this section. Fabrication and finishing are covered in the next section.

Blow Molding. Blow molding rapidly produces thin-walled, hollow thermoplastic parts. There are two different processes in common use: extrusion blow molding (EBM), and injection blow molding (IBM).

Extrusion blow molding is the most common of the processes, and constitutes approximately 75% of total production. It is a hybridization of basic extrusion and molding techniques and rapidly produces thin-walled, hollow thermoplastic parts in the following sequence. 1) The extruder produces a thin, hollow cylinder of molten plastic (called a "parison"). 2) The parison is then captured between the chilled halves of a split mold. 3) Compressed air is introduced and forces the molten parison to expand against the walls of the mold. 4) The formed part is cooled and released from the mold. Flash forms when molten material is forced between the mold halves, and it can be reclaimed and reused. Systems are comprised of a hopper where the material is prepared, an extruder, an extruder head incorporating a die, a clampable mold, and an air-injector capable of pressure of 100 PSI or greater. The process is excellent for producing thin-walled parts such as bottles and other containers, automobile heater ducts, and packaging units. Materials suitable for the process include polyethylene (LDPE and HDPE) polyvinyl chloride (PVC), polypropylene (PP), polyethylene terephthalate (PET), polycarbonate (PC), polyolefins (PO), and polyamide (PA).

Injection blow molding incorporates the principles of injection molding and blow molding. It uses a three-stage process on a rotary indexing table. At the first stage, the parison is preformed. At the second, the preform is blown into the finishing container, and the finished part is removed in stage three. Much higher rates of production can be achieved with EBM, and tooling costs are lower. However, with IBM there is no flash to trim (because the molds are designed to extremely tight tolerances), and wall thickness control is better, which results in less scrap. IBM also provides better surface finishes and superior distribution of material, resulting is more reliable formation of critical neck areas. Materials that can be used include LDPE, HDPE, PP, PC, PET, polystyrene (PS), styrene acrylonitrile copolymer (SAN), polyethylene-vinyl acetate (EVA), polyurethane (PU), polyacrylonitrile (PAN), and polyethylene naphthalate (PEN).

A recent and growing method is *sequential blow molding* (SBM). The perceived advantage is that, unlike EBM, it will be able to shape sharply three-dimensionally shaped parts, while simultaneously reducing flash and pinch lines. This is achieved partially through the use of programmable manipulators or six-axis robots that are used to manipulate the extruded parison. Improved wall thickness control is also claimed through the use of a radial wall thickness system developed by Krupp Kautex. These machines are currently in limited use.

Casting. Casting differs from other processing methods in that external pressure is not required. Gravity and heat settle and harden the mass. Both thermoplastic and thermoset resin liquids or melts can be poured into casting molds to set to a solid by curing with catalysts, at room or higher temperatures (thermoset resins), or by chilling (thermoplastic resins). Casting of film is performed on a continuously moving turntable or belt, or by precipitation in an aqueous chemical bath. In the turntable or belt process, the liquid resin is spread to the desired thickness and, as it dries, the film is stripped off. The most extensively used casting materials are the epoxies, phenolics, polyesters, stryenes, acrylics, and silicones. Mostly, the process is used for rods, tubes, cylinders, sheets, and slabs for further fabrication into parts, and for encapsulation of electrical components. Some advantages of casting are: better optical and strength properties can be achieved with casting than with extrusion or molding; molds may be made from low cost materials such as wood, plaster of paris, or sand-resin formulations; materials can be compounded with reinforced fillers prior to casting. Disadvantages are: high cost due to the amount of hand labor required. Incomplete removal of air bubbles can lead to high scrap rates; relatively few liquid plastics are available for casting.

Calendering. This process is restricted to thermoplastic materials—primarily PVC, but cellulosics, and styrenes, ABS, and EVA can also be calendered—and is used to produce continuous sheets and film. It is also used for applying plastic coatings to textiles, paper, or other supporting material. In the process, the plastic material is softened by heat, plasticizer, and possibly some solvent, and is then fed between a series of several pairs of large heated revolving rollers that squeeze the material between them into the desired

Table 3. Primary Methods of Plastics Processing. *(Source, MIL-HDBK-755.)*

Process Method	Description	Remarks
Blow Molding	Shaping a thermoplastic material into a hollow form by forcing material into a closed mold by internal air pressure. Normally initiated with a form made by extrusion; it is now possible to injection mold the initial form.	High production rate process for manufacturer of hollow items such as bottles, tanks, and drums; uses relatively inexpensive molds.
Casting	Forming a solid part by pouring a liquid resin into a mold and removing the part following curing or solidification.	Practical for small production where inexpensive tooling is employed; primarily used for encapsulation of components or for parts having thick sections.
Compression Molding	Principally used for thermoset parts formed by placing material into an open mold and curing the part by use of heat and pressure after the material is confined with a plunger.	Principally for thermosetting materials of simple shapes having heavy cross sections, high impact fillers, or large deep-draw areas. Cycles are relatively slow, and finishing of parts is required.
Extrusion	A process for making continuous forms by forcing a thermoplastic material in the plastic state through an orifice or die. Thermosets are now somewhat adaptable to extrusion.	Limited to the continuous, low-cost production of rods, tubes, sheets, or other profile shapes; capable of extruding solid, foamed shapes.
Injection Molding	A basic molding process wherein a heat-softened thermoplastic material is forced under pressure into a closed mold. Upon cooling and solidification, the part is ejected. Now widely used as a process method for thermosets and for thermoplastic forms.	A major, high production rate process used for manufacture of intricate shapes. Process maintains good dimensional accuracy. Mold costs are relatively high.
Reaction Injection Molding	A process that involves the mixing of two liquid components, injecting the liquid stream into a closed mold at relatively low pressure, and removing the part following cure.	An injection-molding-type process used for the manufacture of large, solid, or cellular parts. Low pressure and ambient temperature materials eliminate the need for expensive machinery.
Rotational Molding	The forming of a hollow part from thermoplastic resin within a closed mold by heating, rotating, and cooling the material for subsequent removal.	Useful for manufacture of hollow forms with practically no limit to size and shape. Either rigid or flexible parts can be made with inexpensive tooling.
Thermoforming	The forming of thermoplastic sheet material into a three-dimensional shape by heat-softening the sheet and forcing it to conform to the shape of the mold by pressure or vacuum; followed by cooling.	Limited to simple, three-dimensional shapes for signs, trays, cups, domes, and packaging. Tooling is relatively inexpensive.
Transfer Molding	A basic molding process for thermosets; part is formed by transferring the molding compound, which has been heated in a loading well, under pressure into a closed mold where curing takes place.	A process used exclusively for thermoset materials in the manufacture of intricate shapes and parts with fragile inserts. Process maintains good dimensional accuracy.

thickness. The space between the last series of revolving rollers controls the thickness of the end item, film, or sheet. Thicknesses range from 0.00315 to 0.315 in. (0.08 to 0.80 mm) as rolled from the calender, but can be reduced to as thin as 0.00118 in. (0.03 mm) after stretching. Widths can be as wide as 118 inches (3,000 mm) with existing machinery. Tolerance control is very good with this process, and patterned or textured effects are obtainable at low cost.

Compression Molding. In compression molding, a plastic material, usually partially preformed, is placed in a heated mold cavity. The mold is then closed and heat and high pressure are applied to the plastic, thereby forcing the plastic to fill the mold cavity while simultaneously undergoing a chemical reaction that cross-links the polymer chains and hardens the plastic into a finished form. Low cost, quality parts are produced with compression molding. Normally, thermosetting resins in the form of sheet molding compounds (SMC) or bulk loading compounds (BMS), diallyl phthalate (DAP), melamines, phenolics, ureas, silicones, or epoxies are shaped with this process, but glass mat thermoplastics (GMT) such as Azdel, and long fiber thermoplastics (LFT) can also be compression molded. While some materials can be compression molded at room temperature, the cycle is lengthy, so the process is usually performed in the temperature range of 250 to 400° F (121 to 204° C). Compression pressures in the range of 5,000 PSI (34.5 MPa) are normal, but both temperature and pressure will vary with material. The process is usually used to form large parts such as housings, furniture, vehicle body panels, appliance cabinets, and washing machine agitators. There is little waste and finishing costs are low, but intricate parts with undercuts or delicate details are not practical. Tolerances closer than ± 0.005 in. (0.127 mm) are difficult to achieve.

Extrusion. In extrusion, plastic granules or powder are fed through a hopper to a heated plasticizing cylinder where they are melted, then driven by rotating screws through a shaped die that determines the shape of the extruded part. The process is used for the fabrication of pipe, flexible tubing, rods, sheets, tubes, bars, house siding, and other shapes. Single screw extruders are most common. They normally use pellets, screw rotation rates range from 20 to 200 rpm, and screw shapes are varied depending on material application. Twin screw units are sometimes counter-rotating and usually use powders. They incorporate screw geometries that can produce back pressure, knead the material, and affect the volume of the extrusion. Materials commonly used in extrusion are PVC, ABS, polystyrene, thermoplastic elastomers, nylons, and polyolefins.

Extrusion, Blown Film. This process is used in the production of plastic film. Polyethylene (low- to high-density) are the polymers most often used, but other resins such as polypropylene and polycarbonate are also used. In the process, plastic particles are fed into a hopper, melted, and then metered at a controlled rate through an extruder to a ring shaped blown film die that releases the melt in a bubble form. An air stream is fed into the bubble as it is pulled (usually vertically, though some machines are horizontally inclined) by nip or pinch rollers. Before the bubble passes through the rollers, it is flattened by a collapsing frame. The film then is "pinched" to size by the rollers and carried to a spool where it is accumulated on a large roll. Blown films are used for packaging, including food coverings, and in agricultural and building materials. A somewhat similar system, known as *Flat Web Extrusion*, is a continuous system for forming plastics into belt shaper sheet or film. It employs a calender takeoff system for producing sheet and film in thicknesses ranging from 0.007 to over 0.500 in. (0.178 to over 12.7 mm).

Injection Molding. Injection molding is capable of producing complex three-dimensional parts to high quality and tolerance standards. The process is characterized by the fact that the molding mix is preheated to a temperature high enough for it to become a quasi-liquid. When the material reached the desired temperature, a reciprocating screw forces the melt through a nozzle into a controlled temperature closed mold. Basic machines use the screw to force the melt into the mold, but machines intended for high production rates use a plunger device to hasten the injection. Pressures of up to 30,000 PSI (207 MPa) are used to insure that the mold is fully filled. It is common to use a multicavity mold to form several parts simultaneously. The cavities are joined by sprue and runners that fill with scrap which can usually be returned to the process. Thermoplastics are most commonly used in injection molding, but some thermosets (epoxy, melamine, DAP, and others) containing high volumes of fillers to reduce cost and shrinkage are also used. Another development in the field is *Powder Injection Molding* (PIM) for the manufacture of metallic or ceramic moldings. With the exception of aluminum, almost any metal that can be powderized can be used in PIM. Polyolefin wax mixtures and polyacetals are normally used as binders. Typical applications for plastic injection molded parts include cases for appliances, handles, knobs, gears, plumbing and hardware, steering wheels, vehicle grills, fan blades, and fasteners. **Table 4** provides process guidelines for injection molding.

Reaction Injection Molding. Reaction injection molding (RIM) requires the high pressure impingement mixing and polymerization of two or more reactive liquid

Table 4. Guidelines for Injection Molding of Selected Polymers.

Polymer	Processing Temperature		Pre-Dry Time		Mold Temperature		Molding Shrinkage
	°F	°C	°F/hours	°C/hours	°F	°C	
ABS	392-500	200-260	158-176/2	70-80/2	122-176	50 to 80	0.4-0.7
ABS foamed	392-500	200-260	158-176/2	70-80/2	50-104	10 to 40	0.4-0.7
ASA	428-500	220-260	158-176/2-4	70-80/2-4	122-185	50 to 85	0.4-0.7
CA	356-428	180-220	176/2-4	80/2-4	104-176	40 to 80	0.4-0.7
CAB	356-428	180-220	176/2-4	80/2-4	104-176	40 to 80	0.4-0.7
CP	374-446	190-230	176/2-4	80/2-4	104-176	40 to 80	0.4-0.7
EVA	266-464	130-240	-	-	50-122	10 to 50	0.8-2.2
PB	392-554	200-290	-	-	50-140	10 to 60	1.5-2.6
PBT	446-536	230-280	230/4	120/4	104-176	40 to 80	1.0-2.2
PC	518-716	270-380	230-248/4	110-120/4	176-248	80 to 120	0.6-0.7
PE-HD	392-572	200-300	-	-	50-140	10 to 60	1.5-3.0
PE-HD foamed	392-500	200-260	-	-	50-68	10 to 20	1.5-3.0
PE-LD	320-518	160-270	-	-	68-140	20 to 60	1.0-3.0
PES	608-734	320-390	320/5	160/5	212-320	100 to 160	0.6
PET	500-572	260-300	230/4	120/4	266-302	130 to 150	1.6-2.0
PMMA	374-554	190-290	158-212/2-6	70-100/2-6	104-194	40 to 90	0.3-0.8
POM	356-446	180-230	230/2	110/2	140-248	60 to 120	1.5-2.5
PP	392-572	200-300	-	-	68-194	20 to 90	1.3-2.5
PP foamed	392-554	200-290	-	-	50-68	10 to 20	1.5-2.5
PPE/PS	500-590	260-310	212/2	100/2	104-230	40 to 110	1.5-2.5
PS	338-536	170-280	-	-	50-140	10 to 60	0.4-0.7
PSU	644-734	340-390	230/5	120/5	212-320	100 to 160	0.4-0.7
PVC-P	320-374	160-190	-	-	68-140	20 to 60	0.7-3.0
PVC-U	338-410	170-210	-	-	68-140	20 to 60	0.4-0.8
SAN	392-500	200-260	185/2-4	85/2-4	122-176	50 to 80	0.4-0.6
SB	374-536	190-280	-	-	50-176	10 to 80	0.4-0.7

Notes: U = unplasticized P = plasticized

components, the injection of the mixture into a closed mold to form a solid or foamed part. Polyurethanes are normally used in the process, but other potential materials are nylon, epoxy, and polyester. Three RIM processes are usually identified. 1) Standard RIM does not use fillers or reinforcements. 2) Reinforced RIM (known as RRIM) uses fillers such as glass fiber or mica. 3) Structural RIM (SRIM) injects the material onto and around a glass fiber preform encased in a mold to produce a high strength and durable part. RIM is used to produce vehicle body and instrument panel parts, wheel covers, and seat frames. Parts are typically light in weight, have good impact strength, and can be produced in a wide variety of shapes and sizes.

Rotational Molding. Rotational molding is a relatively simple process used to manufacture hollow objects from thermoplastics and, to a lesser extent, thermoset materials. The solid or liquid polymer is placed in a mold which is first heated, then cooled while being rotating around two perpendicular axes simultaneously. In the initial phase of the heating process, a porous skin is formed on the mold surface which gradually melts to form a homogeneous layer of uniform thickness. When molding a liquid material, the liquid tends to flow and coat the mold surface until the gel temperature of the resin is reached. The mold is then cooled with forced air, water spray, or both. The mold is then opened, the finished part removed, and the mold is recharged for the following cycle. About 80% of the materials used in rotational molding are low- to high-density and cross-linkable polyethylenes (LDPE, LLDPE, HDPE). The other 20% is a mix of polyvinyl chloride, polycarbonate, polypropylene, and polyesters. Early emphasis on rotational molding was in the toy industry, and almost every plastic hobbyhorse and squeeze toy is made with the process. It is also used to produce refuse containers, boat and auto parts, and chemical storage tanks.

Thermoforming. There are many techniques for thermoforming plastics, and seven of them will be discussed here. Each of the techniques involves the heating of plastic sheet or film material until it becomes limp, and then causing the material to slump over a mold or form. Vacuum, air, or mechanical pressure is used to create close conformity. Sheet thermoforming is used to produce such items as luggage cases, hot tubs, refrigerator door liners, and vehicle interior panels. Typical materials used are ABS, PVC, PE, PP, PC, polysulfone (PSU), and styrene acrylonitrile copolymer (SAN).

Straight vacuum forming. The sheet is clamped in a stationary frame, then heated, and the frame is clamped to the mold cavity. A vacuum is then applied through holes in the mold cavity to draw the plastic sheet into contact with the inner mold surfaces. It is then chilled and the part removed.

Drape vacuum forming. In this process, the heated sheet is draped over the male form, and a vacuum is used to pull it down onto the form. Relief maps and embossed and texture mats are made with this process.

Male form forced above sheet. The heated sheet is secured in a frame and a male plug is descended from above by means of hydraulic pressure. Partial forming to the shape of the male plug takes place, and then vacuum applied through holes in the plug draws the sheet around the plug. The formed sheet is then chilled.

Vacuum snap-back forming. The heated plastic sheet is positioned over a cavity and partially pulled into the cavity by the application of a vacuum. A male plug is then moved to a predetermined position and the vacuum is released, which allows the stretched sheet to snap back against the male plug. Vacuum is then applied through holes in the plug. The formed sheet is then chilled.

Plug and ring forming. The heated sheet is placed over a ring and clamped into position. A male plug, mounted above, is forced into the sheet, stretching it to conform to the shape of the plug. The formed sheet is then chilled.

Air pressure forming. Many variations are used to make use of air pressure (higher than that available from a vacuum) for forming. In many of these operations, this is accomplished by placing the mold in an autoclave where it can be heated by placing vacuum on one side of the heated sheet, and high pressure on the other. In one instance, the heated sheet is first clamped into position above the cavity of the mold. A cored plug then pushes the heated sheet into the cavity and tightly seals the mold. Air pressure is then introduced through holes in the plug, pushing the sheets against the sides of the female mold. Holes in the latter allow the air to escape from the underside of the sheet. In another case, the sheet is clamped over the female cavity mold, heated to the softening point, and then air is applied on the underside to blow the sheet into a hemispherical shape of the desired size. Vacuum is then applied within the mold cavity, inverting the blister and sucking it into contact with the surfaces of the female mold where it is hardened.

Matched metal mold forming. This process is somewhat similar to compression molding. A heated sheet of plastic resin impregnated fabric or compound thermoset resin is placed between matched male and female dies. Hydraulic pressure is then applied on the two parts of the mold to compress the sheet within the opening between the two halves. The sheet is then cooled by water coils embedded in the metal molds.

Transfer molding. Transfer molding is a combination of injection and compression molding. The plastic (thermosetting resins) is first melted in a heated cylinder, then fed by plunger through sprues and runners to a closed mold where it is compressed into the desired shape and polymerization is completed. Transfer molding differs from compression molding as follows: in compression molding the molding material is charged directly to the mold, but in transfer molding it is first heated to melt temperature in a separate chamber. The essential difference between transfer and injection molding is that in transfer molding, only enough material is heated at each cycle to fill the mold cavities, whereas in injection molding a reservoir of molten plastic is maintained in the heat cylinder. In fact, transfer molding is usually chosen when the designated resin cannot be injection molded. Production rates are high, and dimensional accuracy is good. Molds are usually elaborate, and often encapsulate sensitive electronic components, but waste is comparatively high because runners and sprues cannot be reused. Epoxies and phenolics are common materials, but some elastomeric materials can also be processed. The process is popular for making semiconductors and resistors.

Machining and fabrication of plastics

Practically all plastics, except for the very softest, can be machined with conventional machinery. However, allowances must be made for the greater heat sensitivity of plastics (compared to metals), and for the greater thermal expansion of plastics. Because plastics are poor heat conductors, they sustain high heat build-up during machining, which, in the case of thermoplastics, can cause softening and heat distortion. The high cutting speeds used with metals should be avoided. Instead, a light cut and slow feed are recommended. Coolants or cutting fluids can be used: plain or soapy water, cool air, and less frequently mixtures of water and oil are permissible. Tools should be designed to provide adequate clearance so as to clear away plastic chips as fast as they are formed.

Because thermosetting plastics do not suffer to the same degree from heat distortion, their machining is more straightforward. Coolants may be required to keep surfaces from burning, but chip removal can be done with a vacuum hose as chips are reduced to powder form by machining.

Plastics are routinely machined by grinding, turning, sawing, milling, routing, drilling, reaming, and tapping. When turning plastics, positive geometry inserts with ground peripheries are recommended to help reduce material build-up on the cutting tool. Rake angles of 0 to 15°, and clearance angles of 5 to 15° are appropriate. Cutting speeds average 655 to 1,640 sfm (200 to 500 smm) with a rate of advance of 0.0040 to 0.020 in./rev (0.1 to 0.5 mm/rev).

For drilling, low helix drills having two flutes and a point angle in the range of 70° to 110° are well suited for drilling holes up to one-inch in diameter in plastics, and peck drilling is preferred as it will allow for the removal of swarf. For larger holes, a low helix drill with a 110° point angle and 9° to 16° lip clearance is recommended, and a pilot hole one-half inch in diameter should be drilled before it is enlarged to full size. Cutting speeds for drilling range from 160 to 325 sfm (50 to 100 smm), and rate of advance should be approximately 0.0040 to 0.014 in./rev (0.1 to 0.3 mm/rev).

Sawing (circular and band) speeds of approximately 10,000 sfm (3,000 smm) with a feed rate of 0.0040 to 0.014 in/tooth can be used for most materials. The tooth rake angle should be 0 to 15°, and the clearance angle 10 to 15° for circular saws. For band saws, the rake angle should be 0 to 5°, and the clearance angle 30 to 40°. Plastics are also easily cut by lasers. A 250 W CO_2 laser can typically cut 5/32 inch (4 mm) thick thermoplastics at speeds ranging from 40 to 85 inches (1,000 to 2,100 mm) per minute.

Plastic actually dulls the edges of cutting tools more rapidly than metal, so frequent resharpening is usually required. Tool steel grade tools are normally satisfactory for plastics, but glass reinforced materials may often require tougher tools.

Many plastics can be welded with hot gas (usually nitrogen) equipment. The temperatures should be 600 to 640° F (316 to 338° C) for high-density polyethylene and polystyrene, 640 to 660° F (338 to 349° F) for polypropylene, and 660 to 700° F (349 to 371° C) for acetal. Filler rods should be of the same, or similar, material as the weldment, and range from $^1/_{16}$ to $^3/_{16}$ inch (1.5875 to 4.7625 mm) in diameter. Round rods are preferable, and the diameter should be constant through the length of the rod. Good plastic welds resemble those produced on metal by electric arc welding. No preparation is required for filet welds. For butt welds, the two pieces should be welded as in arc welding, but at a smaller angle (60 to 80°) and with no sharp edges. A sealing run on the reverse side will assure higher tensile strength in butt welds. In plastics welding, a simple bonding process takes place between filler rod and parent material, since only the actual mating surfaces melt. Other parts, such as the center of the rod, remain relatively unaffected and rigid. This allows for the exertion of slight pressure on the rod to force it into the weld and thereby combining the melted surfaces into one homogeneous mass. To assure uniform coalescence, the surfaces and the filler rod should be preheated before insertion into the weld. Overheating must be avoided as it will degrade the material. Tensile strength for butt welds may be as much as 90% of the parent tensile strength. Filet weld strength is somewhat lower, but can be strengthened with reinforcements.

Lasers, which are commonly used to cut plastics, can also be used for welding. CO_2 and Nd:YAG lasers are both used to produce good welds, especially with robot systems. Microwave welding is another recent development. It involves placing a conducting polymer between the sections to be welded and then placing the assembly in a microwave field, and then applying pressure as the mated pieces cool. This system can be employed on three-dimensional joints and assemblies and offers exciting potential in the vehicle and appliance industries.

Several other methods for joining plastics with heat are used for special applications. Most call for heating the material with metal plates, or electric current, or friction to sufficiently melt thermoplastics for joining. With the exception of spin (friction) welding, which is restricted to circular sections, these methods cannot approach the tensile strength of joining with the methods described above. They can, however, be useful for joining panels tightly enough to provide leak free joints.

After fabrication or molding, most plastic parts require some form of finishing, which may consist of removing tool marks, mold flashing, or polishing. Care in machining will minimize the amount of finishing required. Thermoplastics such as the acetates, butyrates, and acrylics that have been dulled by buffing, sanding, or polishing, can often achieve a high gloss finish by dipping them in acetone, butyl acetate, ethylene dichloride, or other suitable solvent. Wiping them with these solvents can achieve the same result.

Compression mold design and construction

Modern, intricate mold design is often performed with computer aided design technology, as there are many variables to be considered. The first consideration is whether the mold will have manual, semiautomatic, or automatic operation. Automatic molds are operator-free, but require sophisticated controls. Operator involvement varies with semiautomatic molds, but manual versions require the operator to handle all phases of operation. Injection and transfer molds are almost universally designed for semi-automatic or automatic operation, but molds used in compression molding vary from the simplest hand-operated types to fully automatic molds.

Well-designed, durable molds are essential to any successful molding operation. The length of the run, the rate of production, and the accuracy of the detail and dimension of the part are some of the major factors to be considered. Additional consideration must be given to the following variables. 1) The type of plastics material being used. This is essential for predetermining the bulk factor and mold shrinkage. 2) The size of the press, especially the size of the platens and the maximum space between them. 3) Molding pressure to be used. 4) Determining the dividing point between the mold halves, and where the parting line will appear on the molded part. 5) Size and shape of the part, which will determine the number of cavities per mold. 6) Inclusion of any inserts (usually metal) that will require provisions for side-core pins and unscrewing threaded pins. 7) Heat requirements for proper spacing of heating coils or wires needed to rapidly and uniformly heat the mold. 8) Ejector assembly for removing molded parts.

Molds for compression molding consist of a cavity, which forms the outside shape of the part, and a plunger which informs the inside shape. Usually, the lower half is secured to the stationary platen, and the upper half (normally the plunger) is secured to a moving platen that is led by alignment guide pins when it is lowered. The material is loaded in the cavity and is compressed and shaped by the plunger upon closing. The material is confined within the area between the force and cavity while it hardens under heat and pressure for the time required to permanently form the part. Provisions for heating the mold can be made directly in the mold, or by use of heated platens mounted on the press. Ejection pins or forced air are usually used to remove the part from the mold.

Generally, there are five basic designs used in compression molding.

Positive type. The plunger descends into the filled cavity. The pressure is almost entirely directed at the material, but tight clearances of 0.0015 to 0.005 in. (0.04 to 0.13 mm) between the cavity and the plunger allow very little material to escape. The material that does escape becomes flash that must be removed from the part. Flash can be controlled by careful measurement of the amount of mold material placed in the cavity. An additional disadvantage of this type of mold is that the vertical flash excreted from the mold leads to eventual degradation of the cavity sides, making part ejection difficult.

Flash type. The procedure is similar to the positive type, except that the flash is forced out of the lands between the cavity and plunger at the top of the cavity. Flash is horizontal, and the flash does not inflict damage to the mold. This method is normally used on parts with shallow thickness such as dinnerware plates.

Semi-Positive type. As the two halves of the mold begin to close, flash is allowed to escape. Then, as full pressure is reached, clearance between the plunger and cavity is minimized, resulting in very thin and relatively easily removed vertical flash at the parting line. This mold combines the advantage of free material flow (flash mold) with the ability to produce dense parts (positive mold).

Land Plunger type. Very similar in appearance to semi-positive molds, but the land does not contact the mold. It is stopped by external landing bars approximately 0.001 in. (0.025 mm) before contact. Flash escapes through the gap, leaving a horizontal flash at the parting line.

Split Wedge type. These molds are more complicated and are used when the part has projections or undercuts. The cavity is in two or more parts, with the inside surface of the part on one part and the outside on the other. As the mold closes, the outside impression presses toward the inside impression through clamping action. This is accomplished with spring loaded pins or other mechanical sliding devices.

Fiber Reinforced Resins

Carbon fibers are produced by the pyrolysis (a chemical change induced by thermal action) of organic precursor fibers such as rayon, polyacrylonitrile (PAN), and pitch in an inert atmosphere. The term carbon fiber is often used interchangeably with “graphite,” but carbon fibers and graphite fibers differ in the temperature at which the fibers are made and heat treated, and in the amount of carbon produced. Carbon fibers are typically carbonized at 3,400 to 5,440° F (1,900 to 2,000° C), and assay at more than 99% elemental carbon. Important carbon and other reinforcement fibers, as well as the resin materials used in composite materials, are discussed in this section. Much of the material in this section, and most of the property values, are from MIL-HDBK-17-3E.

Reinforcement fiber materials

Aramid. Kevlar ™ aramid fiber, an organic fiber with high tensile modulus and strength, was introduced by the Du Pont Company in the 1970s. It was the first organic fiber to be used as reinforcement in advanced composites. Today, it is used in various structural parts including reinforced plastics, ballistics, tires, high performance vehicle components, ropes, cables, asbestos replacement, coated fabrics, and protective apparel. Aramid fiber is manufactured by extruding a polymer solution through a spinneret. It is available in many forms, including continuous filament yarns, rovings, chopped fiber, spun-laced sheet, and thermoformable composite sheets.

Important generic properties of aramid fibers include low density, high tensile strength, high tensile stiffness, low compressive properties (nonlinear), and exceptional toughness characteristics. Its density is 0.052 lb/in.^{3} (1.44 gr/cm^3), which is about 40% lower than glass and about 20% lower than commonly used carbons. Aramids do not melt—they decompose at 900° F (500° C). Tensile strength of various aramid yarns varies from 500 to 600 ksi (3.4–4.1 GPa). The nominal coefficient of thermal expansion is 3×10^{-6} in./in/°F (-5×10^6 m/m/°C) in the axial direction. Aramid fibers, being aromatic polyamide polymers, have high thermal stability and dielectric and chemical properties. Their excellent ballistic performance and general damage tolerance is derived from fiber toughness, both in fabric and composite form.

Composite systems reinforced with aramid have excellent vibration dampening characteristics, and they resist shattering upon impact. Continuous service temperatures range from –33 to + 390° F (–36 to + 200° C). These composites are ductile under compression or flexure, but ultimate strength in this condition is lower than glass or carbon composites. Composite systems reinforced with aramid are resistant to fatigue and stress rupture, and epoxy reinforced aramid (60% fiber volume) specimens under tension/tension fatigue testing survive 3,000,000 cycles at 50% of their ultimate stress. Thermoplastic composites reinforced with aramid have recently been developed. These systems have exhibited mechanical properties comparable to those of thermoset systems, plus offer lower processing costs and the potential for bonding and repair. Properties are shown below.

Nominal Composite Properties Reinforced with Aramid Fiber (60%)					
		Thermoset (epoxy)		*Thermoplastic*	
Tensile Property	*Units*	*Unidirectional*	*Fabric* [1]	*Unidirectional*	*Fabric* [1]
Modulus	Msi (GPa)	11 (68.5)	6 (41)	10.5–11.5 (73–79)	5.1–5.8 (35–40)
Strength	ksi (GPa)	200 (1.4)	82 (0.56)	180–200 (1.2–1.4)	77–83 (0.53–0.57)

Note: [1] Fabric aramid volume 40%.

Aramid fibers are available in many forms with different fiber modulus. Kevlar ™ 29 has the lowest modulus and highest toughness, and is used primarily in ballistics, protective apparel, ropes, tires, etc., and in composites where maximum impact and damage tolerance is critical and stiffness is less important. Kevlar ™ 49 is predominately used in reinforced plastics—in both thermosetting and thermoplastic resin systems. It is also used as core in fiber optic cable, high pressure rubber hoses, conveyor belts, etc. Recently, Kevlar ™ 149 with an ultra-high modulus has become available. Nominal properties for these fibers are given below.

		Type of Kevlar ™		
Tensile Property	*Units*	*29*	*49*	*149*
Modulus	Msi (GPa)	12 (83)	18 (124)	25 (173)
Strength	ksi (GPa)	525 (3.6)	525–600 (3.6–4.1)	500 (3.4)

Aramid fiber is available in various weights, weave patterns, and constructions: from very thin (0.0002 in. [0.005 mm]), lightweight (275 gr/m^2), to thick (0.026 in. [0.66 mm]), heavy (2.8 gr/m^2 roving). Woven fiber prepreg is the most common form used in thermoset composites. Chopped aramid fiber is available in lengths of 6 mm to 100 mm, with the shorter lengths being used primarily to reinforce thermoset, thermoplastic, and elastomeric resins in automotive brake and clutch linings, gaskets, and electrical parts. Aramid short fibers can be processed into spun-laced and wet-laid papers that are useful for surfacing veil, thin-printed wiring boards, and gasket material. Uniform dispersion of aramid short fiber in resin formulations is achieved through special mixing methods and equipment.

Glass. Although the rapid evolution of carbon and aramid fibers has resulted in stronger and lighter materials, glass composite products have prevailed in certain applications. Cost per weight or volume, chemical or galvanic corrosion resistance, electrical properties, and availability of many product forms are among their advantages. Coefficient of thermal expansion and modulus properties may be considered disadvantages when compared to carbon composites. Compared to aramid composites, the tensile properties of glass are not as good, but ultimate compression, shear properties, and moisture absorption properties are superior.

Typical glass compositions used for reinforcement are electrical/Grade "E" glass, a calcium aluminoborosil mica composition with an alkali content of less than 2%; and chemical resistant “C” glass made from soda-lime-borosilicates and high strength S-2 glass (a low alkali magnesi-silicate composition). Glass roving products (untwisted) type yarns are most often directly finished with the final coupling agents during the filament manufacturing step. Other finishes include Volan finishes, with the most recognized variant, Volan A, providing good wet and dry strength properties when used with polyester, epoxy, and phenolic resins. Saline finishes are also often used on epoxy. Most of these finishes are formulated to enhance laminate wet-out, and some also provide high laminate clarity or good composite properties in aqueous environments. Although other finishes are used in combination with matrix materials other than epoxy, these finishes may have proprietary formulations or varied designations relative to the particular glass manufacturer or weaver.

There are many forms of glass products. The continuous filament forms are continuous rovings, yarn for fabrics or braiding, mats, and chopped strand. They are available with a variety of physical surface treatments and finishes, but most structural applications utilize fabric, roving, or rovings converted to unidirectional tapes. Perhaps the most versatile fiber type used to produce glass fiber forms is “E” glass, identified for electrical applications. “E” glass is available in eight or more standard diameters, ranging from 1.4 to 5.1 mils (3.5 to 13 micrometers), which facilitates very thin product forms. The “S” glasses are identified as such to signify high strength. S-2 glasses are available in only one filament diameter, but S-2 rovings are available in yields of 250, 750, and 1,250 yards per pound (500, 1,500, and 2,500 m/kg).

		Type of Glass	
Property	*Units*	*E*	*S-2*
Density	lb/in^3 (gr/cm^3)	0.094 (2.59)	0.089 (2.46)
Tensile Strength	ksi (MPa)	500 (34,450)	665 (45,820)
Modulus of Elasticity	Msi (GPa)	10.5 (72.35)	12.6 (86.81)
Percent Elongation	–	4.8	5.4
Coeff. Thermal Expan.	10^6 in./in./°F (10^6 m/m/°C)	2.8 (5.1)	1.3 (2.6)
Softening Point	°F (°C)	1,530 (832)	1,810 (988)
Annealing Point	°F (°C)	1,210 (654)	1,510 (821)

Boron. Boron fiber is unmatched for its combination of strength, stiffness, and density. The tensile modulus and strength of boron fiber are 60 % 10^6 PSI and 0.52 % 10^6 PSI (40 GPa and 3,600 MPa), respectively. Thermal conductivity and thermal expansion are both low—the coefficient of thermal expansion is 2.5–3.0 % 10^{-6}/°F (4.5–5.4 % 10^{-6}/°C). Boron is available as a cylindrical fiber in two nominal diameters, 4 and 5.6 mils (10 and 14 micrometers) which have a density of 0.0929 and 0.0900 lb/in.^{3} (2.57 and 2.49 gr/cm^3), respectively. Available in filament or epoxy matrix prepreg form, boron fiber has been used for aerospace applications requiring high strength and/or stiffness, and as reinforcement in selected sporting goods.

Typical End-use Properties of Unidirectional Boron/Epoxy Laminate (60% Volume)			
Property		*Units*	*Value*
Moduli	Tensile, Longitudinal	ksi (MPa)	30 (207)
	Tensile, Transverse		2.7 (19)
Strength	Tensile, Longitudinal		192 (1,323
	Tensile, Transverse		10.4 (72)
	Compressive, Longitudinal		353 (2,432)

Alumina. Continuous polycrystalline alumina fiber is ideally suited for the reinforcement of a variety of materials including plastics, metals, and ceramics. It is prepared as continuous yarn containing a nominal 200 filaments, and supplied in bobbins containing continuous filament yarn, and aluminum/aluminum and aluminum/magnesium plates. Alumina staple is also available for short fiber reinforcement. Its high modulus of 55 Msi (380 GPa) is comparable to that of boron and carbon. Average filament strength is 200 ksi (1,379 MPa) minimum. Since alumina is a good insulator, it can be used in applications where conductive fibers cannot. Alumina, in continuous form, offers many advantages for composite fabrication including ease of handling, the ability to align fibers in desired directions, and filament winding capability. The fact that alumina is an electrical insulator, combined with its high modulus and compressive strength, make it of interest for polymer matrix composite applications. For example, alumina/epoxy and aramid/epoxy hybrid composites reinforced with alumina and aramid fibers have been fabricated and are of interest for radar transparent structures, circuit boards, and antenna supports.

Nominal Properties of Alumina			
Property	*Value*	*Property*	*Value*
Composition	>99% α–Al_2O_3	Filaments/yarn	200, nominal
Melting Point	3,713 °F (2,045 °C)	Tensile Modulus	55 Msi (385 GPa)
Filament Diameter	0.8×10^{-3} in. (20 μm)	Tensile Strength	200 ksi (1,379 MPa) min.
Length/Weight	~ 4.7 m/gr	Density	0.14 lb/in.^{3} (3.9 (gr/cc)

Silicon Carbide. Silicon fibers are produced with a nominal 0.0055 in. (140 micrometer) filament diameter and are characteristically found to have high strength, modulus, and density. Practically all silicon carbide monofilament fibers are currently produced for metal composite reinforcement. Alloys employing aluminum, titanium, and molybdenum have been produced. General processing for epoxy, bisimide, and polyimide resin can be either a solvated or solventless film impregnation process, with cure cycles equivalent to those for carbon or glass reinforced products. Organic matrix silicon carbide impregnated products may be press, autoclave, or vacuum bag oven cured. Lay-up on tooling proceeds as with carbon or glass composite products with all bleeding, damming, and venting as required for part fabrication. General temperature and pressure ranges for the cure of the selected matrix resins used in silicon carbide products will not adversely affect the fiber morphology. Silicon carbide ceramic composites engineered to provide high service temperatures (in excess of 2,640° F [1,450° C]) are unique in several thermal properties.

Nominal Properties of Alumina Composite (50–55% Volume)					
Property		*Value*	*Property*		*Value*
Moduli	Tensile, Axial	30–32 Msi (210–220 GPa)	Strength	Tensile, Axial	80 ksi (600 MPa)
	Tensile, Transverse	20–22 Msi (140–150 GPa)		Tensile, Transverse	26–30 ksi (130–210 MPa)
	Shear	7 Msi (50 GPa)		Shear	12–17 ksi (85–120 GPa)
Fatigue–Axial Endurance Limit		10^7 cycles at 75% static ultimate	Thermal Conductivity *		22–29 Btu/lbm-°F (38–50 J/m-s-°C)
Average	Axial	4.0 μin./in./°F (7.2 μm/m/°C)	Specific Heat *		0.19–0.12 Btu/lbm-°F (0.8–0.5 J/gr-°C)
Thermal Exp.	Transverse	11 μin./in./°F (20 μm/m/°C)	Density		0.12 lbm/in.^{3} (3.3 gr/cc)

* Thermal conductivity and specific heat at 68–750 ° F (20–400° C).

Quartz. Quartz fiber is very pure (99.95%) fused silica glass fiber. It is produced as continuous strands consisting of 120 to 240 individual filaments of 9 micron nominal diameter. The single strands are twisted and plied into heavier yarns. Quartz fibers are generally coated with an organic binder containing a silane coupling agent that is compatible with many resin systems. Quartz rovings are continuous reinforcements formed by combining a number of 300 2.0 zero twist strands. End counts of 8, 12, and 20 are available with yields ranging from 750 to 1,875 yards/pound. Quartz fibers are also available in the form of chopped fiber in cut lengths of $^1/_8$ inch to 2 inches (3 to 50 mm). Quartz fiber nomenclature is the same as for "E" or "S" glass except that the glass composition is designated by the letter Q.

Quartz fibers with a filament tensile strength of 850 ksi (5,900 MPa) have the highest strength to weight ratio, exceeding virtually all other high temperature materials. Therefore, quartz fibers can be used at temperatures much higher than "E" or "S" glass, with service temperatures up to 1,920° F (1,050° C) possible. Quartz fibers do not melt or vaporize until the temperature exceeds 3,000° F (1,650° C), and the fibers retain virtually all characteristics and properties of solid quartz up to failure. The fibers are chemically stable, but should not be used in environments where strong concentrations of alkalies are present.

Properties of Quartz Fiber			
Property	*Value*	*Property*	*Value*
Specific Gravity	2.20	Coef. Thermal Exp.	0.3 10^{-6} in./in./°F (0.54 10^{-6} cm/cm/°C)
Density	0.0795 lb/in.^{3} (2.20 gr/cc)	Thermal Conductivity	0.80 Btu/hr/ft/°F (0.0033 Cal/sec/cm/°C)
Tensile Strength	870 ksi (5,900 GPa)	Specific Heat	1.80 Btu/lb/°F (7,500 J/kg/°C)
Modulus	10 ksi (72 MPa)	Dielectric Constant	10 GHz, 75°F (24°C)

Typical Properties for Quartz Epoxy (laminate resin content 32–33.5% by weight)		
Property	*Room Temperature*	*1/2 Hour at 350° F (180° C)*
Tensile Strength	74.9–104 ksi (516–717 MPa)	65.4–92.9 ksi (451–636 MPa)
Tensile Modulus	3.14–4.09 Msi (21.7–28.2 GPa)	2.83–3.67 Msi (19.5–25.3 (GPa)
Flexure Strength	95.5–98.9 ksi (658–682 MPa)	53.9–75.9 ksi (372–523 MPa)
Flexure Modulus	3.27–3.46 Msi (22.5–23.8 GPa)	2.78–3.08 Msi (19.2–21.2 GPa)
Compressive Strength	66.4–72.4 ksi (459–499 MPa)	42.6–49.9 ksi (294–344 MPa)
Compressive Modulus	3.43–3.75 Msi (23.6–25.9 GPa)	3.10–3.40 Msi (21.4–23.4 GPa)
Specific Gravity	1.73–1.77	

Typical Properties for Quartz Polyimide (laminate resin content 36.2% by weight)		
Property	*Room Temperature*	*1/2 Hour at 350° F (180° C)*
Tensile Strength	79.1–105 ksi (545–724 MPa)	–
Tensile Modulus	3.9 Msi (27 GPa)	–
Flexure Strength	93.7–102 ksi (646–703 MPa)	62.4–68.3 ksi (430–471 MPa)
Flexure Modulus	3.2 Msi (22 GPa)	2.6–2.8 Msi (18–19 GPa)
Compressive Strength	67–67.4 ksi (462–465 MPa)	38.6–45.2 ksi (266–312 MPa)
Compressive Modulus	3.5–3.7 Msi (24–26 GPa)	2.8 Msi (19 GPa)

Reinforcement resin materials—thermosets

Resin is a generic term used to designate the polymer, polymer precursor material, and/or mixture or formulation thereof with various additives or chemically reactive compounds. The resin, its chemical composition, and its physical properties fundamentally affect the processing, fabrication, and ultimate properties of composite materials.

Epoxy. The term epoxy is a general description of a family of polymers that are based on molecules that contain epoxide groups. Epoxies are widely used in resins for prepregs and structural adhesives. Their advantages are high strength and modulus, low levels of volatiles, excellent adhesion, low shrinkage, good chemical resistance, and ease of processing. Their major disadvantages are brittleness and the reduction of properties in the presence of moisture. The processing or curing of epoxies is slower than for polyester resins, and the cost of the resin is also higher than for the polyesters. Processing techniques include autoclave molding, filament winding, press molding, vacuum bag molding, resin transfer molding, and pultrusion. Curing temperatures vary from room temperature to approximately 350° F (180° C). Higher temperature cures generally yield greater temperature resistance in the cured product.

Polyester (thermosetting). Generally, for a fiber reinforced resin system, the advantage of a polyester is its low cost and its ability to be processed quickly. Common processing methods include matched metal molding, wet lay-up, press (vacuum bag) molding, injection molding, filament winding, pultrusion, and autoclaving. Polyesters can be formulated to cure more rapidly than phenolics during the thermoset molding process. While phenolic processing, for example, is dependent on a time/temperature relationship, polyester processing is primarily dependent on temperature. Depending on the formulation, polyesters can be processed from room temperature to 350° F (180° C). If the proper temperature is applied, a quick cure will occur. Compared to epoxies, polyesters process more easily and are much tougher, whereas phenolics are more difficult to process and are brittle, but they do have higher service temperatures.

Phenolics. Phenolics, in general, cure by condensation through the off-gassing of water. The resulting matrix is characterized by both chemicals and thermal resistance, as well as hardness, plus low smoke and toxic degradation products. Phenolics are often called either resole or novolac resins. The basic difference between the two is that the novolacs have no methylol groups, and therefore they require an extension agent of paraformaldehyde, hexamethyleneetetraamine, or additional formaldehyde as a curative. Since the additives have higher molecular weights and viscosities than either parent material, they allow either press or autoclave cure and allow relatively high temperature free-standing postcures.

Bismaleimide. This type of thermosetting resin only recently became available commercially in prepreg tapes, fabrics, rovings, and sheet molding compound. The physical form of bismaleimide (BMI) resin depends on the requirements of the final application. At room temperature, their form can vary from a solid to a pourable liquid. For aerospace prepregs, stick resins are required. The advantages of BMI resins are best discussed in comparison to epoxy resins. Emerging data suggests that BMIs are versatile resins with many applications in the electronic and aerospace industries. Their primary advantage over epoxy resins is their high glass transition temperature, which is in the range of 500–600° F (260–320° C). Glass transition temperatures for high temperature epoxies are generally less than 500° F. Another BMI advantage is high elongation with the corresponding high service temperature capabilities. BMI resins are processed with essentially the same methods used for epoxies—they are suited for standard autoclave processing, injection molding, resin transfer molding, and sheet molding compounds, among others. The processing time is similar to epoxies, but to prepare for high service temperatures a free-standing postcure is required. At present, there are no room temperature curing BMIs. Due to their recent introduction, these resins are not as widely available as others, and costs are generally higher.

Polyimides. Polyimide matrix composites excel in high temperature environments where their thermal resistance, oxidative stability, low coefficient of thermal expansion, and solvent resistance benefit the design. Their primary uses are circuit boards and hot engine and aerospace structures. A polyimide may be either a thermoset resin or a thermoplastic. The thermoplastic variety will be discussed later, but, because partially cured thermoset polyimides containing residual plasticizing solvents can exhibit thermoplastic behavior, it is difficult to state with certainity that a particular polyimide is indeed a thermoset or a thermoplastic. Polymides, therefore, represent a transition between these two polymer classifications.

Most polyimide resin monomers are powders (some bismaleimides are an exception), and solvents are added to the resin to enable impregnation of unidirectional fiber and woven fabrics. Commonly, a 50/50 by weight mixture is used for fabrics, and a 90/10 by weight high solids mixture is used to produce a film for unidirectional fiber and low areal weight fabric prepregs. Solvents are further used to control prepreg handling qualities, such as tack and drape. Most of the solvents are removed in a drying process during impregnation, but total prepreg volatiles contents range between 2 and 8% by weight. Polyimides require high cure temperatures, usually in excess of 550° F (288° C). Consequently, normal epoxy composite consumable materials are not usable, and steel tooling becomes a necessity. Polyimide bagging and release films, such as Kapton and Upilex, replace the lower cost nylon bagging and polytetrafluoroethylene (PTFE) release films common to epoxy composite processing. Fiberglass fabrics must be used for bleeder and breather materials rather than polyester mat materials.

Reinforcement resin materials—thermoplastics

Semi-crystallines. These materials are so named because a percentage of their volume consists of a crystalline morphology. The remaining volume has a random molecular orientation (called amorphous). A partial list of semi-crystalline thermoplastics includes polyethylene, polypropylene, polyamides, polyphenylene sulfide, polyetheretherketone, and polyarylketone. The inherent speed of processing, ability to produce complicated, detailed parts, excellent thermal stability, and corrosion resistance have enabled them to become established in the automotive, electronic, and chemical processing industries. Processing speed is, in fact, the primary advantage of thermoplastic materials. However, thermoplastic impregs are typically boardy and do not exhibit the tack and drape of thermosets although drapeable forms are available that have commingled (interlaced together) thermoplastic and reinforcing fibers. High performance thermoplastics are slightly more expensive than equivalent performance epoxies, and tooling costs may be higher as well. However, final part cost may be less due to decreased processing time, and the ability to reprocess molder thermoplastics.

Some semi-crystalline thermoplastics possess properties of inherent flame resistance, superior toughness, good mechanical properties at elevated temperatures and after impact, and low moisture absorption—factors that have led to their wide use in the aerospace and automotive industries. Their primary disadvantage is the lack of a design database.

Amorphous polymers. The majority of thermoplastic polymers are composed of a random molecular orientation and are termed amorphous. This group includes polysulfone, polyamide-imide, polyphenylsulfone, polyphenylene sulfine sulfone, polyether sulfone, polystyrene, polyetherimide, and polyarylate. Amorphous thermoplastics are available in many forms, including films, filaments, and powders. Combined with reinforcing fibers, they are also available in injection molding compounds, compression moldable random sheets, unidirectional tapes, woven prepregs, etc. The fibers used are primarily carbon, aramid, and glass.

The primary advantages of amorphous thermoplastics in continuous fiber reinforced composites are potential low cost process at high production rates, high temperature capability, good mechanical properties before and after impact, and chemical stability. A service temperature of 350° F (177° C), and toughness two or three times that of conventional thermoset polymers are typical, and cycle times in production are less than for thermosets since no chemical reaction occurs during the forming process. Disadvantages are similar to those of semi-crystalline thermoplastics, including a lack of an extensive design database and reduced compression properties compared to 350° F (177° C) cure thermosets. Solvent resistance, which is good for semi-crystalline thermoplastics, is a concern for most amorphous ones. Processing methods include stamp molding, thermoforming, autoclave molding, roll forming, filament winding, and pultrusion. The high melting temperatures require process temperatures ranging from 500 to 700° F (260 to 370° C). Thermal expansion differences between the tool and the thermoplastic material should be addressed due to the high processing temperatures. Forming pressures range from 100 PSI (0.7 MPa) for thermoforming, to 5,000 PSI (34.4 MPa) for stamp molding.

Amorphous thermoplastics are used in many applications including the medical, communication, transportation, chemical processing, electronic, and aerospace industries. The majority of applications use unfilled and short-fiber forms—some uses for the unfilled polymers include cookware, power tools, business machines, corrosion resistant piping, medical instruments, and aircraft canopies. Uses for short-fiber reinforced forms include printed circuit boards, transmission parts, under-hood automotive applications, electrical connectors, and jet engine components.

Cure and consolidation processes

Vacuum bag molding. In vacuum bag molding, the lay-up is cured under pressure generated by drawing a vacuum in the space between the lay-up and a flexible sheet placed over it and sealed at the edges. The reinforcement is generally placed in the mold by hand lay-up using prepreg or wet resin. High flow resins are preferred for vacuum bag molding. The following steps are followed. 1) Place composite material for part into mold. 2) Install bleeder and breather material. 3) Place vacuum bag over part. 4) Seal bag and check for leaks. 5) Place tool and part in oven and cure as required at elevated temperature. 6) Remove part from mold. Parts fabricated using vacuum bag oven cure have lower fiber volumes and higher void contents. This is a low cost method of fabrication and uses low cost tooling for short production runs.

Autoclave cure. This process uses a pressurized vessel to apply both pressure and heat to parts that have been sealed in a vacuum bag. Generally, autoclaves operate at 10 to 300 PSI (10 to 30 kPa) and up to 800° F (420° C). Heat transfer and pressure application to the part is achieved by circulation (convection) of pressurized gas (usually air, nitrogen, or carbon dioxide) with the autoclave. Composite materials that are typically processed in autoclaves include adhesives, reinforced matrix (epoxy, bismaleimide, etc.) laminates and reinforced matrix laminates. In the case of the thermoset resin systems, the cure cycle is developed to induce specific chemical reactions within the polymer matrix by exposing the material to elevated temperatures while simultaneously applying vacuum and pressure to consolidate individual plies and compress voids. The cure cycle and vacuum bagging procedure affect such cured product characteristics as degree of cure, glass transformation temperature, void content percentage, cured resin content/fiber volume, residual stress, dimensional tolerances, and mechanical properties.

Pultrusion. Pultrusion is an automated process for the continuous manufacture of composites with a constant cross sectional area. It can be dry, employing prepreg thermosets or thermoplastics; or wet, where the continuous fiber bundle is resin-impregnated in a resin bath. In pultrusion, the material is drawn through a heated die that is specially designed for the shape being made. The key elements in the process consist of a reinforcement delivery platform, resin bath (for wet pultrusion), preform dies, a heated curing die, a pulling system, and a cutoff station. In general, the following process is used. 1) Reinforcements are threaded through the reinforcement delivery station. 2) The fiber bundle is pulled through the resin bath (wet pultrusion) and die preforms. 3) A strap is used to initiate the process by pulling the resin impregnated bundle through the preheated die. 4) As the impregnated fiber bundle is pulled through the heated die, the die temperature and pulling rate are controlled so that the cure to the product (or thermosets) is completed prior to exiting the heated die. 5) The composite parts are cut off by sawing when they reach the desired length.

Resin transfer molding and *thermoforming* are also used for fiber reinforced resins. These processes were discussed in the earlier section on "Process Technology."

Wood Products

Commercially important woods

Native species of trees are divided into two classes: hardwoods which have broad leaves; and softwoods which have scale-like leaves (such as cedars), or needle-like leaves (such as pines). Native softwoods, with the exception of cypress, tamarack, and western larch, are evergreen. Softwoods are also called "conifers," because all native species bear cones of one kind or another. The terms hardwood and softwood are not descriptively exact. Some softwoods, such as southern yellow pine and Douglas fir, are harder than some hardwoods, such as basswood and cottonwood.

Measuring wood products

The standard unit of measurement in the U.S.A. for wood products is the board foot. Lumber less than one inch thick is based on surface measure, and some types of finish lumber, such as moldings and trim, are sold by the linear foot. Common commercial quantity units are "per 1,000 feet board measure" or "per MBF," or "per 100 lineal feet," or "per CLF," and prices are generally quoted in terms of these quantity units. A board foot represents the quantity of lumber contained in a board that is one inch thick, twelve inches wide, and one foot long, or its cubic equivalent. For example, a board one inch thick, six inches wide, and two feet long is also one board foot. It is important to note that, in practice, the board foot calculations for lumber are based on its nominal thickness, nominal width, and actual length. Nominal sizes are not always actual sizes.

Table 1 provides the actual board foot content of various sizes of boards and dimension lumber. The cross-sectional sizes given are "nominal," meaning the sizes commonly used for rough lumber. This Table can be used to compute board footage if the number of pieces needed is known. Thus, 1,000 2 × 4s, each twelve feet long, would contain 8,000 board feet of lumber.

Board measure can be computed mathematically. Multiply the number of pieces by the nominal thickness in inches by the nominal width in inches by the length in feet, then divide the result by twelve.

Pieces × thickness (inches) × width (inches) × length (feet) ÷ 12 = Board Feet.

To calculate surface measure, in square feet, the thickness is not required.

Pieces × width (inches) × length (feet) ÷ 12 = Square Feet.

For items sold by the lineal foot, calculate the total lineal feet on the basis of a board foot estimate. When the thickness and width of the product is such that they must be produced from lumber nominally one inch thick and wide, or more, use the following formula.

(Board Foot measure × 12) ÷ (thickness (inches) × width (inches) = Lineal Feet.

Mechanical properties of wood

Wood can be considered an orthotropic material, which means that it has unique and independent mechanical properties in three mutually perpendicular axes: longitudinal, radial, and tangential. As shown in *Figure 1*, the longitudinal axis, *L*, is parallel to the grain; the radial axis, *R*, is normal to the growth rings (perpendicular to the grain in the radial direction); and the tangential axis, *T*, is perpendicular to the grain but tangent to the growth rings.

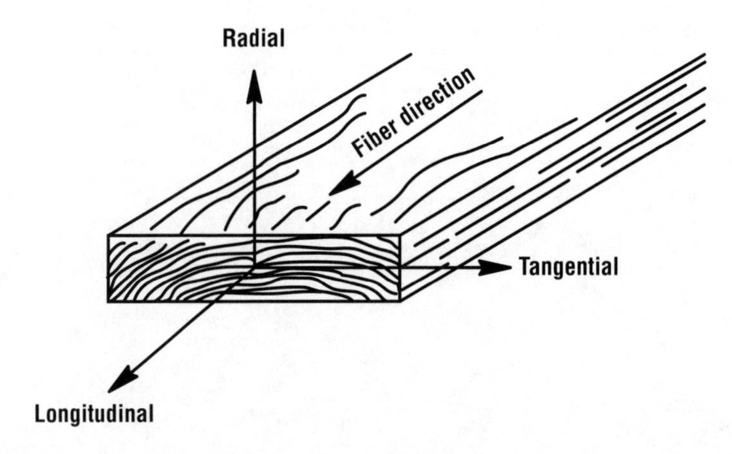

Figure 1. Three principal axes of wood.

Elastic Properties

Twelve constants (nine of which are independent) are needed to describe the elastic behavior of wood. These are three moduli of elasticity, E; three moduli of rigidity, G; and six Poisson's ratios, μ. The moduli of elasticity and the Poisson's ratios are related as follows.

$$\mu_{ij} \div E_i = \mu_{ji} \div E_j, \quad i \neq j \; i,j = L, R, T.$$

Elasticity implies that deformations produced by low stress are completely recoverable after loads are removed. When loaded to higher stress levels, plastic deformation or failure occurs. Wood's three moduli of elasticity, denoted by E_L, E_R, and E_T, are, respectively, the elastic moduli along the longitudinal, radial, and tangential axes. Average values of E_R, and E_T for samples from selected species (along with modulus of rigidity—see below) are shown in **Table 2** as ratios with E_L. It should be remembered that elastic ratios, as well as the elastic constants themselves, vary within and between species and with moisture content and specific gravity. The test results in this section were obtained either from "green" specimens, or from specimens with approximately 12% moisture content.

When a member is loaded axially, the deformation perpendicular to the direction of the load is proportional to the deformation parallel to the direction of the load, and the ratio of the transverse to axial strain in Poisson's ratio. The various ratios are denoted by μ_{LR}, μ_{RL}, μ_{LT}, μ_{TL}, μ_{RT}, and μ_{TR}. The initial letter of the subscript indicates the direction of applied stress, and the second letter the direction of lateral deformation. For example, μ_{LR} indicates deformation along the radial axis caused by stress along the longitudinal axis. Average values of Poisson's ratio for selected species are shown in **Table 3**. Poisson's ratio varies between species and is affected by moisture content and specific gravity.

The modulus of rigidity (shear modulus) for wood indicates its resistance to deflection caused by shear stresses. The three moduli of rigidity, denoted by G_{LR}, G_{LT}, and G_{RT}, are the elastic constants in the *LR*, *LT*, and *RT* planes, respectively. For example, G_{LR} is the modulus of rigidity based on shear strain in the *LR* plane, and shear stresses in the *Lt* and *RT* planes. Average values of shear moduli for samples of selected species, expressed as ratios with E_L, are given in **Table 2**. As with the modulus of elasticity, the modulus of rigidity will vary between species and with moisture content and specific gravity.

Table 1. Lumber Footage by Lengths, Widths, and Thicknesses. *(Source, MIL-HDBK-7B.)*

Nominal Cross Section (inches)	Footage Tally (in Board Feet) for Length of—					
	8 Feet	10 Feet	12 Feet	14 Feet	16 Feet	18 Feet
1 × 3	2	2 1/2	3	3 1/2	4	4 1/2
1 × 4	2 2/3	3 1/3	4	4 2/3	5 1/3	6
1 × 6	4	5	6	7	8	9
1 × 8	5 1/3	6 2/3	8	9 1/3	10 2/3	12
1 × 10	6 2/3	8 1/3	10	11 2/3	13 1/3	15
1 × 12	8	10	12	14	16	18
1 1/2 × 4	4	5	6	7	8	9
1 1/2 × 6	6	7 ½	9	10 1/2	12	13 1/2
1 1/2 × 8	8	10	12	14	16	18
1 1/2 × 10	10	12 1/2	15	17 1/2	20	22 1/2
1 1/2 × 12	12	15	18	21	24	27
2 × 4	5 1/3	6 2/3	8	9 1/3	10 2/3	12
2 × 6	8	10	12	14	16	18
2 × 8	10 2/3	13 1/3	16	18 2/3	21 1/3	24
2 × 10	13 1/3	16 2/3	20	23 1/3	26 2/3	30
2 × 12	16	20	24	28	32	36
3 × 4	8	10	12	14	16	18
3 × 6	12	15	18	21	24	27
3 × 8	16	20	24	28	32	36
3 × 10	20	25	30	35	40	45
3 × 12	24	30	36	42	48	54
4 × 4	10 2/3	13 1/3	16	18 2/3	21 1/3	24
4 × 6	16	20	24	28	32	36
4 × 8	21 1/3	26 2/3	32	37 1/3	42 2/3	48
4 × 10	26 2/3	33 1/3	40	46 2/3	53 1/3	60
4 × 12	32	40	48	56	64	72
6 × 6	24	30	36	42	48	54
6 × 7	28	35	42	49	56	63
6 × 8	32	40	48	56	64	72
6 × 10	40	50	60	70	80	90
6 × 12	48	60	72	84	96	108
7 × 9	42	52 1/2	63	73 1/2	84	94 1/2
8 × 8	42 2/3	43 1/3	64	74 2/3	85 1/3	96
8 × 10	53 1/3	66 2/3	80	93 1/3	106 2/3	120
8 × 12	64	80	96	112	128	144

Table 2. Elastic Ratios for Selected Species (Approximately 12% Moisture Content). See Text for Explanation. *(Source, Wood Handbook, U.S.D.A.)*

Species	E_T/E_L	E_R/E_L	G_{LR}/E_L	G_{LT}/E_L	G_{RT}/E_L
Hardwoods					
Ash, White	0.080	0.125	0.109	0.077	-
Balsa	0.015	0.046	0.054	0.037	0.005
Basswood	0.027	0.066	0.056	0.046	-
Birch, yellow	0.050	0.078	0.074	0.068	0.017
Cherry, black	0.086	0.197	0.147	0.097	-
Cottonwood, eastern	0.047	0.083	0.076	0.052	-
Mahogany, African	0.050	0.111	0.088	0.059	0.021
Mahogany, Honduras	0.064	0.107	0.066	0.086	0.028
Maple, sugar	0.065	0.132	0.111	0.063	-
Maple, red	0.067	0.140	0.133	0.074	-
Oak, red	0.082	0.154	0.089	0.081	-
Oak, white	0.072	0.163	0.086	-	-
Sweet gum	0.050	0.115	0.089	0.061	0.021
Walnut, black	0.056	0.106	0.085	0.062	0.021
Yellow-poplar	0.043	0.092	0.075	0.069	0.011
Softwoods					
Baldcypress	0.039	0.084	0.063	0.054	0.007
Cedar, northern white	0.081	0.183	0.210	0.187	0.015
Cedar, western red	0.055	0.081	0.087	0.086	0.005
Douglas-fir	0.050	0.068	0.064	0.078	0.007
Fir, subalpine	0.039	0.102	0.070	0.058	0.006
Hemlock, western	0.031	0.058	0.038	0.032	0.003
Larch, western	0.065	0.079	0.063	0.069	0.007
Pine					
Loblolly	0.078	0.113	0.082	0.081	0.013
Lodgepole	0.068	0.102	0.049	0.046	0.005
Longleaf	0.055	0.102	0.071	0.060	0.012
Pond	0.041	0.071	0.050	0.045	0.009
Ponderosa	0.083	0.122	0.138	0.115	0.017
Red	0.044	0.088	0.096	0.081	0.011
Slash	0.045	0.074	0.055	0.053	0.010
Sugar	0.087	0.131	0.124	0.113	0.019
Western white	0.038	.078	0.052	0.048	0.005
Redwood	0.089	0.087	0.066	0.077	0.011
Spruce, Sitka	0.043	0.078	0.064	0.061	0.003
Spruce, Engelmann	0.059	0.128	0.124	0.120	0.010

Table 3. Poisson's Ratios for Selected Species (Approximately 12% Moisture Content). See Text for Explanation. *(Source, Wood Handbook, U.S.D.A.)*

Species	μ_{LR}	μ_{LT}	μ_{RT}	μ_{TR}	μ_{RL}	μ_{TL}
Hardwoods						
Ash, White	0.371	0.440	0.684	0.360	0.059	0.051
Aspen, Quaking	0.489	0.374	-	0.496	0.054	0.022
Balsa	0.229	0.488	0.665	0.231	0.018	0.009
Basswood	0.364	0.406	0.912	0.346	0.034	0.022
Birch, yellow	0.426	0.451	0.697	0.426	0.043	0.024
Cherry, black	0.392	0.428	0.695	0.282	0.086	0.048
Cottonwood, eastern	0.344	0.420	0.875	0.292	0.043	0.018
Mahogany, African	0.297	0.641	0.604	0.264	0.033	0.032
Mahogany, Honduras	0.314	0.533	0.600	0.326	0.033	0.034
Maple, sugar	0.424	0.476	0.774	0.349	0.065	0.037
Maple, red	0.434	0.509	0.762	0.354	0.063	0.044
Oak, red	0.350	0.448	0.560	0.292	0.064	0.033
Oak, white	0.369	0.428	0.618	0.300	0.074	0.036
Sweet gum	0.325	0.403	0.682	0.309	0.044	0.023
Walnut, black	0.495	0.632	0.718	0.378	0.052	0.035
Yellow-poplar	0.318	0.392	0.703	0.329	0.030	0.019
Softwoods						
Baldcypress	0.338	0.326	0.411	0.356	-	-
Cedar, northern white	0.337	0.340	0.458	0.345	-	-
Cedar, western red	0.378	0.296	0.484	0.403	-	-
Douglas-fir	0.292	0.449	0.390	0.374	0.036	0.029
Fir, subalpine	0.341	0.332	0.437	0.336	-	-
Hemlock, western	0.485	0.423	0.442	0.382	-	-
Larch, western	0.355	0.276	0.389	0.352	-	-
Pine						
Loblolly	0.328	0.292	0.382	0.362	-	-
Lodgepole	0.316	0.347	0.469	0.381	-	-
Longleaf	0.332	0.365	0.384	0.342	-	-
Pond	0.280	0.364	0.389	0.320	-	-
Ponderosa	0.337	0.400	0.426	0.359	-	-
Red	0.347	0.315	0.408	0.308	-	-
Slash	0.392	0.444	0.447	0.387	-	-
Sugar	0.356	0.349	0.428	0.358	-	-
Western White	0.329	0.344	0.410	0.334	-	-
Redwood	0.360	0.346	0.373	0.400	-	-
Spruce, Sitka	0.372	0.467	0.435	0.245	0.040	0.025
Spruce, Engelmann	0.422	0.462	0.530	0.255	0.083	0.058

Strength Properties

The mechanical properties that are most commonly measured and represented as strength properties for design include modulus of rupture in bending, maximum stress in compression perpendicular to grain, and shear strength parallel to grain. Additional measurements are often made to evaluate work in bending, impact bending strength, tensile strength perpendicular to grain, and hardness. These properties are shown in **Table 4** and are defined below.

Modulus of rupture reflects the maximum load carrying capacity of a member in bending, and is proportional to maximum moment borne by the specimen. Modulus of rupture is an accepted criterion of strength, but it is not a true stress because the formula by which it is computed is valid only to the elastic limit.

Work to maximum load in bending demonstrates the material's ability to absorb shock with some permanent deformation and injury to the specimen. Work to maximum load is a measurement of the combined strength and toughness of wood under bending stresses.

Impact bending is measured in the impact bending test in which a hammer of given weight is dropped upon a beam from successively increased heights until rupture occurs, or the beam deflects 6 inches (152 mm) or more. The height of the maximum drop, or the drop that causes failure, is a comparative value that represents the ability of wood to absorb shocks that cause stresses beyond the proportional limit.

Compressive strength parallel to grain. This is the maximum stress sustained by a compression parallel to grain specimen having a ratio of length to least dimension of less than 11.

Compressive stress perpendicular to grain. Reported as stress at proportional limit, there is no clearly defined ultimate stress for this property.

Shear strength parallel to grain is the ability to resist internal slippage of one part upon another along the grain. Values presented on the Tables are average strength in radial and tangential shear planes.

Tensile strength perpendicular to grain is a measurement of the resistance of wood to forces acting across the grain that tend to split a member. Values presented are the average of radial and tangential observations.

Hardness is defined as the resistance to indentation using a modified Janka hardness test, measured by the load required to embed a 0.444 inch (11.28 mm) ball to one-half its diameter. Values presented are the average of radial and tangential penetrations.

Tensile strength parallel to grain is the maximum tensile stress sustained in the direction parallel to grain, and is not represented on the Table because little information is available on the tensile strength of various species of clear wood parallel to grain. The modulus of rupture is considered to be a conservative estimate of tensile strength for clear specimens (this is not true for lumber).

Working qualities of wood

The ease of working wood with hand tools generally varies directly with the specific gravity of the wood (specific gravities are given in **Table 4**). The lower the specific gravity, the easier it is to cut the wood with a sharp tool. A wood species that is easy to cut does not necessarily develop a smooth surface when it is machined. Consequently, tests have been conducted on many domestic hardwoods to evaluate their machining qualities (see **Table 5**).

Machining evaluation tests are not available for many imported woods, but three major

(Text continued on p. 362)

Table 4. Strength Properties of Selected Commercial Woods Grown in the United States[a] (English Units). *(Source, Wood Handbook, U.S.D.A.)*

Common Species Names	Moisture Content	Specific Gravity[b]	Static Bending			Impact Bending (in.)	Compression Parallel to Grain (lbf/in²)	Compression Perpendicular to Grain (lbf/in²)	Shear Parallel to Grain (lbf/on²)	Tension Perpendicular to Grain (lbf/in²)	Side Hardness (lbf)
			Modulus of Rupture (lbf/in²)	Modulus of Elasticity[c] (× 10⁶ lbf/in²)	Work to Maximum Load (in-lbf/in³)						
Hardwoods											
Alder, Red	Green	0.37	6,500	1.17	8.0	22	2,960	250	770	390	440
	12%	0.41	9,800	1.38	8.4	20	5,820	450	1,080	420	590
Ash											
Black	Green	0.45	6,000	1.04	12.1	33	2,300	350	860	490	520
	12%	0.49	12,600	1.60	14.9	35	5,970	760	1,570	700	850
Blue	Green	0.53	9,600	1.24	14.7	-	4,180	810	1,540	-	-
	12%	0.58	13,800	1.40	14.4	-	6,980	1,420	2,030	-	-
Green	Green	0.53	9,500	1.40	11.8	35	4,200	730	1,260	590	870
	12%	0.56	14,100	1.66	13.4	32	7,080	1,310	1,910	700	1,200
Oregon	Green	0.50	7,600	1.13	12.2	39	3,510	530	1,190	590	790
	12%	0.55	12,700	1.36	14.4	33	6,040	1,250	1,790	720	1,160
White	Green	0.55	9,500	1.44	15.7	38	3,990	670	1,350	590	960
	12%	0.60	15,000	1.74	16.6	43	7,410	1,160	1,910	940	1,320
Aspen											
Bigtooth	Green	0.36	5,400	1.12	5.7	-	2,500	210	730	-	-
	12%	0.39	9,100	1.43	7.7	-	5,300	450	1,080	-	-
Quaking	Green	0.35	5,100	0.86	6.4	22	2,140	180	660	230	300
	12%	0.38	8,400	1.18	7.6	21	4,250	370	850	260	350
Basswood, American	Green	0.32	5,000	1.04	5.3	16	2,220	170	600	280	250
	12%	0.37	8,700	1.46	7.2	16	4,730	370	990	350	410
Beech, American	Green	0.56	8,600	1.38	11.9	43	3,550	540	1,290	720	850
	12%	0.64	14,900	1.72	15.1	41	7,300	1,010	2,010	1,010	1,300

(Continued)

Table 4. *(Continued)* **Strength Properties of Selected Commercial Woods Grown in the United States[a] (English Units).** *(Source, Wood Handbook, U.S.D.A.)*

Common Species Names	Moisture Content	Specific Gravity[b]	Static Bending			Impact Bending (in.)	Compression Parallel to Grain (lbf/in²)	Compression Perpendicular to Grain (lbf/in²)	Shear Parallel to Grain (lbf/on²)	Tension Perpendicular to Grain (lbf/in²)	Side Hardness (lbf)
			Modulus of Rupture (lbf/in²)	Modulus of Elasticity[c] (× 10⁶ lbf/in²)	Work to Maximum Load (in-lbf/in³)						
Hardwoods											
Birch											
Paper	Green	0.48	6,400	1.17	16.2	49	2,360	270	840	380	560
	12%	0.55	12,300	1.59	16.0	34	5,690	600	1,210	-	910
Sweet	Green	0.60	9,400	1.65	15.7	48	3,740	470	1,240	430	970
	12%	0.65	16,900	2.17	18.0	47	8,540	1,080	2,240	950	1,470
Yellow	Green	0.55	8,300	1.50	16.1	48	3,380	430	1,110	430	780
	12%	0.62	16,600	2.01	20.8	55	8,170	970	1,880	920	1,260
Butternut	Green	0.36	5,400	0.97	8.2	24	2,420	220	760	430	390
	12%	0.38	8,100	1.18	8.2	24	5,110	460	1,170	440	490
Cherry, Black	Green	0.47	8,000	1.31	12.8	33	3,540	360	1,130	570	660
	12%	0.50	12,300	1.49	11.4	29	7,110	690	1,700	560	950
Chestnut, American	Green	0.40	5,600	0.93	7.0	24	2,470	310	800	440	420
	12%	0.43	8,600	1.23	6.5	19	5,320	620	1,080	460	540
Cottonwood											
Balsam, Poplar	Green	0.31	3,900	0.75	4.2	-	1,690	140	500	-	-
	12%	0.34	6,800	1.10	5.0	-	4,020	300	790	-	-
Black	Green	0.31	4,900	1.08	5.0	20	2,200	160	610	270	250
	12%	0.35	8,500	1.27	6.7	22	4,500	300	1,040	330	350
Eastern	Green	0.37	5,300	1.01	7.3	21	2,280	200	680	410	340
	12%	0.40	8,500	1.37	7.4	20	4,910	380	930	580	430

(Continued)

Table 4. *(Continued)* **Strength Properties of Selected Commercial Woods Grown in the United States[a] (English Units).** *(Source, Wood Handbook, U.S.D.A.)*

Common Species Names	Moisture Content	Specific Gravity[b]	Static Bending			Impact Bending (in.)	Compression Parallel to Grain (lbf/in²)	Compression Perpendicular to Grain (lbf/in²)	Shear Parallel to Grain (lbf/on²)	Tension Perpendicular to Grain (lbf/in²)	Side Hardness (lbf)
			Modulus of Rupture (lbf/in²)	Modulus of Elasticity[c] (× 10⁶ lbf/in²)	Work to Maximum Load (in-lbf/in³)						
Hardwoods											
Elm											
American	Green	0.46	7,200	1.11	11.8	38	2,910	360	1,000	590	620
	12%	0.50	11,800	1.34	13.0	39	5,520	690	1,510	660	830
Rock	Green	0.57	9,500	1.19	19.8	54	3,780	610	1,270	-	940
	12%	0.63	14,800	1.54	19.2	56	7,050	1,230	1,920	-	1,320
Slippery	Green	0.48	8,000	1.23	15.4	47	3,320	420	1,110	640	660
	12%	0.53	13,000	1.49	16.9	45	6,360	820	1,630	530	860
Hackberry	Green	0.49	6,500	0.95	14.5	48	2,650	400	1,070	630	700
	12%	0.53	11,000	1.19	12.8	43	5,440	890	1,590	580	880
Hickory, Pecan											
Bitternut	Green	0.60	10,300	1.40	20.0	66	4,570	800	1,240	-	-
	12%	0.66	17,100	1.79	18.2	66	9,040	1,680	-	-	-
Nutmeg	Green	0.56	9,100	1.29	22.8	54	3,980	760	1,030	-	-
	12%	0.60	16,600	1.70	25.1	-	6,910	1,570	-	-	-
Pecan	Green	0.60	9,800	1.37	14.6	53	3,990	780	1,480	680	1,310
	12%	0.66	13,700	1.73	13.8	44	7,850	1,720	2,080	-	1,820
Water	Green	0.61	10,700	1.56	18.8	56	4,660	880	1,440	-	-
	12%	0.62	17,800	2.02	19.3	53	8,600	1,550	-	-	-

(Continued)

Table 4. *(Continued)* **Strength Properties of Selected Commercial Woods Grown in the United States[a] (English Units).** *(Source, Wood Handbook, U.S.D.A.)*

Common Species Names	Moisture Content	Specific Gravity[b]	Static Bending			Impact Bending (in.)	Compression Parallel to Grain (lbf/in²)	Compression Perpendicular to Grain (lbf/in²)	Shear Parallel to Grain (lbf/on²)	Tension Perpendicular to Grain (lbf/in²)	Side Hardness (lbf)
			Modulus of Rupture (lbf/in²)	Modulus of Elasticity[c] (× 10⁶ lbf/in²)	Work to Maximum Load (in-lbf/in³)						
Hardwoods											
Hickory, True											
Mockernut	Green	0.64	11,100	1.57	26.1	88	4,480	810	1,280	-	-
	12%	0.72	19,200	2.22	22.6	77	8,940	1,730	1,740	-	-
Pignut	Green	0.66	11,700	1.65	31.7	89	4,810	920	1,370	-	-
	12%	0.75	20,100	2.26	30.4	74	9,190	1,980	2,150	-	-
Shagbark	Green	0.64	11,000	1.57	23.7	74	4,580	840	1,520	-	-
	12%	0.72	20,200	2.16	25.8	67	9,210	1,760	2,430	-	-
Shellbark	Green	0.62	10,500	1.34	29.9	104	3,920	810	1,190	-	-
	12%	0.69	18,100	1.89	23.6	88	8,000	1,800	2,110	-	-
Honeylocust	Green	0.60	10,200	1.29	12.6	47	4,420	1,150	1,660	930	1,390
	12%	-	14,700	1.63	13.3	47	7,500	1,840	2,250	900	1,580
Locust, Black	Green	0.66	13,800	1.85	15.4	44	6,800	1,160	1,760	770	1,570
	12%	0.69	19,400	2.05	18.4	57	10,180	1,830	2,480	640	1,700
Magnolia											
Cucumbertree	Green	0.44	7,400	1.56	10.0	30	3,140	330	990	440	520
	12%	0.48	12,300	1.82	12.2	35	6,310	570	1,340	660	700
Southern	Green	0.46	6,800	1.11	15.4	54	2,700	460	1,040	610	740
	12%	0.50	11,200	1.40	12.8	29	5,460	860	1,530	740	1,020

(Continued)

Table 4. *(Continued)* **Strength Properties of Selected Commercial Woods Grown in the United States[a] (English Units).** *(Source, Wood Handbook, U.S.D.A.)*

Common Species Names	Moisture Content	Specific Gravity[b]	Static Bending			Impact Bending (in.)	Compression Parallel to Grain (lbf/in²)	Compression Perpendicular to Grain (lbf/in²)	Shear Parallel to Grain (lbf/on²)	Tension Perpendicular to Grain (lbf/in²)	Side Hardness (lbf)
			Modulus of Rupture (lbf/in²)	Modulus of Elasticity[c] (× 10⁶ lbf/in²)	Work to Maximum Load (in-lbf/in³)						
Hardwoods											
Maple											
Bigleaf	Green	0.44	7,400	1.10	8.7	23	3,240	450	1,110	600	620
	12%	0.48	10,700	1.45	7.8	28	5,950	750	1,730	540	850
Black	Green	0.52	7,900	1.33	12.8	48	3,270	600	1,130	720	840
	12%	0.57	13,300	1.62	12.5	40	6,680	1,020	1,820	670	1,180
Red	Green	0.49	7,700	1.39	11.4	32	3,280	400	1,150	-	700
	12%	0.54	13,400	1.64	12.5	32	6,540	1,000	1,850	-	950
Silver	Green	0.44	5,800	0.94	11.0	29	2,490	370	1,050	560	590
	12%	0.47	8,900	1.14	8.3	25	5,220	740	1,480	500	700
Sugar	Green	0.56	9,400	1.55	13.3	40	4,020	640	1,460	-	970
	12%	0.63	15,800	1.83	16.5	39	7,830	1,470	2,330	-	1,450
Oak, Red											
Black	Green	0.56	8,200	1.18	12.2	40	3,470	710	1,220	-	1,060
	12%	0.61	13,900	1.64	13.7	41	6,520	930	1,910	-	1,210
Cherrybark	Green	0.61	10,800	1.79	14.7	54	4,620	760	1,320	800	1,240
	12%	0.68	18,100	2.28	18.3	49	8,740	1,250	2,000	840	1,480
Laurel	Green	0.56	7,900	1.39	11.2	39	3,170	570	1,180	770	1,000
	12%	0.63	12,600	1.69	11.8	39	6,980	1,060	1,830	790	1,210
Northern Red	Green	0.56	8,300	1.35	13.2	44	3,440	610	1,210	750	1,000
	12%	0.63	14,300	1.82	14.5	43	6,760	1,010	1,780	800	1,290
Pin	Green	0.58	8,300	1.32	14.0	48	3,680	720	1,290	800	1,070
	12%	0.63	14,000	1.73	14.8	45	6,820	1,020	2,080	1,050	1,510
Scarlet	Green	0.60	10,400	1.48	15.0	54	4,090	830	1,410	700	1,200
	12%	0.67	17,400	1.91	20.5	53	8,330	1,120	1,890	870	1,400
Southern Red	Green	0.52	6,900	1.14	8.0	29	3,030	550	930	480	860
	12%	0.59	10,900	1.49	9.4	26	6,090	870	1,390	510	1,060

(Continued)

Table 4. *(Continued)* **Strength Properties of Selected Commercial Woods Grown in the United States[a] (English Units).** *(Source, Wood Handbook, U.S.D.A.)*

Common Species Names	Moisture Content	Specific Gravity[b]	Static Bending			Impact Bending (in.)	Compression Parallel to Grain (lbf/in²)	Compression Perpendicular to Grain (lbf/in²)	Shear Parallel to Grain (lbf/on²)	Tension Perpendicular to Grain (lbf/in²)	Side Hardness (lbf)
			Modulus of Rupture (lbf/in²)	Modulus of Elasticity[c] ($\times 10^6$ lbf/in²)	Work to Maximum Load (in-lbf/in³)						
Hardwoods											
Oak, Red											
Water	Green	0.56	8,900	1.55	11.1	39	3,740	620	1,240	820	1,010
	12%	0.63	15,400	2.02	21.5	44	6,770	1,020	2,020	920	1,190
Willow	Green	0.56	7,400	1.29	8.8	35	3,000	610	1,180	760	980
	12%	0.69	14,500	1.90	14.6	42	7,040	1,130	1,650	-	1,460
Oak, White											
Bur	Green	0.58	7,200	0.88	10.7	44	3,290	680	1,350	800	1,110
	12%	0.64	10,300	1.03	9.8	29	6,060	1,200	1,820	680	1,370
Chestnut	Green	0.57	8,000	1.37	9.4	35	3,520	530	1,210	690	890
	12%	0.66	13,300	1.59	11.0	40	6,830	840	1,490	-	1,130
Live	Green	0.80	11,900	1.58	12.3	-	5,430	2,040	2,210	-	-
	12%	0.88	18,400	1.98	18.9	-	8,900	2,840	2,660	-	-
Overcup	Green	0.57	8,000	1.15	12.6	44	3,370	540	1,320	730	960
	12%	0.63	12,600	1.42	15.7	38	6,200	810	2,000	940	1,190
Post	Green	0.60	8,100	1.09	11.0	44	3,480	860	1,280	790	1,130
	12%	0.67	13,200	1.51	13.2	46	6,600	1,430	1,840	780	1,360
Swamp	Green	0.60	8,500	1.35	12.8	45	3,540	570	1,260	670	1,110
Chestnut	12%	0.67	13,900	1.77	12.0	41	7,270	1,110	1,990	690	1,240
Swamp White	Green	0.64	9,900	1.59	14.5	50	4,360	760	1,300	860	1,160
	12%	0.72	17,700	2.05	19.2	49	8,600	1,190	2,000	830	1,620
White	Green	0.60	8,300	1.25	11.6	42	3,560	670	1,250	770	1,060
	12%	0.68	15,200	1.78	14.8	37	7,440	1,070	2,000	800	1,360

(Continued)

Table 4. *(Continued)* **Strength Properties of Selected Commercial Woods Grown in the United States[a] (English Units).** *(Source, Wood Handbook, U.S.D.A.)*

Common Species Names	Moisture Content	Specific Gravity[b]	Static Bending			Impact Bending (in.)	Compression Parallel to Grain (lbf/in²)	Compression Perpendicular to Grain (lbf/in²)	Shear Parallel to Grain (lbf/on²)	Tension Perpendicular to Grain (lbf/in²)	Side Hardness (lbf)
			Modulus of Rupture (lbf/in²)	Modulus of Elasticity[c] (× 10⁶ lbf/in²)	Work to Maximum Load (in-lbf/in³)						
Hardwoods											
Sassafras	Green	0.42	6,000	0.91	7.1	-	2,730	370	950	-	-
	12%	0.46	9,000	1.12	8.7	-	4,760	850	1,240	-	-
Sweetgum	Green	0.46	7,100	1.20	10.1	36	3,040	370	990	540	600
	12%	0.52	12,500	1.64	11.9	32	6,320	620	1,600	760	850
Sycamore, American	Green	0.46	6,500	1.06	7.5	26	2,920	360	1,000	630	610
	12%	0.49	10,000	1.42	8.5	26	5,380	700	1,470	720	770
Tanoak	Green	0.58	10,500	1.55	13.4	-	4,650	-	-	-	-
	12%	-	-	-	-	-	-	-	-	-	-
Tupelo											
Black	Green	0.46	7,000	1.03	8.0	30	3,040	480	1,100	570	640
	12%	0.50	9,600	1.20	6.2	22	5,520	930	1,340	500	810
Water	Green	0.46	7,300	1.05	8.3	30	3,370	480	1,190	600	710
	12%	0.50	9,600	1.26	6.9	23	5,920	870	1,590	700	880
Walnut, Black	Green	0.51	9,500	1.42	14.6	37	4,300	490	1,220	570	900
	12%	0.55	14,600	1.68	10.7	34	7,580	1,010	1,370	690	1,010
Willow, Black	Green	0.36	4,800	0.79	11.0	-	2,040	180	680	-	-
	12%	0.39	7,800	1.01	8.8	-	4,100	430	1,250	-	-
Yellow-Poplar	Green	0.40	6,000	1.22	7.5	26	2,660	270	790	510	440
	12%	0.42	10,100	1.58	8.8	24	5,540	500	1,190	540	540

(Continued)

Table 4. *(Continued)* **Strength Properties of Selected Commercial Woods Grown in the United States[a] (English Units).** *(Source, Wood Handbook, U.S.D.A.)*

Common Species Names	Moisture Content	Specific Gravity[b]	Static Bending			Impact Bending (in.)	Compression Parallel to Grain (lbf/in²)	Compression Perpendicular to Grain (lbf/in²)	Shear Parallel to Grain (lbf/on²)	Tension Perpendicular to Grain (lbf/in²)	Side Hardness (lbf)
			Modulus of Rupture (lbf/in²)	Modulus of Elasticity[c] ($\times 10^6$ lbf/in²)	Work to Maximum Load (in-lbf/in³)						
Softwoods											
Baldcypress	Green	0.42	6,600	1.18	6.6	25	3,580	400	810	300	390
	12%	0.46	10,600	1.44	8.2	24	6,360	730	1,000	270	510
Cedar											
Atlantic White	Green	0.31	4,700	0.75	5.9	18	2,390	240	690	180	290
	12%	0.32	6,800	0.93	4.1	13	4,700	410	800	220	350
East. Redcedar	Green	0.44	7,000	0.65	15.0	35	3,570	700	1,010	330	650
	12%	0.47	8,800	0.88	8.3	22	6,020	920	-	-	-
Incense	Green	0.35	6,200	0.84	6.4	17	3,150	370	830	280	390
	12%	0.37	8,000	1.04	5.4	17	5,200	590	880	270	470
North. White	Green	0.29	4,200	0.64	5.7	15	1,990	230	620	240	230
	12%	0.31	6,500	0.80	4.8	12	3,960	310	850	240	320
Port-Orford	Green	0.39	6,600	1.30	7.4	21	3,140	300	840	180	380
	12%	0.43	12,700	1.70	9.1	28	6,250	720	1,370	400	630
West. Redcedar	Green	0.31	5,200	0.94	5.0	17	2,770	240	770	230	260
	12%	0.32	7,500	1.11	5.8	17	4,560	460	990	220	350
Yellow	Green	0.42	6,400	1.14	9.2	27	3,050	350	840	330	440
	12%	0.44	11,100	1.42	10.4	29	6,310	620	1,130	360	580
Douglas-Fir[d]											
Coast	Green	0.45	7,700	1.56	7.6	26	3,780	380	900	300	500
	12%	0.48	12,400	1.95	9.9	31	7,230	800	1,130	340	710
Interior West	Green	0.46	7,700	1.51	7.2	26	3,870	420	940	290	510
	12%	0.50	12,600	1.83	10.6	32	7,430	760	1,290	350	660
Interior North	Green	0.45	7,400	1.41	8.1	22	3,470	360	950	340	420
	12%	0.48	13,100	1.79	10.5	26	6,900	770	1,400	390	600
Interior South	Green	0.43	6,800	1.16	8.0	15	3,110	340	950	250	360
	12%	0.46	11,900	1.49	9.0	20	6,230	740	1,510	330	510

(Continued)

Table 4. *(Continued)* **Strength Properties of Selected Commercial Woods Grown in the United States[a] (English Units).** *(Source, Wood Handbook, U.S.D.A.)*

Common Species Names	Moisture Content	Specific Gravity[b]	Static Bending			Impact Bending (in.)	Compression Parallel to Grain (lbf/in²)	Compression Perpendicular to Grain (lbf/in²)	Shear Parallel to Grain (lbf/on²)	Tension Perpendicular to Grain (lbf/in²)	Side Hardness (lbf)
			Modulus of Rupture (lbf/in²)	Modulus of Elasticity[c] (× 10⁶ lbf/in²)	Work to Maximum Load (in-lbf/in³)						
Softwoods											
Fir											
Balsam	Green	0.33	5,500	1.25	4.7	16	2,630	190	662	180	290
	12%	0.35	9,200	1.45	5.1	20	5,280	404	944	180	400
Calif. Red	Green	0.36	5,800	1.17	6.4	21	2,760	330	770	380	360
	12%	0.38	10,500	1.50	8.9	24	5,460	610	1,040	390	500
Grand	Green	0.35	5,800	1.25	5.6	22	2,940	270	740	240	360
	12%	0.37	8,900	1.57	7.5	28	5,290	500	900	240	490
Noble	Green	0.37	6,200	1.38	6.0	19	3,010	270	800	230	290
	12%	0.39	10,700	1.72	8.8	23	6,100	520	1,050	220	410
Pacific Silver	Green	0.40	6,400	1.42	6.0	21	3,140	220	750	240	310
	12%	0.43	11,000	1.76	9.3	24	6,410	450	1,220	-	430
Subalpine	Green	0.31	4,900	1.05	-	-	2,300	190	700	-	260
	12%	0.32	8,600	1.29	-	-	4,860	390	1,070	-	350
White	Green	0.37	5,900	1.16	5.6	22	2,900	280	760	300	340
	12%	0.39	9,800	1.50	7.2	20	5,800	530	1,100	300	480
Hemlock											
Eastern	Green	0.38	6,400	1.07	6.7	21	3,080	360	850	230	400
	12%	0.40	8,900	1.20	6.8	21	5,410	650	1,060	-	500
Mountain	Green	0.42	6,300	1.04	11.0	32	2,880	370	930	330	470
	12%	0.45	11,500	1.33	10.4	32	6,440	860	1,540	-	680
Western	Green	0.42	6,600	1.31	6.9	22	3,360	280	860	290	410
	12%	0.45	11,300	1.63	8.3	23	7,200	550	1,290	340	540
Larch, Western	Green	0.48	7,700	1.46	10.3	29	3,760	400	870	330	510
	12%	0.52	13,000	1.87	12.6	35	7,620	930	1,360	430	830

(Continued)

Table 4. *(Continued)* **Strength Properties of Selected Commercial Woods Grown in the United States[a] (English Units).** *(Source, Wood Handbook, U.S.D.A.)*

Common Species Names	Moisture Content	Specific Gravity[b]	Static Bending			Impact Bending (in.)	Compression Parallel to Grain (lbf/in²)	Compression Perpendicular to Grain (lbf/in²)	Shear Parallel to Grain (lbf/on²)	Tension Perpendicular to Grain (lbf/in²)	Side Hardness (lbf)
			Modulus of Rupture (lbf/in²)	Modulus of Elasticity[c] (× 10⁶ lbf/in²)	Work to Maximum Load (in-lbf/in³)						
Softwoods											
Pine											
East. White	Green	0.34	4,900	0.99	5.2	17	2,440	220	680	250	290
	12%	0.35	8,600	1.24	6.8	18	4,800	440	900	310	380
Jack	Green	0.40	6,000	1.07	7.2	26	2,950	300	750	360	400
	12%	0.43	9,900	1.35	8.3	27	5,660	580	1,170	420	570
Loblolly	Green	0.47	7,300	1.40	8.2	30	3,510	390	860	260	450
	12%	0.51	12,800	1.79	10.4	30	7,130	790	1,390	470	690
Lodgepole	Green	0.38	5,500	1.08	5.6	20	2,610	250	680	220	330
	12%	0.41	9,400	1.34	6.8	20	5,370	610	880	290	480
Longleaf	Green	0.54	8,500	1.59	8.9	35	4,320	480	1,040	330	590
	12%	0.59	14,500	1.98	11.8	34	8,470	960	1,510	470	870
Pitch	Green	0.47	6,800	1.20	9.2	-	2,950	360	860	-	-
	12%	0.52	10,800	1.43	9.2	-	5,940	820	1,360	-	-
Pond	Green	0.51	7,400	1.28	7.5	-	3,660	440	940	-	-
	12%	0.56	11,600	1.75	8.6	-	7,540	910	1,380	-	-
Ponderosa	Green	0.38	5,100	1.00	5.2	21	2,450	280	700	310	320
	12%	0.40	9,400	1.29	7.1	19	5,320	580	1,130	420	460
Red	Green	0.41	5,800	1.28	6.1	26	2,730	260	690	300	340
	12%	0.46	11,000	1.63	9.9	26	6,070	600	1,210	460	560
Sand	Green	0.46	7,500	1.02	9.6	-	3,440	450	1,140	-	-
	12%	0.48	11,600	1.41	9.6	-	6,920	836	-	-	-
Shortleaf	Green	0.47	7,400	1.39	8.2	30	3,530	350	910	320	440
	12%	0.51	13,100	1.75	11.0	33	7,270	820	1,390	470	690
Slash	Green	0.54	8,700	1.53	9.6	-	3,820	530	960	-	-
	12%	0.59	16,300	1.98	13.2	-	8,140	1,020	1,680	-	-

(Continued)

Table 4. *(Continued)* **Strength Properties of Selected Commercial Woods Grown in the United States[a] (English Units).** *(Source, Wood Handbook, U.S.D.A.)*

Common Species Names	Moisture Content	Specific Gravity[b]	Static Bending			Impact Bending (in.)	Compression Parallel to Grain (lbf/in²)	Compression Perpendicular to Grain (lbf/in²)	Shear Parallel to Grain (lbf/on²)	Tension Perpendicular to Grain (lbf/in²)	Side Hardness (lbf)
			Modulus of Rupture (lbf/in²)	Modulus of Elasticity[c] (× 10⁶ lbf/in²)	Work to Maximum Load (in-lbf/in³)						
Softwoods											
Pine											
Spruce	Green	0.41	6,000	1.00	-	-	2,840	280	900	-	450
	12%	0.44	10,400	1.23	-	-	5,650	730	1,490	-	660
Sugar	Green	0.34	4,900	1.03	5.4	17	2,460	210	720	270	270
	12%	0.36	8,200	1.19	5.5	18	4,460	500	1,130	350	380
Virginia	Green	0.45	7,300	1.22	10.9	34	3,420	390	890	400	540
	12%	0.48	13,000	1.52	13.7	32	6,710	910	1,350	380	740
West. White	Green	0.35	4,700	1.19	5.0	19	2,430	190	680	260	260
	12%	0.38	9,700	1.46	8.8	23	5,040	470	1,040	-	420
Redwood											
Old-growth	Green	0.38	7.500	1.18	7.4	21	4,200	420	800	260	410
	12%	0.40	10,000	1.34	6.9	19	6,150	700	940	240	480
Young-growth	Green	0.34	5,900	0.96	5.7	16	3,110	270	890	300	350
	12%	0.35	7,900	1.10	5.2	15	5,220	520	1,110	250	420
Spruce											
Black	Green	0.38	6,100	1.38	7.4	24	2,840	240	739	100	370
	12%	0.42	10,800	1.61	10.5	23	5,960	550	1,230	-	520
Engelmann	Green	0.33	4,700	1.03	5.1	16	2,180	200	640	240	260
	12%	0.35	9,300	1.30	6.4	18	4,480	410	1,200	350	390
Red	Green	0.37	6,000	1.33	6.9	18	2,720	260	750	220	350
	12%	0.40	10,800	1.61	8.4	25	5,540	550	1,290	350	490
Sitka	Green	0.37	5,700	1.23	6.3	24	2,670	280	760	250	350
	12%	0.40	10,200	1.57	9.4	25	5,610	580	1,150	370	510
White	Green	0.33	5,000	1.14	6.0	22	2,350	210	640	220	320
	12%	0.36	9,400	1.43	7.7	20	5,180	430	970	360	480

(Continued)

Table 4. *(Continued)* **Strength Properties of Selected Commercial Woods Grown in the United States[a] (English Units).** *(Source, Wood Handbook, U.S.D.A.)*

Common Species Names	Moisture Content	Specific Gravity[b]	Static Bending			Impact Bending (in.)	Compression Parallel to Grain (lbf/in²)	Compression Perpendicular to Grain (lbf/in²)	Shear Parallel to Grain (lbf/on²)	Tension Perpendicular to Grain (lbf/in²)	Side Hardness (lbf)
			Modulus of Rupture (lbf/in²)	Modulus of Elasticity[c] (× 10⁶ lbf/in²)	Work to Maximum Load (in-lbf/in³)						
Softwoods											
Tamarack	Green	0.49	7,200	1.24	7.2	28	3,480	390	860	260	380
	12%	0.53	11,600	1.64	7.1	23	7,160	800	1,280	400	590

Notes:

[a] Results of tests on small clear specimens in the green and air-dried conditions. Definition of properties: impact bending is height of drop that causes complete failure, using 0.71-kg (50-lb) hammer; compression parallel to grain is also called maximum crushing strength; compression perpendicular to grain is fiber stress at proportional limit; shear is maximum shearing strength; tension is maximum tensile strength; and side hardness is hardness measured when load is perpendicular to grain.

[b] Specified gravity is based on weight when ovendry and volume when green or at 12% moisture content.

[c] Modulus of elasticity measured from a simply supported, center-loaded beam, on a span depth ratio of 14/1. To correct for shear deflection, the modulus can be increased by 10%.

[d] Coast Douglas-fir is defined as Douglas-fir growing in Oregon and Washington State west of the Cascade Mountains summit. Interior West includes California and all counties in Oregon and Washington east of, but adjacent to, the Cascade summit; Interior North, the remainder of Oregon and Washington plus Idaho, Montana, and Wyoming; and Interior South, Utah, Colorado, Arizona, and New Mexico.

Table 5. Machining and Related Properties of Selected Domestic Hardwoods. *(Source, Wood Handbook, U.S.D.A.)*

Type of Wood[a]	Planning: Perfect Pieces (%)	Shaping: Good to Excellent Pieces (%)	Turning: Good to Excellent Pieces (%)	Boring: Good to Excellent Pieces (%)	Mortising: Fair to Excellent Pieces (%)	Sanding: Good to Excellent Pieces (%)	Steam Bending: Unbroken Pieces (%)	Nail Splitting: Pieces Free From Complete Splits (%)	Screw Splitting: Pieces Free From Complete Splits (%)
Alder, Red	61	20	88	64	52	-	-	-	-
Ash	75	55	79	94	58	75	67	65	71
Aspen	26	7	65	78	60	-	-	-	-
Basswood	64	10	68	76	51	17	2	79	68
Beech	83	24	90	99	92	49	75	42	58
Birch	63	57	80	97	97	34	72	32	48
Birch, paper	47	22	-	-	-	-	-	-	-
Cherry, black	80	80	88	100	100	-	-	-	-
Chestnut	74	28	87	91	70	64	56	66	60
Cottonwood [b]	21	3	70	70	52	19	44	82	78
Elm, soft [b]	33	13	65	94	75	66	74	80	74
Hackberry	74	10	77	99	72	-	94	63	63
Hickory	76	20	84	100	98	80	76	35	63
Magnolia	65	27	79	71	32	37	85	73	76
Maple, bigleaf	52	56	80	100	80	-	-	-	-
Maple, hard	54	72	82	99	95	38	57	27	52
Maple, soft	41	25	76	80	34	37	59	58	61
Oak, red	91	28	84	99	95	81	86	66	78
Oak, white	87	35	85	95	99	83	91	69	74
Pecan	88	40	89	100	98	-	78	47	69

[a] Commercial lumber nomenclature.
[b] Interlocked grain present.

(Continued)

Table 5. *(Continued)* **Machining and Related Properties of Selected Domestic Hardwoods.** *(Source, Wood Handbook, U.S.D.A.)*

Type of Wood[a]	Planning: Perfect Pieces (%)	Shaping: Good to Excellent Pieces (%)	Turning: Good to Excellent Pieces (%)	Boring: Good to Excellent Pieces (%)	Mortising: Fair to Excellent Pieces (%)	Sanding: Good to Excellent Pieces (%)	Steam Bending: Unbroken Pieces (%)	Nail Splitting: Pieces Free From Complete Splits (%)	Screw Splitting: Pieces Free From Complete Splits (%)
Sweetgum [b]	51	28	86	92	58	23	67	69	69
Sycamore [b]	22	12	85	98	96	21	29	79	74
Tanoak	80	39	81	100	100	-	-	-	-
Tupelo, water [b]	55	52	79	62	33	34	46	64	63
Tupelo, black [b]	48	32	75	82	24	21	42	65	63
Walnut, black	62	34	91	100	98	-	78	50	59
Willow	52	5	58	71	24	24	73	89	62
Yellow-poplar	70	13	81	87	63	19	58	77	67

[a] Commercial lumber nomenclature.
[b] Interlocked grain present.

factors other than density can affect production of smooth surfaces during wood machining: interlocked and variable grain; hard mineral deposits; and reaction wood—particularly tension wood in hardwoods. Interlocking grain is characteristic of a few domestic species and many tropical species, and it presents difficulty in planning quartersawn boards unless attention is paid to feed rate, cutting angles, and sharpness of cutters. Hard deposits in cells, such as calcium carbonite and silica, can have a pronounced effect on all cutting edges. This dulling effect becomes more pronounced as the wood is dried to normal inservice requirements. Tension wood can cause fibrous and fuzzy surfaces, and can be very troublesome on species of lower density. Reaction wood can also be responsible for the pinching effect on saws as a result of stress relief. The pinching can result in burning and dulling of saw teeth.

Fasteners for wood

Nails

Nails, of many types, forms, and shapes, are the most common fasteners for wood construction. Nails in use must resist withdrawal and/or lateral loads, and their resistance is affected by the species of wood and the type of nail. Sizes of readily available nails are given in **Table 6**. Because international nail producers do not always adhere to the dimensions in the Table, it is advisable to specify length and diameters when purchasing nails.

The resistance of a nail to withdraw from a piece of wood is dependent on the density of the wood, the diameter of the nail, and the depth of penetration. For bright common wire nails driven into the side grain of seasoned wood, or unseasoned wood that remains wet, the maximum withdrawal load can be obtained with the following equation.

For withdrawal load in pounds $p = 7{,}850 \times G^{5/2} \times D \times L$

For withdrawal load in Newtons $p = 54.12 \times G^{5/2} \times D \times L$

where p = maximum load in pounds or Newtons
G = specific gravity of the wood at 12% moisture content (see **Table 4**)
D = diameter of nail in inches or millimeters
L = depth of penetration of the nail in inches or millimeters.

The resistance of nails to withdrawal is generally greatest when they are driven perpendicular to the grain of the wood. When the nail is driven parallel to the wood fibers (into the end of the piece), withdrawal resistance in the softer woods drops by 25 to 50% of the resistance obtained when the nail is driven perpendicular to the grain. The difference between side- and end-grain withdrawal loads is less for dense woods than for softer woods. Withdrawal resistance is also affected by other factors including the type of nail point, type of shank, time the nail remains in the wood, surface coatings of the nail, and moisture content changes in the wood. Another factor to consider when nailing wood is that the less dense species do not split as readily as the denser ones, thus offering the opportunity to increase the diameter, length, and number of nails to compensate for the wood's lower resistance to nail withdrawal. Withdrawal resistance in plywood is approximately 15 to 30% less than solid wood of the same thickness, primarily because the fiber distortion is less uniform in plywood.

Wood Screws

The common types of wood screws have flat, oval, or round heads, as shown in *Figure 2*. For flush surfaces, flat heads are essential, but round head screws are required when countersinking is objectionable. The root diameter for most screws averages about

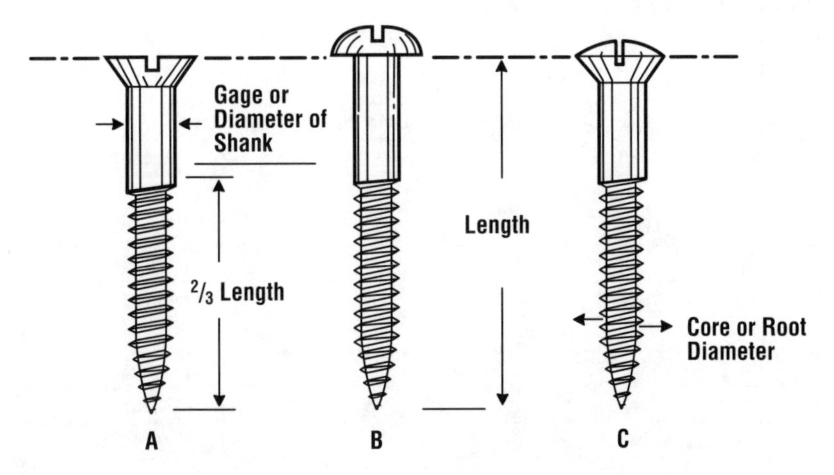

Figure 2. Common wood screws: A) flathead, B) roundhead, C) ovalhead.

two-thirds the shank diameter. Resistance of wood screw shanks to withdrawal varies directly with the square of the specific gravity of the wood. Within limits, the withdrawal load varies directly with the depth of penetration of the threaded portion and the diameter of the screw, provided that the screw does not fail in tension. Failure in tension results when the screw's strength is exceeded by the withdrawal strength from the wood, and the limiting length to cause a tension failure decreases as the density of the wood increases because the withdrawal strength of the wood increases with density. The longer lengths of standard screws are therefore superfluous in dense hardwoods.

Ultimate test values for withdrawal loads of wood screws inserted into the side grain of seasoned wood may be obtained with the following equation.

For withdrawal load in pounds $p = 15{,}700 \times G^2 \times D \times L$

For withdrawal load in Newtons $p = 108.25 \times G^2 \times D \times L$

where p = maximum load in pounds or Newtons
G = specific gravity of the wood at 12% moisture content (see **Table 4**)
D = shank diameter of screw in inches or millimeters
L = depth of penetration of the threaded part of the screw in inches or millimeters.

This equation is applicable when the prebored screw lead hole has a diameter of about 70% of the root diameter of the threads in softwoods, and about 90% in hardwoods. The equation values are applicable to the screw sizes listed in **Table 7**, and when lengths and gages are outside the indicated limits, the actual values are likely to be less than the equation values. The withdrawal loads of screws inserted in the end grain of wood are somewhat erratic, but when splitting is avoided, they should average 75% of the load sustained by screws inserted in the side grain.

Withdrawal resistance of tapping screws (screws that have threads the full length of the shank–commonly known as sheet metal screws) is in general about 10% greater than that for wood screws of comparative diameter and length of threaded portion. The ratio between the withdrawal resistance of tapping screws and wood screws varies from 1.16 in denser woods such as oak, to 1.05 in lighter woods such as redwood.

Table 6. Dimensions of Commonly Available Nails.

Size	Gauge	Length		Diameter	
		inch	mm	inch	mm
Bright Common Wire Nails (Penny Nails)					
6d	11-1/2	2	50.8	0.113	2.87
8d	10-1/4	2-1/2	63.5	0.131	3.33
10d	9	3	76.2	0.148	3.76
12d	9	3-1/4	82.6	0.148	3.76
16d	8	3-1/2	88.9	0.162	4.11
20d	6	4	101.6	0.192	4.88
30d	5	4-1/2	114.3	0.207	5.26
40d	4	5	127.0	0.225	5.72
50d	3	5-1/2	139.7	0.244	6.20
60d	2	6	152.4	0.262	6.65
Smooth Box Nails					
3d	14-1/2	1-1/4	31.8	0.076	1.93
4d	14	1-1/2	38.1	0.080	2.03
5d	14	1-3/4	44.5	0.080	2.03
6d	12-1/2	2	50.8	0.098	2.49
7d	12-1/2	2-1/4	57.2	0.098	2.49
8d	11-1/2	2-1/2	63.5	0.113	2.87
10d	10-1/2	3	76.2	0.128	3.25
16d	10	3-1/2	88.9	0.135	3.43
20d	9	4	101.6	0.148	3.76

Size	Length		Diameter	
	inch	mm	inch	mm
Helical and Annularly Threaded Nails				
6d	2	50.8	0.120	3.05
8d	2-1/2	63.5	0.120	3.05
10d	3	76.2	0.135	3.43
12d	3-1/4	82.6	0.135	3.43
16d	3-1/2	88.9	0.148	3.76
20d	4	101.6	0.177	4.50
30d	4-1/2	114.3	0.177	4.50
40d	5	127.0	0.177	4.50
50d	5-1/2	139.7	0.177	4.50
60d	6	152.4	0.177	4.50
70d	7	177.8	0.207	5.26
80d	8	203.2	0.207	5.26
90d	9	228.6	0.207	5.26

Table 7. Screw Sizes and Gage Diameters.

Screw Length		Gage Limits	Gage Conversion to Diameter		
inch	mm		Gage Number	Dia. inch	Dia. mm
1/2	12.7	4 to 6	4	0.112	2.84
			5	0.125	3.18
3/4	19.0	4 to 11	6	0.138	3.51
			7	0.151	3.84
1	25.4	4 to 12	8	0.164	4.17
			9	0.177	4.50
1 1/2	38.1	5 to 14	10	0.190	4.83
			11	0.203	5.16
2	50.8	7 to 16	12	0.216	5.49
			14	0.242	6.15
2 1/2	63.5	9 to 18	16	0.268	6.81
			18	0.294	7.47
3	76.2	12 to 20	20	0.320	8.13
			24	0.372	9.45

Lag Screws

Lag screws are normally used where it would be difficult to fasten a bolt, or in instances where a nut on the surface would be objectionable. Commonly available lag screws range from about 0.2 to one inch (5.1 to 25.4 mm) in diameter, and from 1 to 16 inches (25.4 to 406 mm) in length. The thread length varies with screw length and ranges from $^3/_4$ inch (19 mm) in lengths up to 1 $^1/_4$ inch (31.8 mm), to one-half the overall screw length in lengths 10 inches (254 mm) or more. The head on lag screws is hexagonal shaped to provide for tightening by wrench. The equation for withdrawal resistance of lag screws, which follows, is based on the screw material having a tensile yield strength of 45 ksi (310 MPa) and an average ultimate tensile strength of 77 ksi (531 MPa).

For withdrawal load in pounds $p = 125.4 \times G^{3/2} \times D^{3/4} \times L$

For withdrawal load in Newtons $p = 108.25 \times G^{3/2} \times D^{3/4} \times L$

where p = maximum load in pounds or Newtons
G = specific gravity of the wood at 12% moisture content (see **Table 4**)
D = shank diameter of screw in inches or millimeters
L = depth of penetration of the threaded part of the screw in inches or millimeters.

Lag screws, like wood screws, require prebored lead holes, and the diameter of the lead hole for the threaded section varies with the density of the wood. For low density softwoods, such as cedars and white pines, the lead hole should be 40 to 70% of the shank diameter; for Douglas fir and Southern pine, 60 to 75%; and for dense hardwoods, 65 to 85%. The smaller percentage in each range applies to lag screws of the smaller diameters, and the larger percentage to lag screws with larger diameters. A lead hole should also be prebored to the depth of penetration of the shank, equal to the diameter of the shank.

Temperature and Its Effect on Dimensions

The "standard" temperature of 68° F or 20° C was set in the first half of the last century. Initially, it was intended as the ambient temperature for calibrating gages, but it logically spread to be the temperature of record for measurements of all kinds. At 68° F/20° C, part size should be precisely that size given in the tables for specified limits and fits. However, as temperature rises or falls, shaft size changes. **Table 1** for inch sizes and **Table 2** for metric sizes provide length differences that will be encountered at different temperatures. The numbers 1 to 10 across the top of each Table represent coefficients of thermal expansion. If, for example, the coefficient of expansion of a material is 5, the changes experienced by that material at a given temperature, in µin./in. or µm/m, can be found in the column headed 5. If the coefficient needed is above 10, add or multiply values to arrive at the amount of expansion or contraction. If, for example, the coefficient is 13, add the value for 10 to the values for 2 and 1. For reference, the coefficient of thermal expansion is quantified as the coefficient $\times 10^{-6}$. Therefore, the length/degree change for Monel, with a coefficient of 8.7, would be $8.7 \times 10^{-6} = 0.0000087$ in./in. At 78° F, which is ten degrees more than the ambient of 68°, the expansion will be $10 \times .0000087 = .0000870$ in./in., or 87 µin./in. This can be confirmed on the Table by adding the values for 5 (50) and 3 (30) and 0.7 (.7 × 10 [the value for 1] = 7. The total is 87 µin./in. (50 + 30 +7).

Compensation for thermal effects

The ability to meet higher part tolerance is crucial in today's metal cutting business. Manufacturing engineers expect the high accuracy performance of a machining center to yield high accuracy parts. But this is not necessarily the case. Thermal errors introduced during part processing can have a significant impact on accuracy. The following article by Satish Shivaswamy, R&D Technology Analysis Engineer at Cincinnati Machine, investigates thermal errors. It originally appeared in *Modern Machine Shop on line*.

Effects on part accuracy

Part accuracy is affected by various errors, which may be broadly classified into geometric and processing errors. Geometric errors include positioning accuracy, roll, pitch, yaw, straightness, and squareness. Processing errors include tooling, fixturing, NC programming, and thermal errors. Thermal errors play a considerable role in part accuracy. The effects of part temperature and the coefficient of thermal expansion (CTE) of the part material are described in this article, and a simple method to overcome thermal errors is discussed. Keep in mind that inspection errors must also be considered, but are not specifically discussed in this article.

Effects of thermal expansion

Many manufacturing plants require machining materials as varied as titanium, cast iron, steel, brass, aluminum, and magnesium on a single machine. However, the coefficients of thermal expansion (CTE) of these materials are all different. The standards organizations throughout the world have adopted 68° F or 20° C as the standard temperature for the measurement of length. This means that all materials at the standard temperature of 68° F or 20° C will have the same length. At any other temperature, the length will vary according to the CTE of that material. Because the CTEs of various materials are different, the length of similar parts of different materials will be unique unless measured at 68° F or 20° C. **Table 3** shows CTEs for some of the commonly used materials in machining.

Table 1. Length Differences per Inch from Standard for Temperatures 38° to 98° F. *(See Notes.)*

Temp. °F	Coefficient of Thermal Expansion of Material per Degree F × 10^{-6}					
	1	2	3	4	5	10
	Total Change in Length from Standard, Microinches per Inch of Length (μin./in.)					
38	-30	-60	-90	-120	-150	-300
39	-29	-58	-87	-116	-150	-290
40	-28	-56	-84	-112	-145	-280
41	-27	-54	-81	-108	-140	-270
42	-26	-52	-78	-104	-135	-260
43	-25	-50	-75	-100	-130	-250
44	-24	-48	-72	-96	-125	-240
45	-23	-46	-69	-92	-120	-230
46	-22	-44	-66	-88	-115	-220
47	-21	-42	-63	-84	-110	-210
48	-20	-40	-60	-80	-105	-200
49	-19	-38	-57	-76	-100	-190
50	-18	-36	-54	-72	-95	-180
51	-17	-34	-51	-68	-90	-170
52	-16	-32	-48	-64	-85	-160
53	-15	-30	-45	-60	-80	-150
54	-14	-28	-42	-56	-75	-140
55	-13	-26	-39	-52	-70	-130
56	-12	-24	-36	-48	-65	-120
57	-11	-22	-33	-44	-60	-110
58	-10	-20	-30	-40	-55	-100
59	-9	-18	-27	-36	-50	-90
60	-8	-16	-24	-32	-45	-80
61	-7	-14	-21	-28	-40	-70
62	-6	-12	-18	-24	-35	-60
63	-5	-10	-15	-20	-30	-50
64	-4	-8	-12	-16	-25	-40
65	-3	-6	-9	-12	-20	-30
66	-2	-4	-6	-8	-15	-20
67	-1	-2	-3	-4	-5	-10
68	0	0	0	0	0	0
69	1	2	3	4	5	10
70	2	4	6	8	10	20
71	3	6	9	12	15	30
72	4	8	12	16	20	40
73	5	10	15	20	25	50
74	6	12	18	24	30	60
75	7	14	21	28	35	70

(Continued)

Table 1. *(Continued)* **Length Differences per Inch from Standard for Temperatures 38° to 98° F.** *(See Notes.)*

Temp. °F	Coefficient of Thermal Expansion of Material per Degree F × 10^{-6}					
	1	2	3	4	5	10
	Total Change in Length from Standard, Microinches per Inch of Length (μin./in.)					
76	8	16	24	32	40	80
77	9	18	27	36	45	90
78	10	20	30	40	50	100
79	11	22	33	44	55	110
80	12	24	36	48	60	120
81	13	26	39	52	65	130
82	14	28	42	56	70	140
83	15	30	45	60	75	150
84	16	32	48	64	80	160
85	17	34	51	68	85	170
86	18	36	54	72	90	180
87	19	38	57	76	95	190
88	20	40	60	80	100	200
89	21	42	63	84	105	210
90	22	44	66	88	110	220
91	23	46	69	92	115	230
92	24	48	72	96	120	240
93	25	50	75	100	125	250
94	26	52	78	104	130	260
95	27	54	81	108	135	270
96	28	56	84	112	140	280
97	29	58	87	116	145	290
98	30	60	90	120	150	300

Notes. To calculate for coefficients of thermal expansion not given, combine values given to give the sum needed. For example, for a coefficient of expansion of 25, double the "10" column and add the "5" column. For 7, add the values for "5" and "2." Fractional interpolation may be similarly calculated. For examples of how to use this Table, see text. Source, USAS B4.1. (Originally issued as ASA B4.1-1947. Currently issued as ANSI B4.1.)

Table 2. Length Differences per Centimeter from Standard for Temperatures 0° to 40° C. *(See Notes.)*

Temp. °C	Coefficient of Thermal Expansion of Material per Degree C × 10^{-6}					
	1	2	3	4	5	10
	Total Change in Length from Standard, Micron per Meter of Length (μm/m)					
0	-20	-40	-60	-80	-100	-200
1	-19	-38	-57	-76	-95	-190
2	-18	-36	-54	-72	-90	-180
3	-17	-34	-51	-68	-85	-170
4	-16	-32	-48	-64	-80	-160
5	-15	-30	-45	-60	-75	-150

(Continued)

Table 2. *(Continued)* **Length Differences per Centimeter from Standard for Temperatures 0° to 40° C.** *(See Notes.)*

Temp. °C	Coefficient of Thermal Expansion of Material per Degree C × 10^{-6}					
	1	2	3	4	5	10
	Total Change in Length from Standard, Micron per Meter of Length (μm/m)					
6	-14	-28	-42	-56	-70	-140
7	-13	-26	-39	-52	-65	-130
8	-12	-24	-36	-48	-60	-120
9	-11	-22	-33	-44	-55	-110
10	-10	-20	-30	-40	-50	-100
11	-9	-18	-27	-36	-45	-90
12	-8	-16	-24	-32	-40	-80
13	-7	-14	-21	-28	-35	-70
14	-6	-12	-18	-24	-30	-60
15	-5	-10	-15	-20	-25	-50
16	-4	-8	-12	-16	-20	-40
17	-3	-6	-9	-12	-15	-30
18	-2	-4	-6	-8	-10	-20
19	-1	-2	-3	-4	-5	-10
20	0	0	0	0	0	0
21	1	2	3	4	5	10
22	2	4	6	8	10	20
23	3	6	9	12	15	30
24	4	8	12	16	20	40
25	5	10	15	20	25	50
26	6	12	18	24	30	60
27	7	14	21	28	35	70
28	8	16	24	32	40	80
29	9	18	27	36	45	90
30	10	20	30	40	50	100
31	11	22	33	44	55	110
32	12	24	36	48	60	120
33	13	26	39	52	65	130
34	14	28	42	56	70	140
35	15	30	45	60	75	150
36	16	32	48	64	80	160
37	17	34	51	68	85	170
38	18	36	54	72	90	180
39	19	38	57	76	95	190
40	20	40	60	80	100	200

Notes. To calculate for coefficients of thermal expansion not given, combine values given to give the sum needed. For example, for a coefficient of expansion of 25, double the "10" column and add the "5" column. For 7, add the values for "5" and "2." Fractional interpolation may be similarly calculated. For examples of how to use this Table, see text. Source, USAS B4.1. (Originally issued as ASA B4.1-1947. Currently issued as ANSI B4.1.)

Table 3. Coefficient of Thermal Expansion for Commonly Machined Materials.

Material	CTE (α) μin./°F	CTE (α) μm/°C
Glass Scale	4.4 – 5.6	8 – 10
Titanium	4.7	8.4
Cast Iron	5.7	10.5
Stainless Steel	5.5 – 9.6	9.9 – 17.3
Brass	11.3	20.3
Aluminum	13.2	23.8
Magnesium	14.1	25.2

The thermal expansion or contraction in length as a result of a change in temperature of the part may be computed when the CTE of the part is known, and the equation is:

$$\Delta L = L \times \alpha \times \Delta T$$

where: ΔL = Change in length (meters or inches) due to change in temperature
L = Nominal distance (meters or inches)
α = Coefficient of thermal expansion in μm/° C or μin /° F
ΔT = (T - 20) when L is in meters and α is in μm/°C or
ΔT = (T - 68) when L is in inches and α is in μm/° F.

Current compensation methods

Modern machine tools are equipped with a feature called axis compensation. The concept of axis compensation is similar to the pitch error compensation on older NC machines. In axis compensation, the machine movement to a commanded position will be adjusted based on the values entered in a look up table to correct for positioning errors.

Often the machine thermal behavior tracks closely to that of steel or cast iron. Steel ballscrews, cast iron and steel structures, and glass scale systems all typically have CTEs close to that of cast iron or steel. However, wet machining, and machining parts of materials other than steel or cast iron will lead to inaccurate parts. By combining the knowledge of the part temperature and its CTE with the feature of axis compensation, the part accuracy can be significantly improved.

Example process

Consider, for example, a machining center that is equipped with a coolant chiller and a glass scale for the machine's feedback system. Let us consider a hypothetical situation where two bored holes spaced 400 mm apart are machined both in steel and aluminum parts. Based on how the machine's axes are compensated, different results are obtained for distance between the bored holes in the steel and aluminum parts.

Conditions:
Aluminum CTE = 23.8 μm/°C (23.8 % 10^{-6}/°C)
Steel CTE = 11.7 μm /°C
Glass Scale CTE = 8 μm /°C
Air/Glass Scale/Pallet Temperature = 24 °C
Coolant Temperature = 26.1°C
Distance Between the Bored Holes = 400 mm.

Machine compensated using scale temperature (or pallet temperature) and CTE of glass scale:

A. Compensation value generated at the time of machine calibration
   $0.4 \text{ m} \times 8 \times 10^{-6}$ per °C × (24 – 20)°C ≈ 13 microns

B. Aluminum part soaked in coolant is machined
   $0.4 \text{ m} \times 23.8 \times 10^{-6}$ per °C × (26.1 – 20)°C ≈ 58 microns

   Error in aluminum part when measured at 20° C = 13 – 58 = - 45 microns (*SHORT*)

C. Steel part soaked in coolant is machined
   $0.4 \text{ m} \times 11.7 \times 10^{-6}$ per °C × (26.1 – 20)°C ≈ 29 microns

   Error in steel part when measured at 20° C = 13 – 29 = - 16 microns (*SHORT*)

If the axes of the machine were compensated using the CTE of steel and the coolant temperature, the steel part when measured at 20°C would have no error. Inspection of the aluminum part at 20°C will reveal that the distance between the bored holes would be spaced 15 microns (0.0006") short of 400 mm. If the CTE of aluminum were used instead, the steel part would show that the distance between the bored holes is farther apart by 29 microns (0.00114"), while the aluminum part would have no error when measured at 20°C.

Temperatures must remain stable after the compensations are established. Continued temperature changes will cause an error.

Calculations demonstrate that the part accuracy may be enhanced using the CTE of the part and its temperature. To counter the thermal growth of the part, the axis compensation tables in the machine control may be suitably changed to match the dimensional change of the part using the part CTE value and the coolant temperature to which the part temperature will be soaked.

Multiple compensation tables

If the coolant temperature is controlled, it is possible to generate a set of axis compensation tables for various CTE values. This can be done while generating axis compensation data using the laser, and modifying these values for the various CTEs. The operator can then choose and load the appropriate axis compensation file into the control based on the particular material being processed. This process of first order thermal compensation may be used on any type of controller.

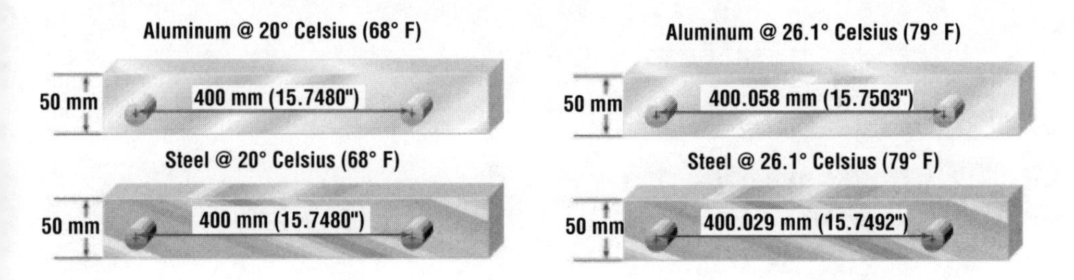

Figure 1. The coefficient of thermal expansion (CTE) of the part material and the coolant temperature play an important role in the final accuracy of the part. The coolant temperature may be controlled to within ±1° C using optional coolant chiller units. The compression values should be generated using the CTE of the part material and the coolant temperature for enhancing part accuracy.

Surface Finish Parameters: Terms and Definitions. *(Source, Norton Company.)*

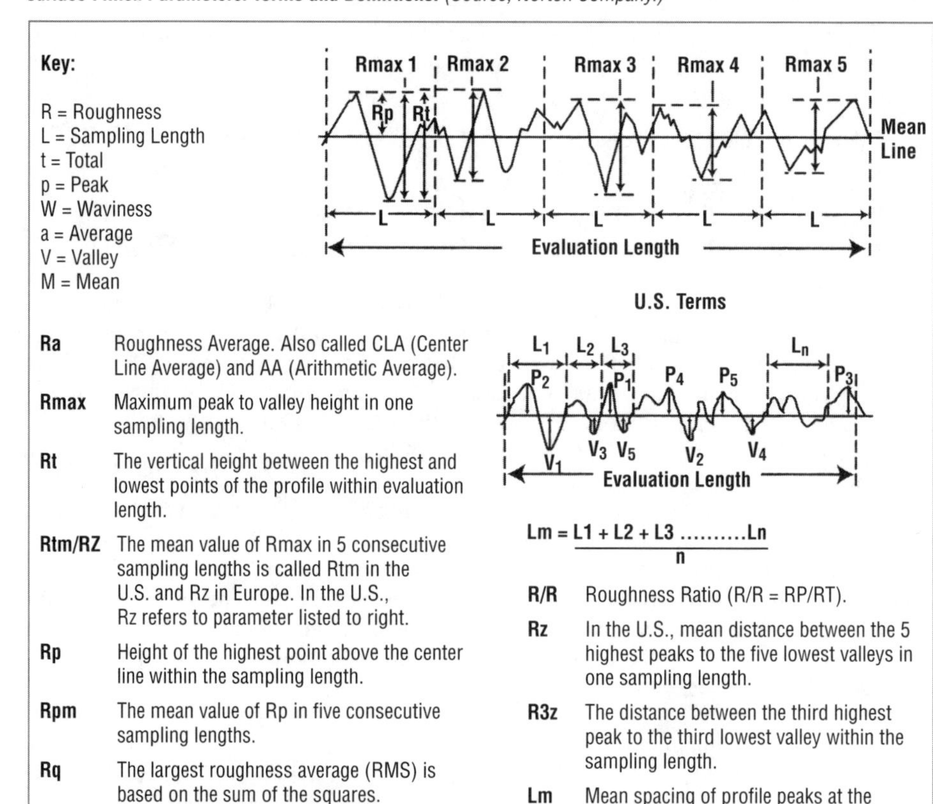

Key:

R = Roughness
L = Sampling Length
t = Total
p = Peak
W = Waviness
a = Average
V = Valley
M = Mean

Ra Roughness Average. Also called CLA (Center Line Average) and AA (Arithmetic Average).

Rmax Maximum peak to valley height in one sampling length.

Rt The vertical height between the highest and lowest points of the profile within evaluation length.

Rtm/RZ The mean value of Rmax in 5 consecutive sampling lengths is called Rtm in the U.S. and Rz in Europe. In the U.S., Rz refers to parameter listed to right.

Rp Height of the highest point above the center line within the sampling length.

Rpm The mean value of Rp in five consecutive sampling lengths.

Rq The largest roughness average (RMS) is based on the sum of the squares.

$$Rz = \frac{P1 + P2 + P3 + P4 + P5 + V1 + V2 + V3 + V4 + V5}{10}$$

$$Lm = \frac{L1 + L2 + L3 \ldots\ldots\ldots\ldots Ln}{n}$$

R/R Roughness Ratio (R/R = RP/RT).

Rz In the U.S., mean distance between the 5 highest peaks to the five lowest valleys in one sampling length.

R3z The distance between the third highest peak to the third lowest valley within the sampling length.

Lm Mean spacing of profile peaks at the mean line.

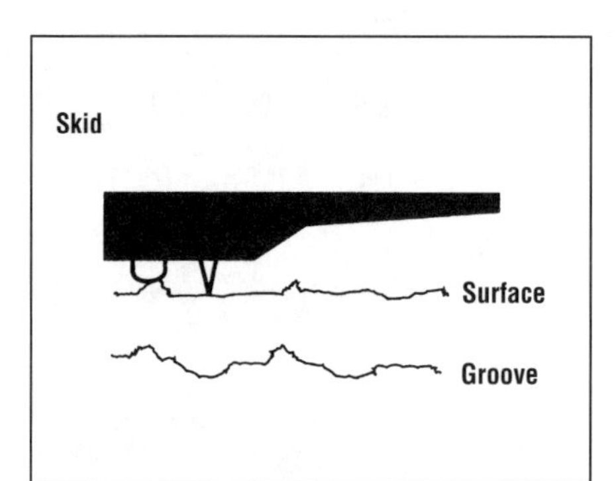

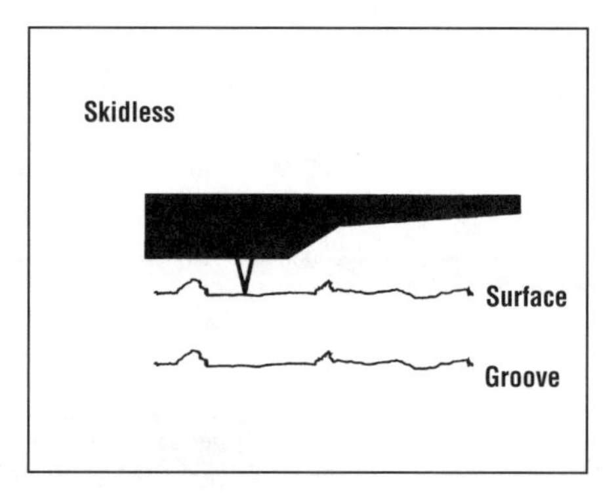

Skid (left) and skidless (right) tracing styluses. (Source, Norton Company.)

Comparison Chart of Surface Roughness Values. *(Source, Norton Company.)*

Rt	Rtm	Ra/CLA/AA		Rq/RMS/Rs	
μm	μm	μm	μin.	μm	μin.
0.06	0.03	0.006	0.2	0.007	0.2
0.08	0.04	0.008	0.3	0.009	0.3
0.1	0.05	0.01	0.4	0.011	0.4
0.12	0.06	0.012	0.5	0.013	0.5
0.15	0.08	0.015	0.6	0.018	0.7
0.2	0.1	0.002	0.8	0.022	0.9
0.25	0.12	0.025	1.0	0.027	1.1
0.3	0.15	0.03	1.2	0.033	1.3
0.4	0.2	0.04	1.6	0.044	1.8
0.5	0.25	0.05	2.0	0.055	2.2
0.6	0.3	0.06	2.4	0.066	2.6
0.8	0.4	0.08	3.2	0.088	3.5
0.9	0.5	0.10	4.0	0.11	4.4
1.0	0.6	0.12	4.8	0.13	5.2
1.2	0.8	0.15	6.0	0.18	7.2
1.6	1.0	0.20	8.0	0.22	8.8
2.0	1.2	0.25	10	0.27	10.8
2.5	1.6	0.3	12	0.33	13.2
3.0	2.0	0.4	16	0.44	17.6
4.0	2.5	0.5	20	0.55	22
5.0	3.0	0.6	24	0.66	26
10	6.0	1.2	48	1.3	52
15	10	2.5	100	2.7	108
30	20	5.0	200	5.5	220
50	40	10	400	11	440
100	80	20	800	22	880
200	160	40	1600	44	1760
400	320	80	3200	88	3500

Note: Due to the nature of measuring surface finishes, comparison values may vary up to 25%.
1 μm = 0.001 mm = 0.000040" = 1 micron. 1 μin. = 0.000001" = 0.254 μm.

Two surface finishes with equal Ra values.

Conversion Factors for Commonly Used Units of Measurement.

To Convert... U.S. System	*To... Metric System*	*Multiply by...*	*To Convert... Metric System*	*To... U.S. System*	*Multiply by...*
Length					
mil	millimeter	0.0254	millimeter	mil	39.37
inch	µm	25,400	mm	inch	0.00003937
inch	millimeter	25.4	millimeter	inch	0.03937
inch	centimeter	2.54	centimeter	inch	0.3937
inch	meter	0.0254	meter	inch	39.37
foot	millimeter	304.8	millimeter	foot	0.0032808
foot	centimeter	30.48	centimeter	foot	0.032808
foot	meter	0.3048	meter	foot	3.2808
yard	millimeter	914.1	millimeter	yard	0.0010936
yard	centimeter	91.44	centimeter	yard	0.010936
yard	meter	0.9144	meter	yard	1.0936
mile	meter	1609.344	meter	mile	0.000621
mile	kilometer	1.609344	kilometer	mile	0.621
Area					
inch^2	millimeter^2	645.16	millimeter^2	inch^2	0.00155
inch^2	centimeter^2	6.4516	centimeter^2	inch^2	0.155
foot^2	millimeter^2	92903.04	millimeter^2	foot^2	0.0000107639
foot^2	centimeter^2	929.0304	centimeter^2	foot^2	0.001076391
foot^2	meter^2	0.0929	meter^2	foot^2	10.76391
yard^2	millimeter^2	836127.36	millimeter^2	yard^2	0.0000011960
yard^2	centimeter^2	8361.2736	centimeter^2	yard^2	0.000119599
yard^2	meter^2	0.83612736	meter^2	yard^2	1.19599
acre	meter^2	4046.8726	meter^2	acre	0.0002471
Volume					
fluidounce	centimeter^3	29.57352956	centimeter^3	fluidounce	0.0338
quart	liter	0.946352946	liter	quart	1.056688
gallon	liter	3.785411748	liter	gallon	0.264172
gallon	meter^3	0.003785412	meter^3	gallon	0.0264172
inch^3	centimeter^3	16.387064	centimeter^3	inch^3	0.06102374
foot^3	centimeter^3	28316.85185	centimeter^3	foot^3	0.000035315
foot^3	liter	28.31685185	liter	foot^3	0.235315
foot^3	meter^3	0.028316852	meter^3	foot^3	35.315
yard^3	liter	764.555	liter	yard^3	0.00130795
yard^3	meter^3	0.764555	meter^3	yard^3	1.30795
inch^3/pound	meter^3/kilogram	0.000036	meter^3/kilogram	inch^3/pound	27.680
foot^3/pound	meter^3/kilogram	0.0624	meter^3/kilogram	inch^3/pound	16.018
Mass					
ounce(avdp.)	gram	28.34952313	gram	ounce	0.03527396
ounce(avdp.)	kilogram	0.028349523	kilogram	ounce	35.27396
pound	gram	453.59237	gram	pound	0.002204622
pound	kilogram	0.45359237	kilogram	pound	2.204623
ton(U.S.short)	metricton	0.09072	metricton	ton(U.S.)	1.1023
Density					
pound/inch^3	gram/centimeter^3	27.67990	gram/centimeter^3	pound/inch^3	0.0361273
pound/foot^3	gram/centimeter^3	0.160185	gram/centimeter^3	pound/foot^3	62.4279744
pound/foot^3	kilogram/meter^3	16.01846	kilogram/meter^3	pound/foot^3	0.0624280

Conversion Factors for Commonly Used Units of Measurement. *(continued)*

To Convert... U.S. System	*To... Metric System*	*Multiply by...*	*To Convert... Metric System*	*To... U.S. System*	*Multiply by...*
		Force			
poundforce	Newton	4.448222	Newton	poundforce	0.2248089
poundfoot	Newtonmeter	1.355818	Newtonmeter	poundfoot	0.7375612
pound/foot	Newton/meter	175.1268	Newton/meter	pound/foot	0.06852178
		Pressure			
pound/inch^2	pascal	6894.76	pascal	pound/inch^2	0.000145
pound/inch^2	megapascal	0.00689476	megapascal	pound/inch^2	145.04
pound/inch^2	bar	0.0689476	bar	pound/inch^2	14.504
atmosphere	pascal	101.325	pascal	atmosphere	0.9869
		Energy			
foot pound force	joule	1.355818	joule	foot pound force	0.737562
BTU	joule	1055.0559	joule	BTU	0.0009478
calorie	joule	4.1868	joule	calorie	0.23885
watt hour	joule	3600	joule	watt hour	0.0002778
		Power			
horsepower	kilowatt	0.7456999	kilowatt	horsepower	1.341022
footpound/minute	watt	0.0225969	watt	footpound/minute	44.25372
		Velocity			
foot/second	meter/second	0.3048	meter/second	foot/second	3.28084
foot/minute	meter/second	0.00508	meter/second	foot/minute	196.8504
foot/second	kilometer/hour	1.09728	kilometer/hour	foot/second	0.911344
mile/hour	kilometer/hour	1.609344	kilometer/hour	mile/hour	0.6213712
knots	kilometer/hour	1.852	kilometer/hour	knots	0.54
		Temperature			
°Fahrenheit	°Celsius	(°F–32)/(1.8)	°Celsius	°Fahrenheit	(1.8°C)+32
°Fahrenheit	°Kelvin	(°F+459.67)/(1.8)	°Kelvin	°Fahrenheit	1.8°K–459.67
°Kelvin	°Celsius	°K–273.15	°Celsius	°Kelvin	°C+273.15